QUANTUM DOTS: OPTICS, ELECTRON TRANSPORT AND FUTURE APPLICATIONS

A comprehensive review of cutting-edge solid state research, focusing on its prominent example – quantum dot nanostructures – this book features a broad range of techniques for fabrication of these nano-structured semiconductors and control of their quantum properties.

Written by leading researchers, the book considers advanced III–V and II–VI semiconductor quantum dots (QDs) realized by self-assembly, lithography and chemical synthesis; novel QD structures in nanowires and graphene; and transport and optical methods for control of single QDs. Significant attention is given to manipulation of single spins and control of their magnetic environment, generation of quantum light emitted by single dots in dielectric cavities, and dots coupled to plasmons in metallic structures. It is a valuable resource for graduate students and researchers new to this field.

ALEXANDER TARTAKOVSKII is a Reader in Experimental Semiconductor Physics in the Department of Physics and Astronomy, University of Sheffield. He has been working in quantum dot research since 2001, recently with a focus on high-sensitivity nano-NMR in ultra-small nuclear spin ensembles in individual quantum dots.

QUANTUM DOTS: OPTICS, ELECTRON TRANSPORT AND FUTURE APPLICATIONS

Edited by

ALEXANDER TARTAKOVSKII
University of Sheffield

CAMBRIDGE
UNIVERSITY PRESS

CAMBRIDGE
UNIVERSITY PRESS

University Printing House, Cambridge CB2 8BS, United Kingdom

Cambridge University Press is part of the University of Cambridge.

It furthers the University's mission by disseminating knowledge in the pursuit of education, learning and research at the highest international levels of excellence.

www.cambridge.org
Information on this title: www.cambridge.org/9781107012585

First published 2012

A catalogue record for this publication is available from the British Library

ISBN 978-1-107-01258-5 Hardback

Contents

Contributors

M. Aßmann Experimentelle Physik 2, Technische Universität Dortmund, D-44221 Dortmund, Germany

G. C. Aers Institute for Microstructural Sciences, National Research Council, Ottawa, Canada, K1A 0R6

N. Akopian Kavli Institute of Nanoscience, Delft University of Technology, 2628CJ Delft, The Netherlands

T. Amand Université de Toulouse, INSA-CNRS-UPS, LPCNO, 135 Avenue de Rangueil, 31077 Toulouse, France

M. Atatüre University of Cambridge, Cavendish Laboratory, J. J. Thomson Avenue, Cambridge CB3 0HE, UK

E. P. A. M. Bakkers Kavli Institute of Nanoscience, Delft University of Technology, 2628CJ Delft, The Netherlands

M. Barkelid Kavli Institute of Nanoscience, Delft University of Technology, 2628CJ Delft, The Netherlands

M. Bayer Experimentelle Physik 2, Technische Universität Dortmund, D-44221 Dortmund, Germany

A. J. Bennett Toshiba Research Europe Limited, 208 Cambridge Science Park, Cambridge CB4 0GZ, UK

L. Besombes Institut Néel, CNRS & Université Joseph Fourier, 25 rue des Martyrs, 38042 Grenoble, France

H. Boukari Institut Néel, CNRS & Université Joseph Fourier, 25 rue des Martyrs, 38042 Grenoble, France

G. Bulgarini Kavli Institute of Nanoscience, Delft University of Technology, 2628CJ Delft, The Netherlands

G. Burkard Department of Physics, University of Konstanz, D-78457 Konstanz, Germany

J. I. Climente Department de Química Física i Analítica, Universitat Jaume I, E-12080 Castelló, Spain

D. Dalacu Institute for Microstructural Sciences, National Research Council, Ottawa, Canada, K1A 0R6

M. F. Doty Department of Materials Science and Engineering, University of Delaware, Newark, Delaware 19716, USA

Y. Dovzhenko Department of Physics, Princeton University, Princeton, New Jersey 08544, USA

B. Eble Institut des NanoSciences de Paris, CNRS-UMR 7588, UPMC Univ. Paris 06, 140 Rue de Lourmel, 75015 Paris, France

D. Englund Department of Electrical Engineering and Department of Applied Physics, Columbia University, New York, NY 10027, USA

K. Ensslin Solid State Physics Laboratory, ETH Zürich, 8093 Zürich, Switzerland

A. Faraon Hewlett Packard Laboratories, 1501 Page Mill Road, Palo Alto, CA 94304, USA

A. Forchel Technische Physik, Physikalisches Institut and Wilhelm Conrad Röntgen Research Center for Complex Material Systems, Universität Würzburg, Am Hubland, D-97074 Würzburg, Germany

A. M. Fox Department of Physics and Astronomy, University of Sheffield, Sheffield S3 7RH, UK

I. Fushman Koshla Ventures, 3000 Sand Hill Road, Building 3, Suite 190, Menlo Park, CA 94025, USA

B. D. Gerardot Heriot-Watt University, Edinburgh EH14 4AS, UK

J. Guettinger Solid State Physics Laboratory, ETH Zürich, 8093 Zürich, Switzerland

R. Heeres Kavli Institute of Nanoscience, Delft University of Technology, 2628CJ Delft, The Netherlands

M. Hocevar Kavli Institute of Nanoscience, Delft University of Technology, 2628CJ Delft, The Netherlands

M. Huefner Solid State Physics Laboratory, ETH Zürich, 8093 Zürich, Switzerland

S. Hughes Department of Physics, Queen's University, Kingston, Canada, K7L 3N6

S. Höfling Technische Physik, Physikalisches Institut and Wilhelm Conrad Röntgen Research Center for Complex Material Systems, Universität Würzburg, Am Hubland, D-97074 Würzburg, Germany

T. Ihn Solid State Physics Laboratory, ETH Zürich, 8093 Zürich, Switzerland

J. G. Keizer COBRA Inter-University Research Institute, Eindhoven University of Technology, P.O. Box 513, NL-5600MB Eindhoven, The Netherlands

P. M. Koenraad COBRA Inter-University Research Institute, Eindhoven University of Technology, P.O. Box 513, NL-5600MB Eindhoven, The Netherlands

O. Krebs CNRS-Laboratoire de Photonique et de Nanostructures, Route de Nozay, 91460 Marcoussis, France

J. Lapointe Institute for Microstructural Sciences, National Research Council, Ottawa, Canada, K1A 0R6

C. Le Gall Institut Néel, CNRS & Université Joseph Fourier, 25 rue des Martyrs, 38042 Grenoble, France

A. Lemaître CNRS-Laboratoire de Photonique et de Nanostructures, Route de Nozay, 91460 Marcoussis, France

C.-Y. Lu University of Cambridge, Cavendish Laboratory, J. J. Thomson Avenue, Cambridge CB3 0HE, UK

A. Majumdar Department of Electrical Engineering, Stanford University, Stanford, CA 94305, USA

X. Marie Université de Toulouse, INSA-CNRS-UPS, LPCNO, 135 Avenue de Rangueil, 31077 Toulouse, France

H. Mariette Institut Néel, CNRS & Université Joseph Fourier, 25 rue des Martyrs, 38042 Grenoble, France

C. Matthiesen University of Cambridge, Cavendish Laboratory, J. J. Thomson Avenue, Cambridge CB3 0HE, UK

K. Mnaymneh Institute for Microstructural Sciences, National Research Council, Ottawa, Canada, K1A 0R6

F. Molitor Solid State Physics Laboratory, ETH Zürich, 8093 Zürich, Switzerland

J. R. Petta Department of Physics, Princeton University, Princeton, New Jersey 08544, USA

P. J. Poole Institute for Microstructural Sciences, National Research Council, Ottawa, Canada, K1A 0R6

A. J. Ramsay Department of Physics and Astronomy, University of Sheffield, Sheffield S3 7RH, UK

M. E. Reimer Kavli Institute of Nanoscience, Delft University of Technology, 2628CJ Delft, The Netherlands

H. Ribeiro Department of Physics, University of Konstanz, D-78457 Konstanz, Germany

C. Schneider Technische Physik, Physikalisches Institut and Wilhelm Conrad Röntgen Research Center for Complex Material Systems, Universität Würzburg, Am Hubland, D-97074 Würzburg, Germany

S. Schnez Solid State Physics Laboratory, ETH Zürich, 8093 Zürich, Switzerland

M. D. Schroer Department of Physics, Princeton University, Princeton, New Jersey 08544, USA

P. Senellart Laboratoire de Photonique et de Nanostructures, CNRS, Route de Nozay, 91460, Marcoussis, France

A. J. Shields Toshiba Research Europe Limited, 208 Cambridge Science Park, Cambridge CB4 0GZ, UK

C. Stampfer Solid State Physics Laboratory, ETH Zürich, 8093 Zürich, Switzerland

R. M. Stevenson Toshiba Research Europe Limited, 208 Cambridge Science Park, Cambridge CB4 0GZ, UK

A. Tartakovskii Department of Physics and Astronomy, University of Sheffield, Sheffield S3 7RH, United Kingdom

V. V. Temnov Institut des Molécules et Matériaux du Mans, UMR CNRS 6283, Université du Maine, 72085, Le Mans cedex, France

B. Urbaszek Université de Toulouse, INSA-CNRS-UPS, LPCNO, 135 Avenue de Rangueil, 31077 Toulouse, France

A. N. Vamivakas University of Cambridge, Cavendish Laboratory, J. J. Thomson Avenue, Cambridge CB3 0HE, UK
J. Vučković Department of Electrical Engineering, Stanford University, Stanford, CA 94305, USA
K. Wang Department of Physics, Princeton University, Princeton, New Jersey 08544, USA
R. J. Warburton Department of Physics, University of Basel, CH4056 Basel, Switzerland
R. L. Williams Institute for Microstructural Sciences, National Research Council, Ottawa, Canada, K1A 0R6
B. J. Witek Kavli Institute of Nanoscience, Delft University of Technology, 2628CJ Delft, The Netherlands
U. Woggon Technische Universität Berlin, Straße des 17. Juni 135, 10623 Berlin, Germany
Y. Zhao University of Cambridge, Cavendish Laboratory, J. J. Thomson Avenue, Cambridge CB3 0HE, UK
V. Zwiller Kavli Institute of Nanoscience, Delft University of Technology, 2628CJ Delft, The Netherlands

Preface

Semiconductor quantum dots (QDs) have been extensively researched in the past 20 years or so. Over this period, the field has been stimulated by various motivating factors from fabrication of low-threshold temperature-insensitive QD lasers to the use of single spins for quantum computing and single dots for medical markers. In the past decade, refinement of fabrication and experimental techniques enabled researchers in the field to routinely use single QDs to access and control single electrons and holes and their spins, and to generate non-classical light. The focus of this book is on control of optical and transport properties of single and few QDs. The remarkable progress in this fast-developing field in the past three to five years is reported.

The term "quantum dot", widely used from late 1980s, usually refers to a semiconductor nano-structure. Typical sizes of a quantum dot range from a few nanometers in colloidal dots (also referred to as nano-crystals) to a few hundred nanometers in lithographically fabricated electrostatic structures, so that on average they contain from 10^3 to 10^6 atoms. The small physical size is the main common characteristic feature of quantum dots made from different materials and using various fabrication methods. It is usually combined with additional methods for electron energy engineering, for example, surrounding the dot with a higher band-gap semiconductor, applying gate-voltage creating a higher potential barrier around the dot, etc. This gives rise to the most important basic property of QDs: the motion of electrons and holes in QDs is suppressed in all three dimensions. This property gave quantum dots the name of zero-dimensional ("0D") structures. A result of the full confinement is the complete quantization or discretization of the energy spectrum of a confined charge carrier. This is in great contrast to higher-dimensionality structures such as quantum wells (QW), where the spectrum is continuous for the particle motion in the plane of the QW, and quantum wires where quantization is present only in two directions, and the third, along the wire, is described by an energy continuum. The quantized energy spectrum and complete localization also resembles the behavior of electrons in atoms, which gave quantum dots the title of "artificial atoms". This description emphasizes the increased isolation of confined electrons and holes from the environment outside the dot, one of the most attractive properties widely exploited in applications ranging from quantum dot lasers for telecommunications to QD-based spin-qubits for quantum computing.

According to the Web of Science, studies of quantum dots have resulted in around 27 000 publications since 1987, with increasing number of publications each year exceeding 2500 per year since 2005. Quantum dot research embraces variety of topics in physics, electrical and electronic engineering, chemistry, material science, biology and medicine, where, according to the demands of a specific application, different types of QD structures are employed. The work of the most prolific authors addresses topics such as QD design and growth, devices based on epitaxial self-assembled dots (lasers, photonic cavities, etc.), colloidal dots as fluorescent markers for biological and medical applications, optics of single epitaxial dots, spin qubits in optically active and gate-defined dots, single electron transport, optics of colloidal dots etc.

This book will present the latest advances in QD research in the past three to five years, embracing some of these most active areas including fabrication, optical properties and electron transport. The material presented in this book is divided in to six parts describing: (1) nanostructure design and structural properties of epitaxially grown quantum dots and nanowires; (2) manipulation of individual quantum states in quantum dots using optical techniques; (3) optical properties of quantum dots in photonic cavities and plasmon-coupled dots; (4) magnetic ions and nuclear spins in a dot; (5) electron transport in quantum dots fabricated by lithographic techniques in III–V semiconductors and graphene; (6) single dots for future telecommunications applications. Below I will briefly review each of these parts.

(1) **Nanostructure design and structural properties of epitaxially grown quantum dots and nanowires.** The three chapters here will review epitaxial growth of III–V quantum dots (Chapter 1), a relatively novel addition to the field – QDs embedded in semiconductor nanowires enabling new flexibility with the material and substrate choice (Chapter 2), and finally ultra-sensitive techniques for structural characterization, delivering information on QD dimensions and material distribution inside nano-structures (Chapter 3).
(2) **Manipulation of individual quantum states in quantum dots using optical techniques.** This part deals with advanced optical methods used for control and characterization of holes and electrons and importantly their spins in QDs. Chapter 4 presents applications of differential transmission for spectroscopy of spin-Λ system, with a focus on manipulation of the hole spin – a candidate for a decoherence free spin qubit, a basic building block for quantum computing. Chapter 5 focuses on resonance fluorescence describing the first application of this techniques to the atomic-like two-level system available in single QDs. Chapter 6 describes coherent control of excitons using ultra-fast optical pulses and photo-current detection in single QDs. This chapter focuses on fidelity of coherent operations and reasons leading to damping of exciton Rabi oscillations. Chapter 7 expands the description of hole properties in QDs presented in Chapter 4 to coupled QD structures, and reports on optical properties of these complex systems measured in photoluminescence of single QD "molecules".

(3) **Optical properties of quantum dots in photonic cavities and plasmon-coupled dots.** Work presented in this part describes how optical properties of QDs are modified when they are inserted in optical cavities or interact with plasmons when positioned in proximity of a metal surface. Chapter 8 reports on efforts for deterministic light–matter coupling using single quantum dots positioned in micropillar cavities fabricated using novel in-situ low-temperature lithography methods. Chapter 9 presents optical investigations carried out on another type of cavities comprising QDs: missing-hole defects in thin membrane photonic crystals made of GaAs. Chapter 10 discusses photon statistics in QD microlasers emission based on a GaAs micropillar design. Chapter 11 addresses a different type of QDs, the only example of colloidal dots, or nano-crystals, in this book. Taking advantage of their free surface, nano-crystals coupled to plasmon excitations in metallic nano-structures can easily be realized. Optical studies of such hybrid structures are presented in this chapter.

(4) **Quantum dot nano-laboratory: magnetic ions and nuclear spins in a dot.** This chapter addresses spin effects related to behavior of nuclei in atoms constituting quantum dots, and spin effects related to transition-metal impurities located inside the dot. These phenomena effectively occur on the length-scales smaller than the physical dimensions of the dot. A QD acts as an enclosure, or a "laboratory", where these nano-scale processes can be probed via the important changes they produce in the properties of the electrons and holes confined in QDs. Chapter 12 considers self-assembled CdTe/ZnTe QDs containing single Mn atoms. Optical manipulation of the spin of a single magnetic atom (Mn) in the solid-state environment is demonstrated. This chapter also provides a brief overview of II–VI QDs fabrication. Chapter 13 considers a challenging goal of doping III–V InGaAs dots with single Mn impurities, and reports on magneto-optics of such dots. Chapter 14 reports on optical manifestation of a variety of spin effects in QDs associated with nuclear spin polarization.

(5) **Electron transport in quantum dots fabricated by lithographic techniques from III–V semiconductors and graphene.** This section of the book touches on a very large field of electron transport in quantum dots fabricated using lithography. GaAs-based structures are considered in Chapter 15, where experiments are reported on electron-spin control in laterally coupled double dots, work inspired by the Loss–DiVincenzo proposal for spin-based quantum computing. The theory of quantum control of the singlet–triplet qubit is outlined in Chapter 16, focusing specifically on the electron nuclear spin interaction in III–V QDs and carbon-based structures. Chapter 17 reports on transport studies of single–layer graphene nano-structures including spectroscopy in magnetic field and scanning-gate microscopy with high spatial resolution.

(6) **Single dots for future telecommunications applications.** This part reports on work where single quantum dots are researched for possible future applications in secure optical communications and linear optics quantum computing. Chapter 18 addresses generation of entangled photon pairs using InGaAs/GaAs quantum dots: the first electrically driven entangled light source, and the first photon pair source where the entanglement of the emitted photons can be activated or deactivated

electrically is considered. Chapter 19 discusses a site-control technique for fabrication of self-assembled InAs/InP quantum dots emitting at telecommunication wavelengths enabling deterministic coupling of single dots to high finesse microcavities. Such devices will form the basis of efficient sources of single photons and entangled photon pairs for telecommunications applications.

Alexander Tartakovskii

Part I

Nanostructure design and structural properties of epitaxially grown quantum dots and nanowires

1

Growth of III–V semiconductor quantum dots

C. Schneider, S. Höfling and A. Forchel

1.1 Introduction

The advanced growth of semiconductor quantum dots (QDs) with high optical quality is one key to realizing novel devices in various research disciplines related to modern semiconductor technologies. Despite recent progress in the top-down lithographic fabrication of single semiconductor QD-like emitters [43], bottom-up fabrication methods are commonly applied for the realization of high quality light emitting QDs [7]. The exploitation of high-density QD arrays as an active material in laser diodes [17] and vertical cavity surface emitting lasers (VCSEL) resulted in a new class of devices featuring lower lasing thresholds and improved device performance (such as temperature stability) in comparison to devices with higher dimensional gain [2, 3].

Spintronics and quantum information processing are intensively studied fields in order to provide complementary or entirely novel routes to a future information technology [12, 5]. Single self-assembled quantum dots (QDs) grown in low-density arrays are promising candidates for realizing functional building blocks in these research fields as they allow to confine single charge carriers or spin carriers while providing a solid-state platform capable of electrical injection or manipulation. Extensive research has resulted in the realization of various devices relying on few or single QDs, for instance quantum light emitters such as single photon sources [23, 36], sources of pairs of entangled photons [1, 45], and few- to single-QD lasers [38]. Moreover, many milestone experiments have been performed with In(Ga)As QDs grown on GaAs, for instance the observation of the Purcell effect [10] or reaching the strong QD exciton–photon coupling regime [26], demonstration of the indistinguishability of photons emitted by a QD [29], or complete control of a single spin by ultrafast optical pulses [25]. A desirable feature of these single QDs is a very high optical quality.

Material engineering and epitaxial growth techniques on GaN [15], GaAs and InP substrates have paved the way to realize optically active QD structures and QD-based devices in a wide spectral range throughout almost the entire visible spectral range up to emission wavelength in the important telecommunication bands at 1.3 μm and 1.55 μm.

In this chapter we concentrate on QD growth on GaAs. Please note that site-selected QD growth on InP substrate for long-wavelength applications (e.g. [8]) will be addressed in

Quantum Dots: Optics, Electron Transport and Future Applications, ed. Alexander Tartakovskii. Published by Cambridge University Press.

chapter 19 of this book. One main focus of the present summarized work lies on the growth of low-density QD arrays emitting in various wavelength regimes. In order to scalably fabricate single QD-relying devices, it is of the utmost interest to achieve a high degree of control over the position of single QDs in dilute arrays with periods exceeding the device dimensions [32, 9]. Therefore we will summarize various complementary attempts to realize site-controlled QD arrays on an GaAs substrate, suitable for device integration.

1.2 Properties of semiconductor quantum dots

1.2.1 Growth of Stranski–Krastanov and site-controlled quantum dots

In this section, we will briefly descibe the processes involved during the growth of QDs in the Stranski–Krastanov (SK) mode as well as on patterened substrates. During the heteroepitaxial growth of various semiconductor materials with slightly differing lattice constants, such as InAs on GaAs (7% lattice mismatch), the grown material at first usually adapts the in-plane lattice constant (Fig. 1.1a) of the substrate and is grown pseudomorphically. During this growth process, strain accumulates in the layer system (Fig. 1.1b). After a certain layer thickness is exceeded the strain in the coherently strained layer eventually relaxes partially, resulting in the formation of islands on the surface in order to minimize the total layer energy (Fig. 1.1c). By applying the right growth conditions these islands can be free of crystal defects and their dimensions can be comparable with the exciton Bohr radius in the semiconductor. This SK growth mode was observed in a number of III–V semiconductor systems, including (Al,Ga)InAs on (Al)GaAs, (Ga)InP on Ga(In)P or InAs on (Ga)InP. The morphology of the QDs can be widely tailored by altering the strain in the material system, e.g. via the composition in ternary QDs, as will be described in Section 1.3.1. Furthermore, the surface migration during QD deposition is another important tool for actively manipulating the morphology and the density of the QDs. In this context, the substrate temperature and the ratio of group III–V elements during sample growth have a wide impact on these properties.

The strain-driven nucleation of QDs in the SK mode is self-organized and hence the QDs are usually randomly distributed over the sample. In order to obtain a higher degree of control over the QD positions, several concepts were developed to grow ordered arrays of QDs. The basic mechanism in most of these concepts is to obtain control over the

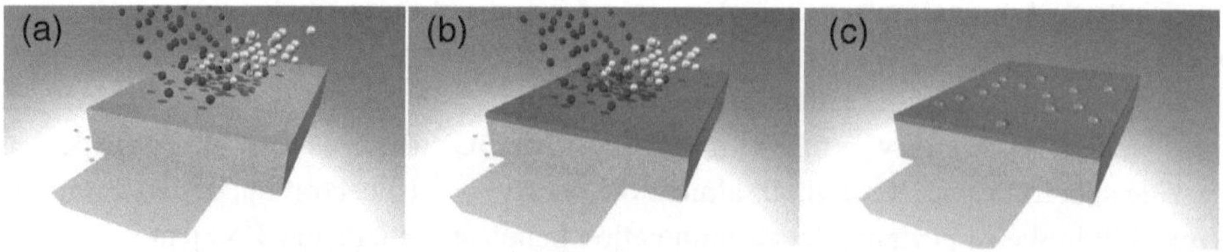

Figure 1.1 Growth of QDs in the Stranski–Krastanov mode. (a) Deposition of the QD material on a GaAs surface. (b) Evolution of a strained wetting layer. (c) Nucleation of coherent islands.

surface binding energies or surface migration length of the adatoms. These adatoms preferentially nucleate at positions with a lower chemical surface potential. Hence, the nucleation of QDs can be actively manipulated by artificially defining crystal edges, nanopits or pyramids on a crystal surface by lithographic techniques before growing the QDs. We refer the reader to reference [31] for a detailed introduction into site-selected QD growth techniques.

1.2.2 Electronic and spectral properties of quantum dots

The unique characteristics of semiconductor QDs originate in pronounced quantization effects and the discrete density of states (DOS) of the quasi zero-dimensional structures. Like single atoms, QDs exhibit a shell structure and the occupation of these shells follows Pauli's principle. A feature that is unique to these 'artificial atoms' is the concept of excitons. These are quasi particles composed of a negatively charged electron in the conduction band and a positively charged hole in the valence band of the semiconductor. Owing to attractive coulomb forces, the electron and the hole form an exciton which can be spatially confined inside the QD due to the energetic band offsets in the heterostructure. The delocalization area of the exciton in bulk material is often described by the exciton Bohr radius, $a = \frac{4\pi\epsilon_0\epsilon_r\hbar^2}{\mu e^2}$, which is usually comparable to the extension of the QD.

The calculated first two single particle eigenenergies of a truncated pyramidal $Ga_{0.5}In_{0.5}As$ QD in the electron and the hole band are shown in Fig. 1.2a, revealing the quantization of the energy levels. Because of the larger effective mass, the lowest energy state of such a system is formed with the heavy hole (HH) in the valence band. In an ideal system, only the indicated transitions between 1s-1HH and 2s-2HH are allowed due to the symmetry of the wavefunctions. This atom-like shell structure can be observed in single QD studies as described in the experimental part of this chapter. Owing to the self-organized fabrication process of the QDs, inhomogeneities in size and material

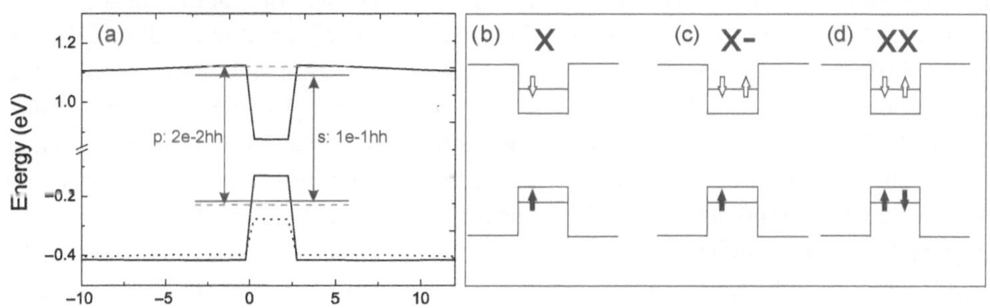

Figure 1.2 (a) Calculated energy structure of a truncated pyramidal QD. Discrete energy levels are formed in the electron and hole band, leading to an atom-like shell structure. (b)–(d) Selected charge and spin configurations in the s-shell, often referred to as exciton X, trion X− (X+ analogous) and biexciton (XX).

composition between different QDs are almost unavoidable. These fluctuations directly manifest themselves in an energy spread of the emission lines up to several tens of millielectronvolts of the QDs, e.g. in a photoluminescence (PL) experiment. Therefore, the inhomogeneous broadening of the ensemble PL signal recorded from a large number of QDs gives direct information about the homogeneity of the QDs. In the light of the exploitation of QDs as a gain material in laser structures, the inhomogeneous ensemble broadening is a figure of merit, since a narrow emission spectrum can lead to a higher material gain and hence improved device characteristics. Since the occupation of the QD shells follows Pauli's principle, the ground state of the QD can be occupied by only a maximum of two electrons and two holes of opposite spin. This configuration is often referred to as the biexciton (see Fig. 1.2d), of which the emission energy is usually shifted by coulomb and exchange interaction with respect to the exciton emission energy. Configurations with a single unpaired spin carrier are mainly referred to as trions or charged excitons. This zoo of excitonic complexes is indicated in Fig. 1.2b–d. If more charges are filled into the QD, higher shells are occupied with emission bands at notably higher energies.

1.3 Epitaxial growth of quantum dots on GaAs substrates

In this section, we summarize the basic morphologic and spectral characteristics of QDs grown on GaAs substrate. The influence of material strain and surface migration length on the QD shape and density are addressed exemplarily for the (Ga)InAs/GaAs QD model system. We will furthermore discuss ensemble and single QD emission properties of these QDs. Finally, approaches for achieving high control over the QD positions are reviewed.

1.3.1 Influence of matrial strain on the formation of GaInAs quantum dots

Since material strain is the driving mechanism in the SK growth mode, it has large impact on the QD morphology and therefore on the optical properties. In order to illustrate this, we compare QD samples all containing GaInAs/GaAs QDs with nominal indium contents of 60%, 45%, and 30%. For the $Ga_{0.4}In_{0.6}As$ QDs, 1.4 nm of material was deposited at a surface temperature of 470 °C by molecular beam epitaxy [13]. QDs with an indium content of 45% and 30% were realized by depositing 2.1 nm and 4.5 nm respectively, at 510 °C. The amount of deposited GaInAs material is increased for the QD samples with lower indium content taking account of the larger critical thickness [34].

Figure 1.3 shows scanning electron microscopy (SEM) images of uncapped QD structures. In order to enhance the image contrast, the surfaces were tilted by 70°. The QDs with an indium content of 60% have diameters between 10 nm and 15 nm, and a rather high QD density of about $1\text{–}2 \cdot 10^{11} \frac{1}{\text{cm}^2}$. For single QD applications, the dot density should be much lower. In order to realize dilute arrays of QDs, the substrate temperature for the 45% and 30% In QDs was increased to 510 °C to enhance the migration length [20]. As a result, the QDs with 45% In content are lens-shaped structures with increased diameter

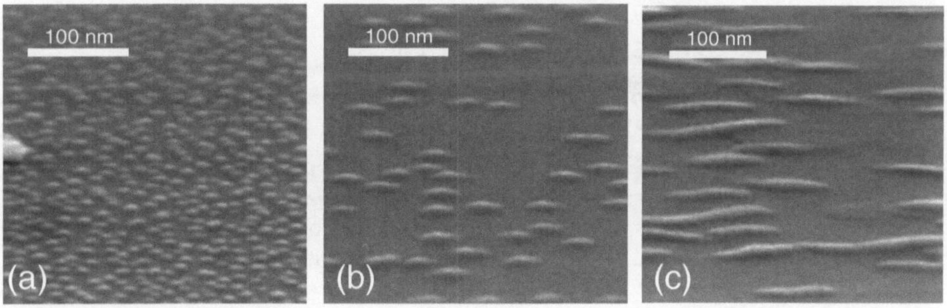

Figure 1.3 Surface SEM images of three uncapped QD samples with different indium contents of (a) 60%, (b) 45% and (c) 30%.

of 20–25 nm and a notably reduced QD density of about $1\text{–}2 \cdot 10^{10} \frac{1}{\text{cm}^2}$. To increase the single QD's volume, the strain of the QD structures was reduced once more by decreasing the In content to 30%. Because of the low-strain condition, the island growth is mainly initiated by crystal steps on the surface. This induces the formation of elongated dot structures in Fig. 1.3c which are preferentially orientated along the $[0\bar{1}1]$ direction with typical lengths of 50–100 nm and widths of about 30 nm as can be extracted from the uncapped structures. The combination of low-density QD growth, the material-related short emission wavelength below 950 nm and large QD volumes resulting in enhanced exciton dipole moments makes these QDs interesting for the study of light–matter coupling.

1.3.2 Quarternary AlGaInAs quantum dots with tailored morphology and emission wavelength

As described above, the morphology of SK QDs is strongly related to the material strain and also to the surface temperature during deposition. The emission wavelength of these QDs cannot be independently tailored in these systems but depend strongly on the material alloy, the QD size and the strain in the islands. In order to independently tailor QD morphology and emission wavelength of the QDs, the growth of quarternary AlGaInAs QDs was established, because the strain mainly depends on the indium fraction while the QD bandgap can be conveniently adjusted by the aluminum vs. gallium fraction [30].

First, we want to clarify the influence of the aluminum content on the QD morphology by studying uncapped QDs on top of an $Al_{0.34}Ga_{0.66}As$ surface. The pronounced dependence of the QD density and the lateral size on the Al and the In content is summarized in Fig. 1.4a, and b for two sample series with a different indium content of 47% and 60%, and varying aluminum contents. It can be clearly seen that, for both indium contents, the QD surface density increases and the average QD diameter decreases with increasing aluminum content.

The fact that the QD density and the average QD size show similar behavior for both moderate and high indium content indicate that these quantities are mainly determined

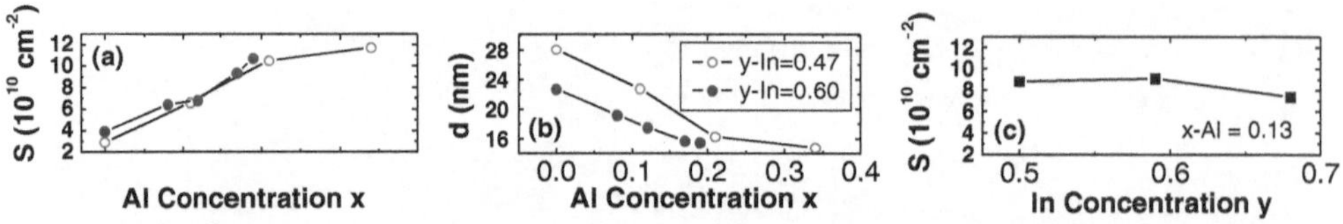

Figure 1.4 (a) QD surface density S and (b) average lateral QD size d for $Al_xGa_{1-x-y}In_yAs$ QDs with two different indium contents ($y = 0.47$ and 0.60) as a function of aluminum content x. (c) QD surface density S of $Al_{0.13}Ga_{0.87-y}In_yAs$ QDs as a function of indium content y.

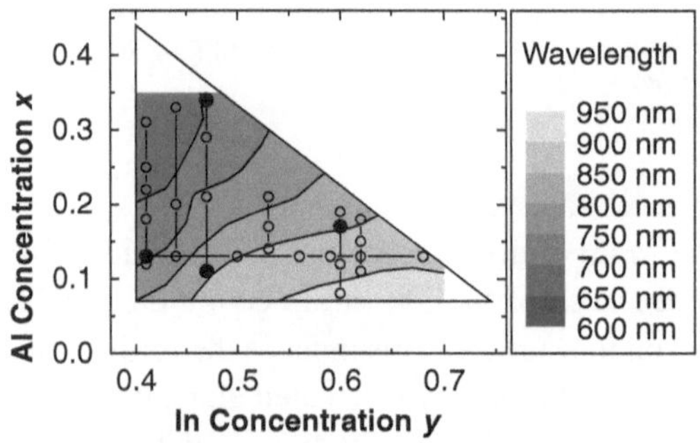

Figure 1.5 QD emission wavelength plotted vs. indium and aluminum content. The emission wavelength is encoded in the gray scale of the graph.

by the aluminum content. Figure 1.4c, which shows the dependence of the QD surface density S on the indium content for a constant aluminum content of $x = 0.13$ supports this conclusion, since only minor changes in the QD density could be observed for indium contents between 50% and 68%. We will now focus on the effect of the aluminum content in the QDs on the emission wavelength of the QD ensembles.

In order to provide electronic confinement in the QDs, the structures were integrated in an $Al_{0.34}Ga_{0.66}As$ matrix for spectroscopic investigations. Figure 1.5 shows an interpolated contour plot of the PL wavelengths (at $T = 8$ K) of 32 QD samples examined as a function of the indium and aluminum contents. Some data points of the systematically acquired grid used to create the figure are indicated in the plot. Apparently the wavelength of the quarternary AlGaInAs QDs can be widely tuned in a range between 660 nm and 940 nm by varying the aluminum and indium contents. Furthermore, it is possible to identify many different QD compositions that would result in the same emission wavelength. Along these iso-wavelength lines it is possible to choose between QD ensembles with morphological properties best suited for the intended application [30].

1.3.3 Morphologic and optical properties of low density InAs quantum dots

The growth of QD arrays with dilute areal densities is a pre-requisite for exploiting the unique properties of the nanostructures in devices relying on single QDs. Typically, the structures that are used in these applications have sizes of several hundred nanometers up to several microns. Therefore it is highly desireable to integrate QD arrays with densities well below $1 \cdot 10^{10} \frac{1}{\mathrm{cm}^2}$ in order to facilitate the observation of single QD related effects. Considering adsorband binding energies and the ability to migrate on a GaAs surface, the most promising candidates to realize very dilute arrays of QDs is the binary InAs material system. During the QD deposition, the morphology of the structures is mainly determined by the surface migration. Hence, a more detailed investigation of the influence of the surface temperature and deposited amount of material is essential. Since the growth rate also has a strong influence on the QD morphology, the surface migration length during QD growth was enhanced by lowering the InAs growth down to $0.008 \frac{\mathrm{nm}}{\mathrm{second}}$. The investigated QD samples were realized by the deposition of 1 nm InAs at substrate temperatures of 510 °C, 513 °C and 517 °C during QD deposition. Figures 1.6a–c show SEM images of the according uncapped QDs. The QD density reduces from a value of $2.34 \cdot 10^{10} \frac{1}{\mathrm{cm}^2}$ for 510 °C down to a value of $5.23 \cdot 10^{9} \frac{1}{\mathrm{cm}^2}$ at 517 °C. Additionally, the lateral extension of the QD increases from 32.7 nm to 39.5 nm when increasing the surface temperature (see Fig. 1.6d).

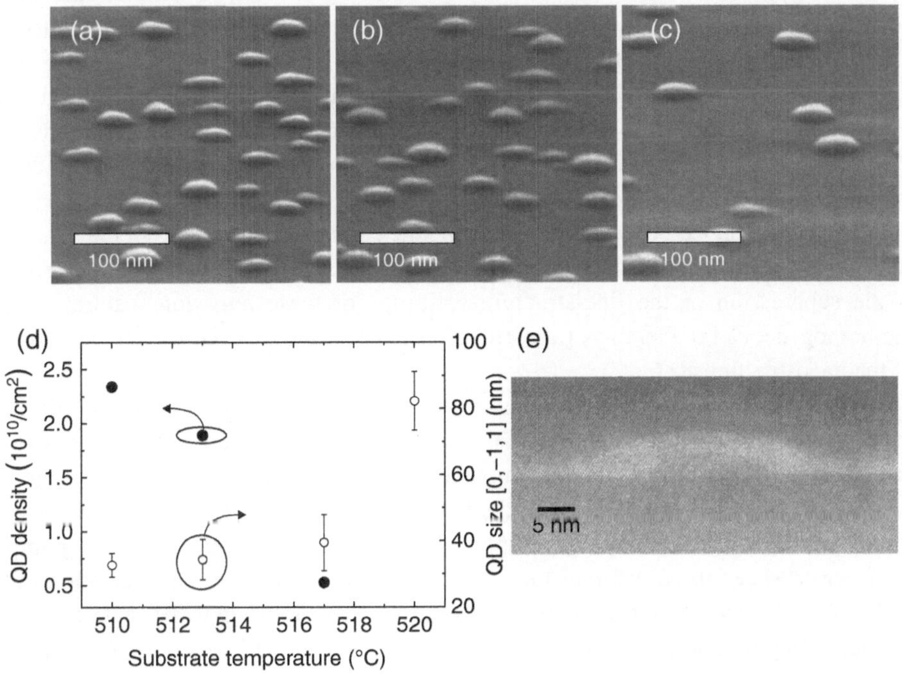

Figure 1.6 SEM images of uncapped InAs QDs grown at substrate temperatures of (a) 510 °C, (b) 513 °C and (c) 517 °C. (d) The QD density decreases with increasing substrate temperature, accompanied by larger lateral dimensions. (e) Scanning transmission electron micrograph of a single InAs QD overgrown by GaAs. The QD features a smooth lens shape with an approximate height of 5 nm on top of a thin wetting layer.

Because it is known that the morphology of QDs can be strongly affected by the overgrowth with GaAs material, additional scanning transmission electron micrographs (STEM) were carried out on buried InAs QDs with a low area density (Fig. 1.6e). The cross-sectional image of the buried QD again reveals the lens shape of the QD, with a QD height of about 5 nm and a baselength of 30 nm on top of a thin wetting layer.

Besides tailoring the morphology of the QDs, a high degree of emission control of the QDs is crucial in order to achieve complex devices and building blocks for QD-based quantum information technology.

- Single QD-based efficient single photon sources. In order to improve the efficiency of the single photon extraction process, QDs are often integrated into optical microcavities. On spectral resonance, the QD emission rate can be significantly enhanced via the Purcell effect, and furthermore the emission can be directed into a collimation lens. In order to realize resonance between the QD and the cavity, the emission wavelength of the QDs needs to be matched to the resonance frequency of the device. This implies the necessity of accurately controlling the emission wavelength of the QDs during growth. Furthermore, since the sensitivity of commercial silicon-based single-photon detectors continuously increases in the spectral range from 1 μm down to wavelengths around 600 nm, the QD emission wavelength should be chosen appropriately. Additionally, it is also interesting to realize QD-based single-photon emitters in the spectral range of the dispersion and absorption minima of optical fibers, namely at wavelengths of 1.3 μm and 1.55 μm [40]. Furthermore, in order to enhance the emission rate of the QDs themselves, it's promising to circumvent long-living dark states in the QDs, for instance by doping the QDs to preferentially create charged excitons [39].
- Polarization entangled photon pairs. Utilizing the cascaded emission from the XX and X states of a single QD, polarization-entangled photons can be generated. One pre-requisite is the suppression of the fine-structure splitting of these emission features (within the homogeneous broadening of the emission lines), caused by structural asymmetries of the nanostructures [36].
- Spin-Qubits. Single spin carriers (electrons or holes) which are spatially confined in a QD are potential candidates for stationary quantum bits (Qubit), and therefore the basic building blocks for quantum computers [18]. Owing to the localization of the spin carriers in InAs QDs effects of spin dephasing via phonon coupling and therefore the loss of information can be notably suppressed. Furthermore, resonant optical pumping of the trion states in an external magnetic field can facilitate the intitialization of the Qubit on a timescale of a few nanoseconds [25] and a ratio of up to 10^5 gate operations per dephasing time has already been demonstrated. For these kinds of experiments, the charged trion state of the QDs especially needs to be controlled.

Since the emission properties of the QDs (especially the emission wavelength) strongly depend on the morphology and alloy [19, 30], it can be challenging to optimize the QD morphology and emission properties independently via the growth conditions and material composition. As discussed in the previous section, very low areal densities of QD ensembles can especially be realized with binary InAs/GaAs QDs. Owing to the rather small InAs bandgap of about 0.415 eV the emission wavelength of these QDs is typically longer than 1 μm. However, the emission wavelength of low-density InAs QDs can be strongly influenced by the overgrowth procedure and widely shifted in a spectral range between 1.3 μm and 900 nm.

In order to realize InAs QDs with short emission wavelengths, one can apply the partial capping and annealing (PCA) technique. In this growth procedure, the InAs QDs are overgrown by a few nanometers of GaAs, and subsequently the uncapped QD tips are desorbed during a growth interruption under arsenic background pressure [44] before growth is completed. During the in-situ annealing procedure, the shape of the QDs is converted into truncated pyramids or flattened lenses, while the QD height can be controlled very accurately via the height of the GaAs cap. Furthermore, during the annealing step, thermal intermixing of the QD and the GaAs matrix can occur, which also results in an even shorter emission wavelength. Figure 1.7a shows ensemble low-temperature (10 K) PL spectra of four samples with different GaAs cap thickness (2, 3, 4 nm and standard InAs QDs without PCA).

The influence of the thickness of the cap on the emission wavelength is evident, and a strong blueshift occurs with decreasing cap thickness. At moderate and high excitation powers, emission from the QD s-shell as well as the first excited p-shell is observed. The ground-state emission strongly shifts from 1150 nm without the PCA technique down to 940 nm for a 2 nm cap. Furthermore, the splitting between the s- and p-states reduces from 58 meV without PCA down to 30 meV for the 2 nm cap.

In contrast, capping QDs by strain and band-gap reducing GaInAs layers can shift the emission wavelength of InAs QDs to larger wavelength into the important regimes around

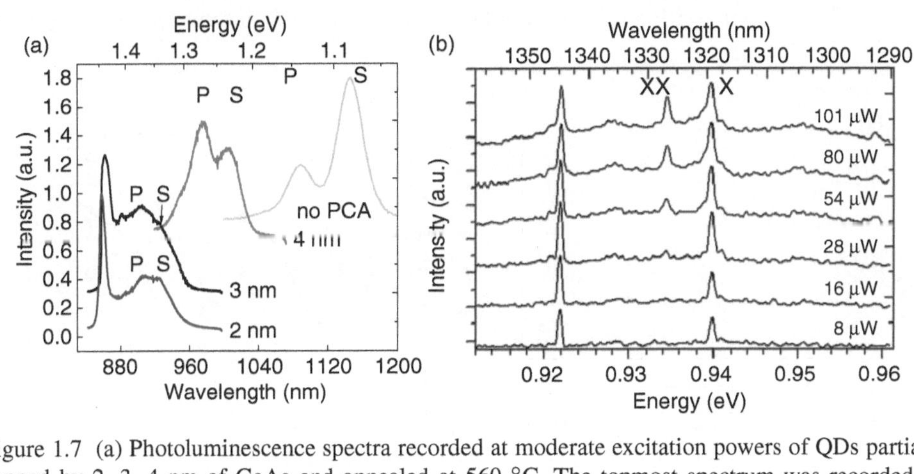

Figure 1.7 (a) Photoluminescence spectra recorded at moderate excitation powers of QDs partially capped by 2, 3, 4 nm of GaAs and annealed at 560 °C. The topmost spectrum was recorded on reference InAs QDs realized without the PCA step. (b) Single QD investigation of an InAs QD capped by a GaInAs–GaInNAs sequence emitting at 1.3 μm.

1.3 μm for fiber communications. It has been demonstrated that a GaInAs cap that is several nanometers thick can preserve the QD height and simultaneously reduces the electronic confinement in the QD, both resulting in a spectral redshift [11]. A further spectral redshift compared to GaInAs capped InAs QDs can be obtained by adding dilute amounts of nitrogen into the QD material or the surrounding barrier. This furthermore reduces either the QD band-gap or the barrier height around the QDs. Even though the intensity of the QD emission is typically reduced by introducing nitrogen, engineering of the QD cap can be used to maintain a high QD quality. Namely, opening the nitrogen source after deposition of several nanometers of GaInAs still results in a strong spectral redshift, but bright QD emission. This is demonstrated in Fig. 1.7b on a QD emitting around 1.3 μm. The narrow single QD emission line exhibits signatures of X and XX with a binding energy of about 5 meV [40].

For the important application of QDs in single-photon sources, the trion state of the QD plays a major role since the creation of dark states is suppressed in charged excitons. Furthermore, it was demonstrated that the intrinsic angular momentum of the excess electron or hole can carry the information in quantum computing schemes [25]. Therefore, it is interesting to realize QDs with strongly pronounced trionic features. In a relatively straightforward approach, a single layer of silicon atoms is introduced 10 nm below the QDs with a layer density comparable to the QD density. Since Si is an electron donator in GaAs it can statistically charge the QDs, and the emission signature of the QDs can reveal trionic features.

Figure 1.8a shows a typical spatially resolved QD spectrum at varying excitation powers. At moderate excitation powers, three emission lines from the QD dominate the spectrum (X (932.3 nm), X− (935.87 nm) and XX (936.22 nm)) accompanied by two emission features at ≈ 934.7 nm (most likely from a neighbouring QD). The power-dependent intensities of the X ($I \propto P^{0,90}$) and XX ($I \propto P^{1,69}$) attributed emission lines suggest the excitonic and biexcitonic origin, with a XX binding energy of 4.45 meV. In the spectral analysis, it

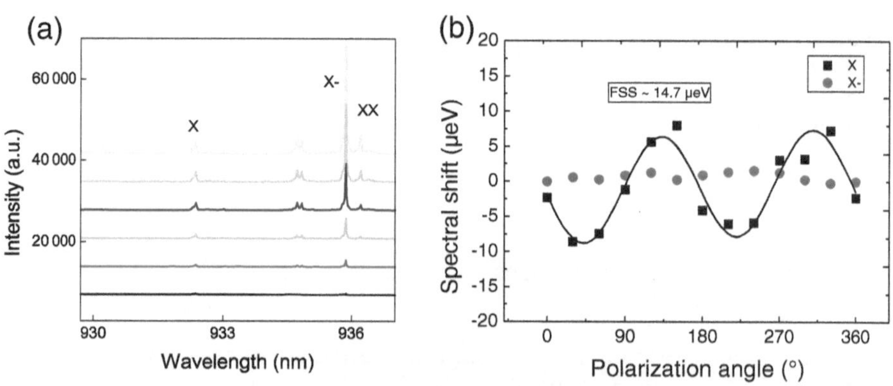

Figure 1.8 (a) Power-dependent μPL spectra of a single InAs QD statistically doped with an additional electron. (b) Polarization-resolved measurement of the X and X− attributed emission line.

is worth noting that the X− attributed emission line is much more intense than the neutral exciton. In order to further support that the X− related emission stems from a trion and not rather from a bright excitonic emission from another QD, polarization-resolved photoluminescence studies were carried out. The recombination of the X state should show some fine structure caused by the exchange interaction between electron and hole with the hamiltonian for the electron spin S and the hole angular momentum J (see e.g. [6]):

$$H_{Ex} = - \sum_{i=x,y,z} a_i J_i S_i + b J_i^3 S_i . \tag{1.1}$$

In the trion configuration the electrons in the conduction band are in a singlet configuration with a complete angular momentum of 0, resulting in a vanishing exchange interaction. Figure 1.8b shows polarization-dependent measurements on both the X and the X− related emission feature: the strongly pronounced oscillation of the X line indicate a fine structure splitting of 14.7 μeV which is comparable with data published elsewhere [33]. In contrast, the oscillation of the X− related line is suppressed, strongly suggesting trionic origin. These results imply that the statistical doping of the QDs can efficiently generate trionic states in the QDs while the optical quality of the QDs (in terms of linewidth and brightness) remains excellent.

1.3.4 Site-controlled quantum dot growth

The growth techniques that have been discussed in this chapter so far were all based on self-organization of QDs in the SK growth mode. For various applications in the field of cavity quantum electrodynamics, nanophotonics, nanoelectronics and quantum information processing, precisely controlling the QD position is of high interest. The absolute position control over arrays of QDs requires a combination of site-controlled QD growth, and furthermore the capability to retrieve the QDs after overgrowth [32]. From a technological point of view, the performance of both fabrication steps with a sufficiently high accuracy is very challenging, but it also paves the way for the scalable fabrication of spatially coupled single QD-resonator devices.

Within this section, we review recent progress in the growth of single site-controlled QDs (SCQDs) grown on (100) oriented GaAs substrate, which is most commonly used for the realization of high-quality photonic devices. The focus is set on the realization of dilute arrays of SCQDs with lattice periods larger than typical device dimensions, since this is a key requirement for the integration of single QDs in quantum devices. The SCQD growth is based on pre-patterning the substrate by etching small pits into the GaAs surface; these pits lower the chemical potential and hence enforce QD nucleation at the site. The possibility of post-growth device alignment is given by the definition of lithographic alignment marks via optical lithography and etching prior to QD growth.

In order to optimize the morphology of SCQDs, it is crucial to monitor the change in surface morphology after overgrowth of the nanoholes. As we will discuss later in this section, the vertical distance between the SCQD and the nucleation centre is of the utmost

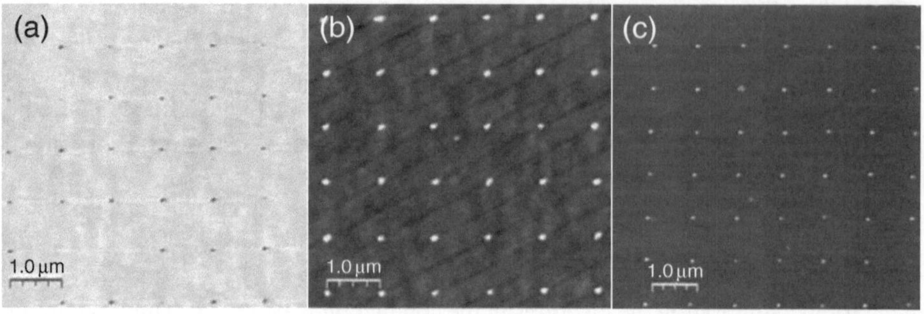

Figure 1.9 Stacked growth of SCQDs on large periods. (a) After growth of a 8 nm thick GaAs buffer layer, an InAs seeding layer is deposited, slightly filling the nanoholes. (b) After deposition of another 10 nm GaAs/AlGaAs/GaAs barrier, the SCQDs are grown. (c) Image of a third SCQD layer separated by 10 nm from the second layer.

importance for the optical quality of the SCQD. In order to maintain a high occupancy of single QDs per nucleation site while increasing the vertical distance between the etched crystal spot and the SCQD, stacked QD growth is a promising approach. In such a growth scheme, the position control of the QDs is maintained by vertical strain coupling between the QD layers.

Figure 1.9 shows atomic force microsopy (AFM) images of three uncapped samples in order to visualize the impact of strain coupling on the formation of ordered SCQD arrays. In Figure 1.9a, a 8 nm thick GaAs buffer and a 0.75 nm thick InAs layer were deposited on top of dry-etched nanoholes at a growth temperature of about 530 °C, resulting in some material accumulation in the nanoholes without forming optically active QDs. The importance of this first InAs layer is shown in Fig. 1.9b: in this sample, the infilled holes are capped by 2 nm of GaAs, and after a 2 minute annealing step at 560 °C, a 10 nm thick GaAs/AlGaAs/GaAs layer sequence is deposited. On top of this layer, 1 nm of InAs is deposited, resulting in the accurate formation of single SCQDs on each site due to the vertical strain coupling.

The stacked QD growth could also be extended to three layers of QDs, as shown in Fig. 1.9c. The third QD layer, which is strain coupled to the second QD layer (separated by a 10 nm thick GaAs/AlGaAs/GaAs sequence), also exhibits excellent QD long-range ordering.

Regarding single QD studies and investigations of coupled QD-resonator systems, SCQDs with a high optical quality are tremendously important and extensively sought after. If the vertical distance between the SCQD and the etch site is too small, the QD linewidth, which is a good indication of the optical quality, is strongly broadened due to defect induced spectral diffusion [4, 32]. The effect of linewidth broadening should therefore strongly depend on the vertical distance between the QD and the etch site and furthermore on the etch technique. The results on single QD investigations are summarized in Fig. 1.10a, where statistical evaluations of the single QD linewidth are shown

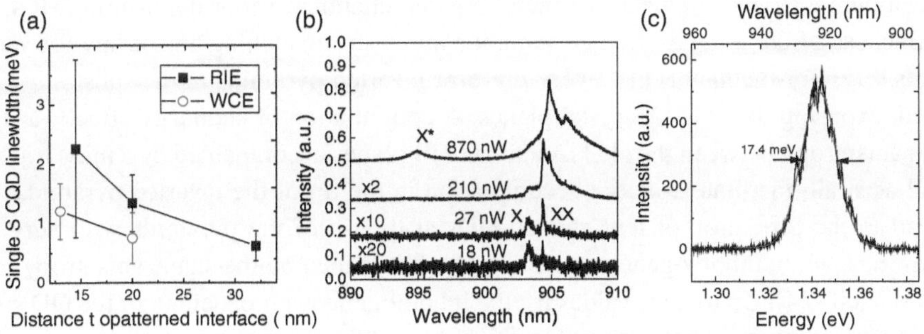

Figure 1.10 Optical properties of SCQDs. (a) The linewidth of single SCQD related emission strongly depends on the etching procedure, and furthermore on the vertical distance between QD and nanohole. (b) μPL investigation of a single SCQD: the QD features a narrow linewidth of 100 μeV, and emission from biexcitonic and p-shell related transitions. (c) Ensemble PL spectrum of an array of SCQDs with a 200 nm pitch.

with respect to the QD–nanohole distance for wet etched (WCE) and dry etched (RIE) nanoholes. As expected, for both kinds of etch techniques the average linewidth decreases when the spacing is increased (up to a factor of 4 for QDs grown on RIE-defined nanoholes, a factor of 2 for WCE-defined nanoholes), and the more gentle WCE technique generally results in improved QD emission characteristics. While the average single QD linewidth still exhibits average linewidth values of about 600 μeV even in the gently WCE-etched samples, some SCQDs with linewidths in the range of 100 μeV could be identified [32]. A power series in a μPL experiment performed on a single SCQD grown on WCE nanoholes and an InAs seeding layer is shown in Fig. 1.10b. The power dependence of the lines reveal emission from excitonic and biexcitonic origin, while at higher pump powers the first excited state from the QD occurs at 1.385 eV. At very low pump powers, the FWHM of the single QD XX line is as small as 100 μeV, which is highly encouraging for further optimization steps. Indeed, recent progress in the growth of high-purity SCQDs has allowed for the observation of very narrow emission lines down to 43 μeV [14] in similar QD structures. Additionally, the directed nucleation of QDs can lead to an improved QD homogeneity as a consequence of a defined amount of material that can contribute to the QD nucleation on each site. Thus, the ensemble broadening of SCQDs grown under optimized conditions can be strongly reduced compared to SK QDs typically exhibiting inhomogeneous linewidth above 30 meV. Figure 1.10c shows an ensemble PL spectrum of a single layer of SCQDs, grown in a similar manner to the uncapped structures in Fig. 1.9b but with a 200 nm pitch and wet chemically defined nanoholes. The ensemble exhibits an inhomogeneous broadening of only 17.4 meV, which is an important step towards the realization of QD structures suitable for spectral and spatial deterministic device integration.

An interesting and promising alternative to SCQD growth on (100) oriented substrate is the growth of QDs in inverted pyramids etched into a (111)B oriented GaAs surface, which has been demonstrated by means of metal organic chemical vapor deposition (MOCVD) [31]. Because of the unique crystal orientation and a strong etching anisotropy for various solvents (e.g. Br_2-methanol) [11], very uniform inverted pyramids in the substrate can be realized, exposing the (111)A crystal planes. A combination of capilarity effects, a strong growth anisotropy between the (111)A and (111)B planes accompanied by a mass transport as well as strain facilitate material accumulation in the tip of the inverted pyramids. This can lead to the formation of lens-shaped QDs at the tip of the pyramids with very high uniformities, while inhomogeneities are mainly determined by the shape anisotropy of the pyramid itself [24]. Up to now, values for the inhomogeneous broadening of the QD s-shell as small as 0.8 meV have been reported, which is more than an order of magnitude narrower than values for SK (Ga)InAs QDs or site-controlled (Ga)InAs grown on a (100) substrate. Furthermore, similar QDs with very high optical quality have been demonstrated by QD growth in very large inverted pyramids and a combined back-etching process [22] where single QD linewidth values down to 18 μeV were measured. Another technique which has been applied to grow high-quality SCQDs in the InAs/GaAs system featuring a single QD linewidth below 100 μeV is the growth on SiO_2 nanotemplates [42].

1.3.5 InP quantum dots on GaAs substrates

Since the efficiency of most single-photon detectors is a sensitive function of the emissison wavelength, it is of great interest to realize QDs on a GaAs substrate emitting in the visible spectral range. Owing to the rather large band-gap of InP and a lattice mismatch on a GaAs substrate comparable to the InAs/GaAs system, the growth of a few monolayers of InP on an (Al)GaInP surface (which can be grown lattice matched to the GaAs substrate) results in the formation of SK QDs with emission bands deep in the red spectral range (see [27, 37]). Analogous to the growth of InAs/GaAs QDs, by enhancing the migration length during QD deposition, the sheet densitites can be reduced by orders of magnitude, making these QDs suitable for single-dot investigations [27a]. By variation of either the QD composition (e.g. by adding gallium or aluminum into the QD material) or the barrier height of the QDs, the emission wavelength can be widely tuned by several hundred millielectronvolt from the red into the green emission band [28], covering the spectral ranges of highest sensitivity of commercially available single-photon detectors. The high optical quality of these InP/GaInAs QDs is demonstrated in single QD investigations. Figure 1.11a shows a series of spectra recorded at various excitation powers: narrow linewidth emission features of both the exciton and the biexciton are clearly resolved in this study with a binding energy of 5.9 meV. Owing to the high purity of the single QD features, the good spectral tunability in the visible range and the availability of dilute QD arrays, the InP QD system should be considered as a promising alternative to the more frequently studied InAs/GaAs system. The capability to emit single photons on demand under pulsed optical excitation is proven in the antibunching experiment in Fig. 1.11b, revealing a value of $g^2(\tau = 0) = 0.09$. The possibility of growing site-controlled InP/GaInP QDs with a high degree of long-range ordering with QD pitches exceeding one micrometer was reported recently [27a]. This is

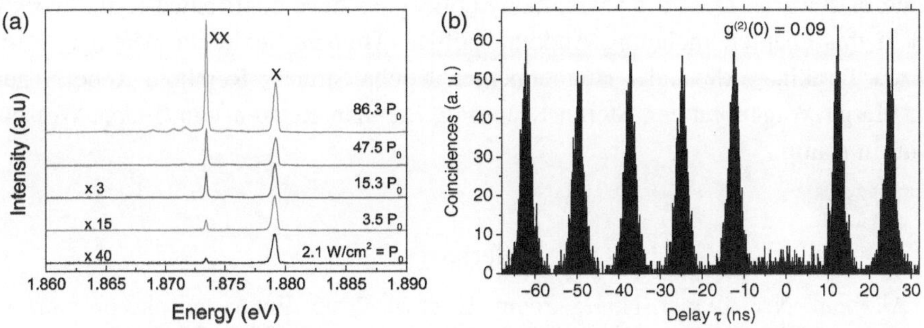

Figure 1.11 μPL images of a single InP QD recorded at various excitation powers. The spectrum reveals clear signatures of X and XX emission features.

highly promising for the deterministic integration of single red emitting QDs into photonic resonators.

1.4 Quantum dot growth on InP substrates

For long distance quantum communication via single mode fibers, single QDs emitting at wavelengths around 1.3–1.55 μm are essential. The 1.55 μm transmission band in glass fibers offers small optical losses, yet up to now there are no reports on high-quality QD single-photon emitters on GaAs substrates emitting in this wavelength regime. A promising alternative for achieving high-quality QDs at 1.55 μm is given by the growth of InAs QDs on InP substrate, as described in references [16, 41].

Another way to realize zero-dimenional QD emitters that are widely tunable in the spectral range around 1.55 μm are so-called InAs quantum dash structures grown on $Al_{0.24}Ga_{0.23}In_{0.53}As$, which is lattice matched to the InP substrate. Despite the rather large dimensions of the nanostructures, zero-dimensional electronic confinement was confirmed e.g. in μPL investigations in [21, 35]. For position-controlled growth of QDs on InP substrates we refer readers to Chapter 19.

1.5 Conclusion

The results that have been summarized in this chapter demonstrate that almost the entire visible spectral range, as well as parts of the near infrared spectrum (including the important telecommunication bands), can be covered by high quality QD emitters fabricated by epitaxial techniques on a III–V semiconductor platform. Furthermore, complementary techniques to gain control over the position of QDs are currently being developed in the various material systems to facilitate scalable integration of single QD emitters in micro- and nanoresonator devices.

The authors acknowledge financial support of the Federal German Ministry of Education and Research, the Deutsche Forschungsgemeinschaft, the European Commission and the State of Bavaria. Over the years, many colleagues have contributed to the work presented in this chapter, including Andreas Löffler, Thomas Schlereth, Matthias Lermer, Nadeszda Tarakina, Alexander Huggenberger, Micha Strauss, Reinhold Rödel, Theresa Steinl, Margit Wagenbrenner, Stefan Kremling, Stephan Reitzenstein, Lukas Worschech and Martin Kamp.

References

[1] Akopian, N., Lindner, N. H., Poem, E. *et al.* 2006. Entangled photon pairs from semiconductor quantum dots. *Physical Review Letters*, **96**, 130501.

[2] Arakawa, Y. and Sakaki, H. 1982. Multidimensional quantum well laser and temperature-dependence of its threshold current. *Applied Physics Letters*, **40**, 939–941.

[3] Asada, M., Miyamoto, Y. and Suematsu, Y. 1986. Gain and the threshold of 3-dimensional quantum-box Lasers. *IEEE Journal of Quantum Electronics*, **22**, 1915–1921.

[4] Atkinson, P., Kiravittaya, S., Benyoucef, M., Rastelli, A. and Schmidt, O. G. 2008. Site-controlled growth and luminescence of InAs quantum dots using in situ Ga-assisted deoxidation of patterned substrates. *Applied Physics Letters*, **93**, 101908.

[5] Awschalom, D. D., Loss, D. and Samarth, N. 2002. *Seminconductor Spintronics and Quantum Computation*. Heidelberg: Springer.

[5a] Baumann, V., Stumpf, F., Schneider, C., *et al.* 2012. Site-controlled InP/GaInP quantum dots emitting single photons in the red spectral range. Applied Physics Letters, **100**, 091109.

[6] Bayer, M., Ortner, G., Stern, O. *et al.* 2002. Fine structure of neutral and charged excitons in self-assembled In(Ga)As/(Al)GaAs quantum dots. *Physical Review B*, **65**, 195315.

[7] Bimberg, D., Grundmann, M. and Ledentsov, N. N. 1999. *Quantum Dot Heterostructures*. John Wiley and Sons.

[8] Dalacu, D., Reimer, M. E., Frederick, S. *et al.* 2010. Directed self-assembly of single quantum dots for telecommunication wavelength optical devices. *Laser & Photonics Reviews*, **4**, 283–299.

[9] Gallo, P., Felici, M., Dwir, B. *et al.* 2008. Integration of site-controlled pyramidal quantum dots and photonic crystal membrane cavities. *Applied Physics Letters*, **92**, 263101.

[10] Gerard, J. M., Sermage, B., Gayral, B. *et al.* 1998. Enhanced spontaneous emission by quantum boxes in a monolithic optical microcavity. *Physical Review Letters*, **81**, 1110–1113.

[11] Hartmann, A., Loubies, L., Reinhardt, E. and Kapon, E. 1997. Self-limiting growth of quantum dot heterostructures on nonplanar 111B substrates. *Applied Physics Letters*, **71**, 1314–1316.

[12] Henneberger, F. and Benson, O. 2009. *Seminconductor Quantum Bits*. Pan Stanford Publishing.

[13] Herman, M. A. and Sitter, H. 1996. *Molecular Beam Epitaxy: Fundamentals and Current Status*. Springer.

[14] Huggenberger, A., Heckelmann, S., Schneider, C. *et al.* 2011. Narrow spectral linewidth from single site-controlled In(Ga)As quantum dots with high uniformity. *Applied Physics Letters*, **98**, 131104.

[15] Kako, S., Santori, C., Hoshino, K. *et al.* 2006. A gallium-nitride single-photon source operating at 200 K. *Nature Materials*, **5**, 887–892.

[16] Kawaguchi, K., Ekawa, M., Kuramata, A. *et al.* 2004. Fabrication of InAs quantum dots on InP(100) by metalorganic vapor-phase epitaxy for 1.55 μm optical device applications. *Applied Physics Letters*, **85**, 4331–4333.

[17] Kirstaedter, N., Ledentsov, N. N., Grundmann, M. *et al.* 1994. Low-threshold, large T-O injection-laser emission from (InGa)As quantum dots. *Electronics Letters*, **30**, 1416–1417.

[18] Ladd, T. D., Jelezko, F., Laflamme, R. *et al.* 2010. Quantum computers. *Nature*, **464**, 45–53.

[19] Löffler, A., Reithmaier, J. P., Forchel, A. *et al.* 2006. Influence of the strain on the formation of GaInAs/GaAs quantum structures. *Journal of Crystal Growth*, **286**, 6–10.

[20] Ma, W. Q., Nötzel, R., Schonherr, H. P. and Ploog, K. H. 2001. Shape transition of coherent three-dimensional (In,Ga)As islands on GaAs(100). *Applied Physics Letters*, **79**, 4219–4221.

[21] Mensing, T., Worschech, L., Schwertberger, R., Reithmaier, J. P. and Forchel, A. 2003. Magneto-optical investigations of single self-assembled InAs/InGaAlAs quantum dashes. *Applied Physics Letters*, **82**, 2799–2801.

[22] Mereni, L. O., Dimastrodonato, V., Young, R. J. and Pelucchi, E. 2009. A site-controlled quantum dot system offering both high uniformity and spectral purity. *Applied Physics Letters*, **94**, 223121.

[23] Michler, P., Kiraz, A., Becher, C. *et al.* 2000. A quantum dot single-photon turnstile device. *Science*, **290**, 228.

[24] Mohan, A., Gallo, P., Felici, M. *et al.* 2010. Record-low inhomogeneous broadening of site-controlled quantum dots for nanophotonics. *Small*, **6**, 1268–1272.

[25] Press, D., Ladd, T. D., Zhang, B. Y. and Yamamoto, Y. 2008. Complete quantum control of a single quantum dot spin using ultrafast optical pulses. *Nature*, **456**, 218–221.

[26] Reithmaier, J. P. 2008. Strong exciton–photon coupling in semiconductor quantum dot systems. *Semiconductor Science Technology*, **23**, 123001.

[27] Richter, D., Rossbach, R., Schulz, W. M. *et al.* 2010. Low-density InP quantum dots embedded in $Ga_{0.51}In_{0.49}P$ with high optical quality realized by a strain inducing layer. *Applied Physics Letters*, **97**, 063107.

[27a] Rödel, R., Bauer, A., Kremling, S., *et al.* Density and size control of InP/GaInP quantum dots on GaAs substrate grown by gas source molecular beam epitaxy. *Nanotechnology* **23**, 015605.

[28] Rossbach, R., Schulz, W. M., Reischle, M. *et al.* 2007. Red to green photoluminescence of InP-quantum dots in $Al_xGa_{1-x}InP$. *Journal of Crystal Growth*, **298**, 595–598. Thirteenth International Conference on Metal Organic Vapor Phase Epitaxy (ICMOVPE XIII).

[29] Santori, C., Fattal, D., Vuckovic, J., Solomon, G. S. and Yamamoto, Y. 2002. Indistinguishable photons from a single-photon device. *Nature*, **419**, 594–597.

[30] Schlereth, T. W., Schneider, C., Höfling, S. and Forchel, A. 2008. Tailoring the morphology and emission wavelength of AlGaInAs quantum dots. *Nanotechnology*, **19**, 045601.

[31] Schmidt, O. G. (ed). 2007. *Lateral Alignment of Epitaxial Quantum Dots*. Heidelberg: Springer.

[32] Schneider, C., Huggenberger, A., Sunner, T. *et al.* 2009. Single site-controlled In(Ga)As/GaAs quantum dots: growth, properties and device integration. *Nanotechnology*, **20**, 434012.

[33] Seguin, R., Schliwa, A., Rodt, S. *et al.* 2005. Size-dependent fine-structure splitting in self-organized InAs/GaAs quantum dots. *Physical Review Letters*, **95**, 257402.

[34] Sek, G., Poloczek, P., Ryczko, K. *et al.* 2008. Photoreflectance determination of the wetting layer thickness in the $In_xGa_{1-x}As/GaAs$ quantum dot system for a broad indium content range of 0.3–1. *Journal of Applied Physics*, **100**, 103529.

[35] Sek, G., Podemsk, P., Musial, A. *et al.* 2009. Exciton and biexciton emission from a single InAs/InP quantum dash. *Journal of Applied Physics*, **105**, 086104.

[36] Shields, A. J. 2007. Seminconductor quantum light sources. *Nature Photonics*, **1**, 215–223.

[37] Skiba-Szymanska, J., Chekhovich, E. A., Nikolaenko, A. E. *et al.* 2008. Overhauser effect in individual $InP/Ga_xIn_{1-x}P$ dots. *Physical Review B*, **77**, 165338.

[38] Strauf, S. and Jahnke, F. 2011. Single quantum dot nanolaser. *Laser and Photonics Reviews*, 1–27.

[39] Strauf, S., Stoltz, N. G., Rakher, M. T. *et al.* 2007. High-frequency single-photon source with polarization control. *Nature Photonics*, **1**, 704–708.

[40] Strauss, M., Höfling, S. and Forchel, A. 2009. InAs/GaInAs(N) quantum dots on GaAs substrate for single photon emitters above 1300 nm. *Nanotechnology*, **20**, 505601.

[41] Takemoto, K., Sakuma, Y., Hirose, S. *et al.* 2005. Single InAs/InP quantum dot spectroscopy in 1.3–1.55 μm telecommunication band. *Physica E: Low-dimensional Systems and Nanostructures*, **26**, 185–189. International Conference on Quantum Dots.

[42] Tran, T., Muller, A., Shih, C. K. *et al.* 2007. Single dot spectroscopy of site-controlled InAs quantum dots nucleated on GaAs nanopyramids. *Appl. Phys. Lett.*, **91**, 133104.

[43] Verma, V. B., Stevens, M. J., Silverman, K. M. *et al.* 2011. Photon antibunching from a single lithographically defined InGaAs/GaAs quantum dot. *Optics Express*, **19**, 4182–4187.

[44] Wang, L., Rastelli, A. and Schmidt, O. G. 2006. Structural and optical properties of In(Ga)As/GaAs quantum dots treated by partial capping and annealing. *Journal of Applied Physics*, **100**, 064313.

[45] Young, R. J., Stevenson, R. M., Atkinson, P. *et al.* 2006. Improved fidelity of triggered entangled photons from single quantum dots. *New Journal of Physics*, **8**, 029.

2

Single semiconductor quantum dots in nanowires: growth, optics, and devices

M. E. Reimer, N. Akopian, M. Barkelid, G. Bulgarini,
R. Heeres, M. Hocevar, B. J. Witek, E. P. A. M. Bakkers and V. Zwiller

2.1 Introduction

Quantum dots have proven to be exciting systems to study light–matter interaction [32, 9]. Self-assembled quantum dots obtained by the Stranski–Krastanow growth mode have been the main system to date [32, 9]. Here we introduce a new type of quantum dot embedded in a one-dimensional nanowire. Quantum dots in nanowires offer a range of advantages over strain-driven Stranski–Krastanov quantum dots. In the case of quantum dots in nanowires, the light extraction efficiency can be very high for the quantum dot emission due to a waveguide effect in the nanowire [14, 45], theoretically approaching 100% according to simulations [14]. Since strain is not the driving mechanism during growth, unprecedented material freedom is available to the quantum engineer in the choice of materials for the quantum dot and the barrier material. At the scale of nanowires, both zincblende and wurtzite crystal structures can coexist, opening the door to a new type of confinement based not only on the material composition, but also on the phase of the crystal lattice [1]. The ability to electrically contact a single nanowire implies that all the current injected in a nanowire will flow through a single quantum dot, enabling an efficient interface between single electrons and single photons [33, 44]. In addition, electrostatic gating is highly versatile, allowing for coherent spin manipulation [36], charge state control [54], and the ability to control the exciton–biexciton splitting by an in-plane electric field [43]. Although optically active nanowire quantum dots have been studied only over the past five years, they already stand out as strong competitors to the well established Stranski–Krastanov quantum dots. In this chapter, we review the current state of the art in growth and optical properties of single quantum dots formed by changing either the material or crystal phase along the axial direction of the nanowire. We also present the most recent developments in nanowire quantum dot devices exploiting one-dimensional properties of nanowires for control of electron charge, novel routes to entangled photon generation, efficient conversion of electrons to photons, and single photon detection.

2.2 Nanowire quantum dot growth

Two approaches enable the fabrication of one-dimensional systems such as nanowires: "top-down" and "bottom-up". The "top-down" approach consists of fabricating arrays of

Quantum Dots: Optics, Electron Transport and Future Applications, ed. Alexander Tartakovskii. Published by Cambridge University Press.

nanowires by etching a patterned substrate with reactive ion etching [14]. The "bottom-up" approach consists of growing nanowires via a catalyst-assisted vapour–liquid–solid (VLS) mechanism [63, 34, 37]. The latter approach is now the most popular, as it allows for a large freedom in material growth. Indeed, not only can the stacking of different materials be achieved in the axial and radial directions of the nanowires [59, 64], but also different crystal structures can be obtained along the axis of a nanowire [50, 15, 1]. Here we discuss the growth of quantum dots in nanowires via the VLS growth technique by changing either the material or crystal phase along the axial direction of the nanowire.

2.2.1 Growth mechanisms

Nanowires are usually grown by metal–organic vapor phase epitaxy (MOVPE) [2, 15] or molecular beam epitaxy (MBE) [19, 52]. In preparation to the growth, metallic nanoparticles are deposited on the substrate either using colloids [2] (Figure 2.1c), or aerosols [15]. The nanoparticles can also be fabricated by patterning a thin layer of gold via e-beam lithography [10] or conformal nanoimprint (Figure 2.1d) [38]. In our case, we use gold but different metals can also be used such as aluminium [62] or copper [46]. Once in the reactor, the substrate is heated up. Gold forms an alloy with the substrate material and with the precursors from the gas phase (i.e. Au–In or Au–Si). As the In composition (or Si) in the gold particle increases, the eutectic forms, turning the seed particle into liquid (Figure 2.1a-1). Subsequently, continuously supplying precursor material from the gas phase leads the particle to saturate and to eventually reach supersaturation. Once supersaturation of the metal particle is reached, crystal nucleation starts at the droplet–substrate interface (Figure 2.1a-2). This process results in the nanowire growing layer by layer, thus lifting the metal particle.

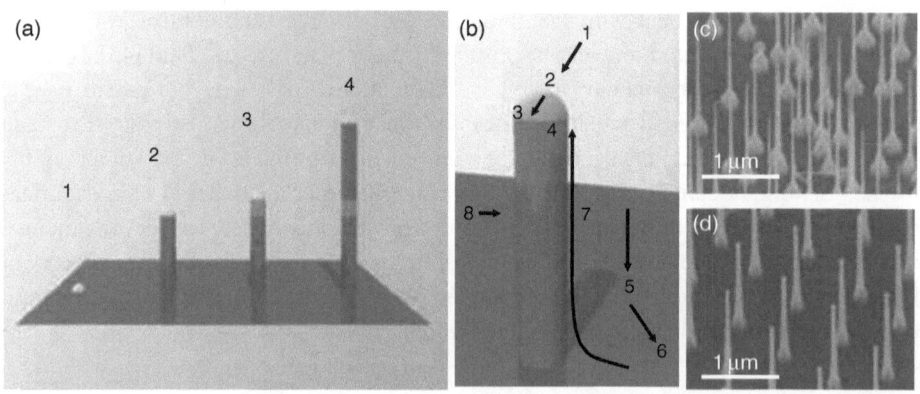

Figure 2.1 (a) The four steps in quantum dot nanowire growth. (b) Relevant processes of the VLS nanowire growth mechanism: 1, precursor transport; 2, precursor dissociation; 3, diffusion through Au; 4, crystal growth; 5, adsorption on substrate; 6, film deposition on substrate; 7, surface diffusion; 8, film deposition on nanowire surface. Example of nanowires grown via (c) colloid catalysts and (d) conformal nanoimprint lithography.

A dynamic equilibrium forms between materials supplied from the gas phase, transport through the droplet and crystal growth at the droplet–nanowire crystal interface (Figure 2.1b). When using MOVPE or chemical vapour deposition (CVD) [49], the material is supplied by metal–organic precursors. In the case of InP, the indium is supplied by tri-methyl-indium (TMI) and the phosphorus is supplied by phosphine (PH_3). In the growth of quantum dot heterostructures, such as InAsP in InP nanowires, an extra precursor, arsine (AsH_3) is supplied during the quantum dot growth (Figure 2.1a-3), and switched off during the remaining part of nanowire growth (Figure 2.1a-4). Shells can also be grown by raising the temperature in order to increase the ratio between radial growth (thin-film growth) and axial growth.

The growth of nanowires by VLS is a complex mechanism that involves many parameters such as the nanowire diameter [23], crystal direction [30], nature of the substrate, substrate temperature [60], total pressure, precursor partial pressure, V–III ratios [15], and impurity concentration, such as dopants [58]. These parameters affect the crystal structure, the morphology and defect density of the nanowires. As a consequence, the band structure and the related physical properties of the nanowires are dependent on the growth parameters.

2.2.2 Nanowire composition and crystal structure engineering

Engineering the crystal structure and composition of the nanowires at the nanoscale is of great interest for many applications, including single photon emission [1], photodetection [55, 44], solar cells and interfacing transport and optics [36]. As discussed in the previous section, control of the crystal structure during nanowire growth is not straightforward and requires a detailed understanding of the growth parameters. Here we present several cases where material and crystal structure is controlled along the axial direction of the nanowire: hetero-material quantum dot; crystal phase quantum dot; defect free nanowires; and twinning superlattices.

Hetero-material quantum dots. Single InAsP quantum dots in InP nanowires are realized during growth by embedding a thin segment of InAsP in an InP nanowire (Figure 2.2a, b). Since InP has a larger bandgap than InAsP, InP is an ideal electronic barrier in the longitudinal direction and a passivation layer in the radial direction. Typical quantum dot heights of 4–10 nm and radial dimensions of 30–40 nm were obtained. To achieve thin segments of InAsP in the InP nanowire during growth, additional arsenic was supplied in the gas phase with TMI for only 1–2 s. In contrast, to achieve a 1 μm long segment of InP, the growth time is approximately 10 min. These types of quantum dots in nanowire devices are useful for electron number control [54], biexciton binding-energy removal for entangled photon pair generation [43], and quantum dot nanowire LEDs [33, 44] (Section 2.4.3). In addition, heterostructure quantum dots can be used as single photon emitters (Section 2.3.1) and for spin memory devices [57], or embedded in LEDs for electrical injection [33].

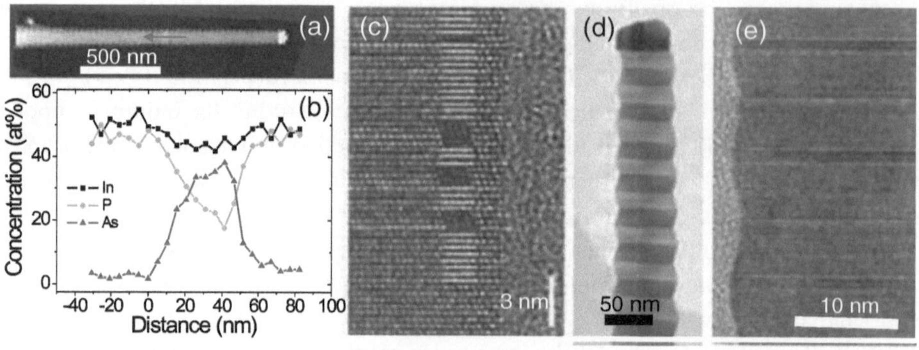

Figure 2.2 (a) InAsP quantum dot in an InP nanowire and (b) EDX scan taken along the arrow in (a). (c) Crystal-phase heterostructures (adapted from Ref. [1]). High-resolution transmission electron microscopy image of a single InP nanowire. White and black lines correspond to atomic planes of wurtzite and zincblende crystal structures, respectively. (d) Twinning superlattices in GaP nanowires imaged by bright field TEM and (e) high-resolution TEM.

Crystal-phase quantum dots. The ability to control the nanowires crystal structure by tuning the growth parameters makes them well-suited for band-gap engineering. For instance, by tuning the growth temperature or gas flow, different crystal structures can be constructed in a nanowire. Crystal-phase quantum dots can be achieved by the insertion of a few monolayers of zincblende into a wurtzite nanowire or vice versa (i.e. embed a thin segment of wurtzite into a zincblende nanowire, as shown in Figure 2.2c). The growth temperature is one of the parameters influencing the crystal structure of the nanowire. In InP nanowires grown by MBE, a low growth temperature (380 °C) is favorable for pure wurtzite InP while the formation of zincblende InP sections is highly probable at higher temperature (420 °C).

Defect-free nanowires. Defect-free nanowires are nanowires grown without any stacking faults or twin planes over a relatively large length ($> 2\,\mu m$) compared to the nanowire diameter. Depending on the composition, the growth parameters are optimized in order to grow the nanowires in a pure hexagonal phase or in a pure cubic phase. However, it is sometimes difficult to reach such a high level of control. As an example, GaP can be grown in a pure hexagonal phase at high temperature; however, even at the lowest growth temperature, GaP currently does not grow in a perfect cubic crystal structure.

Twinning superlattices. Twinning superlattices are common in polar materials such as II–VI and III–V. Twinning superlattices in nanowires are very exciting as direct transitions have been predicted in indirect band-gap materials such as Ge, Si and GaP attributed to the apparition of minibands in the band structure [22]. Twinning superlattices in nanowires are obtained by the utilization of dopants or optimized V/III ratios. Figure 2.2d, e presents a nanowire imaged by TEM featuring alternating dark and light segments, characteristic for a zincblende (ZB) nanowire with twin boundaries, i.e. rotations of the crystal structure by 180° around the nanowire elongation axis. The twinning superlattices nanowires have typical {111} side facets at an angle of 19.5° with the growth direction.

2.3 Optical properties of nanowire quantum dots

We present a brief overview of the optical properties of single quantum dots embedded in individual InP nanowires: either by embedding a thin segment of InAsP in a larger band-gap material, InP (hetero-material quantum dots); or by changing the crystal phase of the nanowire from wurtzite to zincblende (crystal-phase quantum dots). These quantum dots exhibit similar optical behaviour to self-assembled quantum dots embedded in a three-dimensional host matrix, while simultaneously taking advantage of one-dimensional charge transport at the single carrier level (see Section 2.4). Here, we first demonstrate that a single InAsP quantum dot embedded in an individual InP nanowire acts as a single photon emitter and that the electron spin properties can be accessed for implementation of quantum information protocols. We also present the quantum dot photoluminescence (PL) in a magnetic field oriented along three directions, which allows us independently to extract the g-factor along the quantization axis and radial dimension of the nanowire for electrons and holes. This g-factor tensor extraction method is not possible if only the Faraday and Voigt configurations are used. Finally, as an alternative to hetero-material quantum dots in nanowires, we show sub-Poissonian emission from crystal-phase quantum dots defined in an InP nanowire. The measured long exciton lifetime for the crystal-phase quantum dot is indicative of a very high nanowire purity – free of defects or surface states.

2.3.1 Hetero-material quantum dots

Typical photoluminescence (PL) spectra as a function of increasing excitation power for a single InAsP quantum dot in an InP nanowire are shown in Figure 2.3a. The observed behaviour is indicative of single quantum dot emission [6]. At low excitation power, only a single peak at 1.315 eV is observed, attributed to single exciton (X) recombination in the quantum dot s-shell. Increasing the excitation power results in biexciton (XX) emission at 1.4 meV lower energy. The appearance of the biexciton at lower energy is due to the biexciton binding energy, which is dependent on the quantum dot size [67, 47, 48, 42]. Further increase in excitation power results in p-shell emission at approximately 15 meV above the s-shell. This energy difference between the s- and p-shell is determined by the quantum dot radial confinement and is similar to previous reports [54, 43] of quantum dots in nanowires and also to InAs/InP self-assembled quantum dots of comparable size [13].

In the following, we demonstrate single photon emission by filtering the single X line from Figure 2.3a in a standard Hanbury Brown and Twiss setup for photon correlation measurements [18]. A tapered nanowire waveguide design with a single quantum dot

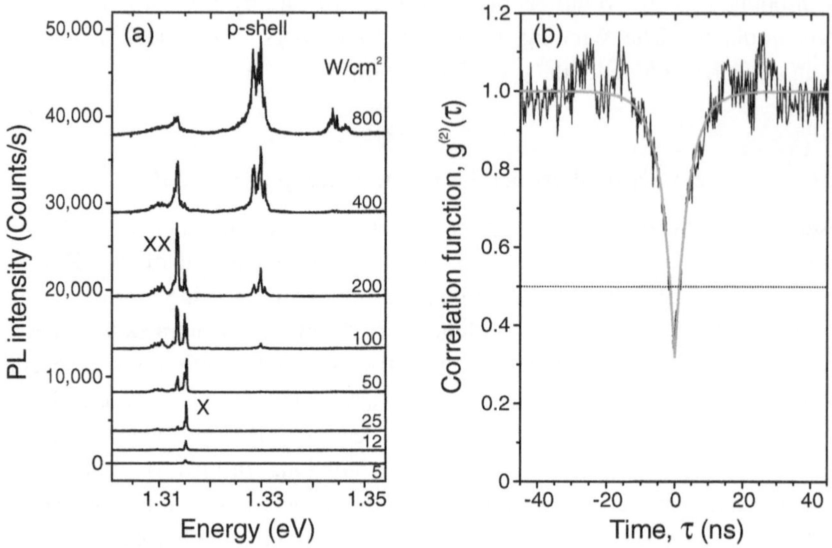

Figure 2.3 (a) Typical power dependent PL of a single quantum dot in a nanowire. The PL spectra are offset along the y-axis for clarity. (b) Second-order correlation function, $g^{(2)}(\tau)$, of single exciton line (X). The anti-bunching dip below 0.5 (dotted line) is characteristic of single photon emission. The extracted fit parameters are the anti-bunching dip of 0.35 and exciton lifetime of 4.1 ± 0.2 ns.

positioned exactly on its axis ensures a high single photon count rate [45]. Figure 2.3b shows the normalized measured count distribution for an excitation power of 50 W/cm^2 where the XX peak is half in intensity as compared to the X peak. We fit the normalized measured count distribution with the second-order correlation function, $g^{(2)}(\tau) = 1 - ae^{-\tau/t_m}$ [8]. Here, a accounts for the background and t_m is the measured lifetime. Accounting for the background radiation in a similar way to Becher *et al.* [8], we extract a measured X lifetime of 4.1 ± 0.2 ns and anti-bunching dip of 0.35. The measured anti-bunching dip below 0.5 for the X line is a signature of single photon emission, indicating that the InAsP quantum dot in an InP nanowire is a single photon emitter.

Electron and hole Landé g-factors in nanowire quantum dots

Application of an external magnetic field to a carrier spin results in Zeeman splitting and therefore generates two well-defined spin states that can be used to define a qubit. The coupling between a carrier spin and magnetic field is determined by the g-factor and is a key parameter for quantum information applications [3, 65, 24, 36, 61]. In certain applications, a large electron and hole g-factor is preferable for high-fidelity spin preparation [3] or coherent spin manipulation [36]. In other cases, a zero exciton g-factor is desirable for tunable polarization insensitive photodetection or coherent spin to photon conversion [61]. Confinement in quantum dots offers a way to tune g-factors value [39, 25] and can be directly controlled by the quantum dot size [24]. Nanowires are ideal systems for controlling the g-factor since the size of a quantum dot can be controlled during the growth.

Here, we have performed a magneto-photoluminescence experiment for a charged exciton line in a nanowire quantum dot. Our setup allows us to apply a magnetic field along two orthogonal directions making it possible to explore the Faraday geometry (Figure 2.4a: magnetic field along the growth direction), the Voigt geometry (Figure 2.4b: magnetic field in plane of the quantum dot), as well as intermediate cases (Figure 2.4c: magnetic field at 60°). This technique reveals all the parameters of the 3D g-factor tensor for both electrons and holes, which has first been demonstrated for self-assembled quantum dots [53].

The measurements presented in Figure 2.4 were fitted with the Hamiltonian, H_{Zeeman}, in a magnetic field corresponding to all three magnetic field orientations [7]. In the Faraday configuration ((Figure 2.4a) we observe the recombination of two bright exciton states

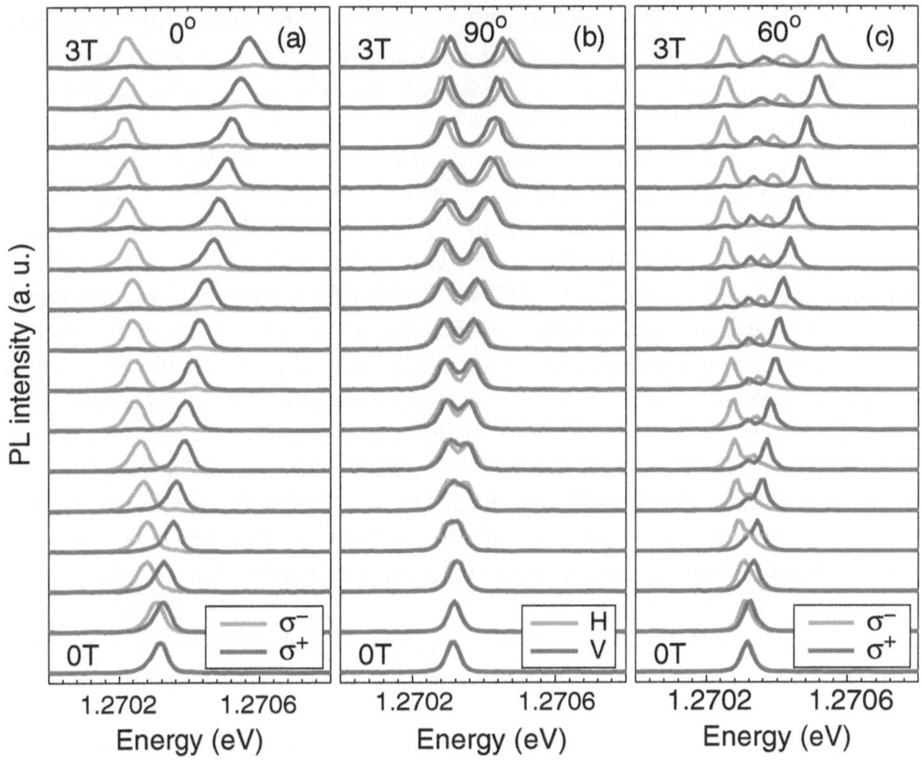

Figure 2.4 A nanowire quantum dot in an external magnetic field. Charged exciton PL in an external magnetic field in the (a) Faraday configuration (0°), and (b) Voigt configuration (90°). (c) Charged exciton PL at an intermediate angle of 60°.

with right (σ^+) and left (σ^-) circular polarization. We therefore measure only the bright exciton g-factor, which is a sum, $g_{ez} + g_{hz}$, of electron and hole g-factors. Application of a magnetic field in-plane of the quantum dot breaks its symmetry and mixes the bright and dark exciton states. Thus, all four optical transitions (two bright plus two dark) become visible (Figure 2.4b) and both the electron $g_{e\perp}$ and the hole $g_{h\perp}$ in-plane g-factors can be extracted. In the measurement at an intermediate angle (60°, Figure 2.4c), all four states are still observed, which is crucial to determine the electron g_{ez} and the hole g_{hz} g-factor in the growth direction.

In this particular dot (diameter $d = 30$ nm, height $h = 10$ nm) we obtain the following values: $|g_{ez}| = 0.74 \pm 0.02$, $|g_{hz}| = 1.24 \pm 0.04$, $|g_{e\perp}| = 1.00 \pm 0.08$ and $|g_{h\perp}| = 0.12 \pm 0.01$. In contrast to quantum wells, the in-plane hole g-factor, $|g_{h\perp}|$ is non-zero implying light hole–heavy hole mixing [26].

2.3.2 Crystal phase quantum dots

In semiconducting nanowires, both zincblende and wurtzite [68] crystal structures can coexist [2, 12, 27, 35]. The band structure difference between the two structures can lead to charge confinement [35]. Here we fabricate and study single quantum dots defined solely by crystal phase in a chemically homogeneous InP nanowire and observe sub-Poissonian emission.

The schematics of a crystal-phase heterostructure is shown in Figure 2.5a. Our heterostructure is formed by growing an InP nanowire with two different crystal phases: wurtzite (WZ) and zincblende (ZB). ZB InP has a smaller band-gap and large negative band offset compared to WZ InP. Hence, the conduction and valence bands of ZB segments are lower in energy compared to WZ parts. Short ZB sections can confine electrons in the growth direction, while simultaneously acting as barriers for holes. Two types of radiative recombination are therefore possible. The first, labeled α, is the direct recombination of electron–hole pairs in the WZ section of the nanowire, and therefore results in a short lifetime. In the second transition, labeled β, electrons confined in ZB segments recombine radiatively with holes in adjacent WZ regions of the nanowire. This indirect electron–hole pair recombination results in longer lifetimes, due to reduced overlap of electron and hole wavefunctions. Optical transitions of type α and β are represented in the PL (Figure 2.5a) and correspond to emission peaks labeled as WZ and ZB/WZ, respectively.

Another case is when a short wurtzite segment is embedded between two short zincblende segments: holes are confined in a wurtzite segment, and are spatially separated from confined electrons in the short zincblende segment. The indirect radiative recombination, labeled γ, results again in longer lifetimes due to reduced overlap of electron and hole wavefunctions.

Time-resolved PL measurements are represented in Figure 2.5b. The spectral line at 1.486 eV, labeled WZ, corresponds to electron–hole recombination of type α in the wurtzite

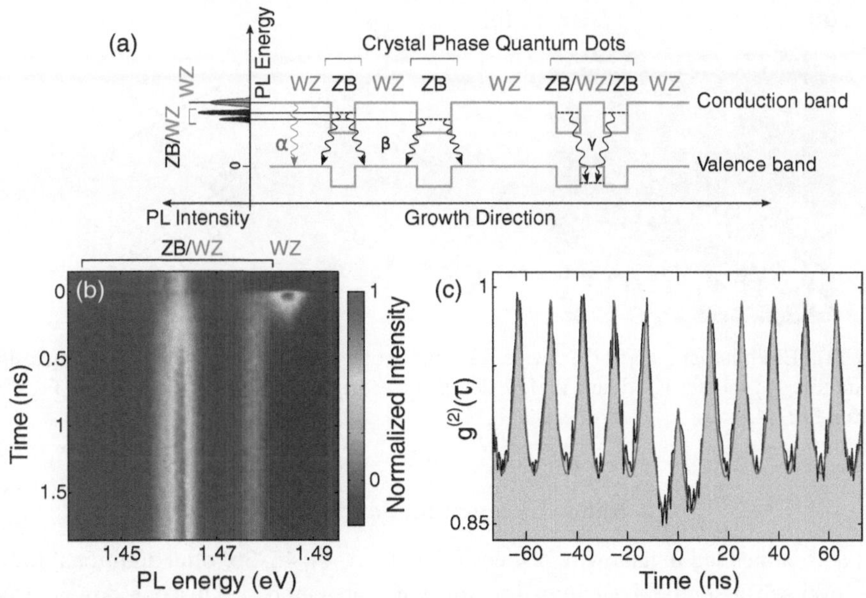

Figure 2.5 Crystal-phase quantum dots (adapted from Ref. [1]). (a) Schematics of crystal-phase heterostructures in an InP nanowire and corresponding photoluminescence from wurtzite (WZ) and zincblende (ZB) crystal phases. α, β, and γ represent possible optical transitions. (b) Time-resolved PL of a single nanowire with crystal-phase heterostructures. (c) Second-order auto-correlation measurement of photons emitted from a few crystal phase QDs under pulsed excitation. The gray line is a fit to the data using the model discussed in [1].

part of the nanowire. We clearly observe a short lifetime for optical transitions of type α. Using a mono-exponential fit to the experimental data, we obtain a lifetime of 120 ± 7 ps (Ref. [1]). Emission peaks at lower energies, labeled ZB/WZ, correspond to optical transitions of type β or γ in crystal phase quantum dots. Only a fraction of their decay traces are captured within a 2 ns time window, indicating much longer lifetimes. The data at negative times represent the part of the PL decay curve that was excited by the previous laser pulse 13 ns earlier. The significant signal measured at these times, the tail-signal, indicates that electrons and holes can remain in short ZB and WZ sections for more than 13 ns. Long lifetimes, measured in our experiments indicate a high nanowire purity of the sample free of defects or surface states [31].

To demonstrate the quantum nature of crystal-phase quantum dots we performed photon correlation measurements under pulsed excitation, as shown in Figure 2.5c. Missing correlation events at time zero indicate that the emission originates from a few single photon emitters. From the peak height at time zero we estimate the number of single photon emitters within the excitation spot to be less than four. To estimate the exciton lifetime, we fit using a model [1] with a few identical single photon emitters. The fit results in an exciton lifetime of 5 ± 1 ns, which is comparable to 3.8 ± 0.2 ns as measured directly [1].

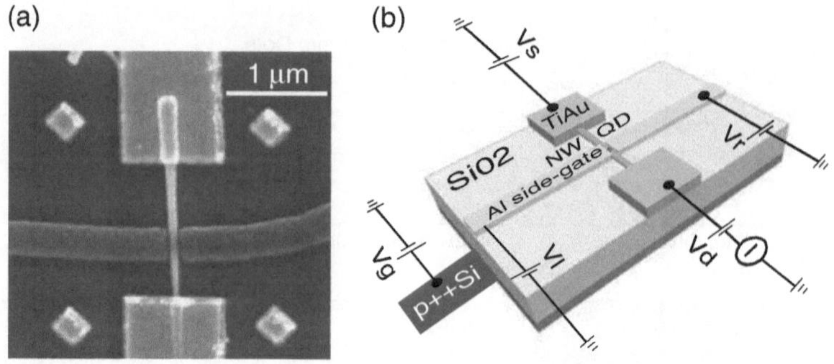

Figure 2.6 (a) SEM micrograph and (b) device schematics of a contacted nanowire containing a single quantum dot. V_s, source; V_d, drain; V_l, left side-gate; V_r, right side-gate; V_g, back-gate. Figures adapted from Ref. [43].

2.4 Nanowire quantum dot devices

Quantum dots embedded in nanowire devices [33, 54, 55, 44, 43, 36] offer additional functionalities over self-assembled quantum dots since they are embedded in a one-dimensional system and not in a three-dimensional host matrix. The one-dimensionality of the nanowire device ensures that all of the current flows through the single quantum dot, as required for an efficient conversion of current into light [33, 44]. These unique electronic properties combined with the possibility to design tapered nanowire waveguides for high extraction efficiencies [14] make these devices desirable for the long-distance transfer of quantum information between stationary qubits [36] and flying qubits [57]. In addition, owing to the small radial dimensions of the nanowires, electrostatic gate geometries are highly versatile allowing for precise control of the electrostatic environment in the nanowire and quantum dot [54, 43]. Recently in InAs nanowires, a gate-defined double quantum dot was created by five closely spaced bottom gates and high-frequency pulses were applied for coherent manipulation of the electron spin with electric fields only [36].

Here, we demonstrate precise control over the electrostatic environment of a single InAsP quantum dot in an individual InP nanowire with two contacts and two lateral gates positioned to an individual nanowire (see Figure 2.6). The two contacts on both ends of the nanowire allows us to control the confined charge in the quantum dot through tunnelling events induced by an applied electric field along the nanowire elongation axis. Combined with an electrostatic back-gate, the electro-chemical potential in the nanowire can be modified independently of the charge state, which tunes the tunnel coupling of the quantum dot with the contacts. The two lateral gates in the vicinity of the quantum dot affects the biexciton binding energy and engineers its removal at finite electric field. Theoretical calculations are presented that are in good qualitative agreement with the observed exciton–biexciton behaviour in a lateral electric field. Removal of the biexciton binding energy is desirable in generating entangled photon pairs without the need to remove the anisotropic exchange splitting of the intermediate exciton states [40, 41, 4, 28, 16, 43].

Finally, we control the chemical doping along the nanowire, thus allowing p–n and p–i–n structures to be fabricated [56, 2] for light emission [33] and detection [66]. We show that light is emitted from the single quantum dot by electrical injection and that the optical quality of the quantum dot emission improves when surrounding the dot material by a small intrinsic section of InP [44]. We also report large multiplication factors in a single Si nanowire avalanche photodiode comprised of p-doped, intrinsic, and n-doped sections, thus opening the possibility of detecting single photons at the nanoscale [44].

2.4.1 Electron charging

Control of the quantum dot charge state is demonstrated in Figure 2.7 by the quantum dot PL as a function of source-drain bias (V_{sd}) for three different back-gate voltages (V_g). In each case, the quantum dot charge state changes from the singly charged exciton X^{1-} containing two electrons and one hole to the neutral exciton (X^0) involving one electron and one hole. The quantum dot is negatively charged at small biases since the Fermi level is pinned at the nanowire surface $\sim$ 200 meV below the InP conduction band [54]. Increasing V_{sd} generates an electric field along the nanowire, which induces tunnel events between the initial and final state of the quantum dot. As the bias is increased, the tunnelling barrier decreases and electrons tunnel out of the quantum dot corresponding to discrete energy steps in the observed optical spectra. At large enough biases, all carriers tunnel out resulting in a quenching of the quantum dot PL and simultaneous increase in resonant photocurrent [54].

In this particular quantum dot, addition of an extra electron results in an observed energy difference in PL of 2 meV for X^{1-} as compared to X^0. This energy difference is due to strong Coulomb interactions present in semiconductor quantum dots and is similar to

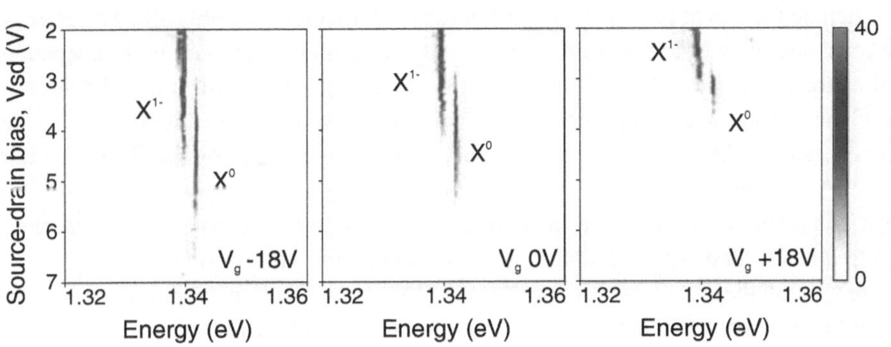

Figure 2.7 Effect of the back-gate potential, V_g on the exciton emission energy for -18 V, 0 V, and 18 V. At $V_g = -18$ V there is a large overlap between X^{1-} and X^0 optical transitions while at $V_g = +18$ V, the transition between X^{1-} and X^0 is abrupt indicating that the tunnel coupling to the contacts is high. At high tunnel coupling, the quantum dot charge state follows the electro-chemical potential in the nanowire. Figures adapted from Ref. [54].

calculations of self-assembled quantum dots of comparable dot diameter [17]. The fact that the X^{1-} PL is at lower energy than X^0 indicates that the electron–hole attraction is stronger than the electron–electron repulsion. Of particular significance is that the tunnel coupling of the quantum dot to the contacts is tuned with a back-gate voltage. At negative back-gate potential ($V_g = -18\,\mathrm{V}$) there is a large overlap observed between X^{1-} and X^0 indicating a low tunneling probability. At low tunnel coupling, the electron charging rate is comparable to the exciton recombination rate ($\Gamma_{charge} \sim \Gamma_{rad}$), which allows for X^0 recombination before an energetically favorable electron tunnels into the quantum dot. In contrast, when the tunnel coupling is high ($\Gamma_{charge} > \Gamma_{rad}$), then the quantum dot charge state can easily follow the electro-chemical potential and the X^{1-} to X^0 transition is more abrupt as observed for $V_g = +18\,\mathrm{V}$ in Figure 2.7. The altered nanowire carrier density by the back-gate changes the charge state of the quantum dot and affects the transparency of the Schottky contacts, thus modifying the tunnel coupling. The reduced quenching voltage of X^0 at $V_g = +18\,\mathrm{V}$ as compared to $V_g = -18\,\mathrm{V}$ confirms the tunnel coupling enhancement. This tunability of the tunnel coupling is crucial for electron spin read-out schemes in which the electron spin is converted into charge [21]. Until this type of nanowire device was demonstrated, the tunnel coupling between the quantum dot energy levels and continuum was fixed during growth for self-assembled quantum dots in a Schottky diode structure [5].

2.4.2 Removal of biexciton binding energy in a lateral electric field

A proposal that circumvents the requirement to remove the anisotropic exchange splitting (AES) and produce entangled photon pairs even in the presence of non-zero AES relies on tuning the biexciton binding energy to zero [40, 41, 28, 43]. In such a scheme, the photons are entangled provided that the timing information of the exciton and biexciton is erased in the measurement setup [4]. Experimental methods to remove the biexciton binding energy to zero include the use of lateral strain [16] or local lateral electric fields [40, 41]. Here, we utilize local lateral electric fields on a single quantum dot in a nanowire with independent control of the charge state [43]. Of particular significance, we demonstrate a large tuning range of the biexciton binding energy of 2 meV in a lateral electric field. Since typical binding energies of the biexciton are within this energy range of 2 meV, the observed tuning range that we demonstrate is desirable in constructing arrays of entangled photon pair sources. The method we utilize has the added benefit that is can be performed dynamically at high frequencies, post-growth and on-chip by simply applying a voltage.

The results of the PL emission for X^0 and $2X^0$ as a function of lateral bias is presented in Figure 2.8a. The difference between the exciton and biexciton emission is known as the biexciton binding energy, $\Delta XX = E_{X^0} - E_{2X^0} = 2V_{eh} - V_{ee} - V_{hh} + \Delta E_{corr}$ [43]. At $V_{lat} = 0\,\mathrm{V}$, ΔXX is $\sim$2 meV. Here, ΔXX is mainly determined by the competition between Coulomb repulsion among electrons (V_{ee}) and holes (V_{hh}), Coulomb electron–hole attraction (V_{eh}), and correlation corrections (ΔE_{corr}) due to higher lying states confined in the quantum dot [28]. At zero electric field, the biexciton can appear

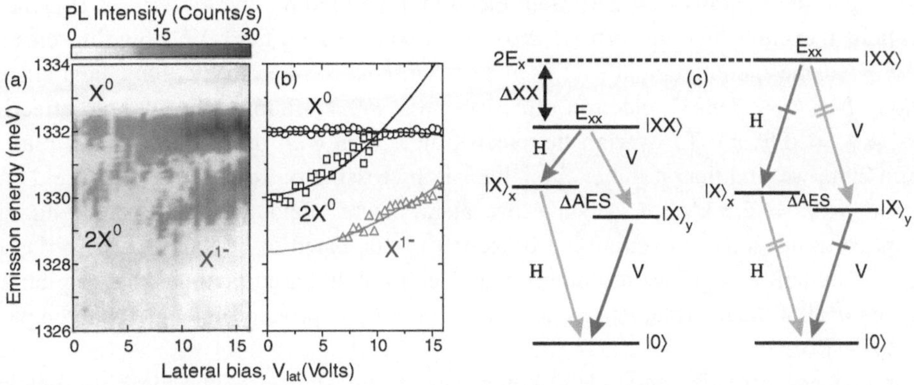

Figure 2.8 (a) Removal of biexciton binding energy in a lateral electric field. (b) Solid lines are fit to peak positions of X^0 (circles), $2X^0$ (squares), and X^{1-} (triangles) optical transitions in a lateral electric field from (a) using realistic system parameters for the quantum dot in a nanowire. See text for fit parameters. (c) Concept to realize entangled photon pairs from the biexciton–exciton radiative cascade through removal of the biexciton binding energy, ΔXX. At $\Delta XX = 0$ (right), two degenerate transitions can be identified, even in the presence of non-zero AES, thus restoring conditions for polarization-entangled photons. H: horizontal polarization; V: vertical polarization; AES: anisotropic exchange splitting. Figures adapted from Ref. [43].

either at higher or lower emission energy depending on the dot size [42]. In cases where ΔXX is positive, the biexciton emission can be engineered to cross the exciton emission at finite electric field [28]. Similar behaviour is observed in Figure 2.8a for increasing electric field (lateral bias). As a function of increasing lateral bias (electric field), dissimilar energy shifts of the exciton and biexciton result in their eventual merging at finite electric field. At a finite bias of $V_{lat} = 10\,\text{V}$, $2X^0$ merges with X^0 and the two peaks cannot be distinguished within the spectral linewidth of the two optical transitions. At the merging of the two peaks, we observe an increase in PL intensity, which suggests that the PL amplitudes of X^0 and $2X^0$ are added.

The small Stark shift observed for X^0 can be understood through the compensation of the single-particle Stark shift and field-induced separation of electron and hole [41]. This compensation of the single-particle Stark shift of X^0 also leads to an observed blue shift of $2X^0$ since there is an additional electron–hole pair building the biexciton state and increased Coulomb interactions [28]. Coulomb interactions of the same type (i.e. V_{ee} and V_{hh}) are not modified in a lateral electric field, while the additional electron–hole Coulomb attraction term, V_{eh}, drives the observed PL for the biexciton to higher energies [28].

Taking realistic parameters for the confinement potential of our quantum dot in a nanowire and measured s-shell, p-shell splitting of 17 meV, we utilize the model of Korkusinski *et al.* for a 2D parabolic potential in a lateral electric field [28]. The results of the electric field dependence for the X^0, $2X^0$, and X^{1-} optical transitions for positive lateral bias are shown in Figure 2.8b. Excellent qualitative agreement is found between the theoretical prediction (solid lines) and experiment (open symbols) when using the following system parameters: zero-field electron–hole binding energy, $V_{eh}^0 = 14.5\,\text{meV}$; zero-field biexciton binding energy, $\Delta_{XX} = 1.9\,\text{meV}$; zero-field X^{1-} binding energy, $\Delta_{X^{1-}} = 3.6\,\text{meV}$; confinement energy of electron, $\hbar\omega_e = 14.3\,\text{meV}$; confinement energy of hole, $\hbar\omega_h = 2.7\,meV$; electron effective mass, $m_e^* = 0.06m_0$; heavy-hole effective mass, $m_h^* = 0.55m_0$. To overlap the electric field axis with the applied lateral bias in Figure 2.8b, we find that at the exciton–biexciton crossing of 10 V, the applied field corresponds to $F = 2.8$ kV/cm. At such low lateral electric fields the oscillator strength of the optical transitions of exciton and biexciton is calculated to be 94%. This overlap of electron and hole is significantly higher than electric field based methods that rely on the AES removal of the intermediate exciton states in order to produce entangled photon pairs [29, 20, 41].

The concept to realize entangled photon pairs from removal of the biexciton binding energy is shown schematically in Figure 2.8c [40, 28, 43]. At zero electric field and in the presence of an anisotropic exchange splitting (ΔAES), each photon emitted in the cascade emits at a different color (Figure 2.8c, left). Thus, the recombination pathway can be distinguished through a photon-energy measurement and the emitted photons are classically correlated. However, tuning the biexciton binding energy to zero in a lateral electric field (Figure 2.8c, right) results in the degeneracy of the cross-colour transitions in the cascade (i.e. $|XX\rangle \rightarrow |X\rangle_x$ and $|X\rangle_y \rightarrow |0\rangle$; $|XX\rangle \rightarrow |X\rangle_y$ and $|X\rangle_x \rightarrow |0\rangle$). In this case, the which-path information cannot be distinguished and the emitted photons are entangled in the polarization degrees of freedom, $|\Psi\rangle = 1/\sqrt{2}(|XX\rangle_V|X\rangle_V + |XX\rangle_H|X\rangle_H)$ [40, 4, 28]. We note that to maintain the conditions of entanglement, the biexciton photon should be delayed with respect to the exciton photon in order to 'erase' the timing information of emitted photons [4]. In order to make this solution viable in the future, improvements to the current PL count rates and access to polarization of the nanowire devices are necessary. Detected PL count rates of quantum dots in nanowires can be drastically improved by designing tapered nanowire waveguides [14] and the polarization information can be accessed if excited and collected along the nanowire growth axis (see Section 2.3.1).

2.4.3 Quantum dot in a nanowire light-emitting diode

We now present our quantum dot nanowire light-emitting diode, which is based on a single InAsP quantum dot embedded within the depletion region of a p-n or p-i-n junction synthesized in an InP nanowire. Growth of nanowire p-n junctions are presented in previous reports [33, 44]. A typical fabricated InP nanowire p-n junction device without the quantum dot is shown in an atomic force microscopy (AFM) image of Figure 2.9a. A step-like change of the nanowire is observed from the n-side to p-side of the p-n junction. A schematic of the band-structure is shown in Figure 2.9b. By applying a voltage across the nanowire p-n junction in forward bias, electrons and holes are injected into the quantum dot and recombine from the exciton ground state (electro-luminescence: EL).

Figure 2.9c compares a typical EL spectrum from a quantum dot in a p-n structure to a typical EL spectrum in a p-i-n structure (Figure 2.9d). In both cases, the quantum dot EL

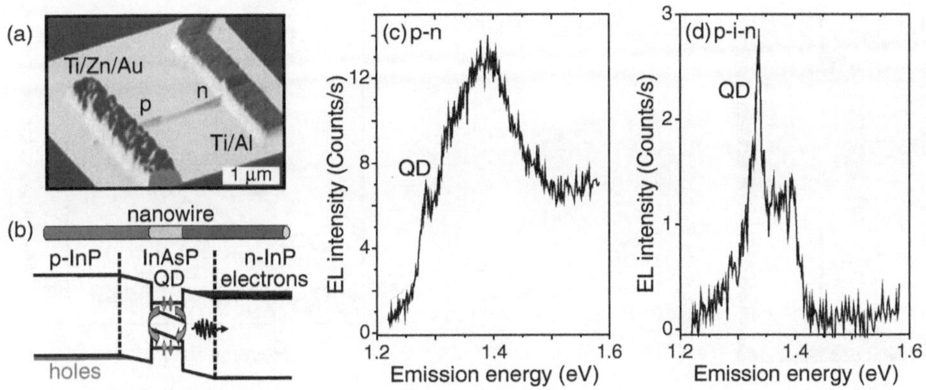

Figure 2.9 (a) AFM image of a single contacted nanowire composed of p- and n-doped sections. (b) Band-structure of a p-n structure with embedded quantum dot within the depletion region. (c) Electro-luminescence (EL) of quantum dot in p–n structure. (d) EL of quantum dot in p–i–n structure. A small intrinsic section of 200 nm is used to improve the optical quality of the quantum dot in the nanowire LED. Figures adapted from Ref. [44].

is observed on top of a broad emission spectrum. However, in cases of the p-i-n structure the optical quality of the quantum dot is improved and the broad emission peak at higher energy is suppressed. We note that the quantum dot EL is broadened compared to PL most likely due to interactions with charges that are in the vicinity of the dot such as dopants in the nanowire [44].

From the EL measurements presented here, we estimate the quantum efficiency of electrons injected to photons generated to be about 10^{-3}. We assume in this calculation a light detection efficiency for the EL measurement of $\sim 0.1\%$. The estimated efficiency is much less than what is expected from a one-dimensional system whereby 100% of the injected current contributes to emitted photons. Further improvements to the p-side contact of the nanowire is required in order to realize the full potential of the one-dimensional system. The material design freedom in nanowire devices makes it possible to overcome this major hurdle in future quantum dot nanowire light-emitting diodes.

2.4.4 Si nanowire avalanche photodiode

High-sensitivity nanoscale photodetectors are a key component in integrated nanophotonic systems and quantum information applications. A photodetector based on an avalanche photodiode (APD) allows for a large photocurrent gain and detection sensitivity down to the single photon regime [44]. Here, we develop an APD from a single Si nanowire composed of p-doped, intrinsic, and n-doped sections (Figure 2.10a). Si is particularly attractive owing to the very low ionization coefficient (i.e. low ionization rate for holes compared to electrons), thus making it possible to make low-noise APDs [11]. Combined with our Si nanowire APD for improved time response, importantly embedding a quantum dot in

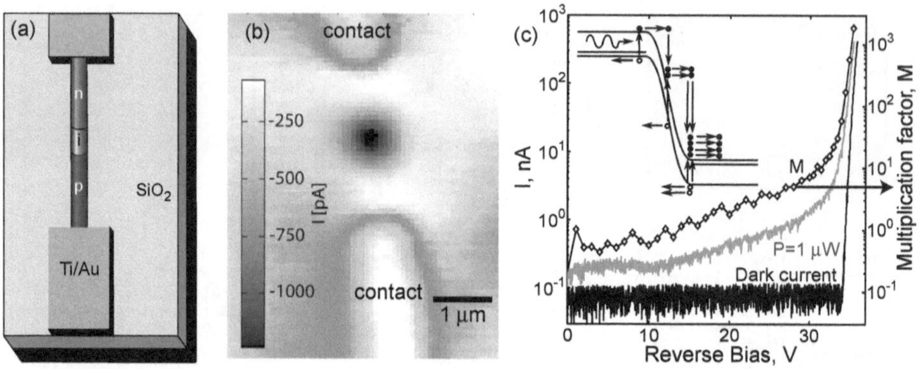

Figure 2.10 (a) Schematic view of a Si nanowire p–i–n structure. (b) A photocurrent image superimposed with a reflection image, indicating that the measured photocurrent originates from the intrinsic region of the Si nanowire. Both images were acquired simultaneously. (c) Left axis: dark current (black) and current under illumination (grey) with a laser power of 1 μW as a function of reverse bias. Right axis: the multiplication factor, M in the nanowire p–n junction depletion region, calculated from Equation (2.1). Inset: schematic of the avalanche process.

the intrinsic region of the Si nanowire is advantageous in constructing an opto-electrical interface in future quantum computers. Here, single photons would be absorbed when in resonance with one of the quantum dot transitions and multiplied within the depletion region of the Si nanowire such that this signal can be detected by external electronics.

An APD consists of a p–i–n junction under high reverse bias, where a photon incident on the photodetector creates an electron–hole pair and initiates the avalanche process (see inset of Figure 2.10c). At large reverse biases above the breakdown field of the diode, an electron entering the intrinsic region is accelerated by the large electric field. Once the electron reaches a large enough energy, upon collision with the lattice, the electron generates an additional electron–hole pair through impact ionization. As the impact ionization process is repeated, the result is a multiplication of electrons and holes and hence, a large current. The ionization threshold energy of Si is derived from the band structure and is 3.6 eV for electrons and 5.0 eV for holes [51].

To confirm that the electron–hole multiplication originates from the intrinsic region of the nanowire, photocurrent imaging is performed with a reverse bias applied to the nanowire APD. In this technique, the laser spot is scanned across a defined area of the sample, while simultaneously measuring the photocurrent and reflection image. By superimposing the two images, the photocurrent can be determined to originate from the intrinsic region of the nanowire, as shown in Figure 2.10b, and any contribution from the contacts can be excluded.

In order to quantify the multiplication in the Si nanowire APD, a multiplication factor is defined as:

$$M = \frac{I_{pc}(V_r) - I_{dark}(V_r)}{I_{pc}(V_{plateau}) - I_{dark}(V_{plateau})}, \tag{2.1}$$

where I_{pc} represents the photocurrent and I_{dark} represents the dark current measured at a certain reverse bias, V_r. In the range of reverse biases where no multiplication of carriers occur, there is a voltage plateau, $V_{plateau}$, and $M = 1$.

In Figure 2.10c, we characterize the photo-response of a single Si nanowire as a function of reverse bias at low temperature (20 K) and extract the multiplication factor using Equation (2.1). We measure with a laser excitation power of 1 μW (spot size = 600 nm) and laser wavelength of 532 nm. At low reverse biases (0 V to 12 V), we observe a plateau for the photocurrent measurement. This regime corresponds to no multiplication of carriers since the electric field across the depletion region is too small to induce impact ionization. In contrast, above 12 V we observe an increase in photocurrent until the reverse breakdown voltage of the diode is reached and an exponential increase of the photocurrent is obtained. The exponential increase in the observed photocurrent is attributed to a multiplication of carriers as the photo-induced electron–hole pairs, which accelerate across the depletion region have enough energy to initiate the avalanche process. In this particular device, we reach a multiplication factor of $\sim$ 3000 for V_r = 35.5 V when using $V_{plateau}$ = 10 V. This multiplication factor that we obtain is two orders of magnitude larger than previously reported on Si nanowire APDs [66].

The multiplication factor that we demonstrate is very promising for reaching the single-photon regime in future experiments from only a single Si nanowire. The challenge remains to operate the nano-APD beyond the reverse breakdown voltage similarly to conventional Si APDs in order to reach multiplication factors necessary for single photon detection.

References

[1] Akopian, N., Patriarche, G., Liu, L., Harmand, J. C. and Zwiller, V. 2010. Crystal phase quantum dots. *Nano Letters*, **10**(4), 1198–1201.

[2] Algra, R. E., Verheijen, M. A., Borgström, M. T. *et al.* 2008. Twinning superlattices in indium phosphide nanowires. *Nature*, **456**, 369–372.

[3] Atatüre, M., Dreiser, J., Badolato, A. *et al.* 2006. Quantum-dot spin-state preparation with near-unity fidelity. *Science*, **312**, 551–553.

[4] Avron, J. E., Bisker, G., Gershoni, D. *et al.* 2008. Entanglement on demand through time reordering. *Phys. Rev. Lett.*, **100**(12), 120501.

[5] Baier, M., Findeis, F., Zrenner, A., Bichler, M. and Abstreiter, G. 2001. Optical spectroscopy of charged excitons in single quantum dot photodiodes. *Phys. Rev. B*, **64**, 195326.

[6] Bayer, M., Stern, O., Hawrylak, P., Fafard, S. and Forchel, A. 2000. Hidden symmetries in the energy levels of excitonic 'artificial atoms'. *Nature*, **405**, 923–926.

[7] Bayer, M., Ortner, G., Stern, O. *et al.* 2002. Fine structure of neutral and charged excitons in self-assembled In(Ga)As/(Al)GaAs quantum dots. *Phys. Rev. B*, **65**(19), 195315.

[8] Becher, C., Kiraz, A., Michler, P. *et al.* 2001. Nonclassical radiation from a single self-assembled InAs quantum dot. *Phys. Rev. B*, **63**(12), 121312.

[9] Bimberg, D., Grundmann, M. and Ledentsov, N. N. 1999. *Quantum Dot Heterostructures*. John Wiley and Sons, Inc.

[10] Borgstrom, M. T., Immink, G., Ketelaars, B., Algra, R. and Bakkers E. P. A. M. 2007. Synergetic nanowire growth. *Nature Nanotechnology*, **2**(9), 541–544.
[11] Capasso, F. 1985. *Semiconductors and Semimetals*. Academic Press.
[12] Caroff, P., Dick, K. A., Johansson, J. *et al.* 2009. Controlled polytypic and twin-plane superlattices in III-V nanowires. *Nature Nanotechnology*, **4**(1), 50–55.
[13] Chithrani, D., Williams, R. L., Lefebvre, J., Poole, P. J. and Aers, G. C. 2004. Optical spectroscopy of single, site-selected, InAs/InP self-assembled quantum dots. *Appl. Phys. Lett.*, **84**(6), 978–980.
[14] Claudon, J., Bleuse, J., Malik, N. S. *et al.* 2010. A highly efficient single-photon source based on a quantum dot in a photonic nanowire. *Nature Photonics*, **4**, 174–177.
[15] Dick, K. A., Thelander, C., Samuelson, L. and Caroff, P. 2010. Crystal phase engineering in single InAs nanowires. *Nano Letters*, **10**(9), 3494–3499.
[16] Ding, F., Singh, R., Plumhof, J. D. *et al.* 2010. Tuning the exciton binding energies in single self-assembled InGaAs/GaAs quantum dots by piezoelectric-induced biaxial stress. *Phys. Rev. Lett.*, **104**(6), 067405.
[17] Ediger, M., Bester, G., Badolato, A. *et al.* 2007. Peculiar many-body effects revealed in the spectroscopy of highly charged quantum dot. *Nature Physics 3*, **3**, 774.
[18] Fox, M. 2006. *Quantum Optics: An Introduction*. Berlin: Oxford University Press.
[19] Fuhrmann, B., Leipner, H. S., Höche, H.-R. *et al.* 2005. Ordered arrays of silicon nanowires produced by nanosphere lithography and molecular beam epitaxy. *Nano Letters*, **5**(12), 2524–2527.
[20] Gerardot, B. D., Seidl, S., Dalgarno, P. A. *et al.* 2007. Manipulating exciton fine structure in quantum dots with a lateral electric field. *Appl. Phys. Lett.*, **90**(4), 041101.
[21] Heiss, D., Jovanov, V., Bichler, M., Abstreiter, G. and Finley, J. J. 2008. Charge and spin readout scheme for single self-assembled quantum dots. *Phys. Rev. B*, **77**(23), 235442.
[22] Ikonic, Z., Srivastava, G. P. and Inkson, J. C. 1995. Optical properties of twinning superlattices in diamond-type and zinc-blende-type semiconductors. *Phys. Rev. B*, **52**(19), 14078.
[23] Johansson, J., Dick, K. A., Caroff, P. *et al.* 2010. Diameter dependence of the wurtzite–zinc blende transition in InAs nanowires. *J. Phys. Chem. C*, **114**(9), 3837–3842.
[24] Kim, D., Sheng, W., Poole, P. J. *et al.* 2009. Tuning the exciton g factor in single InAs/InP quantum dots. *Phys. Rev. B*, **79**(4), 045310.
[25] Kiselev, A. A., Ivchenko, E. L. and Rössler, U. 1998. Electron g factor in one- and zero-dimensional semiconductor nanostructures. *Phys. Rev. B*, **58**(24), 16 353–16 359.
[26] Kiselev, A. A., Kim, K. W. and Yablonovitch, E. 2001. In-plane light-hole g factor in strained cubic heterostructures. *Phys. Rev. B*, **64**(12), 125303.
[27] Koguchi, M., Kakibayashi, H., Yazawa, M., Hiruma, K. and Katsuyama, T. 1992. Crystal structure change of GaAs and InAs whiskers from zinc-blende to wurtzite type. *Jap. J. Appl. Phys.*, **31**, 2061–2065.
[28] Korkusinski, M., Reimer, M. E., Williams, R. L. and Hawrylak, P. 2009. Engineering photon cascades from multiexciton complexes in a self-assembled quantum dot by a lateral electric field. *Phys. Rev. B*, **79**(3), 035309.
[29] Kowalik, K., Krebs, O., Lemaître, A. *et al.* 2007. Monitoring electrically driven cancellation of exciton fine structure in a semiconductor quantum dot by optical orientation. *Appl. Phys. Lett.*, **91**(18), 183104.

[30] Krishnamachari, U., Borgstrom, M., Ohlsson, B. J. *et al.* 2004. Defect-free InP nanowires grown in [001] direction on InP (001). *Appl. Phys. Lett.*, **85**(11), 2077–2079.

[31] Mattila, M., Hakkarainen, T., Lipsanen, H., Jiang, H. and Kauppinen, E. I. 2007. Enhanced luminescence from catalyst-free grown InP nanowires. *Appl. Phys. Lett.*, **90**(3), 033101+.

[32] Michler, P. (ed). 2003. *Single Quantum Dots: Fundamentals, Application and New Concepts*. Topics in Applied Physics, vol. 90. Berlin: Springer-Verlag.

[33] Minot, E. D., Kelkensberg, F., van Kouwen, M. *et al.* 2007. Single quantum dot nanowire LEDs. *Nano Letters*, **7**, 367–371.

[34] Morales, A. M. and Lieber, C. M. 1998. A laser ablation method for the synthesis of crystalline semiconductor nanowires. *Science*, **279**(5348), 208–211.

[35] Murayama, M. and Nakayama, T. 1994. Chemical trend of band offsets at wurtzite/zinc-blende heterocrystalline semiconductor interfaces. *Phys. Rev. B*, **49**(7), 4710–4724.

[36] Nadj-Perge, S., Frolov, S. M., Bakkers, E. P. A. M. and Kouwenhoven, L. P. 2011. Spin–orbit qubit in a semiconductor nanowire. *Nature*, **468**, 1084–1087.

[37] Ohlsson, B. J., Bjork, M. T., Magnusson, M. H. *et al.* 2001. Size-, shape-, and position-controlled GaAs nano-whiskers. *Appl. Phys. Lett.*, **79**(20), 3335–3337.

[38] Pierret, A., Hocevar, M., Diedenhofen, S. L. *et al.* 2010. Generic nano-imprint process for fabrication of nanowire arrays. *Nanotechnology*, **21**(6), 065305.

[39] Pryor, C. E. and Flatté, M. E. 2006. Landé *g* factors and orbital momentum quenching in semiconductor quantum dots. *Phys. Rev. Lett.*, **96**(2), 026804.

[40] Reimer, M. E., Korkusinski, M., Lefebvre, J. *et al.* 2007. Voltage induced hidden symmetry and photon entanglement generation in a single, site-selected quantum dot. arXiv:0706.1075 (unpublished).

[41] Reimer, M. E., Korkusiński, M., Dalacu, D. *et al.* 2008. Prepositioned single quantum dot in a lateral electric field. *Phys. Rev. B*, **78**(19), 195301.

[42] Reimer, M. E., Dalacu, D., Poole, P. J. and Williams, R. L. 2010. Biexciton binding energy control in site-selected quantum dots. *J. Phys.: Conf. Ser.*, **210**, 012019.

[43] Reimer, M. E., van Kouwen, M. P., Hidma, A. W. *et al.* 2011a. Electric field induced removal of the biexciton binding energy in a single quantum dot. *Nano Letters*, **11**(2), 645–650.

[44] Reimer, M. E., van Kouwen, M. P., Barkelid, M. *et al.* 2011b. Single photon emission and detection at the nanoscale utilizing semiconductor nanowires. *J. Nanophoton.*, **5**, 053502.

[45] Reimer, M. E., Bulgarini, G., Akopian, N. *et al.* 2012. Bright single-photon sources in bottom-up tailored nanowires. *Nat. Commun.*, **3**, 737.

[46] Renard, V. T., Jublot, M., Gergaud, P. *et al.* 2009. Catalyst preparation for CMOS-compatible silicon nanowire synthesis. *Nature Nanotechnology*, **4**, 654–657.

[47] Sarkar, D., van der Meulen, H. P., Calleja, J. M. *et al.* 2006. Exciton fine structure and biexciton binding energy in single self-assembled InAs/AlAs quantum dots. *J. Appl. Phys.*, **100**(2), 023109.

[48] Savio, C. Dal, Pierz, K., Ade, G. *et al.* 2006. Optical study of single InAs on $In_{0.12}Ga_{0.88}As$ self-assembled quantum dots: biexciton binding energy dependence on the dots size. *Appl. Phys. B: Lasers and Optics*, **84**(1), 317–322.

[49] Schmid, H., Bjork, M. T., Knoch, J. *et al.* 2008. Patterned epitaxial vapor-liquid-solid growth of silicon nanowires on Si(111) using silane. *J. Appl. Phys.*, **103**(2), 024304.

[50] Spirkoska, D., Arbiol, J., Gustafsson, A. *et al.* 2009. Structural and optical properties of high quality zinc-blende/wurtzite GaAs nanowire heterostructures. *Phys. Rev. B*, **80**(24), 245325.

[51] Streetman, B. 1990. *Semiconductors and Semimetals*. Solid State Electronic Devices. Prentice-Hall, Englewood.

[52] Tchernycheva, M., Cirlin, G. E., Patriarche, G. *et al.* 2007. Growth and characterization of InP nanowires with InAsP insertions. *Nano Letters*, **7**(6), 1500–1504.

[53] Toft, I. and Phillips, R. T. 2007. Hole *g* factors in GaAs quantum dots from the angular dependence of the spin fine structure. *Phys. Rev. B*, **76**(3), 033301.

[54] van Kouwen, M. P., Reimer, M. E., Hidma, A. W. *et al.* 2010a. Single electron charging in optically active nanowire quantum dots. *Nano Letters*, **10**, 1817–1822.

[55] van Kouwen, M. P., van Weert, M. H. M., Reimer, M. E. *et al.* 2010b. Single quantum dot nanowire photodetectors. *Appl. Phys. Lett.*, **97**, 113108.

[56] van Weert, M. H. M., Helman, A., van den Einden, W. *et al.* 2009a. Zinc incorporation via the vapor–liquid–solid mechanism into InP nanowires. *J. Am. Chem. Soc.*, **131**, 4578–4579.

[57] van Weert, M. H. M., Akopian, N., Perinetti, U. *et al.* 2009b. Selective excitation and detection of spin states in a single nanowire quantum dot. *Nano Letters*, **9**, 1989–1993.

[58] van Weert, M. H. M., Helman, A., van den Einden, W. *et al.* 2009c. Zinc incorporation via the vapor–liquid–solid mechanism into InP nanowires. *J. Am. Chem. Soc.*, **131**(13), 4578–4579. doi: 10.1021/ja809871j.

[59] Verheijen, M. A., Immink, G., de Smet, T., Borgström, M. T. and Bakkers, E. P. A. M. 2006. Growth kinetics of heterostructured GaP–GaAs nanowires. *J. Am. Chem. Soc.*, **128**(4), 1353–1359. doi: 10.1021/ja057157h.

[60] Verheijen, M. A., Algra, R. E., Borgström, M. T. *et al.* 2007. Three-dimensional morphology of GaP–GaAs nanowires revealed by transmission electron microscopy tomography. *Nano Letters*, **7**(10), 3051–3055.

[61] Vrijen, R. and Yablonovitch, E. 2001. A spin-coherent semiconductor photo-detector for quantum communication. *Physica E: Low-dimensional Systems and Nanostructures*, **10**(4), 569–575.

[62] Wacaser, B. A., Reuter, M. C., Khayyat, M. M. *et al.* 2009. Growth system, structure, and doping of aluminum-seeded epitaxial silicon nanowires. *Nano Letters*, **9**(9), 3296–3301.

[63] Wagner, R. S. and Ellis, W. C. 1964. Vapor–liquid–solid mechanism of single crystal growth. *Appl. Phys. Lett.*, **4**(5), 89–90.

[64] Wen, C. Y., Reuter, M. C., Bruley, J. *et al.* 2009. Formation of compositionally abrupt axial heterojunctions in silicon–germanium nanowires. *Science*, **326**(5957), 1247–1250.

[65] Xu, X., Sun, B., Berman, P. R. *et al.* 2008. Coherent population trapping of an electron spin in a single negatively charged quantum dot. *Nature Physics*, **4**, 692–695.

[66] Yang, C., Barrelet, C. J., Capasso, F. and Lieber, C. M. 2006. Single p-type/intrinsic/n-type silicon. *Nano Letters*, **6**, 2929–2934.

[67] Young, R. J., Stevenson, R. M., Shields, A. J. *et al.* 2005. Inversion of exciton level splitting in quantum dots. *Phys. Rev. B*, **72**(11), 113305.

[68] Yu, P. Y. and Cardona, M. 2005. *Fundamentals of Semiconductors: Physics and Materials Properties*. Berlin: Springer.

3

Atomic-scale analysis of self-assembled quantum dots by cross-sectional scanning, tunneling microscopy, and atom probe tomography

J. G. Keizer and P. M. Koenraad

3.1 Introduction

Self-assembled quantum dots (QDs) have been at the center of research on the quantum properties of zero-dimensional semiconductor nanostructures. The deep understanding of the physical properties and mechanisms that are active in QDs have allowed for their application in quantum secure single photon communication, quantum processing, etc. This would have been impossible without the progress in the growth control of self-assembled QDs. Nowadays, we can accurately control QD parameters such as height, composition, and strain which determine the optoelectronic and spintronic properties. Several double capping approaches have been developed that allow trimming of the QD height and Sb capping was shown to eliminate QD erosion during capping. Droplet epitaxy is a novel approach to obtain non-strained QDs, which is of great advantage because strain is one of the most complicating factors in understanding and utilizing self-assembled QDs.

In this chapter we will review our recent cross-sectional scanning tunneling microscopy (X-STM) analysis of self-assembled QDs in various hosts and obtained by a range of techniques to control their structural properties. The X-STM technique allows to image atomic scale details in semiconductor structures that are cleaved along a natural cleavage plane perpendicular to the growth direction. Although we have been able to analyze many intricate aspects of the studied QDs by X-STM, this technique is limited by its two-dimensional (2D) nature. Recently, atom probe tomography (APT) has been able to extend its field of application to semiconductor materials. By spatially resolved and mass-selective field-induced ion microscopy, the APT techniques builds a three-dimensional (3D) compositional model of the analyzed semiconductor structure with a near atomic resolution. We finish this chapter with the first structural characterization of a QD structure by a combined analysis with X-STM and APT.

3.2 Specimen preparation

In X-STM the measurements are performed on the (110) natural cleavage plane of zincblende crystals perpendicular to the growth direction that presents a cross-sectional surface

Quantum Dots: Optics, Electron Transport and Future Applications, ed. Alexander Tartakovskii. Published by Cambridge University Press.

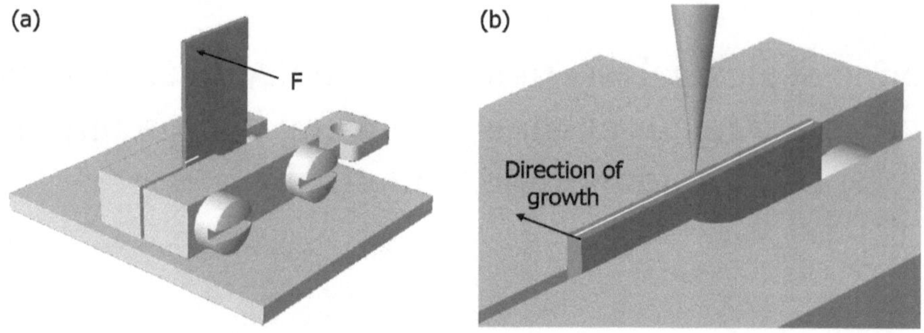

Figure 3.1 (a) The specimen with the embedded nanostructures is clamped vertically in a holder. The specimen is mechanically cleaved under UHV conditions by the application of a force (F). The cleavage is facilitated by a scratch (white line) that is applied to the surface prior to clamping. (b) The X-STM tip and the cleaved specimen with the embedded nanostructure exposed.

of the specimen. Practically, this means that the specimen is mounted vertically in a holder where it is mechanically cleaved, see Figure 3.1a, b. Before mounting, the specimen is first thinned to $\approx 100\,\mu m$ and a scratch of ≈ 1 mm is applied along the (110)-plane to its surface. The thinning and the scratch facilitate the cleavage and ensures the free propagation of the cleave in the (110)-direction. After the specimen is mounted, it is *in situ* heated to remove contaminants such as water and organic sediments. The specimen is then mechanically cleaved under UHV conditions ($p < 5 \times 10^{-11}$ mbar), exposing a contaminant-free surface with large atomically flat terraces.

In APT a needle-shaped specimen that contains the embedded nanostructures is evaporated layer by layer during analysis. The specific needle shape is required to locally enhanced the electrical field and thereby facilitate the evaporation of the constituent atoms. Needle-shaped specimens can be fabricated via the so-called lift-out technique [13] in which a scanning electron microscope, micro manipulators, Pt welding, and focus ion beam milling are used to extract and place a small piece of the original specimen on a Si pillar. With this technique it is possible to produce needle-shaped APT specimens that have a uniform circular cross-section, an apex radius of less than 50 nm, a smooth surface, a taper angle of less than 5°, and a length of several hundred nanometers.

3.3 Quantum dot analysis by X-STM

3.3.1 Shape determination

Embedded nanostructures are, per definition, made out of at least two different materials or crystallographic structures. These materials can be either latticed matched, for example AlGaAs/GaAs, or non-latticed match, for example InAs/GaAs. In the latter case the material will be strained in the vicinity of the nanostructures. Consequently, the surface will relax to minimize its elastic energy after the specimen is cleaved. This is schematically

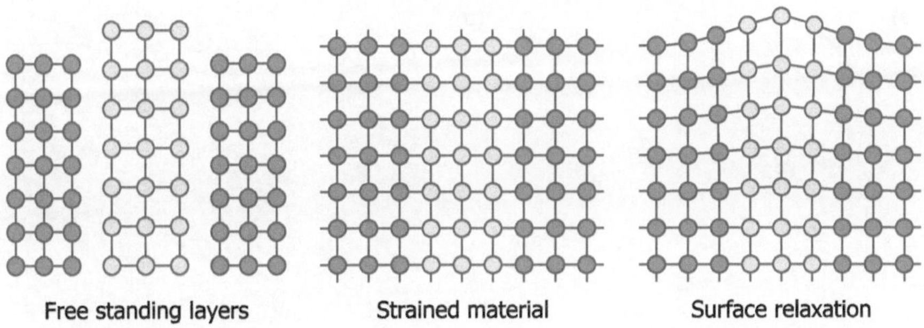

Figure 3.2 Strain relaxation at the cleaved surface of a compressively strained quantum well.

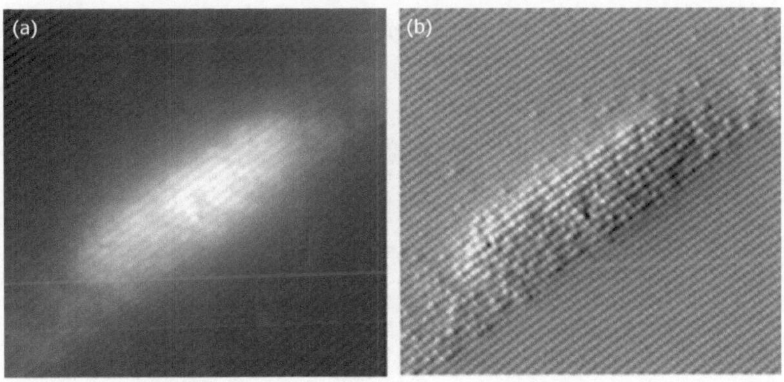

Figure 3.3 (a) A $26 \times 26\,\text{nm}^2$ topographic and (b) current image of an InAs/GaAs QD. Individual In atoms can be distinguished in the current image.

illustrated in Figure 3.2 for a compressively strained quantum well. Typically, the extend of the outward relaxation perpendicular of the surface is in the order of 100 pm and can be observed in careful X-STM measurements. When determining the shape of embedded nanostructures the outward relaxation can obscure the exact outline of the nanostructure, see for example the topographic image of the QD shown in Figure 3.3a. This problem can be circumvented by using the current image, which is a recording of the tunnel current as the tip is scanned across the surface, instead. In the so-called constant current mode, in which the tunnel current is kept constant by means of a feedback loop that controls the z-piezo, the observed contrast in the current image can be thought of as the derivative of the local changes in topography. This is caused by the finite response time of the feedback loop that will result in an (increase) decrease of the current in case of an (under-) overshoot of the tip. As a result, long-scale features in topography (outward relaxation) are filtered out while small-scale features (individual atoms) are accentuated, making the current image ideally suited to determine the exact shape of a nanostructure; see Figure 3.3b.

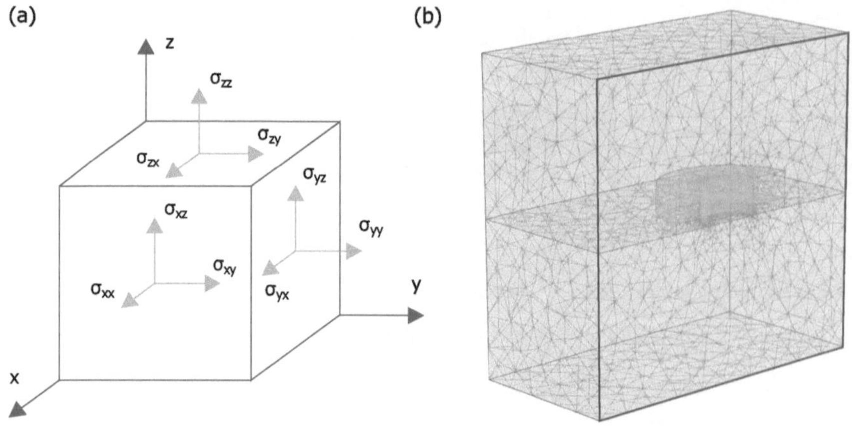

Figure 3.4 (a) Volume element with the components of the stress shown as vectors. (b) Three-dimensional model of a cleaved QD. The cleavage plane is marked with dark lines.

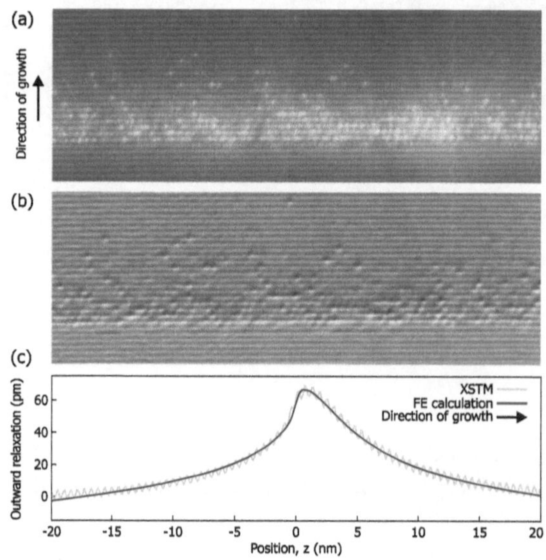

Figure 3.5 (a) A $45 \times 16\,\text{nm}^2$ topographic and (b) current image of an InAs/GaAs WL. (c) Experimental and calculated outward relaxation of the surface. Figures adapted from Ref. [7].

3.3.2 Composition profile

More than just being a nuisance in the size determination of strained nanostructures, the outward relaxation can be put to use to determine the composition of a nanostructure. For this, continuum elastic theory and finite element (FE) calculations are employed. Continuum elastic theory describes the mechanics of elastic solids by relating the stress forces acting on a material to the strain, i.e. the deformation of the solid. Given the stress vectors

σ_{ij} working on a volume element, continuum elastic theory allows the calculation of the deformation, see Figure 3.4a. This is done by solving the stress–strain relation $\sigma = D\epsilon$, in which ϵ is the strain tensor and D the elasticity matrix [12]. The latter contains Young's modulus and Poisson's ratio, both material constants. In FE calculations the solution of the stress–strain relation is evaluated numerically on a dense grid of nodes in the evaluated volume. By building a model of a nanostructure, see for example the model of a cleaved QD as shown in Figure 3.4b, and using it as input for the FE calculations the outward relaxation of the surface can be calculated. The result is checked against the experimentally obtained outward relaxation, after which the model can be adjusted to gain a better match if needed. Depending on the complexity of the nanostructure, typically five iterations in the case of wetting layers (WLs) and up to 100 in the case of QDs are needed to find a matching composition profile. In Figure 3.5, this technique is demonstrated for an InAs/GaAs WL that was modeled with an exponentially decaying In concentration in the growth direction.

3.4 Application to control of quantum dot formation

3.4.1 Double cap

Accurate control of the emission wavelength and size distribution of QDs is of crucial importance for applications in optoelectronic devices. One method of controlling the QD height and reducing the QD height dispersion is based on a discontinuous capping process, where the capping is performed in two sequences [18]. After the growth of the QD, a first capping layer (CL) with a thickness smaller than the QD height is deposited. Next, a growth interruption under phosphorous flux is performed, during which the uncovered part of the QD disappears due to As/P exchange, leading, in principle, to an uniform QD height equal to the thickness of the first CL. The capping of the QDs is then completed by the growth of a second CL. The material system we used to investigates this double cap method, consists of InAs QDs grown on and capped with InGaAsP and InP on a (311)B-substrate [21]. The thickness of the first CL was varied from 1.5 to 5 nm.

Figure 3.6 shows the current image of a QD grown on InP. The QD is found to have a flat top facet. Shallow trenches in the InP buffer layer close to the QD are observed. It is assumed that these trenches are formed immediately after QD formation during the growth interrupt under As pressure. Mass transport from the WL to the QD leads to the formation of a dewetted InP surface on which the As/P exchange reaction occurs. The extra InAs that is formed in this process migrates to the QD, leaving a trench behind. The fact that these trenches were not present in case the QD was grown on a InGaAsP buffer layer, supports this explanation because the As/P exchange is reduced on InGaAsP surfaces [19].

Figure 3.7 shows two topographic images of QDs grown on InP and capped with InGaAsP of which the thickness of the first CL was varied. As the QDs are grown on InP, the trenches are present again, but it is now clear from the figure that this is a global effect affecting the whole WL. The first CL appears as a dark region with inhomogeneous

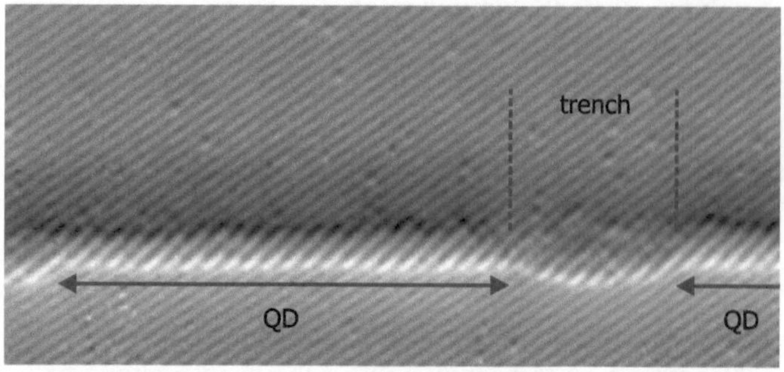

Figure 3.6 A $49 \times 22\,nm^2$ current image of an InAs/InP QD (adapted from Ref. [21]).

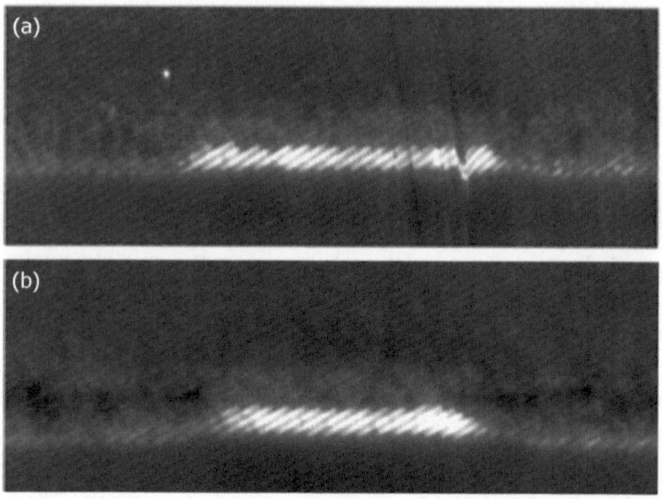

Figure 3.7 Two $55 \times 20\,nm^2$ topographic images of InAs QDs grown on InP and capped with InGaAsP. (a) A 2.0 nm high QD grown with a 2 nm thick first CL and (b) a 2.6 nm high QD grown with a 4 nm thick first CL (adapted from Ref. [21]).

contrast between the WL and the second CL. A bright region in the first CL on top of the QDs is observed, indicating an As and/or In rich region. This is likely due to strain induced composition modulation [19]. Figure 3.7a shows how the height of the QDs is leveled to the thickness of the first CL, namely 2.0 nm. When the thickness of the first CL is increased to 4 nm, see Figure 3.7b, the height of the QDs is less than that of the first CL, in this case 2.6 nm.

The results are summarized in Figure 3.8, in which the average height of the QDs in each material system is plotted as a function of the nominal thickness of the first CL. The solid line represents the ideal situation where the QD height corresponds exactly to the thickness of first CL. The double cap method works well for first CL thicknesses up to 3.5 nm, independently of the material system. For layers thicker than 3.5 nm, a reduction

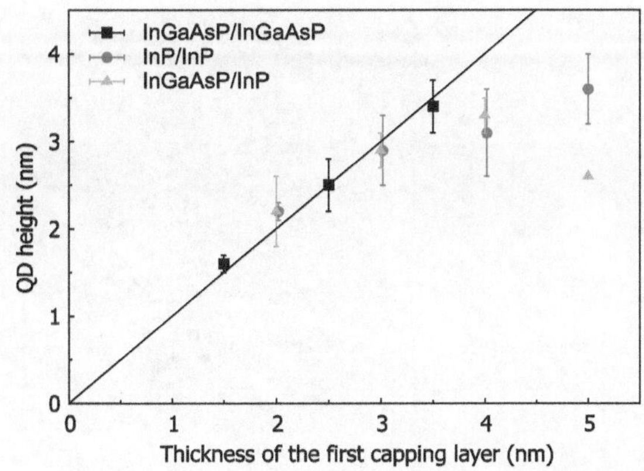

Figure 3.8 Average QD height as a function of first CL thickness for different material systems. The solid line represent the ideal situation where QD height corresponds exactly to the thickness of the first CL (adapted from Ref. [21]).

in the QD height compared to the thickness of the first CL is observed, indicating that the control over the QD height with the double cap process is lost. This can be understood in terms of a transition from a double capping to a classical one-step-capping process when the first CL is thick enough to completely cover the QDs. Once the dots are covered, the growth interrupt has no effect on them anymore.

3.4.2 Indium flush

The technique of indium flush is an adaptation of the double cap method. In this technique surface resident In atoms are desorbed by annealing [25]. The material system we used to investigate this technique consists of InGaAs QD layers grown on a GaAs (001)-substrate [8]. After QD formation at a temperature of 600 °C, the QDs are partially capped with a GaAs layer of which the thickness is varied. Next, the temperature is raised to 650 °C for 30 s, the indium flush step, and lowered again to the nominal growth temperature after which a second GaAs CL is deposited. A conventionally grown QD layer, i.e. without the application of an indium flush step, is included as a reference layer. The height of the QDs in this reference layer is found to vary between 7 nm and 10 nm. Since, the lateral width of all observed QDs is found to be of the order of 60 nm, we can safely assume that none of the QDs is cleaved through their edge and that these values represents the spread in the height distribution of the QDs due to the growth process.

The QD layers are found to be weakly coupled, as expected for a GaAs spacer layer thickness of 30 nm, and moderately strained InGaAs QDs [5], resulting in occasional stacking of the QDs. This is illustrated in Figure 3.9 which shows four QD layers where the QDs are stacked. In this particular case, the thickness of the first CL was varied from 2 nm, 3 nm to 6 nm for the first three QD layers. The last layer is the reference layer. As can be seen,

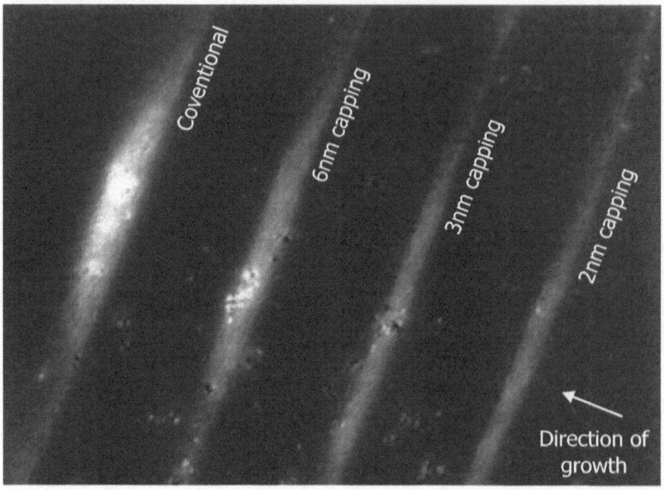

Figure 3.9 A $160 \times 120\,\text{nm}^2$ topographic image of one conventionally grown QD and three QDs grown with an indium flush step incorporated in the growth process. The thickness of the first CL was varied (adapted from Ref. [8]).

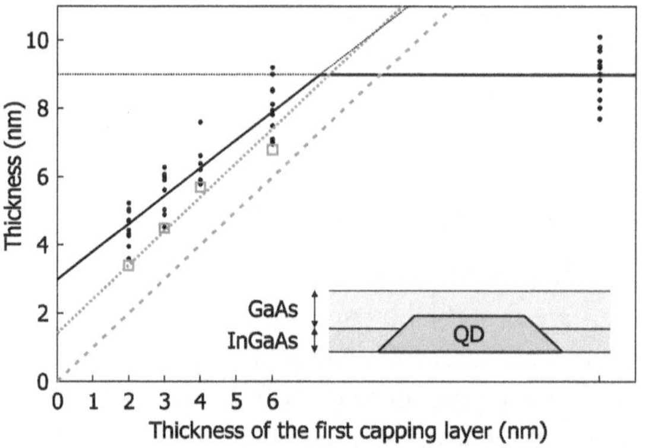

Figure 3.10 The QD height (points) as a function of the thickness of the first CL. The solid line is a linear fit. The dotted line represents the sum of the critical layer thickness for QD formation (5 ML) and the first CL thickness (dashed line). The experimentally determined thickness of the final InGaAs layer connecting the QDs is given by the open boxes (adapted from Ref. [8]).

the application of an indium flush step lowers the QDs and results in flat top facets. This in contrast to the conventionally grown QDs in the reference layer that are higher and lens shaped as previously observed for typical InGaAs QDs [23].

The height of all the observed QDs as a function of the thickness of the first CL is plotted in Figure 3.10. A linear relation is found up to a thickness of ≈ 7 nm, indicated by the solid line. Since increasing the thickness of the first CL beyond 7 nm would make the growth procedure resemble non-interrupted capping, it is expected that the QD height saturates at this value. This is indicated by the horizontal solid line that represents the average height of the conventionally grown QDs. The formation of QDs requires the deposition of a sufficiently thick layer of QD material. In the current case the critical layer thickness is 5 ML [20]. If we add this value to the nominal thickness of the first CL (dashed line) we find that the resulting value (dotted line) is the lower limit of the experimentally observed QD height. Furthermore, all the QDs are found to be a little higher than the experimentally found thickness of the final InGaAs layer connecting the QDs (open boxes). From this we can conclude that the surface was not completely flattened during the indium flush and that the QDs stick out a little afterwards. Still, the results presented above show that an indium flush can be used to gain control over the height of In(Ga)As QDs.

3.4.3 Antimony capping

In the previous sections we have shown that the techniques of double capping and indium flush can be used to control the height of QDs. We now continue with an investigation of antimony (Sb) capping, a technique that can be employed to prevent QD erosion during capping [3]. The material system we used to investigate this technique consists of InAs QD layers grown on a (311)B-oriented InP substrate. After QD formation, a 30 s growth interrupt (GI) under As pressure was performed for all layers. Previously, it has been shown that As/P exchange is limited under these GI conditions [22]. The first QD layer, which is considered as the reference layer, was overgrown with an InP CL. For the second QD layer a growth interrupt under Sb flux (GISb) was performed during 30 s before the growth of the InP CL.

In Figure 3.11a, a typical QD in the reference layer is shown. All the QDs are found to have almost flat top facets. The homogeneity of the contrast within the QDs indicates that they consist of almost pure InAs with minimal intermixing. The average height and width of QDs in the reference layer is found to be 2.0 nm and 25 nm, respectively. To investigate the influence of the growth of the CL on the shape of the QDs we compared the height histograms of uncapped QDs and capped QDs in the reference layer that were deduced from the AFM and X-STM analysis, respectively. The results are shown in Figure 3.12a, b. For the uncapped QDs, a Gaussian distribution centered around 3.3 nm is found, whereas after InP capping the height distribution is truncated at 2.4 nm. The truncated distribution and the flat top facet of InAs/InP QDs are to a large extend the consequence of QD decomposition which is driven by the strain mismatch between the InP CL and the InAs QDs [22].

Figure 3.11b shows an InAs QD for which a 30 s GISb has been performed before the InP CL was grown. The bright spots correspond to Sb atoms. Given the total amount of Sb supplied to the surface and the observed amount of Sb after capping, we conclude that a large part of the Sb is desorbed during overgrowth. Segregation of the small fraction of Sb that gets incorporated in the InP CL is clearly shown. Within the QDs the back diffusion of Sb is negligible and a preferential incorporation of Sb is observed at the outermost layers of the QDs. Again, the QDs appear well defined with minimal intermixing. In Figure 3.12c the height histogram of the QDs as deduced from X-STM measurements is

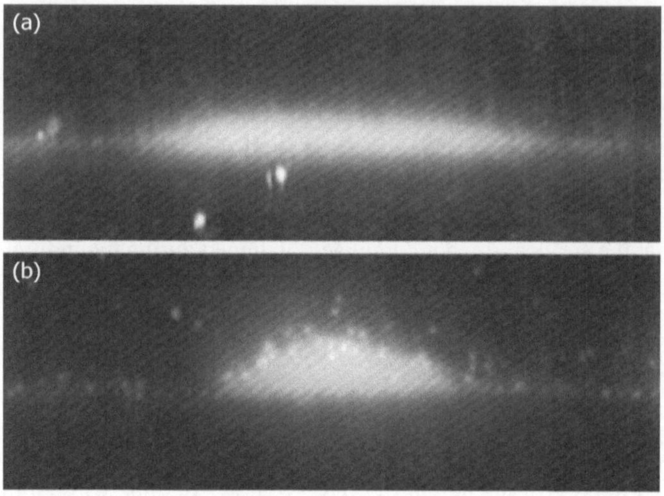

Figure 3.11 Two $60 \times 15\,\mathrm{nm}^2$ topographic images. (a) InAs QD capped with InP after a 30 s GI. (b) InAs QD capped with InP after a 30 s GI + 30 s GISb. The bright spots correspond to Sb atoms (adapted from Ref. [3]).

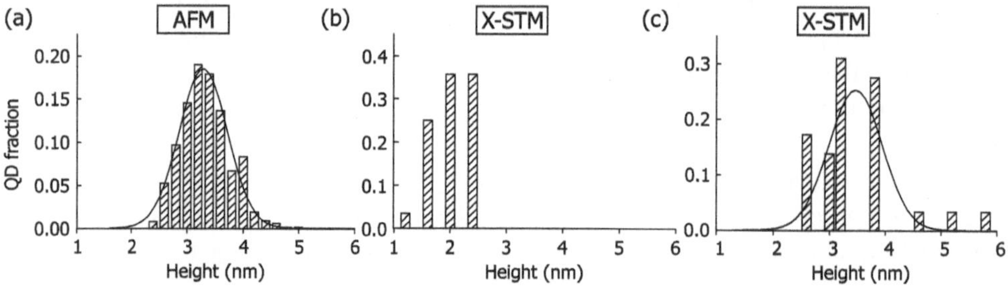

Figure 3.12 QD height distribution of (a) uncapped InAs QDs, (b) with InP capped InAs QDs after a 30 s GI, and (c) with InP capped InAs QDs after a 30 s GI + 30 s GISb (adapted from Ref. [3]).

shown. Apparently, the presence of Sb on the surface has consequences for the QD shape; the mean height and diameter are now 3.5 nm and 21 nm, respectively, corresponding to the dimensions of the uncapped QDs. We can explain the observed shape preservation by the well-documented surfactant effect of Sb atoms [1]. An Sb surfactant can limit the in-plane diffusion of adatoms on the surface. Accordingly, InAs diffusion from the QD apex to the periphery is reduced due to the presence of Sb atoms on the surface. This freezing of mass transport on the growth front results in the preservation of the shape of the uncapped QDs.

3.4.4 *Droplet epitaxy*

Having shown three techniques to control the shape and height of SK-grown QDs, we now turn our attention to droplet epitaxy (DE), which is a method for growing QDs in

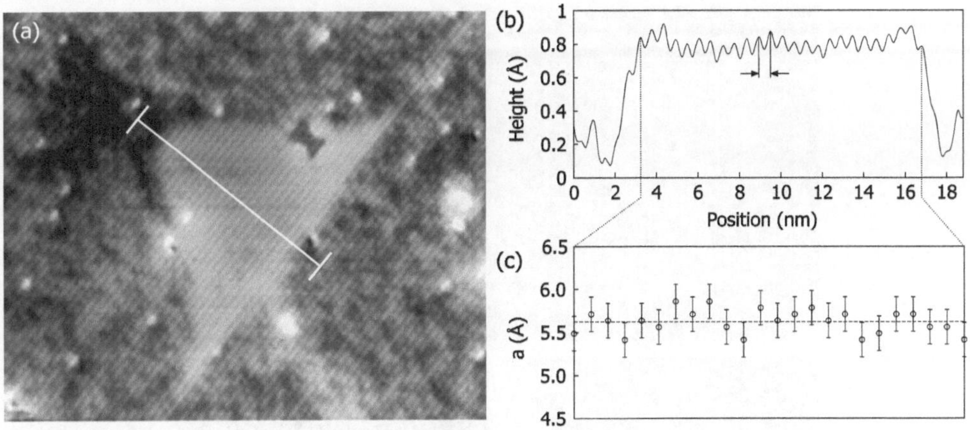

Figure 3.13 (a) A 40 × 34 nm^2 topographic image of a typical GaAs/AlGaAs QD. (b) Average cross-sectional profile and (c) separation between bilayers along the line in (a). Figures adapted from Ref. [9].

lattice-matched material systems. DE involves low-temperature growth of liquid unstrained group III-element droplets that are subsequently crystallized into QDs by incorporation of group V-elements [11]. The material systems we used to investigate this technique are GaAs QD layers grown on (001)- and (311)A-oriented AlGaAs substrates [9, 10]. All the images presented in this section are recorded at high negative bias voltages. At these tunneling conditions and with the color scaling used, dark (light) regions represent AlAs (GaAs)-rich regions.

We start with the QDs grown on a (001)-substrate. A typical QD is shown in Figure 3.13a. As can be seen, the QD is sharply defined by abrupt interfaces. Since AlAs and GaAs are lattice matched materials, the QD is expected to be strain free. This can be checked by taking a cross-sectional profile, see Figure 3.13b. Three distinct regions can be observed. From left to right: an AlAs rich region, the GaAs QD, and the AlGaAs matrix. The height difference between these regions is due to electronic contrast. Importantly, none of the regions show any outward relaxation (see for comparison the outward relaxation induced by the WL in Figure 3.5). To further illustrate that the QD is strain free, the distance between adjacent bilayers along the cross-sectional profile was measured, see Figure 3.13c. There is little deviation from the expected value of 0.565 nm (dashed line), indicating that the QD is indeed strain free. Note that there is an AlAs rich region (dark present on top of the QD. This can be explained by the difference in mobility of Al and Ga adatoms; the Ga adatoms are more mobile and will migrate along the side of the QD during capping while the Al adatoms, which are less mobile, are more likely to remain on top of the QD. The driving force behind the migration of the incoming adatoms is the convex curvature of the growth front at the position of the QDs [26]. Note that this is different from the SK-grown QDs presented in the previous sections were the driving force is strain.

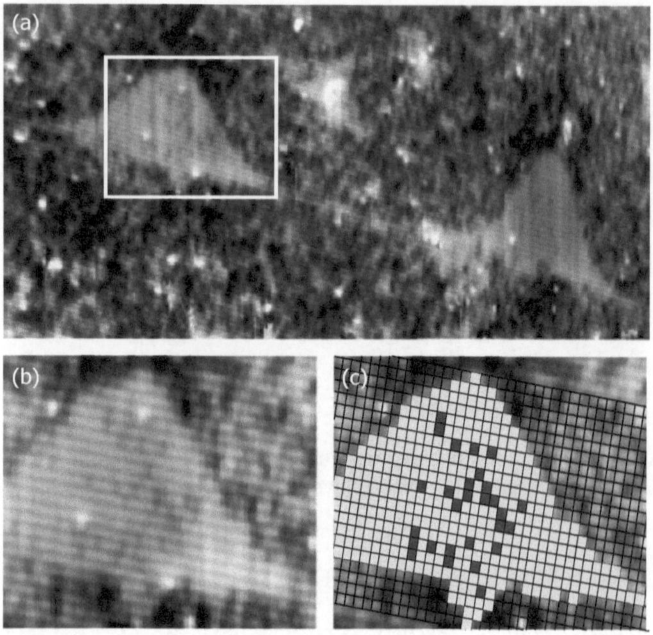

Figure 3.14 (a) A $30 \times 60\,\text{nm}^2$ topographic image of two QDs. (b, c) An atomic grid is overlain on top of a close up of the QD. Al and Ga atoms in the QD are marked by dark and light squares, respectively. Figures adapted from Ref. [9].

In all QDs we observed some degree of intermixing. In Figure 3.14a, two typical QDs are shown. Even without further analysis it is evident that some intermixing of Al has taken place, see dark spots inside the QDs. To make a more quantitative analysis we overlaid a grid with atomic dimensions on top of a close up of the QD that showed the strongest intermixing, see Figure 3.14b, c. On this grid, the positions of the Al and Ga atoms are marked with dark and light squares, respectively. From this analysis we determined that the upper limit of Al intermixing in these QDs is 6%. The buffer layer directly underneath the QD is the most likely source of the Al atoms. It is known that liquid Ga droplets can etch into the buffer layer in a process that is called local etching were the material directly under the liquid droplet is dissolved [24]. The result of this process can be seen in Figure 3.14a, where an GaAs intrusion in the AlGaAs buffer is visible underneath both QDs.

Concerning the shape of the QDs, we notice that the side facets of the observed QDs are not exactly straight. The maximum side facet angles were found to be in the range 34–55° per QD, were the upper limit corresponds to an {111}-facet (54.7°). If we assume that (1) all the QDs are approximately of equal height and (2) the observed height difference is due to the position of the cleavage plane relative to the center of the QD, this result excludes QD shapes with constant facet angles like rectangular (truncated) pyramids [4]. Figure 3.13 shows the highest QD we found. This QD can thus be considered cleaved directly through its center. Consequently, we used this QD profile to generate a three-dimensional model

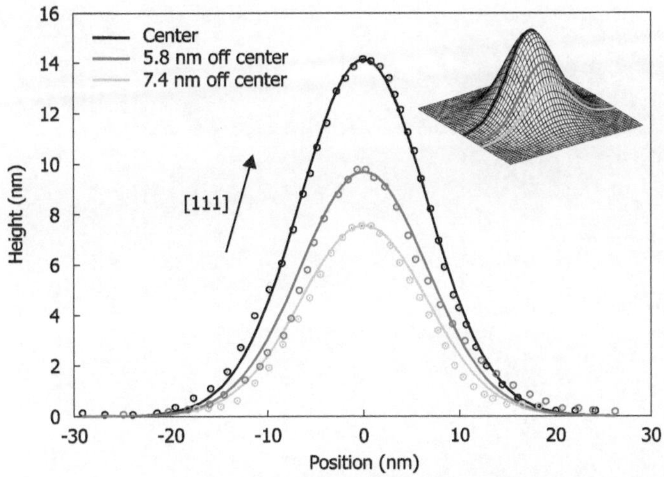

Figure 3.15 Profile of three QDs as measured by X-STM (open circles). A Gaussian function is fitted to the largest QD. The other two QDs are assumed to have the same three-dimensional structure as the largest QD but cleaved off center (see inset). Figures adapted from Ref. [9].

by fitting a Gaussian function, see Figure 3.15, and rotating it around the symmetry axis along the growth direction. Next, we checked whether the profile of all other observed QDs (illustrated for two exemplary QDs) correspond to profiles obtained by cleaving the obtained three-dimensional model at specific distances from the center. As can be seen in Figure 3.15, this is the case. From this we conclude that the observed QDs are Gaussian shaped.

We now turn our attention to QDs grown on a (311)A-surface. In Figure 3.16a, a typical part of the QD layer is shown. Three QDs can be distinguished. A WL connecting the QDs is not observed. Owing to the high density, approximately half of the observed QDs are found to overlap with other QDs. This is illustrated in Figure 3.16c, where the right side of the QD overlaps with another QD. The QDs are found to be bounded by (211)- and (411)-side facets, resulting in a slight elongation of the QDs in the [2-3-3]-direction. Emanating from the top of the QDs are AlAs rich regions (dark). As a guide to the eye the boundary of one of these regions is marked in Figure 3.16a. The formation of these features can again be explained by the difference in mobility of the Ga and Al adatoms. Since Ga adatoms favor the (311)A-surface above the (211)- and (411)-side facets of the QD [2], and given the limited migration length of Al adatoms, this results in AlAs rich regions on top of the QDs. The observation that the AlAs rich regions extend all the way to the GaAs capping layer, and the slight interface irregularities found at this position, indicates that to some extend the original QD shape is retained at the growth front during capping. The lateral position at the growth front of the asymmetric QDs is continuously shifted in the [-233]-direction during overgrowth resulting in a tilt of the AlAs rich regions toward the [011]-direction.

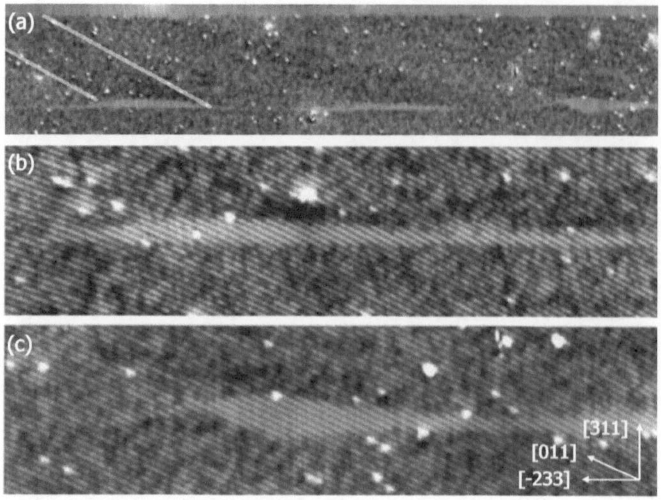

Figure 3.16 (a) A $35 \times 170\,\text{nm}^2$ topographic image of the QD layer. One of the AlAs rich regions emanating from the top of the QDs is marked by two lines. (b) A $16 \times 65\,\text{nm}^2$ topographic image of a typical QD with a smooth and (c) rough bottom AlGaAs/GaAs interface (adapted from Ref. [10]).

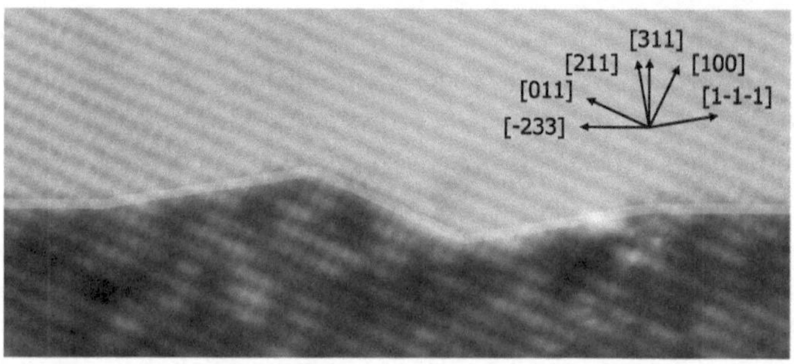

Figure 3.17 A $10 \times 22\,\text{nm}^2$ topographic image of a GaAs QD capped with GaAs. The interface is marked by the transparent white line (adapted from Ref. [10]).

In approximately one third of the observed QDs, a GaAs intrusion in the AlGaAs buffer layer that is paired with an AlGaAs protrusion in the GaAs QD is found at the bottom interface of the QD, see Figure 3.16c. The paired protrusions/intrusions are better resolved in another sample were the QDs are overgrown with a GaAs CL, see Figure 3.17, resulting in a clearer view of the interface. In both samples the protrusions were always found to form in the [-233]-direction relative to the intrusions and are bounded by an (100)- and (211)-surface. The paired protrusions/intrusions are most likely the result of an interface instability induced by the liquid Ga droplet. It is well known that (311)A-surfaces show temperature instabilities that break up the flat surface into facets that have a lower surface energy [17]. We suggest that a similar process happens in the current case, where inside the droplets the (311)A-surface is destabilized by the presence of the liquid Ga and subsequently reconfigures into a more energetically favorable configuration consisting of (100)- and (211)-facets. We would like to stress that this case is different from the QDs grown on an (001)-substrate were local etching is responsible for the interface fluctuations.

3.5 Outlook: atom probe tomography

In the previous sections we showed that X-STM can be used in structural studies of self-assembled QDs. Although a lot of information can be extracted with X-STM, it remains in essence a two-dimensional method that is bound to the cross-sectional surface of a cleaved specimen. To construct a fully three-dimensional model, statistics and FE calculations have to be employed. This is very time consuming, and might not be the most accurate method. In this respect, atom probe tomography (APT) might outperform X-STM in the not to distant future.

APT is the latest evolution of the venerable field emission microscope [15] whereby a field ion microscope is combined with a spatially resolved time-of-flight mass spectroscope, creating a device known as a three-dimensional atom probe [6]. In APT a needle-shaped specimen is placed under a large pulsed voltage causing ions to be emitted from the specimen's apex and accelerated towards a detector screen. The data collected after each voltage pulse can be combined to form a three-dimensional tomographic image of the specimen. This technique works best for materials with a high conductivity, where sub-atomic precision can be achieved, and so traditionally there has been little use for the technique outside of metallic systems. However, the re-emergence of laser based APT has greatly expanded the range of materials that can be analyzed. Where semiconductor materials were previously out of bounds, now the additional thermal excitation from a pulsed laser makes performing APT on these materials feasible. Consequently, pioneering work to study semiconductor nanostructures, such as QDs, has started to be undertaken [16, 7].

The material system we used to investigate the capabilities of APT as a tool to study embedded semiconductor nanostructures consist of a GaAs (001)-oriented specimen that contains five InAs QD layers [7]. In Figure 3.18a, the outline of the needle-shaped specimen and the embedded QD layers are clearly visible. From this figure, it immediately becomes apparent that the technique of APT is of a different nature than X-STM; the APT measurement provides a fully three-dimensional data set. Such three-dimensional data sets can provide a myriad of possibilities to study the QDs. For example, the In concentration through the center of a QD can be readily extracted; see Figure 3.18b, c.

To forge a link between APT and X-STM the QD layers were studied with both techniques. Figure 3.19a, b shows the topographic and current X-STM image of a cleaved QD. Normally, it is not known how the cleavage plane intersects the QDs. However, this particular QD is one of the highest and widest that was observed. Therefore, we can reasonably assume that this particular QD was cleaved through its center. Given the high density, it is not possible to count the number of In atoms in the QD and directly determine an In profile. For this reason the outward relaxation of the cleaved surface is normally used in

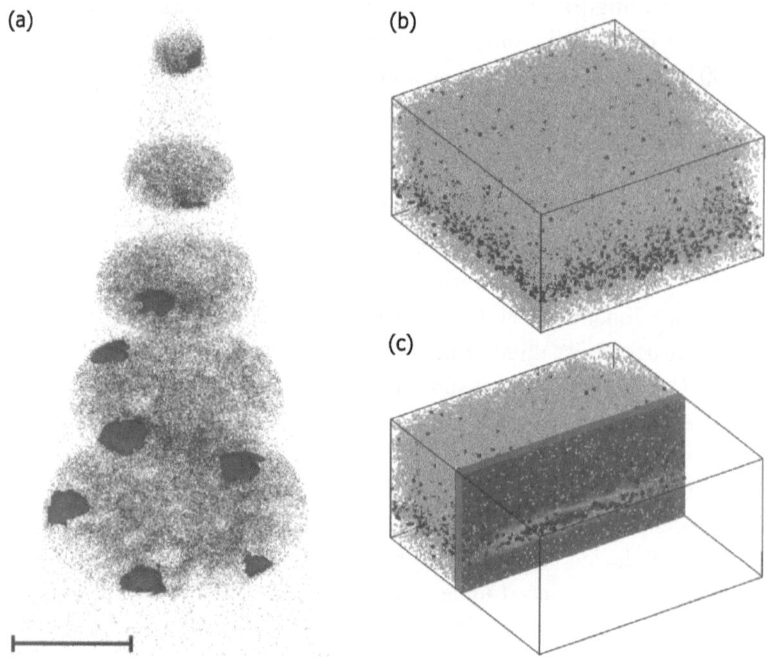

Figure 3.18 (a) Atom map showing only the In atoms. The position of the QDs are marked with 25% In fraction isosurfaces. The scale bar is 50 nm. (b) APT data set of a QD. The In atoms that make up the WL can be distinguished as dark spheres. (c) Taking a cross-section through the data set reveals the core of the QD. A 2D concentration profile has been superimposed. Figures adapted from Ref. [7].

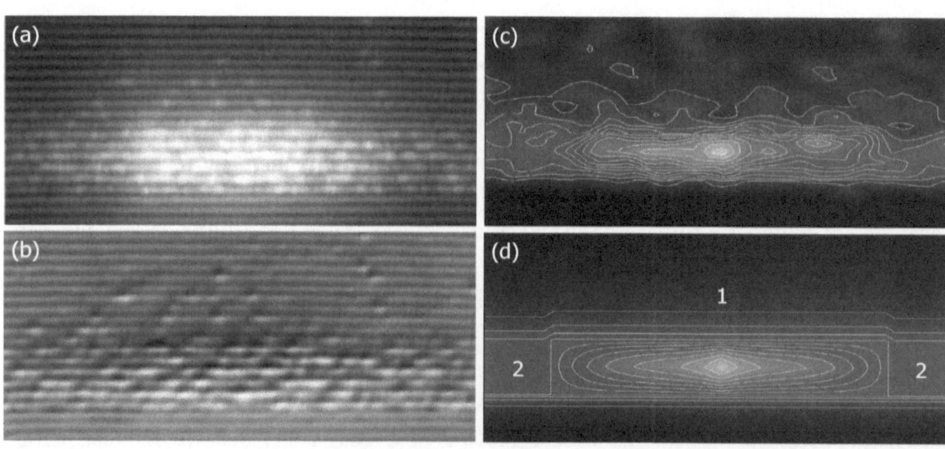

Figure 3.19 Comparative views of two QDs. All images are to the same scale ($25 \times 13\,\text{nm}^2$). (a) Topographic and (b) current X-STM image. (c) contour plot of the In concentration through the center of a QD measured by APT. The contour lines are every 5%, ranging from 0% to 65%. The APT data is used to construct a model (d) that serves as input for the FE calculations. Figures adapted from Ref. [7].

X-STM measurements to assess the In profile. It is, however, difficult to obtain an unique In profile that fits to the observed QD profile. Therefore, an indirect route via APT and FE calculations is followed. The In concentration profile that was extracted from the typical QD shown in Figure 3.18c is used to generate a contour plot of the In concentration profile, see Figure 3.19c. From this an analytical In concentration model is constructed, see Figure 3.19d, that is used as input for the FE calculations. Besides the In profile in the core of the QD, two other regions proved crucial in modeling the QD. First is the exponential decay of the In concentration in the direction of growth above the QD. As much as 16% of the total number of In atoms that constitute the QD reside in this region. Note that in the X-STM images of Figure 3.19a, b, only a few individual In atoms are visible above the QD and that an exponential decay cannot be distinguished. However, the APT data reveals that, together, these atoms form a significant part of the In profile and should thus be included in the model. The second region that is crucial in modeling our QDs is the gradual transition of the QD into the WL. The X-STM and APT images of Figure 3.19a, c show a WL at the sides of the QD that is more substantial than the WL further away from the QDs.

In Figure 3.20, the outward relaxation of the cleaved surface across the center and 7 nm off-center of the QD, as measured by X-STM, is plotted together with the result of the FE calculations. A close match between the measured and the calculated outward relaxation is observed. Here it should be noted that different input models, e.g. the models with a linear In profile, can yield a similar match. However, such models would not resemble the APT data. From this we conclude that care is required when constructing QD models based on X-STM measurements alone. Recently, it has been shown [14] that PL measurements in combination with extensive theoretical modeling and X-STM measurements had to be combined to yield a realistic QD model. However, the subtleties of the decaying In concentration above the QD and the extension of the QD into the WL might be overlooked

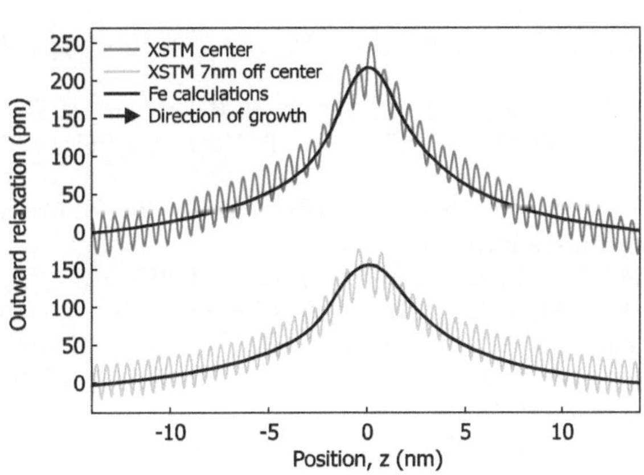

Figure 3.20 Outward surface relaxation along the growth direction through the center and 7 nm off center of the QD shown in Figure 3.19a, b. The X-STM measurements are compared with the results of the FE calculations. Figures adapted from Ref. [7].

in such an approach, and thus some ambiguity still remains. In our view, the abilities of APT in providing a three-dimensional In profile are at the moment unique, and necessary to construct an valid three-dimensional QD model.

3.6 Conclusion

In this chapter we have shown how cross-sectional STM and atom probe tomography can yield a wealth of atomic scale information on the structural and compositional properties of self-assembled quantum dots. We have used these techniques to discuss various double capping methods, including one with an indium flush step, to trim the quantum dot height, various approaches which use Sb to eliminate quantum dot erosion during capping, and droplet epitaxy which allows strain-free quantum dots to be grown.

References

[1] Aivaliotis, P., Wilson, L. R., Zibik, E. A. *et al.* 2007. *Appl. Phys. Lett.*, **91**(1), 013503.
[2] Asano, K., Kangawa, Y., Ishizaki, H. *et al.* 2004. *Appl. Surf. Sci.*, **237**(1–4), 206–212.
[3] Bozkurt, M., Ulloa, J. M. and Koenraad, P. M. 2011. *Semicond. Sci. Technol.*, **26**(6), 064007.
[4] Bruls, D. M., Vugs, J. W. A. M., Koenraad, P. M. *et al.* 2002. *Appl. Phys. Lett.*, **81**(9), 1708.
[5] Bruls, D. M., Koenraad, P. M., Salemink, H. W. M. *et al.* 2003. *Appl. Phys. Lett.*, **82**(21), 3758.
[6] Cerezo, A., Godfrey, T. J. and Smith, G. D. W. 1988. *Rev. Sci. Instrum.*, **59**(6), 862.
[7] Giddings, A. D., Keizer, J. G., Hara, M. *et al.* 2011. *Phys. Rev. B*, **83**, 205308.
[8] Keizer, J. G., Clark, E. C., Bichler, M. *et al.* 2010a. *Nanotechnology*, **21**(21), 215705.
[9] Keizer, J. G., Bocquel, J., Koenraad, P.M. *et al.* 2010b. *Appl. Phys. Lett.*, **96**(6), 062101.
[10] Keizer, J.G., Jo, M., Mano, T. *et al.* 2011. *Appl. Phys. Lett.*, 193112.
[11] Koguchi, N., Takahashi, S. and Chikyow, T. 1991. *J. of Cryst. Growth*, **111**(1–4), 688–692.
[12] Lai, W. M., Rubin, D. and Krempl, E. 2010. *Introduction to Continuum Mechanics.* 4th edn. Butterworth-Heinemann.
[13] Miller, M. K. and Russell, K. F. 2007. *Ultramic.*, **107**(9), 761–766.
[14] Mlinar, V., Bozkurt, M., Ulloa, J. M. *et al.* 2009. *Phys. Rev. B*, **80**(16), 1–7.
[15] Müller, E. W., Panitz, J. A. and McLane, S. B. 1968. *Rev. Sci. Instrum.*, **39**(1), 83.
[16] Müller, M., Cerezo, A., Smith, G. D. W., Chang, L. and Gerstl, S. S. A. 2008. *Appl. Phys. Lett.*, **92**(23), 233115.
[17] Nötzel, R., Däweritz, L. and Ploog, K. 1992. *Phys. Rev. B*, **46**(8), 4736–4743.
[18] Paranthoën, C., Bertru, N., Dehaese, O. *et al.* 2001. *Appl. Phys. Lett.*, **78**(12), 1751.
[19] Poole, P., Williams, R. L., Lefebvre, J. and Moisa, S. 2003. *J. Cryst. Growth*, **257**(1–2), 89–96.
[20] Snyder, C., Mansfield, J. and Orr, B. 1992. *Phys. Rev. B*, **46**(15), 9551–9554.

[21] Ulloa, J. M., Koenraad, P. M., Gapihan, E., Létoublon, A. and Bertru, N. 2007a. *Appl. Phys. Lett.*, **91**(7), 073106.
[22] Ulloa, J. M., Cÿelebi, C., Koenraad, P. M. *et al.* 2007b. *J. Appl. Phys.*, **101**(8), 081707.
[23] Wang, X.-D., Liu, N., Shih, C. K., Govindaraju, S. and Holmes, A. L. 2004. *Appl. Phys. Lett.*, **85**(8), 1356.
[24] Wang, Zh. M., Liang, B. L., Sablon, K. A. and Salamo, G. J. 2007. *Appl. Phys. Lett.*, **90**(11), 113120.
[25] Wasilewski, Z., Fafard, S. and McCaffrey, J. P. 1999. *J. Cryst. Growth*, **201–202**(3), 1131–1135.
[26] Xie, Q., Chen, P. and Madhukar, A. 1994. *Appl. Phys. Lett.*, **65**(16), 2051.

Part II

Manipulation of individual quantum states in quantum dots using optical techniques

4

Studies of the hole spin in self-assembled quantum dots using optical techniques

B. D. Gerardot and R. J. Warburton

4.1 Self-assembled quantum dots as host for spin qubits

A coherent spin in the solid-state would be very attractive for a number of applications. A single spin has an obvious application as a magnetic field sensor; entangled spin states can potentially enhance the sensitivity [54]. An optically active spin is a potential component of a quantum repeater, a technology to extend fibre-based quantum cryptography to large distances [41]. Also, a spin qubit is a potential building block of a quantum information processor [62]. But, applications aside, the targeted investigation of spin coherence in the solid-state is leading to new insights into the microscopic nature of the complex spin environment, allowing some old problems, for instance the central spin problem, to be fruitfully revisited.

The search for spin coherence in the solid-state has led most spectacularly so far to the NV^- centre in diamond whose spin coherence can reach $\sim$1 ms even at room temperature [5]. However, diamond is difficult to process into a real device. Electron and hole spins in III–V semiconductors have yet to achieve the coherence of the NV^- in diamond, but these materials have some considerable advantages. First, quantum dots can be used to confine electron spins to nanometer length scales [47, 61]. The quantum dots can either be defined electrostatically by local depletion of a two-dimensional electron gas, or they can be self-assembled during growth, for instance InAs on GaAs. Second, both a mature heterostructure technology and post-growth nanofabrication can be used to add functionality to the quantum dots.

The advantage self-assembled quantum dots have over electrostatically defined quantum dots is a strong optical transition with an "almost hard" selection rule connecting the individual spin states to exciton states:

$$
\begin{aligned}
|\uparrow\rangle &\overset{\sigma^+}{\longleftrightarrow} |\uparrow\downarrow\Uparrow\rangle \\
|\downarrow\rangle &\overset{\sigma^-}{\longleftrightarrow} |\uparrow\downarrow\Downarrow\rangle
\end{aligned}
\tag{4.1}
$$

where $|\uparrow\rangle$ represents an electron spin up, $|\downarrow\rangle$ represents an electron spin down, $|\uparrow\downarrow\Uparrow\rangle$ represents an exciton (a trion) with two spin-paired electrons and a hole with spin $J_z = +\frac{3}{2}$, and $|\uparrow\downarrow\Downarrow\rangle$ represents a trion with spin $J_z = -\frac{3}{2}$. All spin states are in the $\vec{z}$-basis

Quantum Dots: Optics, Electron Transport and Future Applications, ed. Alexander Tartakovskii. Published by Cambridge University Press.

($\vec{z}$ representing the growth direction), and σ^+ (σ^-) refers to spin angular momentum $+\hbar$ ($-\hbar$) of the photon. The spontaneous radiative decay time from the trion state is just $\sim$1 ns. This strong optical transition allows the spin initialization [2, 40], spin rotation [66, 9, 72] and spin readout [3, 86, 78] all to be carried out optically.

While there are clearly some unresolved issues concerning all these optical techniques, the main limitation so far is arguably the lack of spin coherence. A single electron spin confined to a quantum dot has a time-averaged coherence time of just $T_2^* \sim 10$ ns [66, 9, 84, 58]. The main dephasing mechanism is the effect of noise in the nuclear spins which creates a fluctuating magnetic environment for the electron spin via the hyperfine interaction. A number of approaches are being pursued to tame this dephasing either by polarizing the nuclear spins or by actively reducing the noise.

An interesting alternative to an electron spin is a *hole* spin. The main point is that a pure heavy hole spin experiences the spin noise arising from the nuclei in the growth direction only [33]. This pronounced anisotropy makes it much easier to suppress the effects of the nuclear spins. From an experimental perspective, very little is known presently about hole spins in semiconductor quantum dots. Work so far in realizing the concept of the spin qubit has focused almost exclusively on electron spins. Unfortunately, creating quantum dots electrostatically with p-type material has not been successful – hysteresis and noise in even a simple p-type split-gate has hindered progress [69]. Nevertheless, optically active self-assembled quantum dots can be loaded with a single hole [40, 48, 73, 17, 22]. We present here a snap-shot of results of current optical experiments on holes in InGaAs/GaAs quantum dots.

4.2 Motivating factors for hole spins

A hole spin in a self-assembled quantum dot is potentially coherent. Current theory predicts phonon-limited hole spin relaxation times of order $T_1 \sim 1$ ms [77, 83] and that if the phonon interaction limits the coherence the ideal limit applies, $T_2 = 2T_1$ [43]. The hole spin coherence benefits, first, from strong strain; second, strong quantization; and third, the p-like nature of the underlying atomic states.

Strain drives the self-assembly during semiconductor growth of quantum dots, for instance InAs on GaAs. The resulting quantum dots, typically $\sim$20 nm in diameter and $\sim$5 nm high, are highly strained with a significant uniaxial component along the growth direction. This uniaxial strain splits the heavy hole (HH) – light hole (LH) degeneracy at the Γ-point in the valence band structure, perhaps by 100 meV or more. This energetic "distance" to the light hole states implies that the valence ground state is predominantly heavy hole-like, a conclusion supported by atomistic calculations [11].

Even in strained quantum wells where the HH–LH degeneracy is also lifted by uniaxial strain [80, 64, 81], the hole spin coherence is limited by rapid, typically sub-ns, spin relaxation involving a phonon [24, 63]. However, hole spin relaxation times increase as the temperature decreases as a consequence of hole localization [75, 35], strongly hinting at the potential of quantum dots with particularly strong localization.

In the case of a quantum dot, the nanoscale confinement leads to discrete conduction levels separated by ~20–50 meV, and discrete valence levels separated by ~10–25 meV. Phonon-related electron spin relaxation is relatively straightforward [47, 43]. The spin–orbit interaction admixes the first excited state $|1\downarrow\rangle$ with the ground state $|0\uparrow\rangle$ (and likewise admixes $|1\uparrow\rangle$ with $|0\downarrow\rangle$). The electron–phonon interaction, $H_{\mathrm{e-ph}}$, has a non-zero matrix element $\langle 1|H_{\mathrm{e-ph}}|0\rangle$ and hence the combination of spin–orbit with the electron–phonon interaction provides a relaxation route from $|0\uparrow\rangle$ to $|0\downarrow\rangle$. This relaxation is fast in a quantum well or bulk material. In a quantum dot, the $|1\downarrow\rangle$ admixture to the $|0\uparrow\rangle$ state is suppressed by the large quantization energy separating the $|0\rangle$ and $|1\rangle$ states. Furthermore, at low magnetic field B_{ext}, the phonon density of states at the spin relaxation energy is small, decreasing rapidly with B_{ext}. The net result is that at small magnetic fields, electron spin relaxation times longer than 1 s have been observed [1], with a characteristic reduction as B_{ext}^{-5} as B_{ext} increases [59]. The spin–orbit interaction of a hole spin is considerably more complicated than that of an electron spin [18] but the basic point survives: strong quantization in the quantum dot suppresses spin relaxation via phonons. At magnetic fields of about 1 Tesla, and at a temperature of 4 K, theory predicts that one-phonon processes dominate, with hole spin relaxation times in the ~ millisecond regime [18]. At higher magnetic fields, the hole spin relaxation time decreases; however, at lower magnetic fields, although the one-phonon process becomes increasingly unlikely, two-phonon processes are predicted to take over [77], resulting in a weak magnetic field dependence. For a realistic dot topology, hole spin relaxation times of ~1–20 ms have been predicted in the low field regime [83], an enhancement of many orders of magnitude over spin relaxation times for delocalized holes in quantum wells.

An electron spin in a quantum dot at a small magnetic field can have a very long relaxation time, $T_1 \geq 1$ s, but a rather short time-averaged coherence time, $T_2^* \sim 10$ ns. The origin of the apparent fast dephasing is noise in the nuclear spin bath [65, 32]. A quantum dot electron interacts through the hyperfine interaction with $N \sim 10^5$ nuclear spins, a "central spin problem" with a strong mesoscopic character. The number of nuclear spins is too large to use individual nuclear spins as a quantum resource yet too small for complete cancellation of the hyperfine interaction. Through the contact part of the hyperfine interaction, each nuclear spin acts on the electron spin via an effective magnetic field. We note that for InGaAs, all possible isotopes have non-zero nuclear spins. For N nuclear spins, the effective magnetic fields tend to cancel, but the cancellation is incomplete with the overall field scaling as $1/\sqrt{N}$ [65]. For an InGaAs quantum dot for instance, the overall effective magnetic field (the Overhauser field) is around $B_N \sim 15$ mT. Generally, the Overhauser field fluctuates in the course of an experiment and this leads to apparent dephasing. In a simple view, the components of the contact hyperfine interaction perpendicular to the applied magnetic field represent an electron spin–nuclear spin flip-flop process, a process suppressed by energy conservation by applying a small magnetic field. (The electron spin and nuclear spin gyromagnetic ratios are considerably different of course.) This, along with the suppression of phonon-related spin relaxation, is the origin of the large electron spin T_1 times at small magnetic fields. However, the component of the Overhauser field along the applied field direction cannot be suppressed so simply and, in this simple picture, it is this component which results in the apparent dephasing. Specifically, in an applied magnetic field along, say, the $\vec{z}$-direction, the total field along the applied field direction is $B^z_{ext} + B^z_N$ resulting in random changes in the electron spin precession frequency, equivalently a time-averaged dephasing.

In the case of an electron spin, a number of routes are being pursued in order to limit the dephasing via the hyperfine interaction. One promising approach is to polarize the nuclei dynamically (see Chapter 14 in this book). While polarizing the nuclei completely is probably very challenging, partial polarization can be combined with feedback control to "narrow" the distribution, i.e. reduce the noise [85, 79, 6]. An alternative standpoint is that the nuclear field B_N fluctuates slowly and therefore looks like a dephasing process in a slow experiment (a T_2^* process) but does not represent a true loss of quantum coherence. Equivalently, while the T_2^* may be as short as 10 ns, the true dephasing time T_2 is much longer, tens of μs [6, 70, 44, 45, 12]. Gate operations robust to the slow nuclear wanderings can be implemented starting with the Hahn echo sequence [70] and then adding sophistication [6]. However, these approaches add much complexity even to the simplest spin operation. In a general sense, strongly suppressing electron spin relaxation via the phonons by strongly confining the spin to the nanoscale opens up a new dephasing channel, the hyperfine interaction.

The hole spin has a quite different hyperfine interaction [33, 76]. The main point is that valence states are constructed largely from atomic p states. An atomic p state has the convenient property that the amplitude of the wave function is zero at the location of the nucleus. For a pure valence state in a quantum dot, the wave function amplitude is zero at every nuclear spin in the quantum dot, turning off the contact part of the hyperfine interaction. There is of course admixture of atomic s Bloch states to the atomic p Bloch states through the **k.p**-like interactions, but the large fundamental band gap of the semiconductor implies that this admixture is small. The dipole–dipole part of the hyperfine interaction remains, however [33, 76]. (The dipole–dipole part of the hyperfine interaction is zero of course for a pure conduction state with a zero angular momentum Bloch s state.) For a heavy hole spin, the dipole–dipole hyperfine coupling coefficients have been estimated to be roughly 10% those of the electron spin contact hyperfine interaction [33]. However, there is a crucial difference between the hyperfine interactions of a pure conduction state and a pure heavy hole valence state. A pure heavy hole valence state has spin quantum numbers $J, J_z = \frac{3}{2}, \pm\frac{3}{2}$ in the valence electron basis ($|\Downarrow\rangle$, $|\Uparrow\rangle$ in the hole basis)

$$\begin{aligned} \left|\frac{3}{2}, +\frac{3}{2}\right\rangle &= \frac{1}{\sqrt{2}}\left|(p_x + ip_y)\uparrow\right\rangle \equiv |\Downarrow\rangle \\ \left|\frac{3}{2}, -\frac{3}{2}\right\rangle &= \frac{1}{\sqrt{2}}\left|(p_x - ip_y)\downarrow\right\rangle \equiv |\Uparrow\rangle \end{aligned} \tag{4.2}$$

where p_x and p_y are p orbitals in the plane, along the $\vec{x}$- and $\vec{y}$-directions, respectively. Semi-classically, a valence heavy hole state consists of a circulating electron current either with angular momentum $+\hbar$ and spin up or angular momentum $-\hbar$ and spin down. This means that the magnetic dipole moment of a pure heavy hole state must lie in the

$\vec{z}$-direction. The coupling of this magnetic dipole to each nuclear spin has an Ising-form with terms $\propto J_z I_z$ where I_z is the z-component of the nuclear spin. Equivalently, a pure heavy hole spin "feels" the presence of just a z-component of an Overhauser field. Without an applied magnetic field, this interaction will dephase the heavy hole spin relatively quickly [33]. However, we recall that in the case of an electron spin, it is only the component of B_N along the applied magnetic field direction which results in apparent dephasing. In the case of a heavy hole spin in a reasonably strong in-plane magnetic field, the Ising-like nature of the hyperfine interaction implies that the fluctuations in total magnetic field along the applied field direction are strongly suppressed. A clear prediction for a pure heavy hole state is that the dephasing time T_2^* should increase linearly with in-plane magnetic field [32].

This prediction, that a heavy hole spin becomes highly coherent in an in-plane magnetic field, is very robust for a *pure* heavy hole state. In practice, a pure heavy hole state does not exist. The valence ground state in a real quantum dot will inevitably be an admixture of heavy hole, light hole and spin–orbit split-off Bloch states. For light hole and spin–orbit split-off Bloch states, the basis vectors include not only p_x, p_y but also p_z components such that the dipole–dipole part of the hyperfine interaction is no longer purely Ising-like. Up until recently, it was a completely open question whether the non-heavy hole components in the valence states of *real* quantum dots are sufficiently small for the hole spin to become coherent in an in-plane magnetic field. Recently, strong evidence has emerged in InGaAs quantum dots for, first, relatively slow hole spin relaxation at low magnetic fields and at 4 K, $T_1 \sim 1$ ms [40, 48], and second, a relatively coherent hole spin in an in-plane magnetic field, $T_2^* \sim 1\ \mu$s [17]. The extent to which the hole spin T_2^* can be improved by appropriately designing the quantum dot ("spin engineering") is presently an open question.

Finally, we comment on the possibility of electric field control. The spin–orbit interaction provides a mechanism for an oscillating electric field to drive the electron spin resonance [68]. The spin–orbit interaction of a hole spin is both larger and more complicated than that for an electron spin. This may open the possibility of driving the hole spin resonance rather quickly with electrical pulses [19], and perhaps of performing arbitrary spin rotations electrically.

4.3 Spectroscopy of few-level systems

Laser spectroscopy techniques provide a means to probe spin decoherence. There are two ground states, the spin "up" and "down" states, split by a magnetic field, $|\Uparrow\rangle$ and $|\Downarrow\rangle$ in the case of a hole spin, and two excited states, the trion states, $|\Uparrow\Downarrow\uparrow\rangle$ and $|\Uparrow\Downarrow\downarrow\rangle$ where the basis states are taken along the direction of the applied magnetic field. In the Faraday geometry (magnetic field along the growth direction z and light polarization in the xy-plane), there are two strong "vertical" transitions, $|\Uparrow\rangle \stackrel{\sigma^-}{\longleftrightarrow} |\Uparrow\Downarrow\uparrow\rangle$ and $|\Downarrow\rangle \stackrel{\sigma^+}{\longleftrightarrow} |\Uparrow\Downarrow\downarrow\rangle$, on account of the selection rules. Much weaker "diagonal" transitions exist via heavy hole–light mixing , and via the component of the Overhauser field perpendicular to the applied

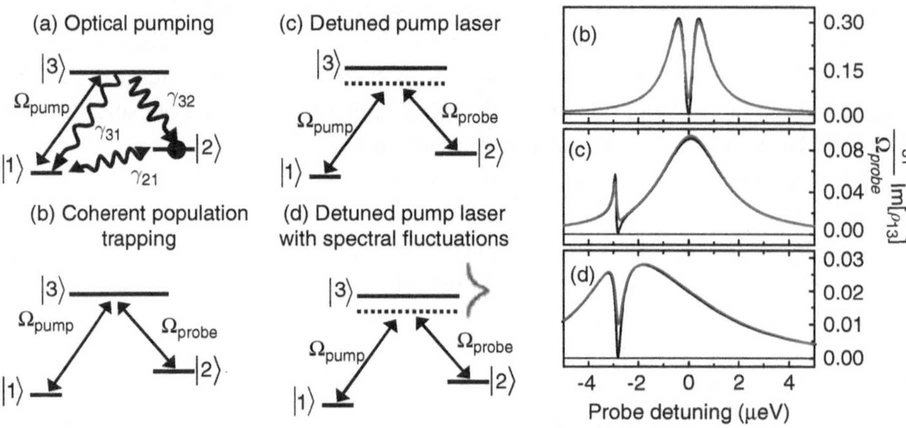

Figure 4.1 In a spin-Λ system, level $|3\rangle$ is dipole coupled to levels $|1\rangle$ and $|2\rangle$, but $|1\rangle$ and $|2\rangle$ are uncoupled. The relaxation channels γ_{ij} are shown as wavy lines. (a) Optical pumping occurs when a coupling field Ω_{pump} is resonant with the $|1\rangle \leftrightarrow |3\rangle$ transition. The relaxation channel γ_{32} projects the system into state $|2\rangle$. The $|2\rangle \leftrightarrow |3\rangle$ transition is off resonance with the driving field, hence high fidelity optical initialization (•) is achieved provided γ_{21} is sufficiently small. (b) A second driving field Ω_{probe} resonant with $|2\rangle \leftrightarrow |3\rangle$ transition frustrates the optical pumping mechanism and the absorption signal completely reappears with the exception of a transparency window due to CPT at zero detuning provided γ_{21} is sufficiently small. (c) CPT with a detuned pump. (d) CPT with a detuned pump plus a spectral wandering of level $|3\rangle$. The probe absorption spectrums in (b)–(d) are calculated using Rabi energies $\hbar\Omega_{\text{pump}} = 1.0$ μeV, $\hbar\Omega_{\text{probe}} = 0.45$ μeV, radiative decay rates $\hbar\gamma_{31} = \hbar\gamma_{32} = \frac{1}{2}\hbar\gamma_r = 0.50$ μeV with a lower level coherence time $T_2 = 1$ μs (light grey lines), $T_2 = 10$ ns (black lines). Here ρ is the density matrix. Spectrum (d) includes a spectral wandering of level $|3\rangle$ with Lorentzian probability distribution with full width 6 μeV.

magnetic field. In the Voigt geometry (both magnetic field and polarization in xy-plane), the "vertical" and "diagonal" transitions have equal optical dipole matrix elements, and this is extremely attractive in order to establish a so-called Λ-system.

In atomic physics, a spin-Λ system is the prototypical system to achieve coherent *optical* manipulation of the ground states. In general, a spin-Λ system consists of three quantum states, $|1\rangle$, $|2\rangle$ and $|3\rangle$, with optical transitions between $|1\rangle \leftrightarrow |3\rangle$ and $|2\rangle \leftrightarrow |3\rangle$ (Fig. 4.1). For a single hole spin, states $|1\rangle$ and $|2\rangle$ correspond to Zeeman-split hole spin states, $|3\rangle$ to an exciton. With this level scheme, "optical pumping" can be exploited to optically initialize the spin, "coherent population trapping" (CPT) can optically create a coherent quantum superposition of the two spin states, and "STIRAP" (Stimulated Raman Adiabatic Passage) can be used to manipulate the spin coherently [10]. A spin-Λ system underpins many applications in quantum information science, including electromagnetic induced transparency [34] and quantum networking [56].

Optical pumping is the process whereby optical excitation of an iso-spin-$\frac{1}{2}$ system produces a population distribution different to the Boltzmann distribution at the environment's temperature. Brossel and Kastler first proposed schemes for optical pumping in atomic gasses in 1949 [16] and successfully implemented them in 1952 [15]. In the framework of quantum information processing, it allows high-fidelity ($> 99\%$) spin initialization. It also permits characterization of spin relaxation dynamics. Optical pumping occurs when a driving field resonantly excites one spin state (state $|1\rangle$ in Fig. 4.1) and a relaxation mechanism projects the system into the spin state orthogonal to the originally excited state (state $|3\rangle$). The orthogonal spin state is shielded from the driving field due to an energy detuning and the polarization. Hence, the absorption signal is related to the spin initialization efficiency. The efficiency of the spin initialization process is determined by the spin relaxation rate (γ_{21}), the driving field Rabi frequency (Ω_{pump}), and the spontaneous emission rate (γ_{32}). The optical pumping mechanism can be frustrated with a second laser, the *repump* laser, resonant with the second transition (Fig. 4.1b).

When both optical transitions of a Λ-system are driven, the system acquires amplitude of the upper state $|3\rangle$ via two pathways. A destructive quantum interference phenomenon is manifest when the frequency difference of the lasers matches the $|1\rangle - |2\rangle$ splitting, the two-photon resonance. The quantum interference creates a ground state coherent superposition: $|\text{Dark}\rangle = \alpha|1\rangle - \beta|2\rangle$, where α and β are determined by the two Rabi frequencies, "pump" and "probe". The creation of a dark state is revealed by a dip in the probe absorption spectrum at the two-photon resonance, an effect called coherent population trapping (CPT), itself the microscopic origin of electromagnetically induced transparency. The extent to which the absorption of the probe laser vanishes is sensitive to the coherence of the ground states, and hence a coherent population trapping experiment on a single spin can probe the quantum coherence of the spin. Specifically, when $\Gamma_3 > \Omega_{\text{pump}} > \Omega_{\text{probe}}$, the probe absorption spectrum acquires a dip with energy width $\hbar\Omega^2_{\text{pump}}/\Gamma_3$ where Γ_3 is the total spontaneous emission rate from state $|3\rangle$. This width sets the sensitivity of the experiment to the decoherence rate γ_{12} of state $|2\rangle$ with respect to $|1\rangle$: for $\gamma_{12} \ll \Omega^2_{\text{pump}}/\Gamma_3$ the signal in the dip goes to zero (Fig. 4.1 light grey curves) but for $\gamma_{12} \gg \Omega^2_{\text{pump}}/\Gamma_3$ the dip is washed out (Fig. 4.1 black curves). Several CPT experiments in semiconductors have struggled to achieve a perfect dip because typically the ground state dephasing rates are too large [84, 37, 36, 71].

4.4 Hole spin sample design

A prerequisite to probing and exploiting any advantageous hole spin attributes is high-fidelity control of the hole population in a quantum dot (QD). A secondary requirement is that the few-level spin system must be sufficiently isolated from the mesoscopic environment. There are three techniques to load a single carrier into a quantum dot deterministically: chemical doping near the QD layer [21, 44, 28, 26], optical generation [59, 29, 73, 42], and controlled carrier tunnelling from a nearby Fermi sea [82, 40]. Chemical doping is achieved by growing the quantum dot layer a few nm above a δ-doped layer. Typically, the nominal doping density is similar to the QD density so that on average one charge from the dopant migrates to the QD. Optical generation of a single hole can be

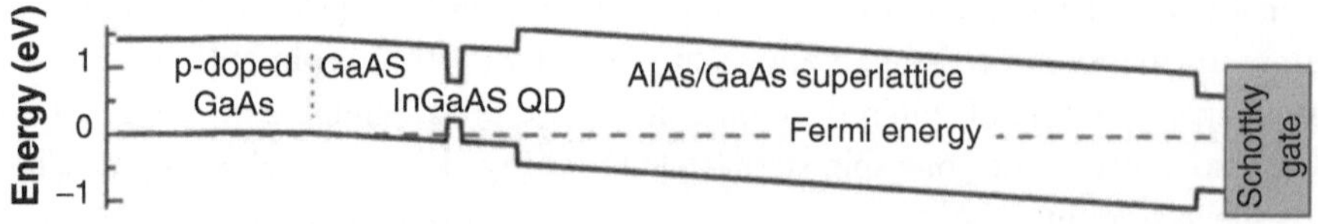

Figure 4.2 The band diagram of the charge-tunable device simulated using a 1D Poisson solver. Holes tunnel through a 25 nm GaAs barrier into the self-assembled InGaAs quantum dots from the *p*-doped back contact. An applied bias between the semi-transparent top gate controls the charge state in the quantum dot.

achieved using an $n-i-$metal photodiode structure with a QD in the intrinsic region. To generate a single hole, a large electric field is used to ionize an optically created neutral exciton, X^0. Both resonant [48, 73, 42] or non-resonant [29] excitation can be used. Owing to the electric field, the electron tunnels quickly ($\tau_{\text{tunnel}} \sim 10$ ps) to the n-doped contact, leaving behind a single hole. We discuss here in detail the third method, deterministic hole charging of a QD embedded in a field-effect structure containing a p-doped back contact.

The samples, grown by molecular beam epitaxy, consist of InGaAs quantum dots tunnel coupled to a grounded back contact (doping 4×10^{18} cm^{-3}) through a 25 nm thick GaAs barrier. The dots are capped with a 10 nm GaAs layer, a 131 nm AlAs/GaAs superlattice (a blocking barrier), a 5 nm GaAs layer, and finally a 4 nm semi-transparent NiCr gate. A bias applied between the Schottky and back contacts creates a vertical electric field in the device (Fig. 4.2). Here we study three samples all identical in structure except for the dopant species in the back gate. Two samples contain different species of acceptor dopants: Be or C. These are compared to a well characterized n-type sample containing Si as a donor dopant (same sample used in refs. [38, 23, 39]). We use non-resonant photoluminescence (PL) to examine the charging dynamics, recombination and Coulomb energies. Resonant laser spectroscopy is then performed to characterize the true lineshape and linewidth of the excitonic transitions.

4.4.1 Photoluminescence spectroscopy

PL spectroscopy is performed on each sample at 4.2 K using a confocal microscope. The PL is excited with an 830 nm laser diode, dispersed with a grating spectrometer, and detected with a Si charge coupled device. Figures 4.3a and b show contour plots of the luminescence as a function of applied bias for the Si-doped and C-doped samples, respectively. In Fig. 4.3a, the excitonic charge is constant over large regions of bias voltage followed by sudden jumps in the emission energy when an electron is added to the dot. The change in emission energy as a carrier is added is a consequence of the Coulomb interaction. Although the Coulomb blockade is well-defined, two charge states are still simultaneously present at the same gate bias, albeit with drastically different intensities. The appearance of multiple emission lines is the result of complications arising on

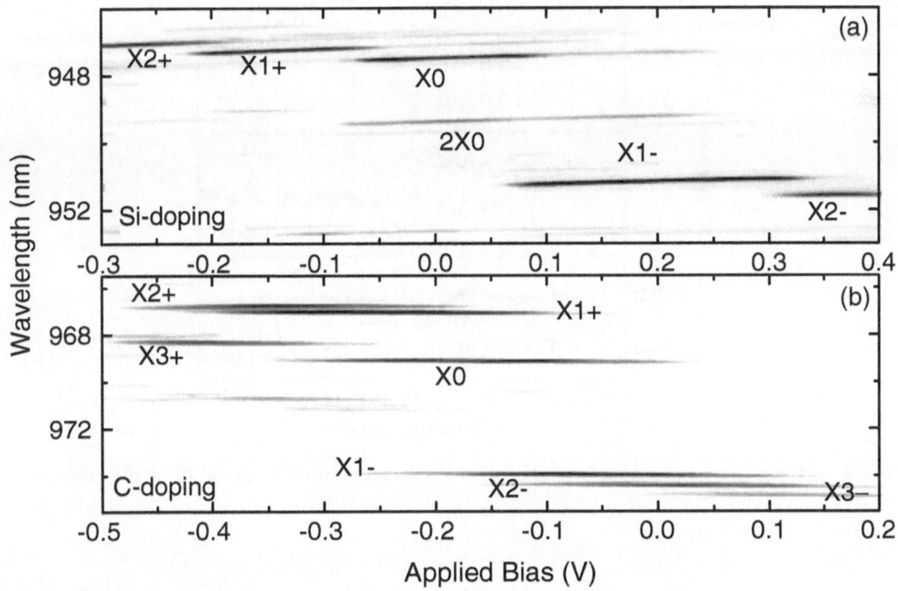

Figure 4.3 Grey-scale intensity contour plots of the photoluminescence wavelength as a function of applied bias for the (a) Si- and (b) C-doped sample, respectively.

non-resonant excitation and the time-integrated nature of the spectra [4]. For the n-type sample with 25 nm tunnel barrier, the tunnelling time is $\tau_{\text{tunnel}} \sim 10$ ps [74], much shorter than typical recombination lifetimes for both the neutral exciton X^0 (0.8 ns) and negatively charged trion X^{1-} (0.9 ns) [23]. However, if the tunnelling time is comparable to or smaller than the recombination lifetime then the quantum dot has insufficient time before recombination to form the charge state with the lowest energy. In this case, overlapping lines occur in the time-averaged PL spectra. This is the situation for tunnelling of holes into a dot though a 25 nm barrier from the *p*-doped back gate (Fig. 4.3b), for which we have determined a hole tunnelling time of $\tau_{\text{tunnel}} \sim 10$ ns [40].

4.4.2 Resonant laser spectroscopy

Once the QD states are characterized in PL, we measure in a transmission geometry the light resonantly scattered by the excitonic dipole in a cryogenic microscope [49, 55]. A glass hemispherical solid immersion lens on top of each sample is used to reduce the spatial resolution of the microscope and increase the signal:noise [38]. Unlike the charging dynamics described for non-resonant excitation, the resonant laser is spectrally specific, interacting at any one time only with only one charge configuration. Hence, absorption into multiple charge states is eliminated for a given applied bias. The charging transitions are close to step-like for each sample, independent of the carrier tunnelling time.

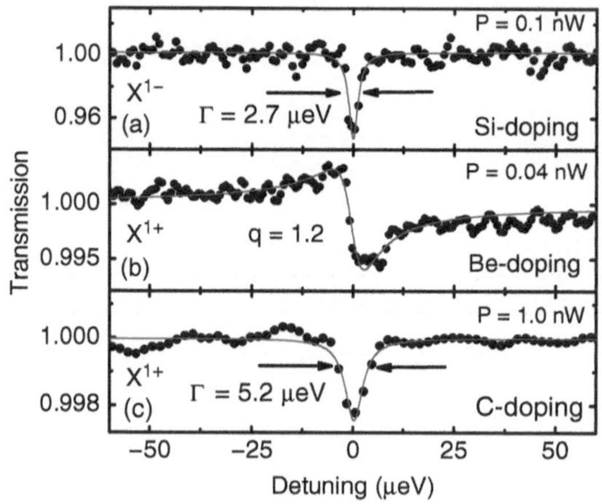

Figure 4.4 Typical resonant excitation spectra for charged excitons from the three different samples (a)–(c).

The absorption of a single-mode electromagnetic field by a two-level atomic transition exhibits a Lorentzian lineshape. Figures 4.4a–c show typical absorption spectra for singly charged QDs from the Si-, Be-, and C-doped samples at zero applied magnetic field, respectively. In each case linear polarization was used to prevent optical pumping [40]. Similar to the absorption spectra for the neutral excitons (not shown), we find that the X^{1-} transitions in the Si-doped sample and the X^{1+} transitions in the C-doped sample display Lorentzian lineshapes, suggesting that these transitions are close to two-level-like in nature. Further evidence of two-level behaviour is the near ideal saturation broadening (saturation curves from the Be- and C-doped samples not shown; see Ref. [38] for the X^0 and X^{1-} saturation curves from the Si-doped sample). However, there are two striking differences between Si- and C-type devices: the amount of signal at resonance (the contrast) is $\sim 10\times$ smaller and the transition linewidths are $\sim 3\times$ larger for the C-doped compared to the Si-doped sample with the identical device structure. We find similar X^0 linewidths for the C- and Be-doped structures. The inhomogeneous broadening is likely caused by spectral fluctuations, slow on the timescale determined by radiative recombination, but fast compared to the measurement integration rate [40, 49].

While the charged transitions in the Si- and C-doped samples exhibit Lorenztian lineshapes, the differential absorption spectrum for the X^{1+} transitions in the Be-doped sample (Fig. 4.4b) displays a dispersive-like lineshape with positive overshoots and zero-crossings, characteristic of a Fano-like quantum interference. Furthermore, the saturation curves for these transitions do not follow the two-level model. Instead, the transitions saturate at much lower power than expected. The Fano effect is a manifestation of quantum interference between two excitation channels: from a discrete ground state to either a continuum of states or a discrete excited state [31]. The coupling of the discrete excited state to the continuum of states was auto-ionization (an Auger-like process) in the original Fano investigation, but can also be tunnelling. The quantum interference is parameterized by the Fano factor, q. Note that $q \to \infty$ corresponds to negligible interference of the two-level transition with the continuum, resulting in a Lorentzian lineshape. Conversely, $q = 1$ signifies a strong quantum interference with the nature, destructive or constructive, changing rapidly as a function of detuning, leading to a dispersive lineshape. Recently, the Fano effect was observed in a self-assembled QD coupled to two-dimensional states in a nearby quantum well [57]. In that case, a discrete ground state (a single electron) was optically coupled both to a discrete excited state (an X^{1-}) and continuum states in the valence band which were tunnel coupled to the confined hole state. It was found that saturation of the discrete two-level transition by an intense laser increased the sensitivity to the continuum states. In addition, q could be tuned by changing the tunnelling dynamics or the pure two-level transition could even be restored by changing the sample structure [57].

The Lorentzian lineshapes and near ideal saturation curves for the Si- and C-doped samples demonstrate that the sample design successfully suppresses the influence of any nearby continuum of states. The only difference between the Be-doped sample and the other two is the dopant, suggesting that for the Be-doped sample, the Be impurities themselves are responsible for the interfering continuum. For MBE growth of Be-doped [001] GaAs, it is difficult to grow sharp interfaces between Be–GaAs and intrinsic GaAs due to significant Be diffusion at typical growth temperatures [67]. Alternatively, C doping with a CBr_4 source offers the ability to obtain high doping concentrations and much sharper interfaces in high quality GaAs [50]. The X^{1+} absorption spectra suggest that Be dopant atoms which have diffused to sites close to the QDs form an electronic continuum at the X^{1+} energy. The microscopic mechanism is unclear. Notably, we see varying Fano visibility from dot to dot.

The results show that, for the current devices, the inhomogeneous linewidth is larger, the absorption contrast smaller, and the carrier tunneling times slower for samples with p-doped back gates compared to n-doped. Nevertheless, resonant laser spectroscopy reveals ideal Coulomb blockade for p-doped structures. A dot dependent non-linear Fano effect is observed for X^{1+} transitions from the Be-doped sample. However, absorption spectroscopy on the X^{1+} transition in the sample with C-doping reveals near ideal two-level behaviour. This sample structure is used in the following sections for probing hole spin dynamics.

4.5 Spin initialization, manipulation, and read-out

For potential applications in quantum information processing, one must be able to initialize, coherently manipulate, and read-out the hole spin state. In general, spin initialization is achieved optically by using a resonant laser while there are various methods for spin read-out: resonant laser spectroscopy [2, 40], optical read-out after a time delay [59, 48], and electrical read-out via photocurrent [73]. We first focus on the application of optical pumping, as described in Section 4.3.

4.5.1 Hole spin initialization in the Faraday geometry

Continuous wave experiments

Optical pumping is typically observed in a spin-Λ system when a magnetic field is applied to Zeeman split the ground states. However, optical pumping can be achieved in a QD charged with a single hole even with degenerate ground states (no externally applied magnetic field). The scheme, depicted in Figs. 4.5a, b along with experimental results, depends on stringent requirements: very clean optical selection rules, slow hole spin relaxation, and an Ising-like interaction between the hole spin and the nuclear spins. The process involves excitation from a ground state ($|\Downarrow\rangle$) with σ^+ polarized light to the excited state $|\Uparrow\Downarrow\downarrow\rangle$. In the excited state, the electron spin experiences a small magnetic field, $B_N \sim 15$ mT [70, 2, 44] and the in-plane component, $B_N^{x,y}$, causes the electron spin to precess with a period of ~ 1 ns. The spin precession is destroyed by spontaneous emission at rate Γ_0, with a characteristic time of ~ 1 ns, resulting in the shelving of the hole spin in the $|\Uparrow\rangle$ state. According to spin selection rules, the σ^+ laser field does not couple the $|\Uparrow\rangle \leftrightarrow |\Uparrow\Downarrow\downarrow\rangle$ transition. Hence, the hole spin is initialized with high fidelity into $|\Uparrow\rangle$ provided T_1^{hole} is sufficiently large. Remarkably, the hole spin is shielded from the laser only by its spin and not by an energy detuning. The spin pumping result at $B_{ext}^z = 0$ T presented in Fig. 4.5 demonstrates that the hole spin relaxation time must be much, much larger than the spin precession time. For the electron spin at fields $B_{ext}^z \leq 0.3$ T, spin pumping is prohibited due to the hyperfine interaction which enables rapid spin relaxation [2]. This is not the case for holes: the high-fidelity hole spin pumping even at $B_{ext}^z = 0$ demonstrates a highly suppressed hole spin hyperfine interaction. Also, rather than a hindrance, the hyperfine interaction between the electron spin and the nuclear spins is beneficial in this scheme as it enables fast hole spin initialization.

Hole spin pumping at small magnetic fields is also highly effective. In the Faraday configuration, the degeneracy of the two optical transitions is lifted by the sum of the electron and hole Zeeman energies. When the σ^+-polarized laser is tuned to the $|\Downarrow\rangle \leftrightarrow |\Uparrow\Downarrow\uparrow\rangle$ resonance, the contrast is very small (see Fig. 4.5) signifying spin pumping into $|\Uparrow\rangle$. As B_{ext}^z increases, the total magnetic field experienced by the electron spin is tilted further from the (x, y)-plane, reducing the electron spin precession rate and hence the spin pumping rate. When $B_{ext}^z \gg B_N^{x,y}$, the total magnetic field lies close to the $\vec{z}$-direction and spin pumping becomes very slow, eventually becoming comparable to the hole spin relaxation rate. In this regime ($B_{ext}^z \sim 3$ T), the transmission signal is fully established. A second feature of the magnetic field dependence is the increasing contrast inequality for excitation with σ^+ relative to σ^- polarization. This dependence arises as the system moves gradually towards thermal equilibrium. The thermal population of $|\Downarrow\rangle$ ($|\Uparrow\rangle$) increases (decreases) with increasing Zeeman splitting and the experimental contrast for σ^+ (σ^-) polarization tends to increase (decrease) at high magnetic fields.

A theoretical analysis of the optical pumping provides a means to extract T_1^{hole}. The master equation for the density matrix describing the 4-level system includes the incoherent processes of spontaneous emission, electron spin relaxation and hole spin relaxation.

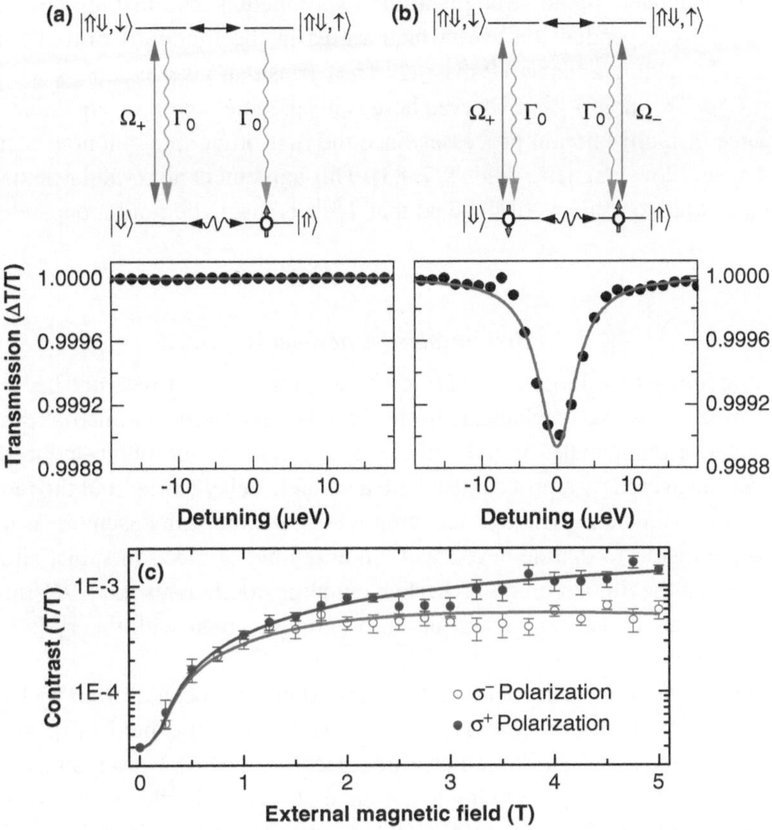

Figure 4.5 The two hole spin states, $|\Uparrow\rangle$ and $|\Downarrow\rangle$, and the two exciton spin states, $|\Uparrow\Downarrow\downarrow\rangle$ and $|\Uparrow\Downarrow\uparrow\rangle$, are depicted with coherent couplings (solid lines) and incoherent couplings (wavy lines). Open (solid) arrows depict hole (electron) spins. The dipole transition from $|\Downarrow\rangle \leftrightarrow |\Uparrow\Downarrow\downarrow\rangle$ is driven on-resonance with a σ^+ polarized laser (Ω_+); electron spin precession in the effective magnetic field generated by the nuclear spins coherently couples the $|\Uparrow\Downarrow\downarrow\rangle$ and $|\Uparrow\Downarrow\uparrow\rangle$ states. (a) With pure σ^+-polarized excitation, spontaneous recombination (rate Γ_0) shelves the population in the $|\Uparrow\rangle$ state and the absorption contrast is negligble. (b) With linear-polarization, the optical pumping is frustrated and the absorption signal reappears. (c) Measured contrast for one-color excitation, either σ^+ or (σ^-) polarization, as a function of external magnetic field applied in the $\vec{z}$-direction. The solid lines represent a fit to the results of the 4-level density matrix calculation using $B_N^{x,y} = 21$ mT and $T_1^{\text{hole}} = 1$ ms. Figures adapted from Ref. [40].

The density matrix is linked to the experiment by computing the QD polarization and scattered field amplitude: the differential transmission signal arises through a destructive interference of the laser and Rayleigh scattered fields. The model is explained in detail in the Supplementary Information of Ref. [40]. All the parameters for the model are either directly measured or known from other experiments apart from T_1^{hole} which is determined by fitting the calculated signal strength to the experimental data in Fig. 4.5c. The fit to determine T_1^{hole} is dependent on B_N, which varies in the literature from 12 to 26 mT [27, 53, 13, 14], yielding a range 0.2 ms $\leq T_1^{\text{hole}} \leq$ 1 ms that gives equally good fits to the data in Fig. 4.5c. The large T_1^{hole} observed here is in agreement with recent theory showing the dominance of multi-phonon processes once the first-order spin–phonon interaction is suppressed at very low magnetic fields [77, 83]. This agreement also validates the assumption in the quantitative analysis of the data that T_1^{hole} is magnetic field independent at low fields.

Time-resolved experiments

Optical spin cooling of a large *ensemble* of hole spins has also recently been achieved using the optical pumping mechanism in the Faraday geometry outlined above [26]. In this experiment, a pump pulse initialized the hole spins and the photo-induced circular dichroism was measured by a probe pulse with a variable delay to read-out the ratio of hole spins still in the initial state. The optical pumping of the hole spin ensemble as a function of B_{ext}^z was compared to that achieved with an ensemble of electron spins, allowing the determination that the dipole–dipole hyperfine coupling coefficients are ~7% those of the electron spin contact hyperfine interaction, in rough agreement with theory [33, 76] (see Section 4.6 below).

Another technique to initialize and read-out spin states has been utilized by Heiss *et al.* to characterize T_1^{hole} of an ensemble as a function of magnetic field [48]. To generate spin polarized holes, a resonant circularly polarized (σ^+ or σ^-) laser optically creates neutral excitons ($|\downarrow\Uparrow\rangle$ or $|\uparrow\Downarrow\rangle$). Owing to the large electric field, the electron rapidly tunnels out of the QD (with $\Gamma_{\text{tunnel}} = 100$ GHz), leaving behind a spin polarized hole ($|\Uparrow\rangle$ or $|\Downarrow\rangle$). Following a delay Δt during which the driving laser is switched off, a forward bias pulse is applied, causing electrons to tunnel back into the QD. The injected electrons are unpolarized, ensuring that when they subsequently recombine with the holes any net polarization of the detected photons is due to a net hole spin polarization. For this experiment, the Zeeman splitting due to B_{ext}^z is much larger than the anisotropic electron–hole exchange interaction. The measured ensemble T_1^{hole} as a function of magnetic field is shown in Fig. 4.6. For comparison, T_1^{electron} for identical QD material and experimental conditions is also included. As B_{ext}^z is decreased from 12 to 1.5 T, T_1^{hole} increases from 8 ± 3 to 270 ± 180 μs. Additionally, the temperature (T) dependence of T_1^{hole} is identical to that found for T_1^{electron} [59], showing a clear T^{-1} dependence. Hence, as for electrons, the dominant relaxation mechanism for hole spins at appreciable magnetic fields is the phonon-spin–orbit interaction [48].

Using the same hole spin initialization sequence as Heiss *et al.*, Godden *et al.* have demonstrated ultra-fast (~ 30 ps) hole spin initialization with high fidelity ($F > 99\%$) in a *single* self-assembled QD [42]. Furthermore, Ramsay *et al.* have added an additional pulse sequence to manipulate coherently a single hole spin and have observed Rabi oscillations between the hole state and the X^{1+} conditional on the hole spin [73].

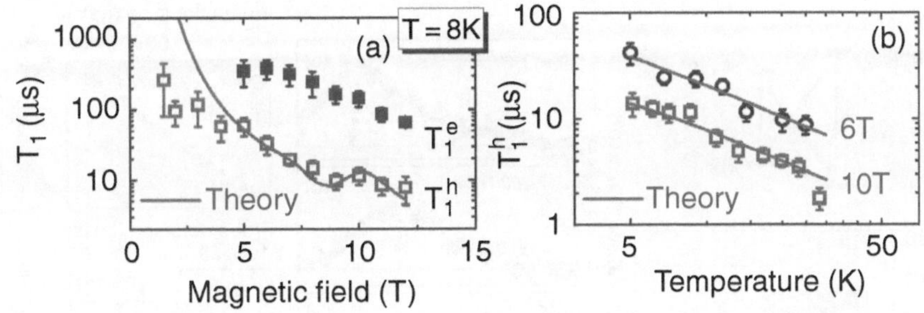

Figure 4.6 (a) A comparison of T_1^{hole} (open symbols) and T_1^{electron} (filled symbols) as a function of B^z_{ext} for identical QD material and experimental conditions. The theory line shows the calculated T_1^{hole} using the one-phonon model from [19]. (b) The temperature dependence of T_1^{hole} at two different magnetic fields. Figure reproduced with permission from [48]. Figures adapted from Ref. [48].

4.5.2 Hole spin coherence in the Voigt geometry

The prediction that pure heavy hole spin states have an Ising-like interaction with nuclear spins in the $\vec{z}$-direction provides a huge stimulus to apply a magnetic field in the Voigt geometry ($B^{x,y}_{ext}$, perpendicular to the growth direction [001]). A second incentive to use an in-plane magnetic field is to create a spin-Λ system. $B^{x,y}_{ext}$ breaks the rotational symmetry of the QD and mixes the spin angular momentum states [7]. In this scenario, each spin eigenstate ($|\Uparrow\rangle$ or $|\Downarrow\rangle$) is dipole coupled to both X^{1+} states ($|\Uparrow\Downarrow\downarrow\rangle$ and $|\Uparrow\Downarrow\uparrow\rangle$). Hence, there are a total of four optical transitions and, once the Zeeman splitting of the exciton states is larger than the linewidths, two separate spin-Λ configurations are established (see Fig. 4.7). Combined, these two driving forces offer a truly exciting prospect: a highly coherent spin-Λ system in the solid-state.

At zero magnetic field and with one linearly polarized laser, resonant laser spectroscopy reveals an X^{1+} absorption peak at the exciton resonance (Fig. 4.4). With $B^x_{ext} > 0$ this resonance disappears through optical pumping. However, scanning with a second laser reveals two "repumping" peaks where the optical pumping is suppressed and a probe absorption signal reappears in the vicinity of two-photon resonance for each spin-Λ system (Fig. 4.7b). Once the two-photon resonance is identified, CPT can be experimentally pursued using very high resolution ($\sim$ 2 MHz) laser scanning. In the example data at B^x_{ext} = 3.0 T there is a pronounced and narrow dip in the probe spectrum. To prove that the dip arises from CPT, the pump laser is detuned (Fig. 4.7c–e). Figure 4.7f shows that the energy shift of the dip in the probe spectrum equals the energy shift of the pump, exactly as expected for a two-photon resonance. There are a number of interesting features in this data. First, the dot is rendered completely transparent at the two-photon resonance. Such a clear destructive interference is only possible with a highly coherent hole spin state. Secondly, the CPT dip clearly survives the broadening of the

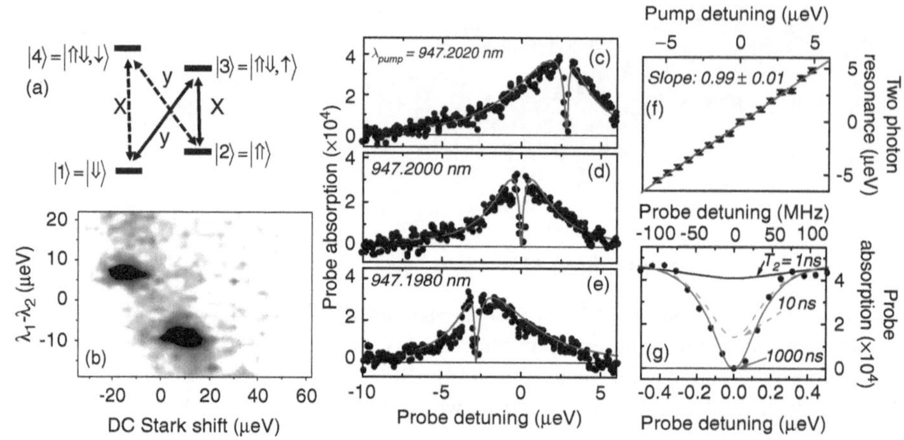

Figure 4.7 (a) Two spin-Λ-systems are established (solid and dashed lines) with an applied field B^x_{ext}. The quantum states of the hole, $|1\rangle$ and $|2\rangle$, are the hole spin eigenvectors in the x-basis, split by the Zeeman effect; the upper levels are X^{1+} excitons consisting of two spin-paired holes and an unpaired electron with spin $\pm\frac{1}{2}$, again in the x-basis. The optical transitions are linearly polarized along either the x- or y-direction. (b) A low-resolution resonant absorption contour plot with $B^x_{ext} = 0.75$ T is obtained by keeping one laser's wavelength constant while scanning the second laser. At each laser two detuning, one spectrum is recorded by sweeping the gate bias. Two resonances are found, corresponding to one spin-Λ-system each. Once the position of a resonance is found, high-resolution spectroscoppy can be performed to observe the CPT dip. (c)–(e) CPT of a single hole spin at 3.0 T and 4.2 K with Zeeman splitting 18.38 μeV. The points show the measured absorption against probe laser detuning for three different pump wavelengths, one blue-detuned from the resonance (c), one close to resonance (d), and one red-detuned (e). The solid lines are fits to the calculated response for pump detunings +2.831 μeV (c), 0.000 μeV (d) and −2.820 μeV (e), $\hbar\Omega_{\text{pump}} = 1.0$ μeV, $\hbar\Omega_{\text{probe}} = 0.45$ μeV, $\hbar\Gamma_{31} = \hbar\Gamma_{32} = 0.50$ μeV, $\hbar\gamma_3 = 0$ (pure dephasing rate of state $|3\rangle$), $\hbar\gamma_2 = 0.000\,67$ μeV, exciton broadening $\Gamma_X = 6.0$ μeV and field scattering ratio $\alpha = 0.01$. (f) The measured detuning of the probe laser at the CPT dip plotted against pump laser detuning with a linear fit. (g) Data in the CPT dip at 2.3 T versus the calculated response (solid lines) for different values of hole spin T_2. The theory uses the same parameters as in (d). Figures adapted from Ref. [17].

exciton resonance which is as large as 6 μeV for this particular dot. Thirdly, for non-zero pump detunings the form of the probe absorption spectrum does not follow the atomic physics model (Fig. 4.1c). Instead of a maximum signal located close to zero probe detuning we find the maximum signal is always close to the CPT dip (Fig. 4.7c–e). Furthermore, for positive pump detunings, the probe absorption falls more rapidly on the positive detuning side than on the negative side, a situation reversed for negative pump detunings.

The CPT dip allows a quantitative statement on the hole spin coherence time T_2^* to be made [52]. To eliminate systematic errors it is first necessary to understand the entire probe spectra (Fig. 4.7c–e). The missing factor in Fig. 4.1c is a description of the broadening of state $|3\rangle$, the exciton. Including sub-ns dephasing in level $|3\rangle$ fails: this simply smears out the curves in Fig. 4.1c. The experiment itself points to the resolution of this problem. The narrow CPT dip amidst the broad exciton resonance suggests that the exciton undergoes spectral fluctuations with little if any effect on the hole spin splitting. Remarkably, once the two-photon resonance condition is satisfied it remains satisfied even as the exciton energy fluctuates. To confirm this view, we convolute the density matrix calculation with a probability distribution in the detunings to describe the fluctuations of the exciton energy, slow relative to radiative recombination but fast relative to the experimental integration. The convolution makes surprisingly large changes to the probe spectrum (Fig. 4.1d) and reproduces all the features in the experiment. We achieve excellent fits to the probe absorption curves (Fig. 4.7c–e), reproducing both the width of the CPT dip and the overall lineshape.

The minimum signal in the dip depends on the hole spin coherence. Figure 4.7g shows the results of the theory for various T_2^* values along with the data from the same dot at $B_{ext}^x = 2.3$ T allowing an immediate and robust conclusion that T_2^* is at least 100 ns. With this data set, $T_2^* \geq 1$ μs with $\sim$ 50% probability. Recalling that the dip width is equal to $\hbar\Omega_{\text{pump}}^2/\Gamma_3$, and that the dip width sets the window of sensitivity on T_2^*, a more concrete statement about T_2^* can be obtained by decreasing $\hbar\Omega_{\text{pump}}$. This is very challenging experimentally. Neverless, recent experiments have been carried out in this regime, confirming that T_2^* is several hundred nanoseconds even at 0.5 T [51]. Notably, this value of T_2^* is considerably larger than has been observed in an ensemble of dots at $B_{ext} = 0$ T [28], possibly a vindication of the positive effect on T_2^* of an in-plane magnetic field. Remarkably, the CPT width in this latest CPT experiment is just 13 MHz (54 neV), ushering semiconductor physics into the realm of high-resolution atomic physics.

4.6 Strength of the hole hyperfine interaction

While the coupling coefficient for the contact hyperfine interaction of a pure heavy-hole spin is zero, the dipole–dipole interaction coupling constant C has been predicted to be ~10% that of the contact hyperfine coupling constant (A) for the electron spin [33, 76]. Three experiments have recently probed this issue in different self-assembled quantum dot samples, and each has obtained similar values [26, 30, 20]. Fallahi *et al.* and Chekhovich *et al.* both used a strong pump laser to achieve strong nuclear spin polarization and then measured the precise shift in energy for each quantum dot state due to the Overhauser field using a weak probe laser. Based on the Overhauser shifts, $\frac{C}{A}$ was determined to be -0.09 to -0.11 [30, 20]. Also, the ratio $\left|\frac{C}{A}\right| = 0.07$ was determined in an ensemble of quantum dots (see Section 4.5.1)[26]. As C is expected to depend on the strength of the heavy-hole confinement as well as the amount of heavy-hole–light-hole mixing, it is striking that each of these experimental measurements shows close agreement. Also, recent experiments by the Yamamoto group [25] have shown that the nuclear feedback for hole spins is dramatically reduced compared to the hysteretic nuclear feedback observed for electron spins during coherent spin manipulation [72, 60]. The validity of the Ising model for the hole hyperfine interaction requires minimal heavy-hole–light-hole mixing (which can be obtained by measuring the degree of circular polarization for the trion [8]). Further experiments are required to elucidate the relationships between the structural dot parameters, the hyperfine interaction, and the spin coherence time.

4.7 Summary and outlook

We have presented recent advances involving a hole spin in a self-assembled quantum dot. Many of the same experimental demonstrations initially accomplished with an electron spin have now been achieved with a hole spin, including: single hole charging (Coulomb blockade), high-fidelity optical spin initialization, ultra-fast spin manipulation, and spin read-out. Furthermore, controlled optical manipulation of interacting hole spins in coupled quantum dots has recently been demonstrated [46]. Long hole spin relaxation times are observed even at zero magnetic field, and, compared to the electron spin, the hole spin experiences a highly suppressed nuclear hyperfine interaction. In the presence of noisy nuclear spins, a clear result is that the hole spin in a quantum dot is considerably more coherent than an electron spin.

However, several open questions and challenges still remain. From a fundamental viewpoint, the limits on the coherence time and the exact dephasing mechanisms remain to be characterized. For instance, the impact of parameters such as the dot geometry, strain, heavy-hole–light-hole mixing and magnetic field on hole spin dephasing is experimentally unexplored. The potential for engineering ultra-long coherence times for the hole spin is intriguing. From a practical perspective, advances in materials or device design are needed to suppress the inhomogeneous broadening of the excitonic transition. Also, while electron spin states are routinely investigated in resonant cavity structures, cavity-quantum electrodynamics with hole spin states are relatively immature. Nevertheless, resolutions to each of these issues should not pose insurmountable obstacles. Hence, exploiting a coherent optical hole-spin-Λ system is particularly exciting both for future fundamental investigations and quantum technologies.

Acknowledgements

We would like to acknowledge and thank our colleagues who have worked with us over the past few years on hole spins in quantum dots, notably Daniel Brunner, Paul A. Dalgarno, Julien Houel, Khaled Karrai, Patrik Öhberg, Pierre M. Petroff and Nick Stoltz.

References

[1] Amasha, S., MacLean, K., Radu, I. P. *et al.* 2008. Electrical control of spin relaxation in a quantum dot. *Phys. Rev. Lett.*, **100**(4), 46803.

[2] Atatüre, M., Dreiser, J., Badolato, A. *et al.* 2006. Quantum-dot spin-state preparation with near-unity fidelity. *Science*, **312**(5773), 551–553.

[3] Atatüre, M., Dreiser, J., Badolato, A. and Imamoglu, A. 2007. Observation of Faraday rotation from a single confined spin. *Nature Physics*, **3**(2), 101–106.

[4] Baier, M., Findeis, F., Zrenner, A., Bichler, M. and Abstreiter, G. 2001. Optical spectroscopy of charged excitons in single quantum dot photodiodes. *Phys. Rev. B*, **64**(19), 195326.

[5] Balasubramanian, G., Neumann, P., Twitchen, D. *et al.* 2009. Ultralong spin coherence time in isotopically engineered diamond. *Nat. Mater.*, **8**(5), 383–387.

[6] Barthel, C., Medford, J., Marcus, C. M., Hanson, M. P. and Gossard, A. C. 2010. Interlaced dynamical decoupling and coherent operation of a singlet–triplet qubit. *Phys. Rev. Lett.*, **105**(26), 266808.

[7] Bayer, M., Ortner, G., Stern, O. *et al.* 2002. Fine structure of neutral and charged excitons in self-assembled In(Ga)As/(Al)GaAs quantum dots. *Phys. Rev. B*, **65**(19), 195315.

[8] Belhadj, T., Amand, T., Kunold, A. *et al.* 2010. Impact of heavy hole–light hole coupling on optical selection rules in GaAs quantum dots. *Appl. Phys. Lett.*, **97**(5), 051111.

[9] Berezovsky, J., Mikkelsen, M. H., Stoltz, N. G., Coldren, L. A. and Awschalom, D. D. 2008. Picosecond coherent optical manipulation of a single electron spin in a quantum dot. *Science*, **320**(5874), 349–352.

[10] Bergmann, K., Theuer, H. and Shore, B. W. 1998. Coherent population transfer among quantum states of atoms and molecules. *Rev. Mod. Phys.*, **70**, 1003–1025.

[11] Bester, G., Nair, S. and Zunger, A. 2003. Pseudopotential calculation of the excitonic fine structure of million-atom self-assembled InGaAs/GaAs quantum dots. *Phys. Rev. B*, **67**(16), 161306.

[12] Bluhm, H., Foletti, S., Mahalu, D., Umansky, V. and Yacoby, A. 2010. Enhancing the coherence of a spin qubit by operating it as a feedback loop that controls its nuclear spin bath. *Phys. Rev. Lett.*, **105**(21), 216803.

[13] Bracker, A., Stinaff, E., Gammon, D. *et al.* 2005. Optical pumping of the electronic and nuclear spin of single charge-tunable quantum dots. *Phys. Rev. Lett.*, **94**(4), 047402.

[14] Braun, P.-F., Marie, X., Lombez, L. *et al.* 2005. Direct observation of the electron spin relaxation induced by nuclei in quantum dots. *Phys. Rev. Lett.*, **94**(11), 116601.

[15] Brossel, J. and Bitter, F. 1952. *Phys. Rev.*, **86**, 308.

[16] Brossel, J. and Kastler, A. 1949. *Compt. Rend.*, **229**, 1213.

[17] Brunner, D., Gerardot, B. D., Dalgarno, P. *et al.* 2009. A coherent single-hole spin in a semiconductor. *Science*, **325**(5936), 70.

[18] Bulaev, D. V. and Loss, D. 2005. Spin relaxation and decoherence of holes in quantum dots. *Phys. Rev. Lett.*, **95**(7), 076805.

[19] Bulaev, D. V. and Loss, D. 2007. Electric dipole spin resonance for heavy holes in quantum dots. *Phys. Rev. Lett.*, **98**(9), 097202.

[20] Chekhovich, E., Krysa, A., Skolnick, M. and Tartakovskii, A. 2011. Direct measurement of the hole–nuclear spin interaction in single InP/GaInP quantum dots using photoluminescence spectroscopy. *Phys. Rev. Lett.*, **106**(2), 027402.
[21] Cortez, S., Krebs, O., Laurent, S. *et al.* 2002. Optically driven spin memory in n-doped InAs-GaAs quantum dots. *Phys. Rev. Lett.*, **89**(20), 207401.
[22] Crooker, S. A., Brandt, J., Sandfort, C. *et al.* 2010. Spin noise of electrons and holes in self-assembled quantum dots. *Phys. Rev. Lett.*, **104**(3), 036601.
[23] Dalgarno, P., Ediger, M., Gerardot, B. *et al.* 2008. Optically induced hybridization of a quantum dot state with a filled continuum. *Phys. Rev. Lett.*, **100**(17), 176801.
[24] Damen, T. C., Vina, L., Cunningham, J. E., Shah, J. and Sham, L. J. 1991. Subpicosecond spin relaxation dynamics of excitons and free-carriers in GaAs quantum-wells. *Phys. Rev. Lett.*, **67**(24), 3432–3435.
[25] De Greve, K., McMahon, P. L., Press, D. *et al.* 2011. Coherent control and suppressed nuclear feedback of a single quantum dot hole qubit. *Nature Physics*, **7**, 872.
[26] Desfonds, P., Eble, B., Fras, F. *et al.* 2010. Electron and hole spin cooling efficiency in InAs quantum dots: the role of nuclear field. *Appl. Phys. Lett.*, **96**(17), 172108.
[27] Dreiser, J., Atatüre, M., Galland, C. *et al.* 2008. Optical investigations of quantum dot spin dynamics as a function of external electric and magnetic fields. *Phys. Rev. B*, **77**(7), 075317.
[28] Eble, B., Testelin, C., Desfonds, P. *et al.* 2009. Hole-nuclear spin interaction in quantum dots. *Phys. Rev. Lett.*, **102**(14), 146601.
[29] Ediger, M., Dalgarno, P. A., Smith, J. M. *et al.* 2005. Controlled generation of neutral, negatively-charged and positively-charged excitons in the same single quantum dot. *Appl. Phys. Lett.*, **86**(21), 211909.
[30] Fallahi, P., Yilmaz, S. and Imamoğlu, A. 2010. Measurement of a heavy-hole hyperfine interaction in InGaAs quantum dots using resonance fluorescence. *Phys. Rev. Lett.*, **105**(25), 257402.
[31] Fano, U. 1961. Effects of configuration interaction on intensities and phase shifts. *Phys. Rev.*, **124**(6), 1866–1878.
[32] Fischer, J. and Loss, D. 2009. Dealing with decoherence. *Science*, **324**(5932), 1277–1278.
[33] Fischer, J., Coish, W. A., Bulaev, D. V. and Loss, D. 2008. Spin decoherence of a heavy hole coupled to nuclear spins in a quantum dot. *Phys. Rev. B*, **78**(15), 155329.
[34] Fleischhauer, M., Imamoglu, A. and Marangos, J. 2005. Electromagnetically induced transparency: optics in coherent media. *Rev. Mod. Phys.*, **77**(2), 633–673.
[35] Fokina, L. V., Yugova, I. A., Yakovlev, D. R. *et al.* M. 2010. Spin dynamics of electrons and holes in InGaAs/GaAs quantum wells at millikelvin temperatures. *Phys. Rev. B*, **81**(19), 195304.
[36] Frogley, M. D., Dynes, J. F., Beck, M., Faist, J. and Phillips, C. C. 2006. Gain without inversion in semiconductor nanostructures. *Nat. Mater.*, **5**(3), 175–178.
[37] Fu, K.-M., Santori, C., Stanley, C., Holland, M. and Yamamoto, Y. 2005. Coherent population trapping of electron spins in a high-purity n-type GaAs semiconductor. *Phys. Rev. Lett.*, **95**(18), 187405.
[38] Gerardot, B. D., Seidl, S., Dalgarno, P. A. *et al.* 2007. Contrast in transmission spectroscopy of a single quantum dot. *Appl. Phys. Lett.*, **90**(22), 221106.
[39] Gerardot, B. D., Brunner, D., Dalgarno, P. A. *et al.* 2009. Dressed excitonic states and quantum interference in a three-level quantum dot ladder system. *New J. Phys.*, **11**(1), 013028.

[40] Gerardot, B. D., Brunner, D., Dalgarno, P. A. *et al.* 2008. Optical pumping of a single hole spin in a quantum dot. *Nature*, **451**(7177), 441–444.

[41] Gisin, N., Ribordy, G. G, Tittel, W. and Zbinden, H. 2002. Quantum cryptography. *Rev. Mod. Phys.*, **74**(1), 145–195.

[42] Godden, T. M., Boyle, S. J., Ramsay, A. J., Fox, A. M. and Skolnick, M. S. 2010. Fast high fidelity hole spin initialization in a single InGaAs quantum dot. *Appl. Phys. Lett.*, **97**(6), 061113.

[43] Golovach, V., Khaetskii, A. and Loss, D. 2004. Phonon-induced decay of the electron spin in quantum dots. *Phys. Rev. Lett.*, **93**(1), 016601.

[44] Greilich, A., Yakovlev, D. R., Shabaev, A. *et al.* 2006. Mode locking of electron spin coherences in singly charged quantum dots. *Science*, **313**(5785), 341–5.

[45] Greilich, A., Shabaev, A., Yakovlev, D. R. *et al.* 2007. Nuclei-induced frequency focusing of electron spin coherence. *Science*, **317**(5846), 1896–9.

[46] Greilich, A., Carter, S. G., Kim, D., Bracker, A. S. and Gammon, D. 2011. Optical control of one and two hole spins in interacting quantum dots. *Nature Photonics*, **5**, 702.

[47] Hanson, R., Kouwenhoven, L. P., Petta, J. R., Tarucha, S. and Vandersypen, L. M. K. 2007. Spins in few-electron quantum dots. *Rev. Mod. Phys.*, **79**(4), 1217.

[48] Heiss, D., Schaeck, S., Huebl, H. *et al.* 2007. Observation of extremely slow hole spin relaxation in self-assembled quantum dots. *Phys. Rev. B*, **76**(24), 241306.

[49] Högele, A., Seidl, S., Kroner, M. *et al.* 2004. Voltage controlled optics of a quantum dot. *Phys. Rev. Lett.*, **93**, 217401.

[50] Hoke, W. E., Weir, D. G., Lemonias, P. J. and Hendriks, H. T. 1994. Carbon tetrabromide carbon doping of molecular beam epitaxial (GaAs) films. *Appl. Phys. Lett.*, **64**(2), 202.

[51] Houel, J. 2012. In press.

[52] Imamoglu, A. 2006. Coherent population trapping in a single-hole-charged quantum dot. *Phys. Stat. Sol. B*, **243**(14), 3725–3729.

[53] Johnson, A. C., Petta, J.R., Taylor, J. M. *et al.* 2005. Triplet–singlet spin relaxation via nuclei in a double quantum dot. *Nature*, **435**(7044), 925–928.

[54] Jones, J. A., Karlen, S. D., Fitzsimons, J. *et al.* 2009. Magnetic field sensing beyond the standard quantum limit using 10-spin NOON states. *Science*, **324**(5931), 1166–1168.

[55] Karrai, K. 2003. Optical transmission and reflection spectroscopy of single quantum dots. *Superlattices Microstruct.*, **33**(5-6), 311–337.

[56] Kimble, H. J. 2008. The quantum internet. *Nature*, **453**(7198), 1023–1030.

[57] Kroner, M., Govorov, A. O., Remi, S. *et al.* 2008. The nonlinear Fano effect. *Nature*, **451**(7176), 311–314.

[58] Kroner, M., Weiss, K., Biedermann, B. *et al.* 2008. Optical detection of single-electron spin resonance in a quantum dot. *Phys. Rev. Lett.*, **100**(15), 156803.

[59] Kroutvar, M., Ducommun, Y., Heiss, D. and Bichler, M. 2004. Optically programmable electron spin memory using semiconductor quantum dots. *Nature*, **432**(November), 81–84.

[60] Ladd, T. D., Press, D., De Greve, K. *et al.* 2010. Pulsed nuclear pumping and spin diffusion in a single charged quantum dot. *Phys. Rev. Lett.*, **105**(10), 107401.

[61] Liu, R. -B., Yao, W. and Sham, L. J. 2010. Quantum computing by optical control of electron spins. *Advances In Physics*, **59**(5), 703–802.

[62] Loss, D. and DiVincenzo, D. P. 1998. Quantum computation with quantum dots. *Phys. Rev. A*, **57**(1), 120–126.

[63] Marie, X., Amand, T., Le Jeune, P. *et al.* 1999. Hole spin quantum beats in quantum-well structures. *Phys. Rev. B*, **60**(8), 5811–5817.

[64] Martin, R. W., Nicholas, R. J., Rees, G. J. *et al.* 1990. Two-dimensional spin confinement in strained-layer quantum-wells. *Phys. Rev. B*, **42**(14), 9237–9240.

[65] Merkulov, I. A. 2002. Electron spin relaxation by nuclei in semiconductor quantum dots. *Phys. Rev. B*, **65**(20), 205309.

[66] Mikkelsen, M. H., Berezovsky, J., Stoltz, N. G., Coldren, L. A. and Awschalom, D. D. 2007. Optically detected coherent spin dynamics of a single electron in a quantum dot. *Nature Physics*, **3**(11), 770–773.

[67] Miller, D. L. and Asbeck, P. M. 1985. Be redistribution during growth of GaAs and AlGaAs by molecular beam epitaxy. *Journal of Applied Physics*, **57**(6), 1816–1822.

[68] Nowack, K. C., Koppens, F. H. L., Nazarov, Yu. V. and Vandersypen, L. M. K. 2007. Coherent control of a single electron spin with electric fields. *Science*, **318**(5855), 1430–1433.

[69] Petta, J. R. 2010. Private communication.

[70] Petta, J. R., Johnson, A. C., Taylor, J. M. *et al.* 2005. Coherent manipulation of coupled electron spins in semiconductor quantum dots. *Science*, **309**(5744), 2180–2184.

[71] Phillips, M., Wang, H., Rumyantsev, I. *et al.* 2003. Electromagnetically induced transparency in semiconductors via biexciton coherence. *Phys. Rev. Lett.*, **91**(18), 1–4.

[72] Press, D., Ladd, T. D., Zhang, B. and Yamamoto, Y. 2008. Complete quantum control of a single quantum dot spin using ultrafast optical pulses. *Nature*, **456**(7219), 218–21.

[73] Ramsay, A. J., Boyle, S. J., Kolodka, R. S. *et al.* 2008. Fast optical preparation, control, and readout of a single quantum dot spin. *Phys. Rev. Lett.*, **100**(19), 197401.

[74] Smith, J. M., Dalgarno, P. A., Warburton, R. J. *et al.* 2005. Voltage control of the spin dynamics of an exciton in a semiconductor quantum dot. *Phys. Rev. Lett.*, **94**(19), 197402.

[75] Syperek, M., Yakovlev, D. R., Greilich, A. *et al.* 2007. Spin coherence of holes in GaAs/(Al,Ga)As quantum wells. *Phys. Rev. Lett.*, **99**(18), 187401.

[76] Testelin, C., Bernardot, F., Eble, B. and Chamarro, M. 2009. Hole spin dephasing time associated with hyperfine interaction in quantum dots. *Phys. Rev. B*, **79**(19), 195440.

[77] Trif, M., Simon, P. and Loss, D. 2009. Relaxation of hole spins in quantum dots via two-phonon processes. *Phys. Rev. Lett.*, **103**(10), 106601.

[78] Vamivakas, A. N., Lu, C. Y., Matthiesen, C. *et al.* 2010. Observation of spin-dependent quantum jumps via quantum dot resonance fluorescence. *Nature*, **467**(7313), 297–300.

[79] Vink, I. T., Nowack, K. C., Koppens, Frank H. L. *et al.* 2009. Locking electron spins into magnetic resonance by electron-nuclear feedback. *Nature Physics*, **5**(10), 764–768.

[80] Warburton, R. J., Sundaram, G. M., Nicholas, R. J. *et al.* 1990. Strain reconstruction of the valence band in $Ga_xIn_{1-x}Sb$–GaSb quantum-wells. *Surf. Sci.*, **228**(1–3), 270–274.

[81] Warburton, R. J., Martin, R. W., Nicholas, R. J., Howard, L. K. and Emeny, M. T. 1991. Valence band spin splitting in strained $In_{0.18}Ga_{0.82}As$/GaAs quantum-wells. *Semicond. Sci. Technol.*, **6**(5), 359–364.

[82] Warburton, R. J., Schaflein, C., Haft, D. *et al.* 2000. Optical emission from a charge-tunable quantum ring. *Nature*, **405**, 926–929.

[83] Wei, H., Gong, M., Guo, G. and He, L. 2010. Atomistic theory of spin relaxation in self-assembled $In_{1-x}Ga_xAs$/GaAs quantum dots at zero magnetic field. *arXiv: 1011.4724* (2010).
[84] Xu, X., Sun, B., Berman, P. R. *et al.* 2008. Coherent population trapping of an electron spin in a single negatively charged quantum dot. *Nature Physics*, **4**(9), 692–695.
[85] Xu, X., Yao, W., Sun, B. *et al.* 2009. Optically controlled locking of the nuclear field via coherent dark-state spectroscopy. *Nature*, **459**(7250), 1105–1109.
[86] Yilmaz, S. T., Fallahi, P. and Imamoglu, A. 2010. Quantum-dot-spin single-photon interface. *Phys. Rev. Lett.*, **105**(3), 033601.

5

Resonance fluorescence from a single quantum dot

A. N. Vamivakas, C. Matthiesen, Y. Zhao, C.-Y. Lu and M. Atatüre

5.1 Introduction

Quantum optics provides a complete description of the resonant interaction of light and a material system. The light scattered as a result of this interaction – the resonance fluorescence – provides a means to monitor and affect the material system's dynamics and also allows for the generation of nonclassical states of the electromagnetic field. Initial theoretical studies characterizing resonance fluorescence addressed fundamental questions regarding the necessity of quantizing the electromagnetic field. This is a particularly illustrative period in optical physics where multiple theories existed that might describe resonance fluorescence – semi-classical optical Bloch equations [1, 2] or quantum electrodynamics [4, 3] – and experimental evidence was necessary to identify the appropriate theoretical framework. Indeed, experiments in 1974 and 1977 investigating the resonance fluorescence lineshape [5, 6] and the photon statistics of the scattered light [7] vindicated the full quantum electrodynamical description of light–matter interaction over other models. In the subsequent 30 years, resonance fluorescence has been used as a high-resolution spectroscopic probe of a variety of physical systems, as a means to investigate quantum coherence in atomic gases and, with the current interest in quantum information science (QIS), as a tool for quantum-state preparation, measurement and control in trapped ions [8].

During this same time period, advances in material science have ushered in an era where it is possible to grow and fabricate solid-state devices at length scales commensurate with the quantum confinement of electrons and holes. It is only recently that resonance fluorescence has been observed from the prototypical artificial nanostructure exhibiting three-dimensional quantum confinement – a single semiconductor quantum dot (QD) [9]. The excitement in accessing a QD's resonance fluorescence is many-fold [10, 11]. First, access to photons resonantly scattered from a QD is prerequisite for many quantum optics experiments. The interest here is that although in many ways a QD exhibits features characteristic of an atom, it is in fact a mesoscopic physical system and we can now experimentally study solid-state quantum optics, i.e. fine and subtle deviations from the simple atomic picture. For example, nuclear spins in the QD or phonon modes in the host matrix lead to signatures in the resonance fluorescence spectrum that provide spectroscopic information about these excitations [12]. Finally, there are multiple areas of application

Quantum Dots: Optics, Electron Transport and Future Applications, ed. Alexander Tartakovskii. Published by Cambridge University Press.

for optically active semiconductor QDs in QIS that benefit from both the coherent coupling with the electromagnetic field and direct access to the QD resonance fluorescence. Charges and spins trapped in a QD are possible quantum bits [13]. Their coherent interface with photons is enabling advances in state preparation and measurement [14]. Further, the single photons emitted by a QD can be used to transmit classical or quantum information between remote locations, prepare nonclassical states of the electromagnetic field [15] or serve as high quality physical bits for a photon-based quantum computer [16].

In the following we first briefly review the theory of resonance fluorescence from an ideal two-level system. Next, we describe different experimental implementations capable of monitoring the resonance fluorescence from a single QD and compare this to a typical atomic physics experimental arrangement. Finally, we outline inturn how resonance fluorescence can be used to investigate solid-state quantum optics and present its use as a tool for solid-state spectroscopy and QIS.

5.2 Resonance fluorescence from a two-level system

There are many excellent books that detail the dynamic and steady-state properties of the interaction of matter as an ideal two-level system and light [17, 18] and below we summarize the main results and illustrate some measurable properties. As illustrated in Fig. 5.1a we assume a monochromatic plane wave (a laser) of frequency ω_L drives the two-level system (TLS). The bare states of the interacting system consist of the TLS ground and excited states and the number of photons in the electromagnetic field mode. For large enough drive field amplitudes, it is more convenient to think in terms of a new coupled basis, right panel of Fig. 5.1a, the dressed states – hybrid matter–light modes [19].

For an optics experiment a useful observable quantity that carries information on the dynamics of the light–TLS interaction is the temporal correlation of the source-field electric field operator and can be obtained by sending the field through a two-path interferometer, e.g. a Michelson or a Mach–Zehnder design. Following the derivation presented [3, 17, 18] the first-order correlation function of the light scattered from a resonantly driven (zero detuning) TLS is

$$g^{(1)}(\tau) = | \rho_{12}^{\infty} |^2 + \frac{\rho_{22}^{\infty}}{2} e^{-\tau/T_2} + \rho_{22}^{\infty} e^{-\frac{\Gamma\tau}{2}} \left(a\cos(\Omega\tau) + b\sin(\Omega\tau)\right), \qquad (5.1)$$

where $\Gamma = 1/T_1 + 1/T_2$ and a and b are constants depending on Γ and $\Omega = \sqrt{\Omega_B^2 - (1/T_1 - 1/T_2)^2/4}$ – the generalized Rabi frequency (Ω_B is the bare Rabi frequency). Note that ρ_{12}^{∞} and ρ_{22}^{∞} are the steady-state density matrix elements. Only in the specific limit, where the Rabi frequency is much smaller than Γ and the dephasing rate $1/T_2$ is much faster than the radiative decay rate, the correlation function reduces to a simple exponential decay. This double-sided exponential form is illustrated in Fig. 5.1b (bottom). For larger laser fields the correlation exhibits oscillations at the generalized Rabi frequency indicative of coherent oscillations between the TLS ground and excited state. This is also plotted in Fig. 5.1b (top).

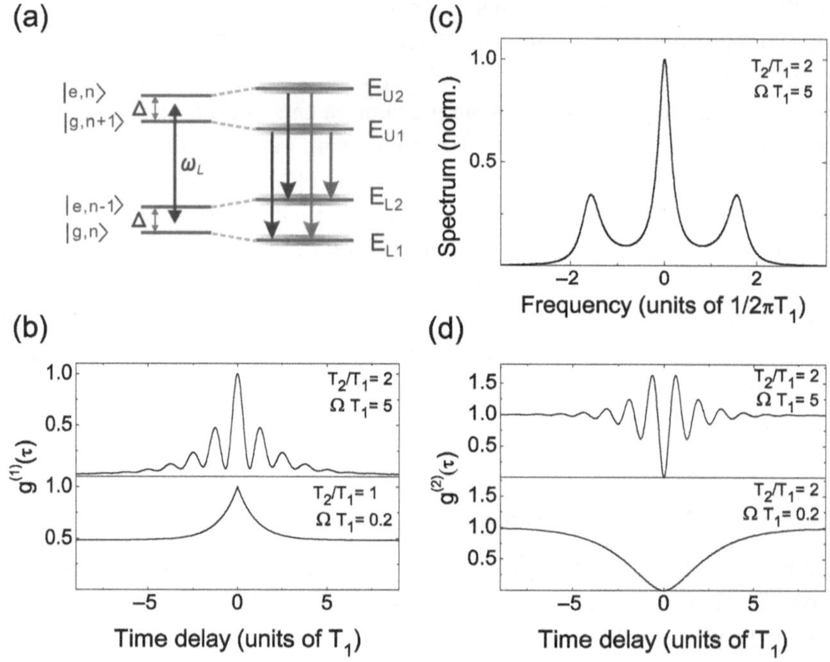

Figure 5.1 Resonance fluorescence from a 2-level system. (a) The electronic structure of a single mode electromagnetic field of angular frequency ω_L interacting with a 2-level system (TLS). The field mode is detuned from the TLS transition by an amount Δ. The left column illustrates the bare system states and the right column illustrates the dressed states – the eigenstates of the field mode–TLS interaction. In the dressed basis there are three distinct transition energies between the upper ($E_{U1/2}$) and lower ($E_{L1/2}$) eigenstates. (b) The resonance fluorescence first-order autocorrelation function $g^{(1)}(\tau)$. The top curve is for an excitation field amplitude well above the saturation Rabi frequency and slow dephasing, while the bottom curve is for a Rabi frequency well below the saturation frequency and faster dephasing. (c) The resonance fluorescence spectrum for an excitation well above the saturation frequency. The three-peaked spectrum is the Mollow triplet. The elastic component of the spectrum has been left out for clarity. (d) The resonance fluorescence second-order autocorrelation function $g^{(2)}(\tau)$. The top curve is for an excitation field amplitude well above the saturation Rabi frequency and the bottom curve is for a Rabi frequency well below the saturation frequency. Both curves exhibit anti-bunching at $\tau = 0$. Figures adapted from Ref. [11].

As a consequence of the Wiener–Khinchin theorem, the Fourier transform of the first-order autocorrelation function is equal to the measured spectrum. In the weakfield coherent limit, $\Omega \ll \Gamma$ and $T_2 \sim 2T_1$, the spectrum is a single peak reflecting the spectral and coherence properties of the driving laser. In the strong field limit, $\Omega \gg \Gamma$, the resonance fluorescence spectrum for a resonantly driven TLS is

$$S(\Delta) \approx \frac{2/T_2}{\Delta^2 + (1/T_2)^2} + \frac{\Gamma/2}{(\Delta+\Omega)^2 + (\Gamma/2)^2} + \frac{\Gamma/2}{(\Delta-\Omega)^2 + (\Gamma/2)^2}, \tag{5.2}$$

where $\Delta = \omega - \omega_L$ is the difference between the measurement and drive laser frequency. This three-peaked spectrum, first predicted in 1969, is now referred to as the Mollow triplet

[3] and plotted in Fig. 5.1c for examplary values of Ω, T_1 and T_2. The appearance of sidebands due to laser-induced coherent oscillations of the population between ground and excited state is perhaps the most striking observable signature of coherent light–matter interaction. An equivalent explanation of the three peak spectrum is the laser-induced dressing of the TLS [19]. Figure 5.1a illustrates the bare and dressed electronic states.

It is also possible to measure the second-order correlation function of the scattered field, expressed as

$$g^{(2)}(\tau) = 1 - e^{-\frac{\Gamma\tau}{2}} \left(\cos(\Omega\tau) + \frac{\Gamma}{2\Omega} \sin(\Omega\tau) \right). \tag{5.3}$$

Observation of photon antibunching in the second order correlation function, ie $g^{(2)}(\tau = 0) = 0$, reveals the quantum behaviour of the measured light field and was first reported in Ref. [7]. In the weak field (small Rabi frequency) limit, as illustrated in Fig. 5.1d for a radiatively broadened TLS, $g^{(2)}(\tau)$ is monotonically increasing from zero, while exhibiting oscillations at the strong field limit (Fig. 5.1d).

In the following sections we will encounter examples where the photons resonantly scattered from a semiconductor QD exhibit the properties just discussed. The main difference to the near ideal example of trapped ions and atoms is that in the case of QDs the transitions are no longer purely radiatively broadened.

5.3 Observation of resonance fluorescence

5.3.1 First realizations in atomic systems

The first experimental investigations of resonance fluorescence focused on transitions in beams of atomic sodium [5, 6]. As discussed previously, the motivation for these initial experiments was to determine which theoretical model correctly described the interaction of laser light with a TLS. The typical apparatus employed to conduct these early experiments is illustrated in Fig. 5.2a [20]. To separate the atomic sodium resonance fluorescence from the excitation laser a crossed-beam geometry is used. The detector measures along a direction that is orthogonal to the mutually orthogonal atomic beam and excitation laser removing laser background from the detector. Details of the sodium level structures and this apparatus can be found in [20]. Since then, extensive studies of resonance fluorescence have been conducted with trapped ions, with particular focus on correlation functions [8].

5.3.2 Experimental approaches to quantum dot resonance fluorescence

The technical challenge when trying to observe resonance fluorescence from a buried QD is the need to reduce the scattered laser background from the host semiconductor material. There are two main approaches, illustrated in Fig. 5.2, that have been adopted to remedy this problem. In the spirit of the original sodium and trapped ion experiments the QD resonance fluorescence can be monitored along a spatial direction that is orthogonal to the

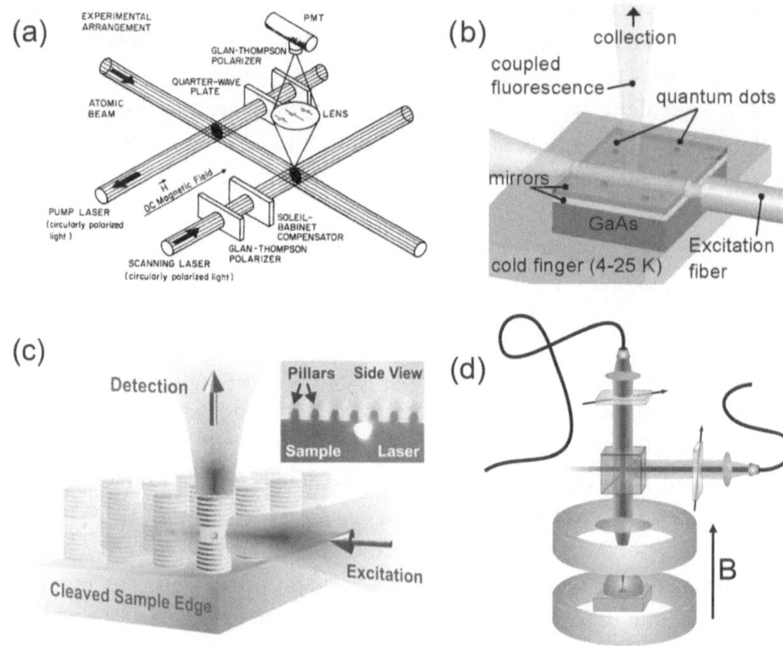

Figure 5.2 Experimental implementations to observe resonance fluorescence. (a) A schematic of the experimental arrangement used to observe resonance fluorescence from atomic sodium. The oven emits a beam of sodium atoms that are first optically prepared to a suitable ground state by a pump laser and subsequently excited by the scanning laser. Resonance fluorescence is observed through a lens in a direction orthogonal to the excitation. (b) In this experimental arrangement, the QDs are embedded in a low-Q Fabry–Pérot cavity. In the spirit of the arrangement presented in (a), the laser excites the QDs through the waveguide in a direction orthogonal to the collection direction. The cavity not only enhances QD fluorescence into the optical miscroscope, but also reduces the laser scatter into the microscope. (c) Resonance fluorescence has also been observed for QDs embedded in micropillar cavities, again using side illumination. High-Q values are possible in this geometry and the orthogonality of excitation and detection reduces the laser background. (d) Our approach to resonance fluorescence observation exploits the laser polarization and QD optical transitions polarization selection rules. Excitation and collection are done through the same lens. The excitation laser is prepared with linear polarization and the collection arm contains a linear polarizer orthogonal to the excitation laser polarizer. In this configuration, background laser suppression of 10^7 is possible. Figures adapted from Refs. [20, 10, 15].

excitation laser path. The two modalities to ensure spatial selectivity involve cavity structures. In Fig. 5.2b [10, 21], the QD layer is embedded in a planar Fabry–Pérot cavity and excitation is accomplished by bringing an optical fiber to the side of the device. The cavity does not affect the lifetime of the QD optical transition, but it does help to direct QD emission along a path orthogonal to the excitation direction and reduce laser background since the excitation is confined to the planar waveguide modes of the cavity. A second approach to spatial selectivity is illustrated in Fig. 5.2c. Here the QDs are fabricated into a pillar microcavity and excitation is again orthogonal to collection [15]. The pillar cavity enhances collection of the QD resonance fluorescence. Further, the Purcell effect can be used to reduce the excited state radiative lifetime. In contrast to the orthogonal excitation–collection geometry, it is also possible to use a combination of polarization selection rules and confocal rejection to suppress the background laser scatter. Figure 5.2d illustrates such an optical system. In this modality the excitation laser is linearly polarized in a direction orthogonal to the collection arm's direction of linear polarization. Since polarization selection rules dictate that the QD fluorescence is either linearly or circularly polarized, it is always possible to find an excitation polarization orientation that can be suppressed by the collection arm's polarization orientation [11]. Each of these two approaches to laser background suppression have enabled a number of experiments leveraging QD resonance fluorescence.

5.3.3 QD resonance fluorescence: solid-state quantum optics

The initial demonstrations of QD resonance fluorescence were from the perspective of solid-state quantum optics in order to identify to what extent a semiconductor QD behaves like a single atom. In Fig. 5.3a the integrated resonance fluorescence from a single QD is plotted [22]. For this particular measurement, the X^{1-} transition (ground state is a single electron; see Fig. 5.5a) of a QD in a Schottky heterostructure [23] is driven by a linearly polarized, frequency- and power-stabilized, singlemode laser. Here, polarization-based background suppression is utilized, where the resonance fluorescence collected back through the same focusing objective passes through a second linear polarizer (orthogonal to the laser polarization) prior to being sent to a single-photon counting avalanche photodiode (APD). The solid circles fit by the curve present an exemplary integrated resonance fluorescence spectrum of the X^{1-} transition as the laser is fixed to a particular frequency and the QD resonance is swept across using the DC Stark shift. The inset in Fig. 5.3a displays APD counts far off resonance, i.e. the background due to laser leakage and APD dark counts, the average of which is indicated by the dashed line. For this measurement, the resonance fluorescence signal-to-background ratio is ~500:1. The number of photons resonantly scattered by the QD transition is sensitive to the drive laser's Rabi frequency. The maximum recorded resonance fluorescence signal, as a function of excitation Rabi frequency, exhibits the well-known saturation behaviour of a TLS. The linescan shown here corresponds to an excitation Rabi frequency of ~0.5 of the saturation Rabi frequency.

More interesting, perhaps, is the spectrum of the resonance fluorescence photons presented in Fig. 5.3b [11]. In this plot, as the Rabi frequency is increased from ~1/10th to eight times the spontaneous emission rate, the resonance fluorescence spectrum is measured (the increased number of counts in the central peak is from residual laser leakage to the detector). At Rabi frequencies larger than the spontaneous emission rate the emergence of sidebands in the fluorescence spectrum are apparent. It is also possible to tune the spectral location of the sideband photon scattering by detuning the laser from the QD

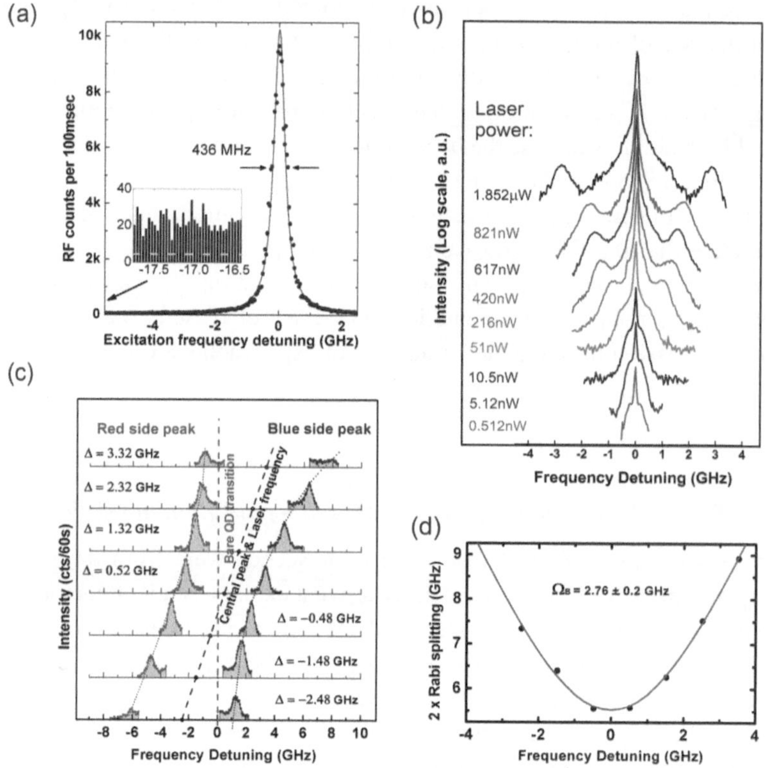

Figure 5.3 Solid-state quantum optics: scattered photon spectrum. (a) Integrated resonance fluorescence (circles) from a single QD. The laser is fixed to a frequency close to resonance and the QD transition is shifted across the laser using the DC Stark shift. Resonance fluorescence is recorded with an APD. The inset shows counts far off resonance. This background is made up of laser leakage and APD dark counts, as indicated by the dashed line. (b) The resonance fluorescence spectrum as the excitation laser power is varied from 0.1 of the Rabi frequency to eight times the Rabi frequency. Prior to detection, the resonance fluorescence is filtered by a 40 MHz Fabry–Pérot spectrum analyzer. The three-peaked spectrum is the Mollow triplet. (c) The resonance fluorescence spectrum as the excitation laser is tuned from blue to red across the QD resonance. The central feature that is contaminated by the laser background has been suppressed for clarity. (d) The extracted sideband Rabi splitting extracted from panel (c). The solid curve is a fit to the data and yields a bare Rabi frequency of $\Omega_B = 2.76$ GHz. Figures adapted from Ref. [11].

transition. Figure 5.3c presents the scattered photon spectrum as the laser is scanned across the transition from lower to higher optical frequency (the middle peak in the spectrum has been suppressed for clarity) [11]. By changing the laser frequency it is possible to vary the spectrum of the photon sideband scattering by 14 GHz. In Fig. 5.3d the extracted Mollow sideband separation as a function of laser detuning is plotted. The measured data fits with the simple theory based on an ideal atom presented in the previous section. We highlight

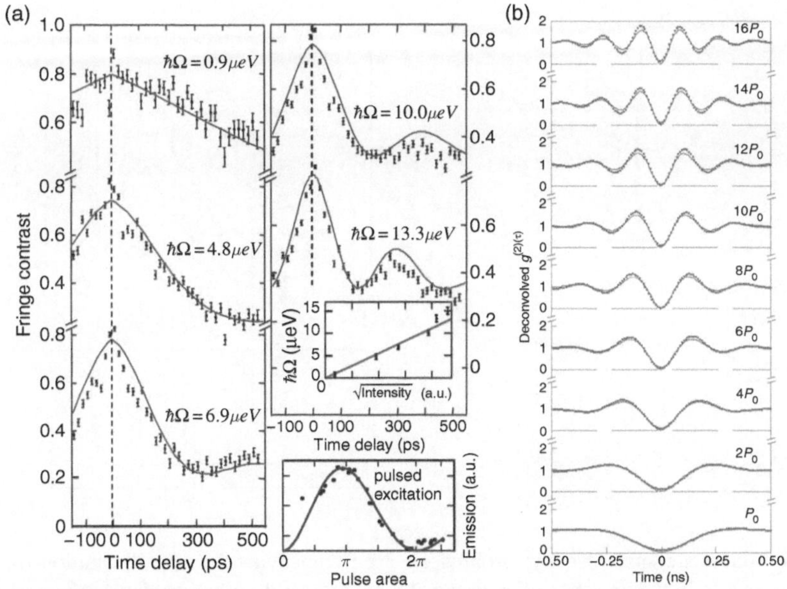

Figure 5.4 Solid-state quantum optics: autocorrelation functions. (a) First-order autocorrelation function as the laser Rabi frequency is varied. The solid curve is a fit to the data and the Rabi energy extracted from the fit for each data set is displayed. Inset 1: Rabi frequency as a function of driving laser intensity. Inset 2: Rabi oscillations in the resonance fluorescence under pulsed excitation as a function of the pulse area. (b) Second-order autocorrelation function of the resonance fluorescence as the laser Rabi frequency is increased (from bottom to top). Background has been subtracted and the instrument response has been deconvolved from the plots. The solid lines are fits to the data using the independently measured values of T_1 = 277 ps and T_2 = 132 ps. Figures adapted from Refs. [10, 21].

the detuned resonance fluorescence spectrum since this is useful tool to probe nuclear spin dynamics and will be discussed in the next section.

The manifestation of Rabi oscillations is not only apparent in the spectral domain, i.e. as sidebands, but also in the temporal domain. In Fig. 5.4a we present a measurement of the QD resonance fluorescence first-order correlation function $g^{(1)}(\tau)$ [10] using a Michelson interferometer. The fringe contrast, defined as the difference between the maximum and minimum of the interference signal divided by their sum, is plotted in Fig. 5.4a with excitation Rabi energies ranging from 0.9 to 13.3 μeV (note the measured linewidth is 2.8 μeV). T_1 and T_2 values are extracted from linewidth measurements in the zero temperature and zero Rabi frequency limit, respectively, to be T_1 = 290 ps and T_2 = 380 ps. In Fig. 5.4a the correlation function exhibits a monotonic decay for small Rabi frequency (0.9 μeV – upper left trace), but exhibits oscillsation at higher Rabi frequencies. The extracted Rabi frequency dependence on laser power, presented below trace 5, confirms a square-root dependence on laser power. Finally, demonstrated in in the lower right trace of Fig. 5.4a is

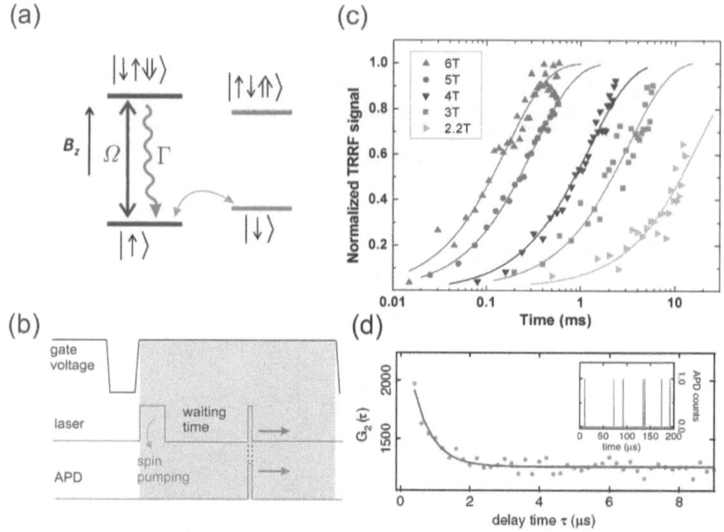

Figure 5.5 Solid-state spectroscopy: probing electron spin dynamics. (a) Illustration of the X^{1-} electronic structure in the presence of an external magnetic field in the Faraday configuration (along the sample growth direction). The ground state consists of a single electron and the excited state is two electrons and one hole. The excited states decay with a spontaneous emission rate Γ and the ground state spin flips with a rate T_{1e}. (b) Illustration of one experimental approach to optically measure T_{1e}. One cycle consists of optically pumping the electron spin to the down state, leaving the system dark for a finite time window and then measuring resonance fluorescence on the spin down correlated transition. (c) Time-resolved resonance fluorescence as a function of the time delay between optically pumping the QD spin and subsequent measurement. The data is for five different magnetic fields and the recovery time of the resonance fluorescence measures T_{1e}. (d) Longtime second-order autocorrelation curve obtained from 60.000 traces such as displayed in the inset. Solid line is an exponential fit with a decay time 540 ± 40 ns, corresponding to the cotunnelling-limited spin lifetime. Inset: A typical time trace recorded from the APD with a 200 ns time resolution. Each pulse arises from the detection of a single photon. Figures adapted from Refs. [26, 28].

a full Rabi cycle in the QD resonance fluorescence under excitation with 8ps long optical pulses.

Higher-order correlation functions of the QD resonance fluorescence have been measured as well [21]. In Fig. 5.4b the second-order correlation function, $g^{(2)}(\tau)$, is presented as the laser Rabi frequency is increased (from the bottom panel to the top panel) where the data are corrected for both background and instrument response. A single pair of values for T_1 (= 277 ps) and T_2 (= 132 ps) is used again to generate reasonable fits for all measured correlation functions in Fig. 5.4b. This is an indication that in this range of Rabi frequencies in this particular experiment excitation induced dephasing plays a minor role (to be compared to [24]) and that the cause of $T_2 < 2T_1$ is spectral diffusion and pure dephasing.

5.3.4 QD resonance fluorescence: solid-state spectroscopy

The development of a technical ability, in this case being able to access the photons from QD resonance fluorescence, presents an opportunity to scrutinize QDs in a new way. Self-assembled solid-state QDs are embedded in a semiconductor matrix and interact with a number of reservoirs that influence the dynamics of confined charges and spins. For example, the roughly 10^5 nuclear spins of an indium arsenide QD are a source of local effective magnetic fields that can interact with the charges confined in a QD via their spin degree of freedom. Also, in QD devices embedded in Schottky diode heterostructures, the buried n^+ layer provides a reservoir of electrons that can interact with electrons confined in the QD via electron cotunnelling – a second-order process that can flip the electron spin (see for a full discussion [25]). In this section we describe how QD resonance fluorescence has served as a unique probe to study both single electron spin dynamics confined in a single QD and the interaction of the QD nuclear spins with the confined charges.

In the theoretical and experimental discussion of resonance fluorescence timescales associated with the excited-state exciton lifetime (T_1) and dephasing (T_2) were introduced. If the ground-state configuration of a QD consists of a single electron, there is an analogous set of timescales associated the spin lifetime and dephasing. Resonance fluorescence in combination with optical pumping has been used as a means to measure the ground state spin lifetime (T_{1e}). The top panel in Fig. 5.5a shows the QD electronic structure when the ground state consists of a single electron with spin projection up or down along the growth direction (parallel to the applied magnetic field and microscope optical axis) and the excited state, the trion or X^{1-}, consists of two electrons and one hole. The experimental approach is illustrated in Fig. 5.5b. The finite magnetic field lifts the spin ground-state degeneracy. When the gate voltage is high, a single electron is loaded into the QD. The application of a laser pulse to the higher energy transition (a few microseconds in duration generated by using the first-order diffraction from an acousto-optic modulator) optically shelves the spin from up to down. The system is left dark for a finite amount of time and then a second laser pulse ($\sim$ microsecond duration) drives the higher energy transition and resonance fluorescence counts are recorded (see rail 2 and 3 in Fig. 5.5b). This cycle is repeated a number of times per dark waiting time to increase the signal. Resonance fluorescence signal is recorded only if the electron spin has flipped during the time window the system is dark and the signal recovery time, as plotted in Fig. 5.5c for magnetic fields between 2.2 T and 6 T, determines the electron spin lifetime T_{1e}. This approach, detailed in Ref. [26], measured spin lifetimes as long as 20 ms and is consistent with phonon assisted spin orbit mediated spin relaxation [27]. A second approach to measuring the spin lifetime is detailed in [28] and uses the decay time of bunching observed in the resonance fluorescence second-order autocorrelation function to extract the spin lifetime. An exemplary data set is presented in Fig. 5.5d. The reduced lifetime, as compared to Fig. 5.5c, is due to the much stronger influence of electron cotunnelling, interaction with the n^+ electron reservoir, on the spin dynamics in this measurement.

In addition to providing information on timescales associated with electron spin dynamics, resonance fluorescence has also been used to understand the interaction of electrons and holes confined in QDs with nuclear spins – one of the main sources of electron spin decoherence. The interplay of electrons and holes confined in a QD with nuclear spins, the n^+ electron reservoir and a resonant laser leads to a number of interesting effects. In particular, the interaction results in optically induced dynamic nuclear spin polarization that exhibits a bidirectional Overhauser shift. Not only does this feedback mechanism result in some remarkable spectroscopic signatures, it also reduces the fluctuations associated with the nuclear spin ensemble. This latter point is critical since these fluctuations are the main source of decoherence for QD electron spin superposition states. A number of recent experiments based on absorption measurements and theoretical works have explored the physics of this complicated interaction. Forgoing much of the details of these works, the interested reader should consult Refs. [29, 30, 31], here we show how QD resonance fluorescence provides a unique probe to investigate the QD–nuclear spin interaction.

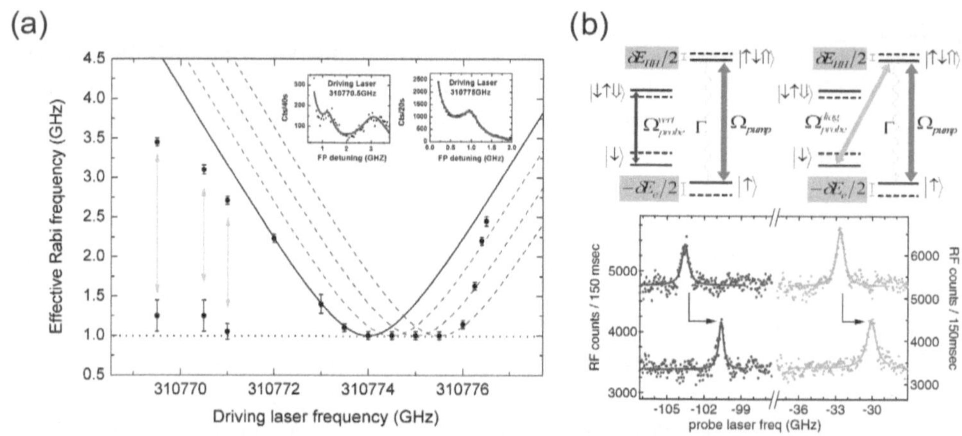

Figure 5.6 Solid-state spectroscopy: probing nuclear spin physics. (a) Mollow triplet sideband splitting as the laser is tuned across the X^0 transition transition. The magnetic field is 6 T with a Rabi frequency of two times saturation. The measured effective Rabi frequency is used to illustrate the expected Mollow sideband detuning dependence without nuclear spin effects for four data points, plotted as solid and dashed curves (compare to Fig. 5.3a). The dashed curves are the expected detuning dependence of the sideband splitting for each of the data points when the nuclear spins shift the resonance to compensate the detuning. Inset: the resonance fluorescence spectrum for two different driving laser frequencies. (b) Top panel: energy level diagrams showing the pump laser, probe laser and 4-level system. The probe laser repumps the spin into the up state by driving either the red trion (left) or the diagonal (right) transition. Dashed (solid) lines indicate the ground and excited energy levels before (after) the polarization of nuclei by the pump laser; δE_{HH} (δE_e) denotes the Overhauser shift seen by a single QD heavy hole (electron). Bottom panel: resonance fluorscence signal recorded as the probe laser is scanned across the diagonal (right) or the red trion (left) transitions. An offset is added to the top two scans for clarity. Before probing, the nuclear spins are prepared by scanning the pump laser to a fixed detuning from the bare transition. Solid lines are Lorentzian fits. Peak positions are shifted due to nuclear spin polarization induced by the pump laser. Figures adapted from Ref. [12].

Figure 5.3d illustrated the Mollow sideband splitting as a function of laser detuning in a regime where nuclear spins have minimal influence on the transition hence the QD can be modelled as an ideal TLS. Figure 5.6a, which plots the splitting of the Mollow sidebands for an uncharged QD at 6T, shows a striking deviation from the expected TLS-behaviour [22]. In this experiment the laser is tuned across the QD transition, from higher energy to lower energy, and the resonance fluorescence spectrum is recorded. The raw spectrum for two laser detunings is shown in the insets. In Fig. 5.6a, as the QD transition is approached from the blue, the sideband splitting decreases much more rapidly than expected from the conventional theory. This suggests that the nuclear spins facilitate the reduction of laser-bare transition detuning as the laser is tuned towards the transition. A second feature in Fig. 5.6a is the four data points across a near 2 GHz excitation window that correspond to the same effective Rabi frequency of 1 GHz (the horizontal dashed line is a guide to the eye). Notably, these measurements reveal a shift of the bare transition energy depending on the frequency of the exciting laser. This confluence of the QD resonance with the exciting laser is referred to as *locking* of the QD transition to the laser frequency or *dragging* of the resonance and shows directly that the broad absorption features observed in Refs. [29, 30] do not arise from the broadening of the QD transition, but rather from locking of the QD transition to the laser via nuclear spin polarization. For each of these four data points, the measured effective Rabi frequency is used to illustrate the expected Mollow sideband detuning dependence, plotted as solid and dashed curves. Once the laser detuning exceeds a particular value the nuclear spins are no longer able to provide suitable feedback and the transition no longer follows the laser as it is detuned from resonance. Third, and finally, as the laser is red detuned from the transition there is a clear bistability in the measured sideband splitting. The appearance of two Mollow sidebands (see inset for the full spectrum) suggests the resonance and hence the nuclear spin ensemble changes between two stable configurations during the course of the measurement. The detuning dependence of QD resonance fluorescence clearly exhibits the rich physics associated with QD electron and/or hole interaction with nuclear spins.

Observation of these nuclear spin effects has initiated a discussion as to the main mechanism leading to the nuclear spin polarization. Specifically, the question is which species – electrons or holes – efficiently polarizes the nuclear spins. Fallahi *et al.* [12] used resonance fluorescence in an elegant two-colour experiment to determine the relative magnitudes of the electron and hole hyperfine interaction strength. The experiment can be understood by considering the illustration in Fig. 5.6b. A strong pump laser is scanned to a precise value of detuning from the bare transition energy of the blue trion transition. This pump laser builds up a precise amount of nuclear spin polarization, dragging the QD resonance along as the laser is scanned away from the bare resonance. Following the strong pump, a second weak probe laser is scanned across either the red trion transition or the diagonal transition linking the red transition ground state and blue transition excited state. On resonance the probe laser enhances resonance fluorescence allowing the resonance location to be determined. The sum and difference of the measured resonances reveal the electron and hole Overhauser shift for the particular pump detuning. Figure 5.6c presents data for two different pump laser detunings at a magnetic field of 4 T in the Faraday configuration. The relative strength of the heavy-hole to electron hyperfine interaction is found to be independent of pump laser detuning and determined to be -0.09 ± 0.02. Very similar results were obtained at the same time by Chekhovich *et al.* in photoluminescence studies [32]. These results are important in the context of using heavy-hole spins in quantum information processing, as nuclear spins are the principal decoherence source for spin qubits.

5.3.5 QD resonance fluorescence: solid-state quantum information science

In this final section we discuss some applications where resonance fluorescence is an enabling technique for advances in solid-state QIS. The utility of resonance fluorescence is that it provides a means to coherently interface photons and QD spins. In the context of photonic QIS this allows for generation of near transform-limited single photons and generation of entangled photons. Further, it provides a means to coherently prepare, control and read-out the orientation of single QD spin.

Some of the early QIS related demonstrations were on the photon statistics of light emitted by a QD. The improvement afforded by resonance fluorescence in the context of photon generation is that coherent excitation avoids any dephasing induced by optically charging the QD environment (although the excitation laser itself can induce dephasing via acoustic phonons [24]). This was discussed with regard to the autocorrelation measurements in Fig. 5.4. A crucial ingredient for using QD generated photons for QIS is the possibility of generating indistinguishable photons. The need for photon indistinguishability is that protocols for linear optics quantum computing or quantum teleportation rely on the two-photon interference of single photon pulses. By using the post-selected resonance fluorescence from a single QD in a pillar microcavity it has been possible to generate near transform limited photon pulses that exhibit a visibility contrast of 90% in Hong–Ou–Mandel two-photon interference experiments [15]. Figure 5.7a confirms under resonant s-shell excitation that the QD emits single photons (as also presented in Fig. 5.4). In Fig. 5.7b (Fig. 5.7c) two-photon interferograms, recorded under cw-resonant s-shell excitation with an asymmetric fibre-based Mach–Zehnder interferometer, are presented for photons impinging on the beam-splitter with orthogonal (parallel) polarizations. The two satellite peaks at 13 ns reflect the fixed path delay between the two arms. The data for distinguishable photons in Fig. 5.7b finds a $g^{(2)}(\tau = 0)_{\perp} = 0.55 \pm 0.01$ whereas in the parallel – indistinguishable configuration – $g_{\parallel}^{(2)}(\tau = 0) = 0.22 \pm 0.01$ in Fig. 5.7c. The two-photon interference visibility $V_{\mathrm{HOM}}(\tau)$, finds a maximum normalized value of 0.6 – limited by the system response. After deconvolving the system response from the recorded data the actual two-photon visibility is found to be 0.90 ± 0.05. These results clearly demonstrate the improvement of the photon's coherence properties using resonance-fluorescence-based photon generation.

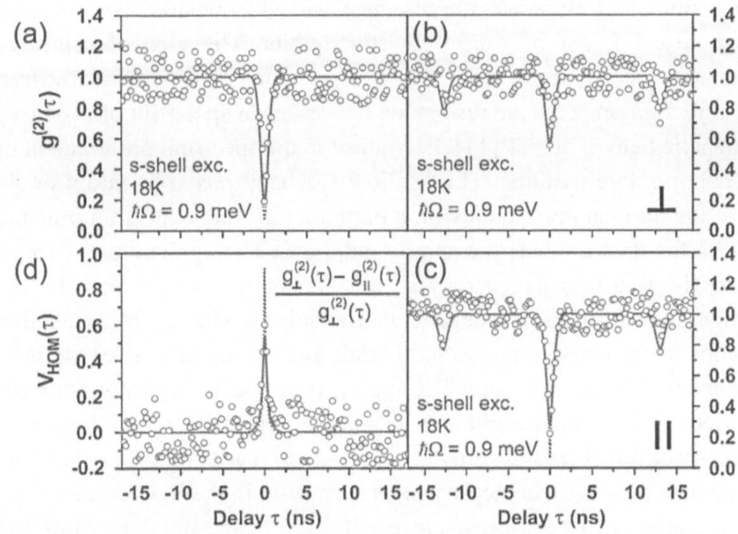

Figure 5.7 Solid-state quantum information science: indistinguishable photon generation: (a) second-order correlation measurement under low-power resonant QD s-shell excitation (18K): $g^{(2)}(0) = 0.08$ (deconvoluted) reflects almost background-free single-photon emission. (b, c) Two-photon interference data under orthogonal and parallel polarizations of the interferometer arms, respectively. (d) Interference visibility obtained from traces (b) and (c). Solid lines in all figures: theoretical fits convolved with the system response ($\delta t = 400 \pm 10$ ps); dotted lines: deconvolved fits, i.e. corrected for δt. Figures adapted from Ref. [15].

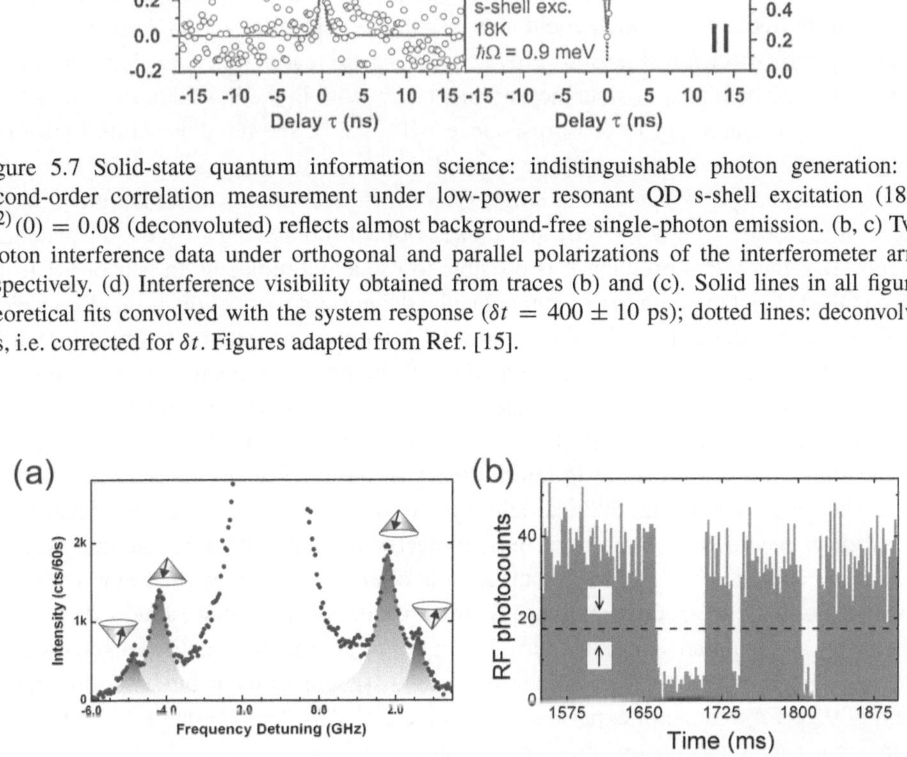

Figure 5.8 Solid-state quantum information science: spin qubit readout. (a) The resonance fluorescence Mollow quintuplet under application of a 50 mT magnetic field (Faraday configuration) and a 1.25 GHz red detuned laser with a power of 1.852 μW corresponding to ~8 times the saturation Rabi frequency. Each sideband splits into a doublet in which each transition is linked to a specific QD ground state spin. (b) Intermittent resonance fluorescence from the host QD exciton of a QD molecule. The time bin per data point is 2 ms. The dashed line shows the optimum threshold to distinguish between spin up and spin down, resulting in a fidelity of 96%. Figures adapted from Ref. [14].

Another application of resonance fluorescence we will consider here is in the optical control and measurement of solid-state spin qubits. The idea of using resonance fluorescence for state measurement is not new and has a long history within the trapped ion community. Figure 5.8a presents the resonance fluorescence spectrum of a singly charged QD under a magnetic field of 50 mT [11]. In contrast to the spectrum presented in Fig. 5.3b, this spectrum exhibits five features. The Mollow sideband has split into a doublet with each member of the doublet correlated with a particular ground state spin orientation (the arrows decorating the data indicate the spin orientation). The application of the magnetic field has broken the two-fold degeneracy of this transition and detuned the laser by a fixed and equal amount from the nondegenerate transitions. The result is a Mollow triplet for each transition with a degenerate central peak and a pair of nondegenerate spectral sidebands correlated with electron spin. The spin-sensitive splitting depends on transition-selective Rabi frequencies which could also be obtained via polarization. Spectral filtering the fluorescence spectrum sidebands provides a means to readout the QD electron spin.

An alternative path to spin-state measurement is to work in the low Rabi frequency limit and tune the measurement laser to resonance with one of the nondegenerate transitions of Fig. 5.5a. The time dynamics of the emitted resonance fluorescence will be correlated with the electron spin dynamics. For single QDs the measurement laser itself induces spin flips on a timescale of $\sim 1\ \mu s$ [26] and makes optical monitoring of the spin dynamics difficult. To remedy this problem it is possible to use a QD molecule (a pair of vertically stacked QDs) [14]. One member of the molecule, the host QD, stores the ground state electron. The second QD in the pair, the readout QD, supports an excitonic transition with a frequency that is determined by the orientation of the host-QD electron spin. To monitor real-time spin dynamics, a cw-laser is tuned to resonance with the readout QD exciton transition correlated with the host-QD electron spin-down orientation. Changes in the host-QD electron spin's orientation result in intermittency in the time resolved resonance fluorescence. The time trace in Fig. 5.8b exhibits exactly this effect. The presence of resonance fluorescence, recorded in 2 ms time bins, indicates the host-QD electron spin is down whereas the absence of signal means the spin is oriented up. Hence time bins where the resonance fluorescence counts are above (below) a suitably chosen threshold – dashed line in the plot – are declared spin down (up). Estimates of the readout fidelity can be obtained from analysing a large sample of time traces (see [14]). Our analysis indicates fidelities in excess of 95% for a range of experimental parameters. This demonstration is the first step towards realizing a feasible quantum spin–photon interface.

5.4 Conclusion

Resonance fluorescence provides state-selective excitation and direct access to the emitted photons – both the elastically and inelastically scattered components. Truly resonant excitation further greatly enhances the coherence of photons emitted by a QD when compared to quasi-resonant or above-bandgap excitation methods.

These features together have enabled the real-time monitoring of electron spin quantum jumps in a QD, revealed the interaction dynamics of electrons and holes with the nuclear spins in spectroscopic studies and hold promise for the realization of a spin–photon interface. Future QD research challenges include entangling distant QD spins via a shared photonic channel, the application of this technique for QD-based quantum metrology and the control and manipulation of mesoscopic aspects of QDs such as nuclear spins. Resonance fluorescence is likely to play a key role in achieving these numerous milestones within the remit of QIS and mesoscopic physics.

References

[1] Crisp, M. D. and Jaynes, E. T. 1969. Radiative effects in semiclassical theory. *Phys. Rev.*, **179**, 1253–1261.

[2] Stroud Jr., C. R. and Jaynes, E. T. 1970. Long-term solutions in semiclassical radiation theory. *Phys. Rev. A*, **1**, 106–121.

[3] Mollow, B. R. 1969. Power spectrum of light scattered by two-level systems. *Phys. Rev.*, **188**, 1969–1975.

[4] Stroud Jr., C. R. 1971. Quantum-electrodynamic treatment of spontaneous emission in the presence of an applied field. *Phys. Rev. A*, **3**, 1044–1052.

[5] Schuda, F., Stroud, Jr., C. R. and Hercher M. 1974. Observation of the resonant Stark effect at optical frequencies. *J. Phys. B*, **7**, L198–L202.

[6] Wu, F. Y., Grove, R. E. and Ezekiel, S. 1975. Investigation of the spectrum of resonance fluorescence induced by a monochromatic field. *Phys. Rev. Lett.*, **35**, 1426–1429.

[7] Kimble, H. J., Dagenais, M. and Mandel, L. 1977. Photon antibunching in resonance fluorescence. *Phys. Rev. Lett.*, **39**, 691–695.

[8] Leibfried, D., Blatt, R., Monroe, C. and Wineland, D. 2003. Quantum dynamics of single trapped ions. *Rev. Mod. Phys.*, **75**, 281–324.

[9] Gammon, D. and Steele, D. G. 2002. Optical studies of single quantum dots. *Physics Today*, 36–41.

[10] Muller, A., Flagg, E. B., Bianucci, P. *et al.* 2007. Resonance fluorescence from a coherently driven semiconductor quantum dot in a cavity. *Phys. Rev. Lett.*, **99**, 187402:1–4.

[11] Vamivakas, A. N., Zhao, Y., Lu, C.-Y. *et al.* 2009. Spin-resolved quantum-dot resonance fluorescence. *Nat. Phys.*, **5**, 198–201.

[12] Fallahi, P., Yilmaz, S. T. and Imamoğlu, A. 2010. Measurement of a heavy-hole interaction in InGaAs quantum dots using resonance fluorescence. *Phys. Rev. Lett.*, **105**, 257402:1–4.

[13] Imamoğlu, A., Awschalom, D. D., Burkard, G. *et al.* 1999. Quantum information processing using quantum dot spins and cavity QED. *Phys. Rev. Lett.*, **83**, 4204–4207.

[14] Vamivakas, A. N., Lu, C.-Y., Matthiesen, C. *et al.* 2010. Observation of spin-dependent quantum jumps via quantum dot resonance fluorescence. *Nature*, **467**, 297–300.

[15] Ates, A., Ulrich, S. M., Reitzenstein, S. *et al.* 2009. Post-selected indistinguishable photons from the resonance fluorescence of a single quantum dot in a microcavity. *Phys. Rev. Lett*, **103**, 167402:1–4.

[16] Knill, E. and LaFlamme, R. 2001. A scheme for efficient quantum computation with linear optics. *Nature*, **409**, 46–52.

[17] Loudon, R. 2000. *The Quantum Theory of Light*. Oxford University Press.

[18] Scully, M. O. and Zubairy, M. S. 1997. *Quantum Optics*. Cambridge University Press.

[19] Tannoudji, C. C., Dupont-Roc, J. and Grynberg, G. 1998. *Atom–Photon Interactions*. Wiley.

[20] Citron, M. L., Gray, H. R., Gabel, C. W. and Stroud, C. R. 1977. Experimental study of power broadening in a two-level atom. *Phys. Rev. A*, **16**, 1507–1511.

[21] Flagg E. B., Muller A., Robertson, J. W. *et al.* 2009. Resonantly driven coherent oscillations in a solid-state quantum emitter. *Nat. Phys.*, **5**, 203–207.

[22] Unpublished data obtained by the authors.

[23] Warburton, R. J., Schein, C., Haft, D. *et al.* 2000. Optical emission from a charge-tunable quantum ring. *Nature* **405**, 926–929.

[24] Ramsay, A. J., Godden, T. M., Boyle, S. J. *et al.* 2010. Phonon-induced Rabi-frequency Renormalization of optically driven single InGaAs/GaAs quantum dots. *Phys. Rev. Lett.*, **105**, 177402:1–4.

[25] Dreiser, J., Atatüre, M., Galland, C. *et al.* 2008. Optical investigations of quantum dot spin dynamics as a function of external electric and magnetic fields. *Phys. Rev. B*, **77**, 075317:1–15.

[26] Lu, C.-Y., Zhao, Y., Vamivakas, A. N. *et al.* 2010. Direct measurement of spin dynamics in InAs/GaAs quantum dots using time-resolved resonance fluorescence. *Phys. Rev. B*, **81**, 035332:1–5.

[27] Khaetskii, A. and Nazarov, Y. 2001. Spin-flip transitions between Zeeman sublevels in semiconductor quantum dots. *Phys. Rev. B*, **64**, 125316:1–6.

[28] Yilmaz, S. T., Fallahi, P. and Imamoğlu, A. 2010. Quantum-dot-spin single-photon interface. *Phys. Rev. Lett.*, **105**, 033601:1–4.

[29] Latta, C., Högele, A., Zhao, Y. *et al.* 2009. Confluence of resonant laser excitation and bidirectional quantum-dot nuclear-spin polarization. *Nat. Phys.*, **9**, 5–5.

[30] Xu, X., Wang, Y., Sun, B. *et al.* 2009. Optically controlled locking of the nuclear field via coherent dark-state spectroscopy. *Nature* **459**, 1105–1109.

[31] Yang, W. and Sham, L. J. 2011. Collective nuclear stabilization by optically excited hole in quantum dot. *arXiv:1012.0060v2*.

[32] Chekhovich, E. A., Krysa, A. B., Skolnick, M. S. and Tartakovskii, A. I. 2010. Direct measurement of the hole–nuclear spin interaction in single InP/GaInP quantum dots using photoluminescence spectroscopy. *Phys. Rev. Lett.*, **106**, 027402:1–4.

6

Coherent control of quantum dot excitons using ultra-fast optical techniques: the role of acoustic phonons

A. J. Ramsay and A. M. Fox

6.1 Introduction

Quantum dots are often referred to as artificial atoms, since they trap carriers in discrete energy-levels due to the nanoscale three-dimensional finite potential energy well they provide. As such, dots exhibit a coherent light–matter interaction that is similar to an atom. This is evidenced by observations of atom–optics phenomena such as Rabi oscillations [43, 26], power broadening [27], Autler–Townes doublet [14, 42], Mollow triplet [42, 8], and coherent population trapping [6]. In this chapter, Rabi rotation measurements are used to examine how an exciton transition deviates from an ideal two-level atom due to its interaction with a reservoir of phonons.

The neutral exciton transition may be regarded as a two-level system, or qubit, composed of the crystal ground-state $|0\rangle$ and a single electron–hole pair $|X\rangle$. The state-vector of a qubit can be described as a pseudo spin-half. When an oscillating electro-magnetic field resonantly excites the two-level transition it drives an oscillation in the population inversion known as a Rabi oscillation. This results from the oscillations of the driving field and the dipole of the two-level system being synchronous, such that in its rotating frame, the driving field acts as a static magnetic field that causes the pseudo-spin to rotate. Coherent control of the pseudo-spin can be achieved by applying well-defined driving fields, enabling the preparation, and manipulation of superposition states. Such coherent control concepts have found widespread use in electron spin, and nuclear magnetic resonance spectroscopy. Coherent control is also widely used to perform quantum logic operations. Examples of driven qubits can be found throughout physics, including: atoms; ion-traps; NV-centers; SQUIDs and quantum dots.

The fidelity of a Rabi oscillation, in other words how closely it matches the ideal case, is limited by decoherence. The simplest models of decoherence consider a qubit interacting with a bath of bosons or half-integer spins, and the case of a driven qubit is important for understanding decoherence in the Rabi regime relevant to the performance of quantum logic operations. In this chapter, a driven qubit coupled to a boson bath is studied by measuring the optically driven Rabi rotations of a neutral exciton transition, which is coupled to a bath of longitudinal acoustic phonons. In this system the energy spectrum of the

Quantum Dots: Optics, Electron Transport and Future Applications, ed. Alexander Tartakovskii. Published by Cambridge University Press.

exciton–phonon interaction is structured on an energy scale comparable to easily attainable Rabi energies, giving rise to a physically rich decoherence dynamics.

In the remainder of this section we will explain why a neutral exciton behaves as a two-level system, explain why a Rabi oscillation occurs, present a Rabi rotation measurement of a quantum dot exciton, and point out how the data deviate from a two-level picture due to the interaction with acoustic phonons.

A neutral exciton driven by a circularly polarized picosecond laser pulse may be regarded as a two-level system; this is justified as follows. The lower energy level is the crystal ground state – an empty dot. The upper energy level is a bright exciton composed of an electron and heavy-hole of total z-component of angular momentum of $m_j = \pm 1/2, \mp 3/2$ respectively. Owing to a small asymmetry in the dot, there is an anisotropic exchange interaction coupling between the exciton spins. However, in the experiments discussed here the fine-structure splitting is small compared with the spectral width of the laser pulse (18 μeV vs 0.2 meV), and the precession of the exciton spin can be neglected on the timescale of the laser pulse [38, 4]. The biexciton can also be neglected, since creation of a biexciton via two-photon absorption is suppressed both by using a laser of circular polarization, and through energy selection, since the 1.9 meV biexciton binding energy is large compared with the 0.2 meV wide laser pulse. What remains, is a two-level system formed by the crystal ground-state $\mid 0\rangle$ and, in the case of a σ_+-polarized laser, a spin-up exciton $\mid X\rangle$.

In the rotating frame of the laser, the control Hamiltonian H_c of the optically driven neutral exciton can be expressed as

$$H_c = \frac{\hbar\Omega(t)}{2}\left[\mid 0\rangle\langle X \mid + \mid X\rangle\langle 0 \mid\right] \equiv \frac{\hbar\Omega(t)}{2}\hat{\sigma}_x, \tag{6.1}$$

where the Rabi energy $\hbar\Omega(t) = \mu E(t)$ is the product of the optical-dipole μ and the envelope of the electric field $E(t)$. Note that $\hat{\sigma}_x$ is the Pauli spin-x operator. When the laser is applied, the energy-eigenstates of H_c are the optically dressed states $\mid \pm\rangle = (\mid 0\rangle \pm \mid X\rangle)/\sqrt{2}$ of energies $\pm\hbar\Omega/2$. If the qubit is initially in the ground-state $\mid 0\rangle$, this is a superposition of the optically dressed states $\mid \pm\rangle$ which accumulate a relative phase-shift $\theta(t) = \int_{-\infty}^{t} \Omega(t')dt'$. In the excitonic basis the beat results in a Rabi oscillation in the population inversion: $\mid \psi(t)\rangle = \cos(\theta(t)/2) \mid 0\rangle + i \sin(\theta(t)/2) \mid X\rangle$. Following excitation by a laser pulse, the final occupation of the exciton state is $\sin^2(\Theta/2)$, where the pulse-area $\Theta = \theta(\infty)$.

Figure 6.1 presents a Rabi rotation measurement taken at 5 K. A circularly polarized Gaussian laser pulse excites the neutral exciton transition on-resonance. The photocurrent, which is proportional to the final exciton population, is measured as a function of the square-root of the time-averaged incident power $\sqrt{P}$. The x-axis is expressed in terms of the pulse-area $\Theta = \int_{-\infty}^{\infty} \Omega(t')dt' = a\sqrt{P}$, where the Rabi frequency of the pulse is described by $\Omega(t) = \frac{\Theta}{2\tau\sqrt{\pi}} \exp(-(\frac{t}{2\tau})^2)$, $\tau = 4$ ps, and a is a measure of the dot–laser coupling. The dot–laser coupling a is deduced from fits to Eqs. (6.3, 6.4). Even at this

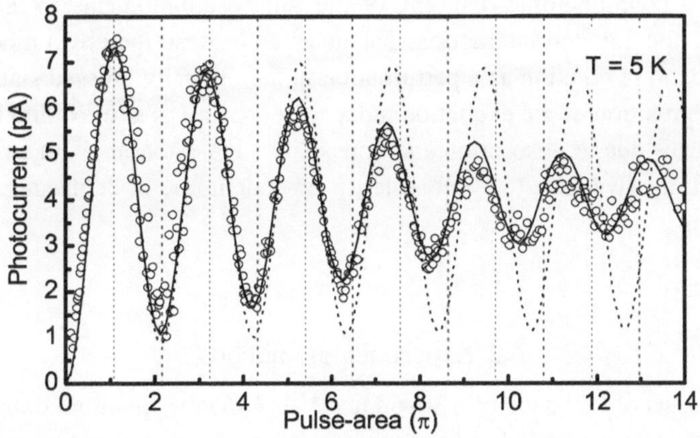

Figure 6.1 A Rabi rotation of the neutral exciton transition. The photocurrent is proportional to the final exciton population, and oscillates with the pulse area which is proportional to the square root of the time-averaged incident power of the single picosecond laser excitation pulse. To help illustrate the aperiodic nature of the oscillations, the vertical lines provide a periodic grid matched to the first two turning points. The dashed line is a calculation using a constant rate of dephasing. The solid line is a fit to data using Eqs. (6.3, 6.4). Figures adapted from Ref. [24].

low temperature, the resulting oscillation is damped with increasing power, and the period decreases with increasing pulse area.

To illustrate the relative unimportance of the conventional exponential-type dephasing, the dashed trace in Fig. 6.1 presents a calculation of the Rabi rotation using a two-level model with a constant, driving-field independent, rate of dephasing $\Gamma_2^* = 0.025\ \text{ps}^{-1}$. The dashed-trace is scaled to best fit the data, and hence the turning points do not occur at integer-π pulse areas. Since the time duration of the pulse is constant there is a fixed damage to the coherence, and the contrast is lost over the first period of the oscillation, but thereafter is almost constant. By comparison, the data are strongly damped. The damping is nonmonotonic, with the rate of damping decreasing with increasing pulse area, and the rotation angle is a nonlinear function of the pulse area, i.e. the period decreases with pulse area. The line trace is calculated by using the model described in Section 6.5. It is this intensity damping that is the subject of this chapter.

The source of the intensity damping observed in measurements of optically driven Rabi rotations in semiconductor quantum dots is a long-standing question in quantum dot physics. In this chapter, experimental evidence identifying bulk longitudinal acoustic phonons as the principal source of driving field dependent dephasing is presented [25, 24]. An understanding of the interplay between the driving field and the phonon environment is important for the realization of high-fidelity quantum logic operations in the high-speed regime. Key signatures of the exciton–phonon interaction in the Rabi regime are observed, including a temperature and driving-field dependent renormalization of the Rabi frequency, and a nonmonotonic damping of the Rabi oscillation due to the finite bandwidth of the exciton–phonon interaction. The data can be described by a model that treats the exciton–phonon interaction as a perturbation in the optically dressed states basis. This leads to an interpretation of the excitation induced dephasing as a relaxation between optically dressed states due to absorption and emission of LA-phonons with an energy equal to the Rabi splitting gives rise to a Rabi-energy dependent rate of dephasing and a shift to the Rabi splitting.

6.2 Experimental methods

The sample consists of low density $\sim$30–60 μm^{-2} InAs/GaAs quantum dots embedded in the intrinsic region of an n-i-Schottky diode structure. The wafer is grown by molecular beam epitaxy and consists of a GaAs substrate with the following layers: 50 nm of n^+-doped GaAs, a 25 nm spacer of i-GaAs, a single layer of InAs/GaAs, 125 nm of i-GaAs, a 75 nm blocking barrier of $Al_{0.3}Ga_{0.7}As$, capped with 5 nm of i-GaAs. The dots were grown by depositing approximately 2.0 ml of InAs at a growth rate of 0.1 mls^{-1} and a temperature of 490°C. Photoluminescence measurements of the unprocessed wafer reveal a Gaussian distribution of dots with an emission centered at 980 nm, and a FWHM of 30 nm. The wafer is processed into 400 $\times$ 200 μm mesas with a semi-transparent titanium top contact and an aluminium shadow-mask. Individual quantum dots are spatially isolated using 400 nm diameter apertures defined by e-beam lithography. The results presented here are from a single dot emitting at 951 nm.

At liquid-helium temperatures, the photodiode is nearly ideal with a ammeter limited dark current of less than 25 fA, and no observable reverse breakdown current for reverse gate voltages of up to 5 V. The diode is connected in series with a Keithley 230 voltage source, and a Keithley 6485 picoammeter. Coaxial cables with a Teflon dielectric are used for improved temperature stability compared with polyethylene, with the current passing through the inner core and a grounded outer core to provide shielding. Care is taken to avoid any ground loops, but otherwise detecting pA currents is relatively straight forward.

A schematic diagram of the experimental setup is shown in Fig. 6.2. The sample is held in a helium bath cryostat, and mounted on a low-temperature piezo-electric *xyz*-stage from Attocube. The dot is excited by a single laser pulse derived from a 100 fs Ti:sapphire laser. The 100 fs laser pulse is passed through a pulse-shaper, consisting of two diffraction gratings and two lenses in a 4f configuration [39]. In other words, it is a double spectrometer in a transmission mode, as shown in the inset of Fig. 6.2. A focal length of 0.5 m is used to achieve high spectral resolution, and the spectrum of the resulting pulse is a convolution of the rectangle arising from the slit, and a Gaussian due to the diffraction limited spot size. In these experiments the pulse-shape is diffraction limited, resulting in a Gaussian pulse of intensity full-width half-maximum of 0.2 meV, corresponding to an electric field envelope of 14 ps FWHM. The slit is motorized, enabling continuous control of the carrier frequency of the pulse.

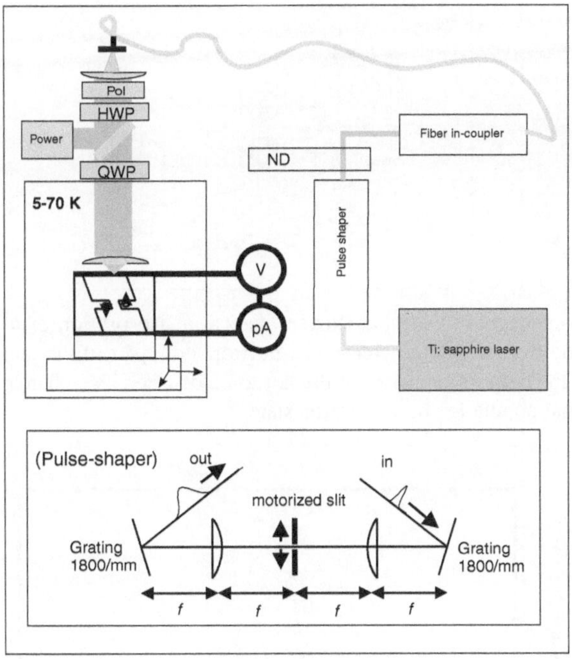

Figure 6.2 Sketch of the experimental setup. The inset shows a close-up of the layout of the pulse-shaper optics. Details in the text.

Proper alignment of the pulse-shaper is important to avoid chirping the pulses, and to avoid beam walk as the frequency of the laser pulse is tuned. This is achieved when the pulse-shaper without the slit has no effect on the pulse, resulting in a good spatial and angular overlap of the spectral components. It is important that the optical components are spaced properly, and that the 1:1 telescope is not used to alter the divergence of the beam. The tilt of the diffraction gratings need to be matched. A useful alignment trick is to place a block in the image plane, allowing only the high- and low-wavelength components of the spectrum to pass; a good overlap of the two beams over a long distance indicates good spatial–spectral overlap. To test the alignment, the pulse spectrum is measured with a spatial filter to test the spatial–spectral overlap.

To control the pulse area, the laser passes through a motorized variable neutral density filter, and is then coupled into a single mode angle polished optical fiber to carry the light to the cryostat. The output of the fiber is collimated, and polarized. A glass slide directs a small portion of the beam to a power meter. The polarization of the laser is then mapped to circular polarization using a quarter waveplate before passing into the sample space of the cryostat. The laser is focused onto the sample using a plano-convex lens of 4.75 mm focal length mounted inside the cryostat. Although the power of the laser incident on the sample fluctuates due to variations in the coupling losses into the fiber, and changes in the birefringence of the fiber occur as a result of variations in the laboratory temperature, the relationship between the pulse area and the detected power is highly robust.

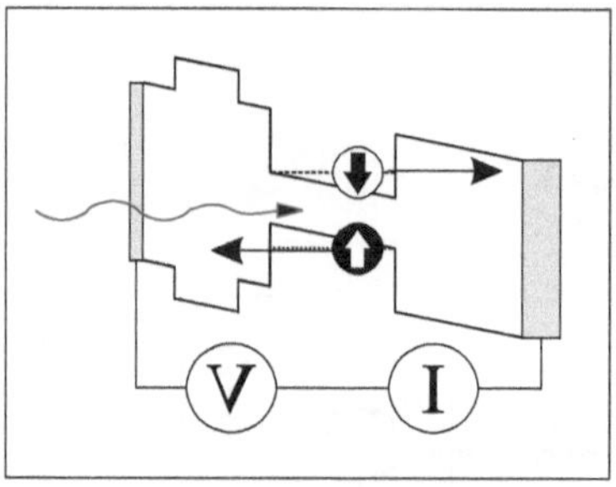

Figure 6.3 Principle of operation of photocurrent detection. Absorption of a laser pulse (wiggly line) resonant with the neutral exciton creates an electron–hole pair in the quantum dot. Under an applied electric field the carriers tunnel from the dot to be detected as a change in the photocurrent proportional to the final population of the exciton state.

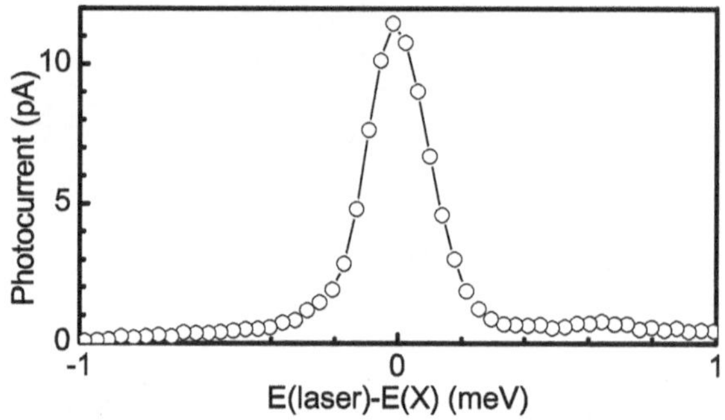

Figure 6.4 A typical photocurrent spectrum of the neutral exciton transition measured at a reverse gate voltage of 0.8 V using a π-pulse.

Photocurrent detection is a quasi-absorption measurement. When a laser is resonant with an optical transition, absorption creates an electron–hole pair in the dot. Under an applied electric field, the carriers tunnel from the dot and are detected as a change in photocurrent, as illustrated in Fig. 6.3. A photocurrent spectrum is measured by monitoring the photocurrent as the frequency of the laser pulse is varied. An example of a photocurrent spectrum for a pulse of pulse area π scanned through the neutral exciton transition is shown in Fig. 6.4. Since the spectral width of the laser pulse is large compared with the linewidth of the exciton transition the Gaussian lineshape is that of the laser pulse.

If the photocurrent detection is 100% efficient then the device acts like an optically triggered single-electron source [43] and the creation of one exciton per pulse results in a current of 12.18 pA for a laser with a 76 MHz repetition rate. However, there are two main factors that limit the detection efficiency: a competition between radiative recombination and electron tunneling rates; and a potentially slow hole tunneling rate that may be comparable to the repetition rate of the laser [15]. As a result, different excitonic species give rise to different sized changes in photocurrent [23, 4, 29].

In addition to the photocurrent originating from the dot, there is a background signal that is proportional to the incident power. The background has been subtracted from all of the data presented. The background is usually attributed to absorption of scattered light by other dots in the mesa [28].

6.3 Candidate mechanisms for the intensity damping

A number of previous reports have found that the intensity damping can be described by a rate of dephasing that is proportional to the square of the Rabi frequency, i.e. $\Gamma_2 = K_2\Omega^2$ [34, 36, 37, 30, 19]. However, as we will discuss below, there are a number of possible mechanisms that can give rise to such behavior.

Ensemble averaging of the dot–laser coupling If there are fluctuations in the dot–laser coupling due to variations in the optics, the time-averaged ensemble of the dot–laser coupling will give rise to a temperature independent damping in the observed Rabi rotation.

Multi-exciton effects If the laser pulse is spectrally broad it will also excite other multi-excitonic transitions, such as the two-photon biexciton transition [29, 5]. This results in a break down of the two-level approximation. A damping of the Rabi rotation occurs as a result of the interplay between the different transitions. In experiments using spectrally broad pulses with a 1.0 meV FWHM the intensity damping has been attributed to the excitation of multiple biexcitonic transitions [22]. Similar effects are discussed theoretically in Ref. [34].

In these experiments, circular polarization is used to suppress the creation of a biexciton. Further suppression is achieved using spectrally narrow laser pulses with a 0.2 meV FWHM, which is small compared to the 1.9 meV biexciton binding energy.

Off-resonant excitation of wetting layer states The wetting layer is a quantum well that naturally accompanies dots grown by Stranski–Krastanov method. The energy-levels of the excited state (p-shell) excitons are often hybridized with the wetting layer forming a broad continuum of states [33]. Villas-Bôas *et al.* [34] have postulated that off-resonant excitation of these wetting layer states would give rise to a rate of dephasing proportional to the Rabi frequency squared. This extrinsic mechanism is likely to be strongest in weakly confined dots, and for p-shell states that are energetically close to the wetting layer. There are experimental reports that attribute intensity damping of Rabi rotations of the p-shell excitons to the wetting layer [36, 37].

The temperature dependence of the wetting layer mechanism would be governed by the thermal occupation of the dot-wetting layer hybrid states, and for the dots used here, one might expect activation energies in the range of 20–40 meV [33].

Phonons The dot is embedded in a GaAs matrix. As such the confined carriers interact with the longitudinal acoustic (LA), transverse acoustic (TA), and longitudinal optical (LO) phonons of the host crystal via the deformation potential, piezo-electric effect, and Fröhlich interactions, respectively. Although non-elastic scattering is suppressed by the carrier confinement, elastic scattering can occur and is one of the dominant sources of pure dephasing for the optical coherence of semiconductor quantum dots and gives rise to a non-exponential dephasing. In photoluminescence experiments, the lineshape of an exciton transition is modified by the exciton–phonon interaction, and resembles a Lorentzian zero-phonon line sitting on top a broad ($\sim$ 1 meV) asymmetric line due to phonon-assisted emission [7, 1]. In the time domain, resonant excitation of the exciton transition results in a non-exponential decay of the excitonic polarization that has been measured in four-wave mixing experiments [2, 3]. The transient has two components. In the first picosecond, a fraction of the coherence is lost as the lattice adjusts to the new charge configuration of the dot. This is followed by an exponential decay, which at low temperature is mostly due to radiative recombination. The Fourier transform of the transient is similar to the lineshape observed in single dot PL, and can be described in terms of the exciton LA-phonon interaction [31]. Unlike the wetting layer, or multi-exciton effects, phonon dephasing cannot be avoided by selecting the right dot, and represents an intrinsic limitation on the quality of excitonic Rabi rotations. Hence theoretical investigations of the intensity damping have mostly focused on the role of phonons, with longitudinal acoustic phonons as the prime suspects [9, 13, 18, 16, 32, 20, 10, 11].

The temperature dependence of the LO-phonon contribution to K_2 would be determined by the thermal occupation of the LO-phonons with an activation energy of around 36 meV.

6.4 Temperature-dependent measurements of Rabi rotations

To identify the source of the intensity damping, a series of Rabi rotation measurements are made versus temperature and are presented in Fig. 6.5. For each temperature the laser is tuned on-resonance with the neutral exciton transition to account for the red-shift. At low temperatures oscillations are observed over the full 14π pulse-area range of the experiments, providing good resolution of the period. As the temperature increases, the damping becomes stronger, and the period decreases. Even at 50 K, at least one period of the Rabi rotation can still be observed. The red lines are fit to the model in Section 6.5. To establish the intensity damping mechanism, the Rabi rotations are fitted to calculations of a two-level system with a rate of dephasing $\Gamma_2 = K_2\Omega^2$. The value of the K_2 time extracted from the data is plotted versus temperature in Fig. 6.6.

As shown in Fig. 6.6, K_2 is proportional to temperature, implying that LA-phonons are responsible for the intensity damping. The gradient $A^{(\Omega^2)} = 12.9\pm0.4$ fs.K^{-1}. To compare this to theory, we note that in Ref. [41] the temperature dependence of the rate of dephasing for a two-level system coupled to a generic three-dimensional boson bath in a high-temperature, low-driving-field regime has been calculated. By substituting values for the carrier-phonon deformation couplings $g_{\mathbf{q}} = q(D_{\mathrm{e}}\mathcal{P}[\psi^{\mathrm{e}}(\mathbf{r})] - D_{\mathrm{h}}\mathcal{P}[\psi^{\mathrm{h}}(\mathbf{r})])\big/\sqrt{2\mu\hbar\omega_{\mathbf{q}}V}$, [17], into Eq. (8.10) of ref. [41] yields:

$$K_2 = \frac{(D_{\mathrm{e}} - D_{\mathrm{h}})^2}{4\pi\mu c_{\mathrm{s}}^5\hbar^2} k_{\mathrm{B}}T \equiv AT, \tag{6.2}$$

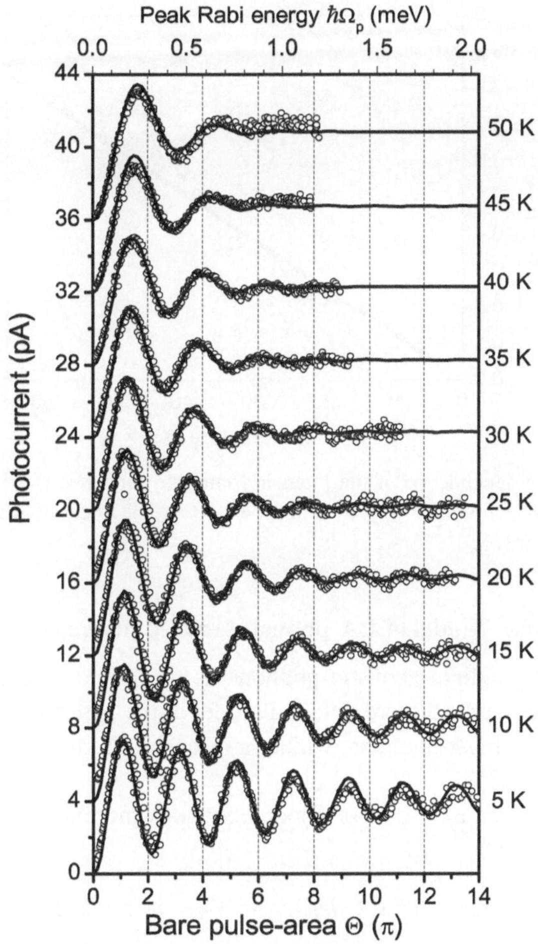

Figure 6.5 Rabi rotation measurement at various temperatures. The solid lines show fits to the LA-phonon model presented in Section 6.5. The top axis indicates the Rabi energy at the peak of the laser pulse. Figures adapted from Ref. [24].

where D_e, D_h are the deformation potential constants of the electron and hole respectively; $\mu = 5.37$ gcm^{-3} is the mass density, and $c_s = 5.11$ nmps^{-1} is the angle-averaged speed of sound in bulk GaAs [17]. $\mathcal{P}[\psi_{e,h}] = \int d^3\mathbf{x}|\psi(x)|^2 e^{i\mathbf{q}\cdot\mathbf{x}}$ are the form factors of the carrier–phonon interaction. In the weak-driving-field regime considered in this section, the form factors are approximately equal to one. From Eq. (6.2) we infer a value of $|D_e - D_h|_{eff} = 9.8$ eV, which is similar to the total hydrostatic deformation potential constant of the GaAs bandgap of -8.5 eV [35]. The close quantitative agreement between the measured gradient and a calculation using literature values for bulk GaAs material parameters implies that other mechanisms must be weak.

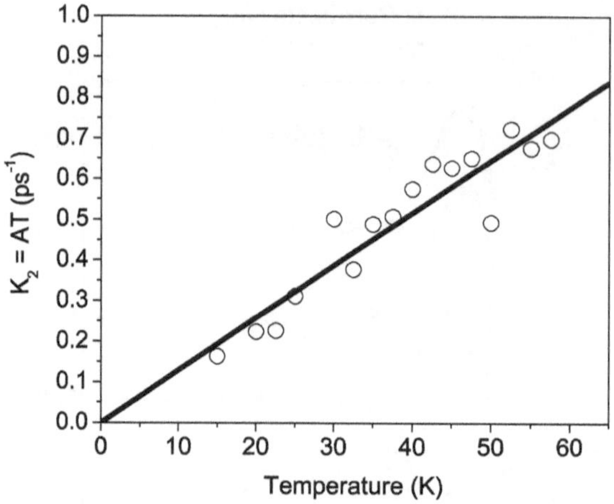

Figure 6.6 Temperature dependence of the intensity damping time K_2. Here K_2 is extracted from fits to data in Fig. 6.5 assuming $\Gamma_2^* = K_2\Omega^2$, and for $\Theta < 6\pi$. The linear dependence is the signature of an LA-phonon mechanism.

6.5 Model of LA-phonon induced dephasing

Having identified longitudinal acoustic phonons as the primary source of the intensity damping, we now summarize a more-detailed model of the optically driven two-level system coupled to a bath of LA-phonons, which is discussed in more detail in ref. [20] and the supplement of Ref. [25] .

The model is expressed by a pair of Bloch equations modified by the carrier–phonon interaction:

$$\dot{s}_y = \{\Omega + \Im[K(\Omega, T)]\}s_z - \{\Gamma_2^* + \Re[K(\Omega, T)]\}s_y \tag{6.3}$$

$$\dot{s}_z = -\Omega s_y. \tag{6.4}$$

The crystal ground and neutral exciton states form a two-level system described by a Bloch vector $\mathbf{s} = (s_x, s_y, s_z)$, where s_x, s_y are the real and imaginary parts of the excitonic dipole in the rotating frame of the laser, and s_z represents the population inversion. The photocurrent is a measure of the final occupation of the upper state: $PC \propto (1 + s_z(\infty))/2$. The initial state is $\mathbf{s} = (0, 0, -1)$. The Bloch equations for a two-level system driven on resonance at a Rabi frequency are modified by a driving field rate of dephasing $\Re[K(\Omega)]$, and a shift to the Rabi rotation frequency $\Im[K(\Omega)]$ due to the fluctuations in the phonon bath. To satisfy causality, $\Re[K(\Omega)]$ and $\Im[K(\Omega)]$ are connected by a Kramers–Kronig relationship. Γ_2^* describes a phenomenological, driving field independent, rate of pure dephasing to account for electron tunneling and other dephasing mechanisms.

Equations (6.3, 6.4) are derived using a weak-coupling Born–Markov approximation [20] that treats the exciton–phonon interaction $H_{e-ph} = \frac{1}{2}(1+\hat{s}_z)\sum_{\mathbf{q}}(g_{\mathbf{q}}b^{+}\mathbf{q}+g_{\mathbf{q}}^{*}b_{\mathbf{q}})$ as a second-order perturbation of the optically dressed states, where $b_{\mathbf{q}}^{+}$, $b_{\mathbf{q}}$ are the creation and

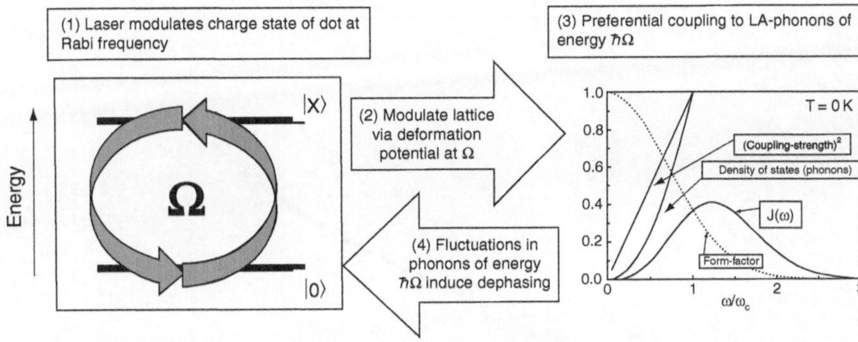

Figure 6.7 An illustration of excitation induced dephasing. (1) Charge state of dot is modulated at Rabi frequency. (2) This modulates the lattice at the Rabi frequency via the deformation potential coupling. (3) Preferential coupling to phonons resonant with the Rabi energy leads to a driving field dependent rate of dephasing. (4) This acts on the exciton transition intensity damping the Rabi rotation.

annihilation operators for an LA-phonon of wave vector $\mathbf{q}$. The exciton–phonon response function $K(\Omega) = \int_0^\infty dt \tilde{K}(t) e^{i\Omega t}$, where:

$$\tilde{K}(t) = \int_0^\infty d\omega J(\omega) \coth(\frac{\hbar\omega}{2k_B T}) \cos(\omega t) \tag{6.5}$$

$$J(\omega) = \sum_{\mathbf{q}} |g_{\mathbf{q}}|^2 \delta(\omega - \omega_{\mathbf{q}}). \tag{6.6}$$

In the following, we assume the LA-phonons have a linear, isotropic dispersion relationship $\omega_{\mathbf{q}} = c_s|\mathbf{q}|$, and that the electron and hole have identical spherical wavefunctions $\psi \propto e^{-x^2/2d^2}$. In this case the form factor can be approximated by $\mathcal{P}(\omega) = e^{-\omega^2/2\omega_c^2}$, where the cut-off energy $\hbar\omega_c = \hbar c_s \sqrt{2}/d$ characterizes the bandwidth of the carrier phonon interaction. If we assume different sized asymmetric wavefunctions for the electron and hole, then for a wide range of parameters, the form factor can still be approximated by a Gaussian with a modified cut-off energy. To avoid over specifying the fitting problem, we therefore treat the wavefunctions as identical, and $J(\omega) \approx \frac{\hbar A}{\pi k_B} \omega^3 e^{-\omega^2/\omega_c^2}$.

A physical interpretation of the model is shown schematically in Fig. 6.7. The laser drives the charge state of the dot at the Rabi frequency, which in turn modulates the lattice via the deformation potential resulting in resonant coupling to phonons of an energy equal to the Rabi energy. This results in a rate of dephasing proportional to the density of states, the coupling strength squared (second-order process), and the thermal occupation of phonons of an energy equal to the Rabi energy. This is described by the driving-field-dependent function $\Re[K(\Omega)]$. In order to satisfy causality, the driving-field-dependent rate of dephasing is accompanied by a driving-field and temperature-dependent shift to the Rabi frequency $\Im[K(\Omega)]$. As the driving-field is increased, the contrast of the Rabi rotations provides a probe of the spectral density of the carrier–phonon interaction at the Rabi energy, providing information on the phononic environment of the dot.

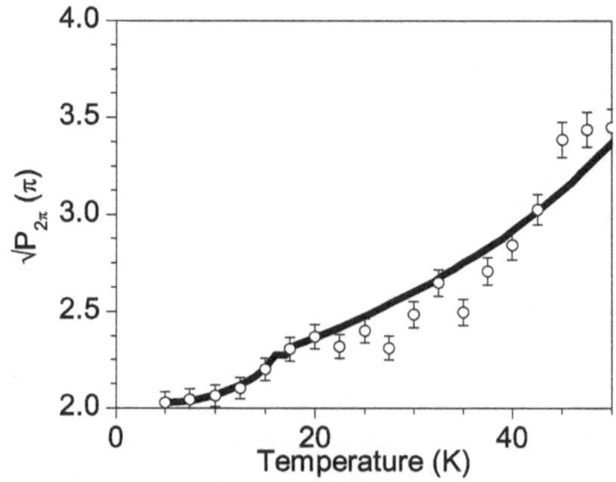

Figure 6.8 Temperature dependence of the period. (o) Square-root of the power needed to reach the first minimum $\sqrt{P}_{2\pi}$, which provides a measure of the period, versus the temperature. A calculation of the pulse area required to reach the first minimum using the carrier–phonon interaction deduced from the fits in Fig. 6.5 is shown by the solid line, assuming the dot–laser coupling a is constant. Figures adapted from Ref. [24].

6.6 Comparison of experiment to model

To extract information on the spectral density of the carrier–phonon interaction we fit the data in Fig. 6.5 to numerical solutions of Eqs. (6.3, 6.4). The A-parameter and cut-off energy $\hbar\omega_c$ are used as global parameters in a fit to the entire temperature dependent set of data. Since the optical field at the dot is not known with any precision, the conversion a between the incident power and the pulse area is unknown, $\Theta = a\sqrt{P}$. Therefore, for each Rabi rotation measurement, the data are scaled in the pulse-area axis to minimize the error at each temperature. $\Gamma_2^* = 0.025$ ps^{-1} is fixed, and A and $\hbar\omega_c$ are deduced from the shape of the Rabi rotations. Note that the shape of each Rabi rotation is described by two fitting parameters and three qualitative features: the strength of the damping, the cut-off in the damping, and the nonlinear dependence of the rotation angle on the driving field. Good fits to data are achieved over the full range of the data, reinforcing the model.

From fits to Fig. 6.5 using the full model, we extract values of $A = 10.8-11.9$ fs K^{-1} and $\hbar\omega_c = 1.37-1.46$ meV. As a result of fitting to the full model, the value of A is slightly revised compared to $A^{(\Omega^2)}$. The cut-off energy $\hbar\omega_c$ corresponds to a spherical [10] wavefunction of $\psi(x) \propto e^{-x^2/2d^2}$, with $d = \sqrt{2}c_s/\omega_c = 3.25 - 3.46$ nm. This corresponds to a probability distribution of $\sim$5.5nm FWHM, which is reasonable for an InAs/GaAs quantum dot.

Having deduced the carrier–phonon interaction from fits to the shape of the Rabi rotations, Fig. 6.8 compares the temperature dependence of the period to the model. The period is parameterized by the square-root of the power needed to reach the first minimum: $\sqrt{P}_{2\pi}$. The solid line is calculated for a fixed value of a using the values of A and $\hbar\omega_c$ deduced from the fits in Fig. 6.5. There is close agreement between the observed period and the calculation. Therefore the observed temperature dependence of the period is fully described by the same carrier–phonon interaction as the damping.

6.7 Outlook

The physics of a driven two-level system coupled to a boson bath is studied by using a laser-driven quantum dot exciton transition coupled to a reservoir of LA-phonons. In the rotating frame of the laser, the driving field couples the crystal ground state and the exciton leading to new eigenstates separated by the Rabi splitting. Phonon mediated relaxation between the Rabi-split optically dressed states results in an additional Rabi-energy-dependent rate of dephasing, and a renormalization of the Rabi splitting. The observed shift in the Rabi energy is the result of the thermal and vacuum fluctuations in the phonon bath, an effect that is analogous to the Lamb shift that occurs in atomic hydrogen due to the interaction with vacuum photons. The Rabi rotation measurements provide a way of probing the spectral density of the exciton–phonon interaction $J(\omega)$, and based on the model, we can deduce that the phonon fluctuations induce a 4.5% decrease in the effective Rabi energy at absolute zero. For Rabi energies of 1 meV, corresponding to a 2π pulse with 4 ps time duration, the phonon-induced dephasing can be fierce, with an effective dephasing time of ≈ 10 ps at zero temperature. This illustrates the importance of designing control fields that minimize the impact of the LA-phonons in order to achieve high-fidelity coherent optical control of excitons, or spins via a charged excitons on a picosecond timescale, as has been explored theoretically in a number of works, for example Refs. [10, 11, 13]. Other driven solid-state qubits should suffer similar dephasing processes, such as superconducting qubits [40], and the optical transitions of NV-centers. A similar system is a dot strongly coupled to a cavity photon interacting with a bath of acoustic phonons, where phonon-assisted cavity feeding effects [12] and a temperature dependent shift in the vacuum Rabi splitting [21] have been observed.

Acknowledgements

The work reviewed in this chapter was the result of team work. We acknowledge experimental assistance from Tim Godden, Stephen Boyle and Achanta Venu Gopal; theory by Erik Gauger, Ahsan Nazir and Brendon Lovett; and growth by Hui-Yun Liu. We acknowledge the advice and support of Maurice Skolnick. The work was funded by the EPSRC UK, the QIPIRC, and the UKIERI.

References

[1] Besombes, L., Kheng, K., Marsal, L. and Mariette, H. 2001. Acoustic phonon broadening mechanism in single quantum dot emission. *Phys. Rev. B*, **65**, 155307.

[2] Borri, P., Langbein, W., Schneider, S. *et al.* 2001. Ultralong dephasing time in InGaAs quantum dots. *Phys. Rev. Lett.*, **87**, 157401.
[3] Borri, P., Langbein W., Woggon U., Stavarache V., Reuter D., and Wieck, A. D. 2005. Exciton dephasing via phonon interactions in InAs quantum dots: Dependence on quantum confinement. *Phys. Rev. B*, **71**, 115328.
[4] Boyle, S. J., Ramsay, A. J., Bello, F. *et al.* 2008. Two-qubit conditional quantum-logic operation in a single self-assembled quantum dot. *Phys. Rev. B*, **78**, 075301.
[5] Boyle, S. J., Ramsay, A. J., Fox, A. M. *et al.* 2009. Beating of exciton-dressed states in a single semiconductor InGaAs/GaAs quantum dot. *Phys. Rev. Lett.*, **102**, 207401.
[6] Brunner, D., Gerardot, B. D., Dalgarno P. A. *et al.* 2009. A coherent single-hole spin in a semiconductor. *Science*, **325**, 70–72.
[7] Favero, I., Cassabois, G., Ferreira, R. *et al.* 2003. Acoustic phonon sidebands in the emission line of single InAs/GaAs quantum dots. *Phys. Rev. B*, **68**, 233301.
[8] Flagg, E. B., Muller, A., Robertson, J. W. *et al.* 2009. Resonantly driven coherent oscillations in a solid-state quantum emitter. *Nature Phys.*, **5**, 203–207.
[9] Förstner, J., Weber, C., Danckwerts, J. and Knorr, A. 2003. Phonon-assisted damping of Rabi oscillations in semiconductor quantum dots. *Phys. Rev. Lett.*, **91**, 127401.
[10] Gauger, E. M., Benjamin, S. C., Nazir, A. and Lovett, B. W. 2008. High-fidelity all-optical control of quantum dot spins: Detailed study of the adiabatic approach. *Phys. Rev. B*, **77**, 115322.
[11] Hodgson, T. E., Viola, L. and D'Amico, I. 2008. Decoherence-protected storage of exciton qubits through ultrafast multipulse control. *Phys. Rev. B*, **78**, 165311.
[12] Hohenester, U., Laucht, A., Kaniber, M. *et al.* 2009. Phonon-assisted transitions from quantum dot excitons to cavity photons. *Phys. Rev. B*, **80**, 201311.
[13] Hohenester, U. and Stadtler, G. 2004. Quantum control of electron–phonon scatterings in artificial atoms. *Phys. Rev. Lett.*, **92**, 196801.
[14] Kamada, H., Gotoh, H., Temmyo, J., Takaghara, T. and Ando, H. 2001. Exciton Rabi oscillation in a single quantum dot. *Phys. Rev. Lett.*, **87**, 246401.
[15] Kolodka, R. S., Ramsay, A. J., Skiba-Szymanska, J. *et al.* 2007. Inversion recovery of single quantum-dot exciton based qubit. *Phys. Rev. B*, **75**, 193306.
[16] Krügel, A., Axt, V. M., Kuhn, T., Machnikowski, P. and Vagov, A. 2005. The role of acoustic phonons for Rabi oscillations in semiconductor quantum dots. *Appl. Phys. B*, **81**, 897–904.
[17] Krummheuer, B., Axt, V. M. and Kuhn, T. 2002. Theory of pure dephasing and the resulting absorption line shape in semiconductor quantum dots. *Phys. Rev. B*, **65**, 195313.
[18] Machnikowski, P. and Jacak, L. 2004. Resonant nature of phonon-induced damping of Rabi oscillations in quantum dots. *Phys. Rev. B*, **69**, 193302.
[19] Melet, R., Voliotis, V., Enderlin, A. *et al.* 2008. Resonant excitonic emission of a single quantum dot in the Rabi regime. *Phys. Rev. B*, **78**, 073301.
[20] Nazir, A. 2008. Photon statistics from a resonantly driven quantum dot. *Phys. Rev. B*, **78**, 153309.
[21] Ota, Y., Iwamoto, S., Kumagai, N. and Arakawa, Y. 2009. Impact of electron-phonon interactions on quantum-dot cavity quantum electrodynamics. *ArXiv, 0908.0788*.
[22] Patton, B., Woggon, U. and Langbein, W. 2005. Coherent control and polarization readout of individual excitonic states. *Phys. Rev. Lett.*, **95**, 266401.

[23] Ramsay, A. J., Boyle, S. J., Kolodka, R. S. *et al.* 2008. Fast optical preparation, control, and readout of a single quantum dot spin. *Phys. Rev. Lett.*, **100**, 197401.

[24] Ramsay, A. J., Godden, T. M., Boyle, S. J. *et al.* 2010a. Phonon-induced Rabi-frequency renormalization of optically driven single InGaAs/GaAs quantum dots. *Phys. Rev. Lett.*, **105**, 177402.

[25] Ramsay, A. J., Achanta, V. G., Gauger, E. M. *et al.* 2010b. Damping of exciton Rabi rotations by acoustic phonons in optically excited InGaAs/GaAs quantum dots. *Phys. Rev. Lett.*, **104**, 017402.

[26] Stievater, T. H., Li, X., Steel, D. G. *et al.* 2001. Rabi oscillations of excitons in single quantum dots. *Phys. Rev. Lett.*, **87**, 133603.

[27] Stufler, S., Ester, P., Bichler, M. and Zrenner, A. 2004. Power broadening of the exciton linewidth in a single InGaAs/GaAs quantum dot. *Appl. Phys. Lett.*, **85**, 4202–4204.

[28] Stufler, S., Ester, P., Zrenner, A. and Bichler, M. 2005. Quantum optical properties of a single $In_xGa_{1-x}As$/GaAs quantum dot two-level system. *Phys. Rev. B*, **72**, 121301(R).

[29] Stufler, S., Machnikowski, P., Ester, P. *et al.* 2006. Two-photon Rabi oscillations in a single $In_xGa_{1-x}As$/GaAs quantum dot. *Phys. Rev. B*, **73**, 125304.

[30] Takagi, H., Nakaoka, T., Watanabe, K., Kumagai, N. and Arakawa, Y. 2008. Coherently driven semiconductor quantum dot at a telecommunication wavelength. *Optics Express*, **16**, 13 949–13 954.

[31] Vagov, A., Axt, V. M., Kuhn, T. *et al.* 2005. Nonmonotonous temperature dependence of the initial decoherence in quantum dots. *Phys. Rev. B*, **70**, 201305.

[32] Vagov, A., Croitoru, M. D., Axt, V. M., Kuhn, T. and Peeters, F. M. 2007. Nonmonotonic field dependence of damping and reappearance of Rabi oscillations in quantum dots. *Phys. Rev. Lett.*, **98**, 227403.

[33] Vasanelli, A., Ferreira, R. and Bastard, G. 2002. Continuous absorption background and decoherence in quantum dots. *Phys. Rev. Lett.*, **89**, 216804.

[34] Villas-Bôas, J. M., Ulloa, S. E. and Govorov, A. 2005. Decoherence of Rabi oscillations in a single quantum dot. *Phys. Rev. Lett.*, **94**, 057404.

[35] Vurgaftman, I., Meyer, J. R. and Ram-Mohan, L. R. 2001. Band parameters for III?V compound semiconductors and their alloys. *J. Appl. Phys.*, **89**, 5815–5875.

[36] Wang, Q. Q., Muller, A., Bianucci, P. *et al.* 2005a. Decoherence processes during optical manipulation of excitonic qubits in semiconductor quantum dots. *Phys. Rev. B*, **72**, 035306.

[37] Wang, Q. Q., Muller, A., Bianucci, P. *et al.* 2006. Internal and external polarization memory loss in single semiconductor quantum dots. *Appl. Phys. Lett.*, **89**, 142112.

[38] Wang, Q. Q., Muller, A., Cheng, M. T. *et al.* 2005b. Coherent control of a V-type three-level system in a single quantum dot. *Phys. Rev. Lett.*, **95**, 187404.

[39] Weiner, A. M. 2000. Femtosecond pulse shaping using spatial light modulators. *Rev. Sci. Instrum.*, **71**, 1929–1960.

[40] Wilson, C. M., Duty, T., Persson, F. *et al.* 2007. Coherence times of dressed states of a superconducting qubit under extreme driving. *Phys. Rev. Lett.*, **98**, 257003.

[41] Würger, A. 1998. Strong-coupling theory for the spin-phonon model. *Phys. Rev. B*, **57**, 347–361.

[42] Xu, X., Sun, B., Berman, P. R. *et al.* 2007. Coherent optical spectroscopy of a strongly driven quantum dot. *Science*, **317**, 929–932.

[43] Zrenner, A., Beham, E., Stufler, S. *et al.* 2002. Coherent properties of a two-level system based on a quantum-dot photodiode. *Nature*, **418**, 612–614.

7

Holes in quantum dot molecules: structure, symmetry, and spin

M. F. Doty and J. I. Climente

7.1 Introduction

The spin projections of single electrons and holes confined in quantum dots (QDs) provide a natural two-level system that can serve as the logical basis for both classical and quantum information processing devices. When two layers of self-assembled InGaAs QDs are grown sequentially, strain propagation causes QDs in the two layers to align along the growth direction. Coherent tunneling of either electrons or holes between the two QDs leads to a variety of Coulomb and spin interactions with possible applications in optoelectronic and logic devices [4, 5, 6, 7, 8, 9, 10, 11, 12, 13, 14, 15]. Because coherent tunneling leads to the formation of delocalized molecular states, these vertically stacked pairs of QDs have come to be known as quantum dot molecules (QDMs). One of the surprising discoveries about QDMs was that the delocalized molecular states have their own unique and tunable properties [16]. In this chapter we review the formation of delocalized molecular states of holes in QDMs and consider how the structure and symmetry of the QDM influence spin properties. Results have been obtained for molecular states charged with one, two, and three holes [9, 13, 17]. We focus here on molecular states occupied by a single hole whose spin projections could serve as the basis for optoelectronic logic devices.

Hole spins were initially discounted for spin-based devices because the complex valence-band interactions were anticipated to degrade spin storage or decoherence times. However, recent experiments have demonstrated that holes have lifetimes and decoherence times competitive with electrons [18, 19, 20]. Moreover, holes have a significantly weaker hyperfine interaction with nuclei, which suppresses a primary decoherence mechanism [21, 22, 23, 24, 25, 26, 27]. As described here, the unique properties of holes also lead to novel interactions that can be used to engineer the symmetry of molecular states and spin mixing interactions [2, 3].

Quantum Dots: Optics, Electron Transport and Future Applications, ed. Alexander Tartakovskii. Published by Cambridge University Press.

7.2 Growth and spectroscopic characterization of QDMs

We briefly summarize the growth and spectroscopic characterization of InAs QDMs, focusing on the concepts required to understand the molecular and spin properties of single holes. A detailed review of QDM growth and spectroscopic characterization can be found in Chapters 2 and 10 of *Single Semiconductor Quantum Dots*, edited by Peter Michler [28].

InAs QDs are formed by Stranski–Krastanov growth during molecular beam epitaxial deposition of InAs on GaAs. Our QDs are truncated to create pancake-shaped QDs of defined thickness [29, 30, 31]. The truncation provides the strongest confinement and controls the center emission wavelength of the QD ensemble. To make QDMs, a barrier layer (usually GaAs) of defined thickness is grown on top of the layer of truncated QDs. The strain induced by the buried InAs QD creates preferential nucleation sites for a second layer of QDs [32, 33, 34, 35, 1] that are again truncated to a specific height. Using this technique, it is possible to grow two dots that are stacked on top of one another in the growth direction with individual control over both the truncation height of each dot and the barrier thickness.

Even if both layers of QDs are grown under nominally identical conditions, strain and asymmetry generally lead to nondegenerate dots with slightly different energy levels. A schematic depiction of a QDM where the dots have slightly different sizes is shown in Fig. 7.1a and can be compared to a cross-sectional scanning tunneling microscope image of such a QDM in Fig. 7.1b. The difference in confined energy levels is shown by the schematic band diagram of Fig. 7.1c. The size asymmetry depicted in Fig. 7.1a causes the top QD to have higher confined energy levels. As a result, embedding this QDM into a diode structure on an n-type substrate allows the hole energy levels to be tuned into resonance *in situ* using an electric field applied in the growth direction (Fig. 7.1d). When the hole energy levels are resonant, holes can tunnel between the two dots. Because the dots have nondegenerate energy levels, the electron energy levels are detuned from resonance at the electric field of maximal hole tunneling.

The inhomogeneous distribution of photoluminescence (PL) energies from an ensemble of QDMs has a full-width half maximum of order 100 meV. This distribution originates primarily in statistical fluctuations in the QD size and degree of alloying that occur during self-assembly. To resolve spin fine structure we cover the sample surface with a metal mask that has small apertures isolating, on average, a single QDM. The sample is illuminated with a continuous wave laser tuned anywhere from above the GaAs bandgap to somewhat below the InAs wetting layer. Optically injected electron–hole pairs relax into the low-energy states of the QDM before recombining to emit a photon. The emitted PL is collected, dispersed with a monochromator, and measured with a charge coupled device (CCD) camera. The PL spectra are sequentially accumulated at various values of the applied electric field and assembled to plot PL intensities as a function of both applied electric field (x axis) and energy (y axis). These spectral maps generally contain PL from several charge states. We will focus here on the molecular properties of the neutral exciton state, which contains only one electron and one hole.

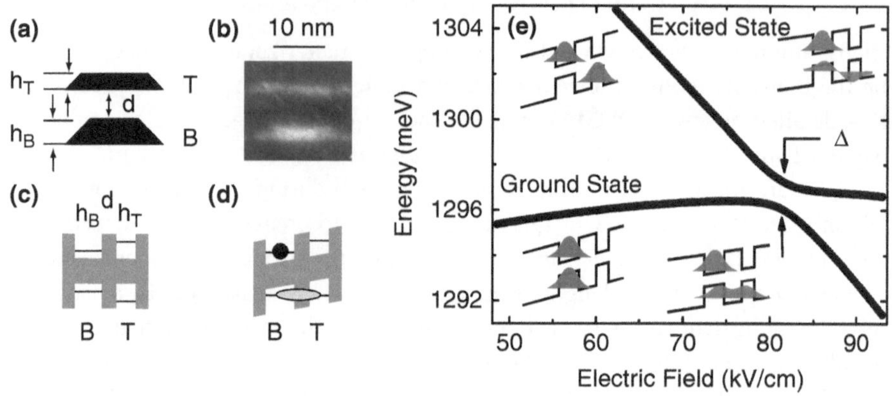

Figure 7.1 (a) Schematic depiction of a QDM with nondegenerate dots, where h_B (h_T) is the height of the bottom (top) dot, and d is the thickness of the barrier. (b) Cross-sectional STM image of such a QDM. (c, d) Schematic band diagram showing the relative energy levels of nondegenerate dots in the absence (c) and presence (d) of a resonant electric field. (e) Energies of observed photoluminescence lines for the neutral exciton show an anticrossing as a function of applied electric field due to holes tunneling through a 2 nm barrier. The QDs are both nominally 2.5 nm high and embedded in an n-type diode. Insets are described in the text. Figures adapted from [1] and [2].

7.3 Anticrossings and the formation of molecular states

The asymmetry in energy levels described in Section 7.2 causes an electron confined in the QDM to rapidly relax into the bottom QD, which is the lowest electron energy state available (Fig. 7.1d). To describe the possible charge configurations for the neutral exciton state, we therefore need only to consider the spatial location of the hole. Note that $\binom{1,0}{1,0}$ labels the case when both the electron and hole are in the bottom dot, and $\binom{1,0}{0,1}$ labels the case when the hole is in the top dot. The PL line emitted as a result of excitonic recombination from the $\binom{1,0}{1,0}$ state is called a 'direct' exciton. Direct excitons have a weak energy dependence on the applied electric field because the electron–hole dipole moment inside a single dot is small. Thus, direct excitons appear in spectral maps as nearly horizontal lines. In contrast, 'indirect' excitons emitted from the $\binom{1,0}{0,1}$ state have a strong energy dependence on applied electric field (F) and appear in spectral maps as diagonal lines. The slope of the line is determined by the thickness of the barrier (d) separating the two dots: $\Delta E = \Delta F e d$, with e being the charge of the electron. The dipole matrix element for PL transitions is proportional to the overlap of the electron and hole wavefunctions and is significantly reduced for indirect exciton states because the electron and hole wavefunctions are centered on different QDs.

In Fig. 7.1e we plot the energy of direct and indirect PL lines emitted by a QDM in which the QDs are separated by a 2 nm GaAs barrier. The direct PL line is the nearly horizontal line at approximately 1296 meV, and the spatial distribution of electron and hole wavefunctions for the direct state is schematically indicated by the lower inset on the left side of Fig. 7.1e. The indirect line is the diagonal line that begins at approximately 1304 meV at 62 kV/cm and decreases to 1292 meV at 92 kV/cm. The spatial distribution of wavefunctions for the indirect state is schematically depicted in the inset at the top left of Fig. 7.1e. Because the ground state after optical recombination contains no electrons or holes, the energy levels of the neutral exciton states map directly onto the energies of PL emitted.

In the absence of interactions between the QDs, the direct and indirect PL lines would be degenerate when the applied electric field tunes the hole energy levels to be equal, approximately 81 kV/cm in Fig. 7.1e. It is clear from Fig. 7.1e that the lines do not cross, but instead anticross. This avoided crossing arises because coherent tunneling forms molecular orbitals delocalized over both dots and the intervening barrier [7]. These molecular states are constructed from the symmetric and antisymmetric combination of the basis states, i.e. $\frac{1}{\sqrt{2}}\left(\binom{1,0}{1,0}+\binom{1,0}{0,1}\right)$ and $\frac{1}{\sqrt{2}}\left(\binom{1,0}{1,0}-\binom{1,0}{0,1}\right)$. The insets on the right side of Fig. 7.1e illustrate the formation of the symmetric and antisymmetric molecular hole states at the anticrossing. The magnitude of the anticrossing, which is the energy separation between the symmetric and antisymmetric states, is indicated by Δ.

We model the anticrossing of the neutral exciton with a matrix Hamiltonian in the basis $\binom{1,0}{1,0}$, $\binom{1,0}{0,1}$. The Hamiltonian is

$$\widehat{H}^{X^0} = \begin{pmatrix} E_0 & -t_{X^0} \\ -t_{X^0} & E_0 - edF \end{pmatrix} \tag{7.1}$$

with E_0 the exciton energy and t_{X^0} the tunneling matrix element. Note that t_{X^0} is determined by the overlap of the hole wavefunctions in the two dots and depends on barrier thickness. The term $-edF$ captures the change in energy with applied electric field (F) for the hole in the top dot separated from the bottom dot by an effective barrier thickness d. A key insight provided by this model is that the states of QDMs evolve continuously between atomic-like and molecular orbitals as a function of applied electric field. Away from resonance (large magnitude of F), the tunnel coupling is much smaller than the Stark shift (edF) and the eigenvalues of Eqn (7.1) are essentially just the energies of the two basis states. This is the atomic-like or Heitler–London limit [36]. On resonance ($F = 0$), the eigenvalues of Eqn (7.1) are $E_0 - t_{X^0}$ and $E_0 + t_{X^0}$. The eigenstates that correspond to these values are, respectively, the symmetric (bonding) and antisymmetric (antibonding) combination of the two basis states. In analogy with natural molecules, one would expect the bonding orbital to have lowest energy. This is the case when t_{X^0} is positive. The formation of anticrossings in PL spectra (e.g. Fig. 7.1e) demonstrates the formation of molecular states and allows the magnitude of t_{X^0} to be measured ($\Delta = 2|t_{X^0}|$). However, the sign of t_{X^0} and the orbital character of the molecular states cannot be determined from data taken at zero magnetic field. To avoid ambiguity, we refer to the lowest-energy molecular state as the molecular ground state and the higher-energy state as the molecular excited state.

As we show in Section 7.4.1, the molecular orbital character of the molecular ground and excited states can be determined using magneto-optical PL spectroscopy.

7.4 Spin projections and Zeeman splitting

We now extend our notation to explicitly list the spin projections of the electron and hole. Electrons have spin $\pm 1/2$, denoted as $\uparrow$ and $\downarrow$. Holes experience a sizable spin–orbit (SO) interaction because they are derived from p-type atomic orbitals of the semiconductor lattice [37]. The SO interaction couples the hole's atomic orbital and spin degrees of freedom to give a total (Bloch) angular momentum $J = 3/2$. The hole states are then described by four-component spinors, with $J_z = \pm 3/2$ projections corresponding to heavy holes (HHs) and $J_z = \pm 1/2$ to light holes (LHs). In single QDs, LHs are shifted up in energy by confinement and strain, and the only significant component of the spinor for low-energy holes is the HH $J_z = \pm 3/2$, which we denote in the pseudo-spin basis $\Uparrow$ and $\Downarrow$. In QDMs, the influence of LH components is not always negligible and can significantly impact the nature of the tunnel coupling. We will return to this point in Section 7.5.

Just as in single QDs, optical selection rules require that recombining electrons and holes have opposite spin projections to conserve angular momentum. There are thus two optically bright direct neutral excitons $\left(\begin{pmatrix}\uparrow,0\\ \Downarrow,0\end{pmatrix}\right.$ and $\left.\begin{pmatrix}\downarrow,0\\ \Uparrow,0\end{pmatrix}\right)$ and two optically dark direct excitons $\left(\begin{pmatrix}\uparrow,0\\ \Uparrow,0\end{pmatrix}\right.$ and $\left.\begin{pmatrix}\downarrow,0\\ \Downarrow,0\end{pmatrix}\right)$. The bright and dark indirect excitons are analogous. When the electron and hole are confined in the same dot (direct exciton), they experience an electron–hole exchange interaction J^{eh}. This interaction breaks the energetic degeneracy of bright and dark exciton states. J^{eh} depends on the overlap of electron and hole wavefunctions and is drastically suppressed when the electron and hole are in separate dots. To include the spin projections and electron–hole exchange interactions, Eqn (7.1) must be expanded. Using the basis:

$$\begin{pmatrix}\downarrow & 0\\ \Uparrow & 0\end{pmatrix} \begin{pmatrix}\downarrow & 0\\ 0 & \Uparrow\end{pmatrix} \begin{pmatrix}\downarrow & 0\\ \Downarrow & 0\end{pmatrix} \begin{pmatrix}\downarrow & 0\\ 0 & \Downarrow\end{pmatrix} \tag{7.2}$$

the neutral exciton (X^0) Hamiltonian can be written as (using $\delta = \mu_B B$):

$$\begin{matrix} J^{eh} + \frac{\delta(g_e + g_{hB})}{2} & -t_{X^0} + \frac{\delta g_{12}}{2} & 0 & -hm \\ -t_{X^0} + \frac{\delta g_{12}}{2} & -edF + \frac{\delta(g_e + g_{hT})}{2} & hm & 0 \\ 0 & hm & -J^{eh} + \frac{\delta(g_e - g_{hB})}{2} & -t_{X^0} - \frac{\delta g_{12}}{2} \\ -hm & 0 & -t_{X^0} - \frac{\mu_B B g_{12}}{2} & -edF + \frac{\delta(g_e - g_{hT})}{2} \end{matrix} \tag{7.3}$$

Equation (7.3) considers only the states with electron spin down. The full matrix is block diagonal and the states with electron spin up are described by a sub-matrix identical to Eqn (7.3) in the basis with all spins flipped. Equation (7.3) contains all interactions that will be discussed in this chapter. To begin, we focus on the energies of each individual basis state, which appear on the diagonal. Here $\binom{\downarrow,0}{\Uparrow,0}$ has energy J^{eh} because of the antiparallel electron and hole spins. Similarly, $\binom{\downarrow,0}{\Downarrow,0}$ has energy $-J^{eh}$ because of the parallel electron and hole spins. Both $\binom{\downarrow,0}{0,\Uparrow}$ and $\binom{\downarrow,0}{0,\Downarrow}$ have an energy $(-edF)$ that depends on applied electric field because they have a single hole in the top dot. The energies of these states do not depend on the relative spins of the electron and hole because the particles are in separate dots. The off-diagonal matrix elements t_{X^0} give the tunnel coupling. Because tunneling is a spin-conserving process, t_{X^0} only couples states with the same hole spin orientation.

At zero magnetic field, all states are at least two-fold degenerate under a flip of all spins.[1] In a non-zero magnetic field along the optical axis (Faraday geometry), the Zeeman splitting between orthogonal spin projections is given by the g factors for the electron and hole. To include the magnetic field in our model we add Zeeman terms that include B (the magnetic field), μ_B (the Bohr magneton), g_e (the longitudinal electron g factor), and g_h (the longitudinal HH g factor) [38]. We use the HH pseudospin projection and assume that its g factor is the same in both dots. This assumption is well supported by experimental measurements of Zeeman splitting when holes are in each of the dots. The Zeeman terms on the diagonal contain the g factor contributions for each basis state, which depend only on the spin orientations and not on the spatial distribution of carriers. Electron spin up (down) corresponds to $+g_e$ $(-g_e)$. Hole spin up (down) corresponds to $-g_h$ $(+g_h)$. The g factors for electrons and holes in InAs QDs are generally negative [38], so states with net electron spin down and hole spin up are at higher energies; g_{12} is the contribution to the g factor from the amplitude of the wavefunction in the barrier, which will be discussed in Section 7.4.1, and hm is a term mixing the spin of holes localized in opposite QDs as we discuss in Section 7.6.

7.4.1 Resonant changes in g factor

One would expect that the application of magnetic fields would result in two copies of the spectral map separated by a constant Zeeman splitting. We observe, however, that the Zeeman splitting is not constant, but depends on the applied electric field. The suppression and enhancement of the Zeeman splitting for the two molecular states is shown in Fig. 7.2a, which corresponds to a QDM with barrier thickness $d = 2$ nm. The resonant change in Zeeman splitting peaks at exactly the electric field of the anticrossing as measured in data taken at zero magnetic field. At this electric field tunnel coupling is

[1] We neglect anisotropic exchange.

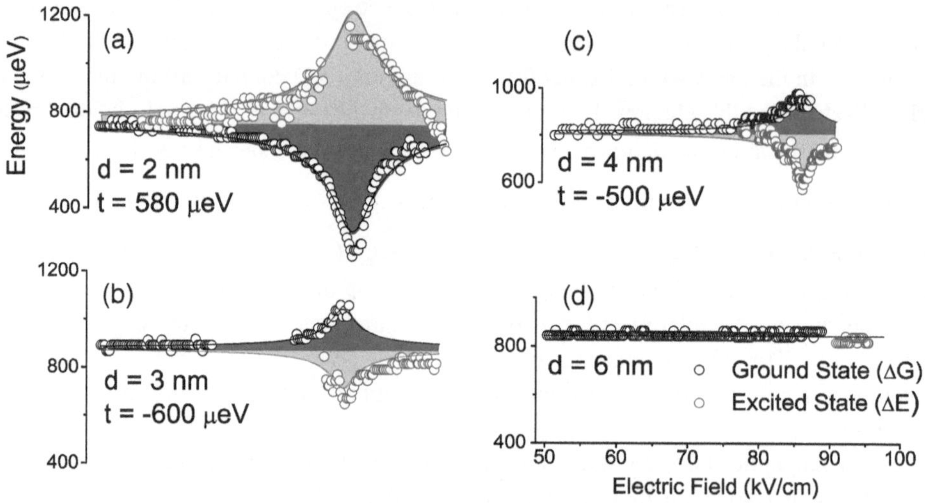

Figure 7.2 Zeeman energy splitting as a function of applied electric field for $B = 6$ T. The QDMs have barrier thicknesses of 2, 3, 4, and 6 nm. The solid curves are calculated with Eqn (7.4) using $(g_e + g_h, h) = (-2.12, 1.32), (-2.57, 0.47), (-2.32, 0.44)$ meV for 2, 3, and 4 nm, respectively. Figure adapted from [2].

maximized and the states have fully molecular character. As we now show, the resonant changes in Zeeman splitting come from barrier contributions to the g factor and allow us to identify the molecular orbital character of the delocalized states formed by tunnel coupling.

The magnitude of Zeeman splitting is directly proportional to the g factor of electrons and holes, as described in Eqn (7.3). Because the g factor depends on material parameters, it is sensitive to the amplitude of the wavefunction in regions comprised of different materials [39, 40]. By measuring far away from any anticrossing, we obtain the electron and hole g factors in the individual InAs dots of the QDM. Obtaining independent values for g_e and g_h requires analysis of spectra taken at several angles of the magnetic field, which is not shown. The values obtained in Fig. 7.2a, $g_e = -0.745$ and $g_h = -1.4$, are consistent with other measurements of InAs dots [38].

At the anticrossing, the formation of molecular states alters the spatial distribution of the hole wavefunction, perturbing the extent of the wavefunction in the GaAs barrier [16]. The g factor for a hole confined in a single InAs QD is negative, but the g factor for holes in bulk GaAs is positive [41]. Resonant changes in the amplitude of the wavefunction in the GaAs barrier consequently have a significant impact on the net hole g factor. Bonding orbitals have a large amplitude in the barrier and thus add a large contribution from the GaAs to the overall hole g factor. This adds a positive component (g_{12}) to the otherwise negative HH g factor and suppresses the Zeeman splitting on resonance. In Fig. 7.2a the low-energy molecular ground states have a suppressed splitting and must therefore have bonding orbital character. In contrast, antibonding states suppress the contributions from the tail of the wavefunction in the barrier, enhance the negative g factor, and result in increased Zeeman splitting on resonance. The increased splitting for the molecular excited states in Fig. 7.2a is a consequence of their antibonding character.

The resonant contribution of the barrier to the g factors of molecular states is captured in the Hamiltonian (Eqn (7.3)) by the term g_{12}. Note that g_{12} is a positive number because it originates in the g factor of a HH in GaAs, though the exact magnitude of g_{12} is expected to differ from the value in bulk GaAs because of additional strain and boundary conditions in the barrier. The sign of g_{12} at each matrix element in the Zeeman Hamiltonian is determined by the spin orientation of the tunneling hole: if the tunneling hole has spin up, the additional matrix element is $-g_{12}$. Equation (7.3) allows us to calculate the energy of each state, including the resonant contributions of the barrier. The calculation results and resonant contribution to the g factor are seen most clearly if we take the difference in energy between Zeeman split molecular states. The Zeeman splitting for the molecular ground state as a function of applied electric field is given by

$$\Delta G(F) = \mu_B B \left(g_e + g_h + \frac{2t_{X0} g_{12}}{\sqrt{e^2 d^2 (F - F_{X0})^2 + 4t_{X0}^2}} \right). \tag{7.4}$$

Here F_{X0} is the electric field of anticrossing, at which the hole states have fully molecular character; t_{X0} and F_{X0} are determined from measured spectral maps of the same QDM at zero magnetic field; g_e and g_h are determined from the Zeeman splitting far from resonance. Because the signs of g_e, g_h and g_{12} are fixed, the sign of t_{X0} determines whether the Zeeman splitting of the ground state is enhanced or suppressed at the resonance. In the 2 nm case, the Zeeman splitting of the molecular ground state decreases at the anticrossing point. Therefore t_{X0} must be positive, which is consistent with the identification that the molecular ground state has bonding character. In contrast, the Zeeman splitting of the molecular excited state is described by an equation identical to Eqn (7.4) except for a minus sign in front of the term containing g_{12} and t_{X0}. A positive t_{X0} therefore leads to an increase in the splitting at resonance, consistent with antibonding character. The fit lines shown in Fig. 7.2a are produced using Eqn (7.4). The excellent agreement between the observed linewidth of the resonant change in g factor and the linewidth calculated using independently measured values of $|t_{X0}|$ provides strong confirmation that the g factor resonance arises from the formation of molecular states with bonding and antibonding orbitals. The Zeeman splitting of the molecular ground and excited states are fit with the same value of $g_{12} = 1.32$. We note that the full Hamiltonian has tunneling matrix elements given by $-t_{X0} \pm \frac{\mu_B B g_{12}}{2}$. This illustrates an important consequence of the resonant barrier contribution to the molecular state g factor: the magnitude of the net tunneling matrix element is spin-dependent. Whether the tunneling rate is enhanced or suppressed depends on the relative signs of t_{X0} and g_{12} [2].

7.5 Antibonding molecular ground states

In Fig. 7.2 we show the resonant increases and decreases in Zeeman splitting for neutral excitons in QDMs with $d = 2, 3, 4$, and 6 nm. Figure 7.2a shows the resonant changes for a QDM with $d = 2$ nm, as discussed above. The theory curves have been filled in to emphasize that the molecular ground state (dark shading) has the resonant decrease in splitting indicative of a bonding orbital. The molecular excited state (light shading) has the resonant increase indicative of an antibonding orbital. As the thickness of the barrier increases (Fig. 7.2b, c, d) the amplitude of the resonance decreases, and is below our noise level for $d = 6$ nm (Fig. 7.2d). This results from the reduction of the amplitude of the wavefunction in the barrier with increasing barrier thickness. There is also a *qualitative* change in the nature of the resonance. When $d = 3$ and 4 nm the molecular ground state (dark shading) shows a resonant *increase* in Zeeman splitting. The resonant increase indicates that the molecular ground state now has antibonding character. This is a remarkable result because antibonding molecular ground states are never observed in natural diatomic molecules or QDMs with resonant electrons.

As discussed in Section 7.3, the energy ordering and symmetry of molecular states in QDMs is given by the off-diagonal matrix element $-t_{X0}$. If t_{X0} becomes negative, the antisymmetric state has lowest energy. In QDMs, t_{X0} becomes negative as the structure of the QDM is changed to include a larger GaAs barrier between the two QDs. This unusual result originates in the spin properties of holes. In QDMs, the low-energy hole spinors are dominated by a single HH pseudospin component. However, the minor LH components, which make up no more than 5% of the spinor weight, have a lighter effective mass and tunnel more efficiently. Consequently LH begin to dominate the tunneling and alter the character of the molecular orbitals for sufficiently thick barriers.

The influence of the minor LH components is apparent from Fig. 7.3. In Fig. 7.3a, we plot the energies of the molecular ground and first excited states for asymmetric QDMs subject to a resonant electric field calculated using a simple one-band, HH effective mass model, which neglects SO interactions. As expected, the energy separation of the bonding (solid line) and antibonding (dashed line) states decreases as a function of increasing barrier thickness and the bonding orbital remains the molecular ground state. In Fig. 7.3b we show the energies of the bonding and antibonding states calculated using a four-band $k \cdot p$ model that includes HH–LH coupling. At $d \sim 1.75$ nm the energies of the bonding and antibonding states cross and the antibonding state becomes the molecular ground state.

In Fig. 7.3c we plot the values for t_0 and t, i.e. half the difference between the state energies given in Figs. 7.3a and 7.3b, respectively. In the absence of SO interaction, the tunneling rate is determined simply by the overlap between the hole orbitals of the individual dots (t_0), which decreases exponentially with increasing barrier thickness at a rate dependent on the HH mass. When the SO interaction is included, there is a correction to the tunneling rate, $t = t_0 - t_{SO}$. This t_{SO} term arises from the small contribution of the LH component of the spinor. The LH component has approximate parity along z opposite to that of the HH component [42]. The LH component therefore adds a small antibonding

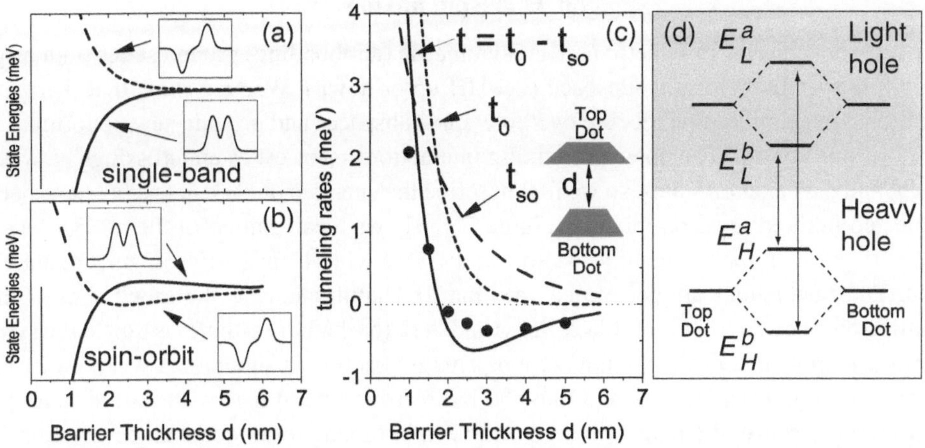

Figure 7.3 (a, b) Energy of the ground and first excited molecular states as a function of d calculated using (a) the single-band effective mass and (b) $k \cdot p$ theory. Scale bars are 5 meV. Insets show the orbital character of the dominant hole spinor component. (c) Tunneling rates of a single hole versus d calculated as described in the text. (d) Schematic depiction of the SO induced mixing between bonding (E^b) and antibonding (E^a) molecular states of the heavy (E_H) and light (E_L) holes. Figure adapted from [2].

(bonding) contribution to the bonding (antibonding) state determined by the dominant HH component, as shown schematically in Fig. 7.3d. The addition of this antibonding component increases the energy of the bonding state and vice versa for the antibonding state. For thin barriers, the t_{SO} correction is small compared to the large t_0, and t remains positive. However, for thicker barriers t_0 decreases faster than t_{SO}, in part because of the LH origin of the latter term. Eventually t_0 becomes smaller than t_{SO}, the net tunneling rate is negative and the antibonding orbital becomes the molecular ground state.

We have verified that this simple four-band $k \cdot p$ approach captures the essential physics of the system by comparison with an atomistic multi-million atom calculation of the hole levels of a QDM described by the $sp^3d^5s^*$ tight-binding model. This approach accounts for strain and changes to the underlying crystal lattice on the atomistic level. The results of this calculation (solid points in Fig. 7.3c) qualitatively match the $k \cdot p$ results. Our theoretical model predicts an antibonding molecular ground state for a barrier thickness $d \geq 1.75$ nm. Experimentally, we find that all examples (7) in the sample with $d = 2$ nm have a molecular ground state with bonding orbital character; in the sample with $d = 4$ nm, all examples (3) have an antibonding ground state. The intermediate case ($d = 3$ nm) has examples of both types of behavior, indicating that the reversal of orbital character occurs near $d = 3$ nm. The small discrepancy with theory most likely arises from details of dot structure and the coexistence of both behaviors at $d = 3$ nm most likely arises from fluctuations. The influence of the split-off band, biaxial strain and position-dependent masses in the $k \cdot p$ description of holes has been shown not to modify the results we present here [43, 44].

7.6 Hole-spin mixing

The resonant changes in g factor and antisymmetric (antibonding) ground states both originate in a specific hole-spin subspace (i.e. HH up or down). We now show that there are additional interactions that mix the two hole-spin subspaces and provide new opportunities for hole-spin control. The hole-spin mixing interactions manifest as anticrossings between bright and dark exciton states, so we first describe the energies of dark states that have been neglected in the discussion thus far. Figure 7.4a shows a calculation of the energy levels and anticrossings for the neutral exciton in a QDM with 4 nm barrier at zero magnetic field. The calculations are performed with matrix Hamiltonians as described above. The splitting between direct bright (black lines) and dark (dashed lines) states is due to the symmetric electron–hole exchange interaction. This interaction is suppressed when electrons and holes are in separate dots and thus the indirect bright and dark states are degenerate [13]. In the absence of a magnetic field or hole spin mixing, orthogonal electron and hole spin projections are degenerate. The two bright exciton spin configurations are thus degenerate and only the electron spin down configuration is labeled in Fig. 7.4a. In the figure, two anticrossings can be observed where direct and indirect excitons would be expected to intersect, one for bright excitons (upper anticrossing) and another for dark excitons (lower anticrossing). In typical experimental spectra, only the bright states are evident.

In Fig. 7.4b we show the calculated energies of the bright states from Fig. 7.4a when a magnetic field of 6 T is applied. The first effect of the magnetic field is a Zeeman splitting of the two bright states that were degenerate in Fig. 7.4a. This is seen most clearly at the edges of Fig. 7.4b: the degenerate lines at +0.1 meV in Fig. 7.4a split and move to −0.3 meV and +0.5 meV. The second effect of the magnetic field is the introduction of a g factor resonance, as described in Section 7.4.1, which is seen most clearly by looking

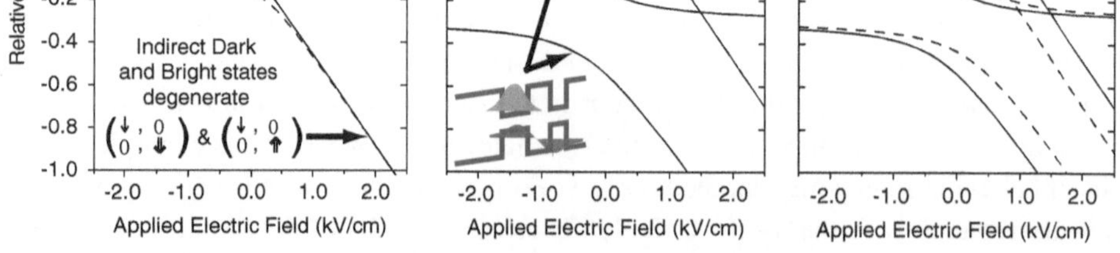

Figure 7.4 (a) Energy levels of bright and dark exciton states in a QDM at B = 0 T. (b,c) Energy levels of bright states (b) and bright and dark states (c) at B = 6 T. Solid (dashed) lines indicate optically bright (dark) exciton states. Figure taken from [3].

at the anticrossings. The degenerate anticrossings in Fig. 7.4a have an anticrossing gap of 214 μeV. In Fig. 7.4b, the anticrossing gap for the lower Zeeman branch expands to approximately 400 μeV, while the anticrossing of the upper Zeeman branch collapses to approximately 30 μeV. Because each anticrossing arises from a single hole-spin subspace, the difference between the two anticrossing energy gaps is equivalent to a spin-dependent tunneling rate [2]. The larger (smaller) magnitude of the lower (upper) anticrossing is due to the antibonding (bonding) nature of the molecular ground (excited) state of the QDM.[2] The combination of the enhanced Zeeman splitting for one molecular branch and suppressed Zeeman splitting for the other branch leads to the different anticrossing energies. In Fig. 7.4c we plot the energies of both the bright and dark exciton states including the Zeeman splitting and the resonant changes in g factor. Note that there are no anticrossings where bright and dark states intersect because they live in different spin subspaces.

We now add a grayscale map to display the optical intensity of the lines calculated in Fig. 7.4c. The optical intensities are calculated by multiplying the eigenvectors by an optical intensity vector that gives unit intensity to the direct transition (electron and hole in the same dot) and a fraction of that intensity to the indirect transition (electron and hole in different dots). Any optical intensity for dark states comes only from mixing with bright states. Figure 7.5a and d display calculated spectra of the anticrossings of the X^0 state at $B = 0$ and 6 T. As expected, the dark states have no optical intensity. These spectra are representative of the observed behavior for the anticrossing of the neutral exciton in most samples. In some samples, however, we observe dark states that gain optical intensity in the vicinity of the anticrossing region. Figures 7.5b and e show experimental data for one such case, corresponding to a QDM where the dots are separated by a 4 nm tunnel barrier. In this example, not only do the dark states gain optical intensity, we also observe new anticrossings between bright and dark states at high magnetic field. In the zero magnetic field case (Fig. 7.5b) the new anticrossing with the dark exciton state is indicated by callout 1. At $B = 6$ T (Fig. 7.5e) a complex pattern of additional anticrossings (callouts 3–6) appear near the electric field of the tunnel resonance. These anticrossings cannot be explained with any of the previously observed QDM properties, including spin-conserving tunneling and molecular g factor resonances. In order to explain them, we must assume the existence of a hole spin mixing that couples bright and dark exciton states, hm in Hamiltonian (7.3).

The calculations in Figs. 7.4 and 7.5a and d are obtained with numerical values determined by the experimental data in Figs. 7.5b and e with the hole-spin mixing term (hm) set to zero. In this case, it is clear that the phenomenological Hamiltonian (Figs. 7.5a and d) does not capture all of the features that appear in the experimental data. However, as shown in Figs. 7.5c and f, all of the dark states and anticrossings in the experimental spectra are explained if we turn on the hole mixing term. Best agreement with the data is obtained when $hm = 92\,\mu\text{eV}$, $t_{X^0} = -20\ \mu\text{eV}$ and all other parameters have the same value as in Figs. 7.5a and d. The signs and locations of the hm terms in the phenomenological matrix

[2] The ground states is antibonding because the barrier thickness is 4 nm.

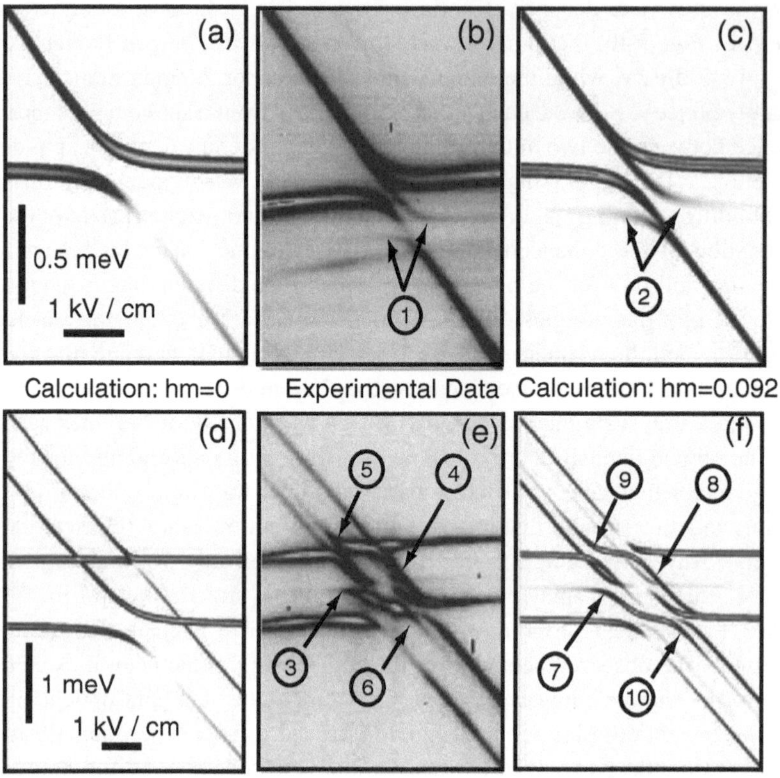

Figure 7.5 Experimental (b) and calculated (a, c) photoluminescence spectral map of the neutral exciton at $B = 0$ T without (a) and with (c) phenomenological hole spin mixing term. (d–f) Spectral maps as in (a–c) with $B = 6$ T. Scales for panels (a–c) are the same and are indicated in panel a, similarly for (d–f). Callout numbers are referenced in the text. Figure adapted from [3].

Hamiltonian (Eqn (7.3)) are derived from the full $k \cdot p$ theory [2]. The explanation of all of the complex anticrossing patterns, both at $B = 0$ and $B = 6$ T, by the inclusion of a single term in the matrix Hamiltonians provides strong phenomenological evidence for the existence of hole-spin mixing [2]

To elucidate the physical origin of the h_m term, we consider the contributions of both HHs and LHs in the Luttinger spinors that properly describe holes. The Luttinger spinor for an ideal QDM with two identical lens-shaped QDs perfectly aligned along the stacking axis (Fig. 7.6a with $O = 0$) is [42, 45]:

$$|F_z, k\rangle = \begin{pmatrix} c_{+\frac{3}{2}}\, f_{m_z}\quad(\mathbf{r})\,|J_z = +\frac{3}{2}\rangle \\ c_{-\frac{1}{2}}\, f_{m_z+1}(\mathbf{r})\,|J_z = +\frac{1}{2}\rangle \\ c_{+\frac{1}{2}}\, f_{m_z+2}(\mathbf{r})\,|J_z = -\frac{1}{2}\rangle \\ c_{-\frac{3}{2}}\, f_{m_z+3}(\mathbf{r})\,|J_z = -\frac{3}{2}\rangle \end{pmatrix}. \tag{7.5}$$

Here $|J_z>$ is the Bloch part of the wavefunction, $f(\mathbf{r})$ is the envelope function, and c_{J_z} is a numerical coefficient that gives the weight of each component. The envelope components of the spinor have the symmetries of the confining potential. Because the ideal QDM has circular symmetry, we can label each of the components by an envelope angular momentum $m_z = 0, \pm 1, \pm 2 \ldots$. The complete Luttinger spinor does not have circular symmetry because it is broken by the valence band SO interaction. The spinor can instead be classified by the total angular momentum $F_z = m_z + J_z$ and the main quantum number k.

As seen in Eqn (7.5), the spinor contains a mixture of $\Uparrow$ ($J_z = +3/2$) and $\Downarrow$ ($J_z = -3/2$) HH components. Despite this mixture, the low-lying hole states of a QD ($|F_z = \pm 3/2\rangle$) are dominated by the HH with $m_z = 0$. In typical InAs QDMs, one HH spin component makes up over 95% of the ground-state weight, with small contributions from the LH components (less than 5%) and the HH component with opposite spin (less than 0.1%). One can then identify the $|F_z = +3/2\rangle$ and $|F_z = -3/2\rangle$ spinors with the $\Uparrow$ and $\Downarrow$ HH states of the single-band description used in Eqn (7.3).

The hm term in Eqn (7.3) implies mixing between $\Uparrow$ and $\Downarrow$ spin configurations, i.e. between $|F_z = +3/2\rangle$ and $|F_z = -3/2\rangle$. In circular QDMs these two states do not interact in spite of the HH–LH coupling, because of the different symmetry. However, the total angular momentum (F_z) symmetry can be removed by structural distortions that break the circular symmetry of the QDM. When the F_z symmetry is broken, strong mixing of states with $F_z = 3/2$ and $F_z = -3/2$, mediated by LH states, leads to the appearance of hole-spin mixing [3]. From Hamiltonian (7.3), one can see that hm is present only in the matrix elements coupling hole states localized in opposite QDs; hm does not appear in matrix elements coupling states with opposite HH spins localized in the same QD. The agreement of this model with the experimental data suggests that the individual QDs are sufficiently symmetric to maintain F_z symmetry. The relevant structural distortions are therefore believed to originate in the lateral offset of the two QDMs along the stacking axis, as schematically depicted in Fig. 7.6a. Cross-sectional STM studies of QDMs (Fig. 7.6c) indicate that offsets along the stacking axis are not uncommon. Offsets of the magnitude

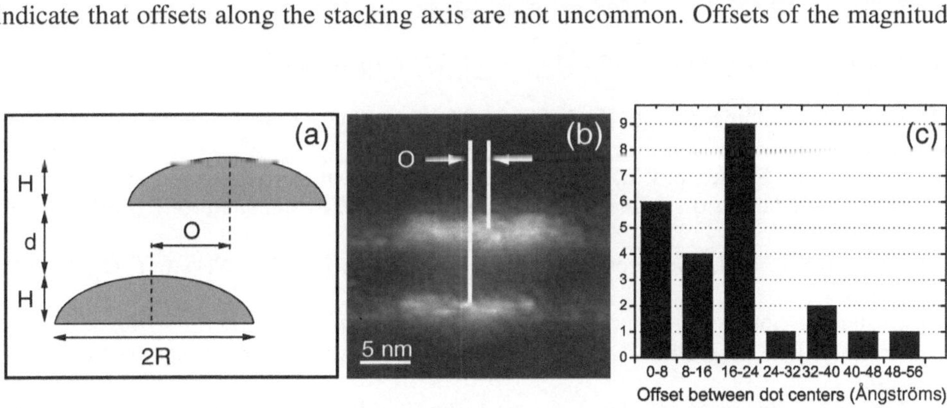

Figure 7.6 (a) Schematic depiction of lens-shaped QDs. In an ideal QDM, misalignment (O) is zero. (b) Cross-sectional STM image of vertically stacked InAs QDs showing misalignment. (c) Histogram of measured offsets between QD centers in a sample of 24 QDMs. Figure adapted from [3].

necessary to explain the large hole spin mixing presented in Fig. 7.5 ($O \sim 5$ nm) are, however, relatively rare, as shown in Fig. 7.6d.

7.7 Summary

In this chapter we have reviewed the current understanding of the properties and interactions of holes confined in QDMs. The QDMs are constructed of two InAs QDs separated by a GaAs barrier and are grown using molecular beam epitaxy. The lowest confined energy levels in each dot are generally not at equal energy, but can be tuned into resonance by an applied electric field. On resonance, coherent tunneling of holes between the two QDs leads to the formation of molecular states. The molecular states have unique spin properties because the different spatial extent of the molecular wavefunctions introduce different barrier contributions to the hole g factor. As a result, bonding orbitals manifest a resonant decrease in Zeeman splitting, while antibonding states manifest a resonant increase in Zeeman splitting. When the structure of the QDM is changed by increasing the thickness of the GaAs barrier region, the SO interaction leads to a surprising reversal of the orbital character of the molecular states, i.e. formation of a molecular ground state with antibonding orbital character. When the symmetry of the QDM is broken, SO interactions mediated by LH states mix orthogonal hole spin projections in different dots, leading to the appearance of new anticrossings between bright and dark exciton states. These surprising properties of holes in QDMs originate in the complex nature of valence-band states and the interplay between LH and HH components of the hole spinor. This complexity makes understanding the properties of holes in QDMs challenging, but also provides new tools for spin control [3].

References

[1] Bracker, A. S., Scheibner, M., Doty, M. F. *et al. Applied Physics Letters* **89**, 233110 (2006).
[2] Doty, M. F., Climente, J. I., Korkusinski, M. *et al. Physical Review Letters* **102**, 47401 (2009).
[3] Doty, M. F., Climente, J. I., Greilich, A. *et al. Physical Review B* **81**, 035308 (2010).
[4] Lyanda-Geller, Y. B., Reinecke, T. L. and Bayer, M. *Physical Review B* **69**, 161308 (2004).
[5] Villas-Bôas, J. M., Govorov, A. O. and Ulloa, S. E. *Physical Review B* **69**, 125342 (2004).
[6] Bester, G., Shumway, J. and Zunger, A. *Physical Review Letters* **93**, 47401 (2004).
[7] Krenner, H. J., Sabathil, M., Clark, E. C. *et al. Physical Review Letters* **94**, 57402 (2005).
[8] Ortner, G., Bayer, M., Geller, Y. L. *et al. Physical Review Letters* **94**, 157401 (2005).
[9] Stinaff, E. A., Scheibner, M., Bracker, A. S. *et al. Science* **311**, 636–639 (2006).
[10] Krenner, H. J., Clark, E. C., Nakaoka, T. *et al. Physical Review Letters* **97**, 76403 (2006).

[11] Jaskolski, W., Zielinski, M., Bryant, G. W. and Aizpurua, J. *Physical Review B* **74**, 195339 (2006).

[12] Ponomarev, I. V., Scheibner, M., Stinaff, E. A. *et al. Physica Status Solidi B-Basic Solid State Physics* **243**, 3869–3873 (2006).

[13] Scheibner, M., Doty, M., Ponomarev, I. *et al. Phys. Rev. B* **75**(24) (2007).

[14] Maialle, M. Z. and Degani, M. H. *Physical Review B* **76**, 115302 (2007).

[15] Szafran, B., Peeters, F. and Bednarek, S. *Physical Review B* **75**, 115303 (2007).

[16] Doty, M. F., Scheibner, M., Ponomarev, I. V. *et al. Physical Review Letters* **97**, 197202 (2006).

[17] Doty, M. F., Scheibner, M., Bracker, A. S. and Gammon, D. *Physical Review B* **78**, 115316 (2008).

[18] Gerardot, B. D., Brunner, D., Dalgarno, P. A. *et al. Nature* **451**, 441–444 (2008).

[19] Heiss, D., Schaeck, S., Huebl, H. *et al. Physical Review B* **76**, 241306 (2007).

[20] Brunner, D., Gerardot, B. D., Dalgarno, P. A. *et al. Science* **325**, 70–72 (2009).

[21] Eble, B., Krebs, O., Lemaitre, A. *et al. Physical Review B* **74**, 81306 (2006).

[22] Testelin, C., Bernardot, F., Eble, B. and Chamarro, M. *Physical Review B* **79**(19), 195440 (2009).

[23] Xu, X., Yao, W., Sun, B. *et al. Nature* **459**(7250), 1105–1109 (2009).

[24] Bulaev, D. V. and Loss, D. *Physical Review Letters* **95**, 76805 (2005).

[25] Burkard, G., Seelig, G. and Loss, D. *Physical Review B* **62**, 2581 (2000).

[26] Khaetskii, A. V., Loss, D. and Glazman, L. *Physical Review Letters* **88**, 186802 (2002).

[27] Chekhovich, E., Krysa, A., Skolnick, M. and Tartakovskii, A. *Physical Review Letters* **106**, 027402 (2011).

[28] Michler, P., editor. *Single Semiconductor Quantum Dots*. Springer, Berlin, (2009).

[29] Wasilewski, Z. R., Fafard, S. and McCaffrey, J. P. *Journal of Crystal Growth* **201**, 1131–1135 (1999).

[30] Gong, Q., Offermans, P., Notzel, R., Koenraad, P. M. and Wolter, J. H. *Applied Physics Letters* **85**, 5697–5699 (2004).

[31] Costantini, G., Rastelli, A., Manzano, C. *et al. Physical Review Letters* **96**, 226106 (2006).

[32] Goldstein, L., Glas, F., Marzin, J. Y., Charasse, M. N. and Leroux, G. *Applied Physics Letters* **47**, 1099–1101 (1985).

[33] Xie, Q., Madhukar, A., Chen, P. and Kobayashi, N. P. *Physical Review Letters* **75**, 2542 (1995).

[34] Solomon, G. S., Trezza, J. A., Marshall, A. F. and Harris, J. S. *Physical Review Letters* **76**, 952–955 (1996).

[35] Ledentsov, N. N., Shchukin, V. A., Grundmann, M. *et al. Physical Review B* **54**, 8743–8750 (1996).

[36] Fazekas, P. *Lecture Notes on Electron Correlation and Magnetism*. World Scientific (1999).

[37] Luttinger, J. M. and Kohn, W. *Physical Review* **97**, 869–883 (1955).

[38] Bayer, M., Stern, O., Kuther, A. and Forchel, A. *Physical Review B* **61**, 7273–7276 (2000).

[39] Salis, G., Kato, Y., Ensslin, K. *et al. Nature* **414**, 619 (2001).

[40] Poggio, M., Steeves, G. M., Myers, R. C. *et al. Physical Review B* **70**, 121305(R) (2004).

[41] Snelling, M. J., Blackwood, E., McDonagh, C. J., Harley, R. T. and Foxon, C. T. B. *Physical Review B* **45**, 3922–3925(R) (1992).

[42] Climente, J. I., Korkusinski, M., Goldoni, G. and Hawrylak, P. *Physical Review B* **78**, 115323 (2008).
[43] Chwiej, T. and Szafran, B. *Physical Review B* **81**, 75302 (2010).
[44] Planelles, J., Climente, J., Rajadell, F. *et al.* *Physical Review B* **82**, 155307 (2010).
[45] Rego, L. G. C., Hawrylak, P., Brum, J. A. and Wojs, A. *Physical Review B* **55**, 15 694–15 700 (1997).

Part III

Optical properties of quantum dots in photonic cavities and plasmon-coupled dots

8

Deterministic light–matter coupling with single quantum dots

P. Senellart

In 1946, E. M. Purcell predicted that the radiative lifetime of an emitter is not an intrinsic property but can be modified by structuring the surrounding electromagnetic field [36]. By inserting a semiconductor quantum dot (QD) in an optical cavity, one can accelerate or inhibit its spontaneous emission. In the present article, we show that the QD spontaneous emission can be deterministically controlled to fabricate bright sources of quantum light.

8.1 QDs in cavities: basics, motivation, first demonstrations

8.1.1 Light–matter coupling

We note f the ground state of the QD and e its excited state. For a cavity mode close to resonance with the QD optical transition, we consider only the states with 0 or 1 photon in the cavity mode. The states $|e, 0\rangle$ and $|f, 1\rangle$ are coupled through light–matter interaction, with a constant g, where $\hbar g = |\langle e, 0|\vec{E}\vec{d}|f, 1\rangle|$, with $\vec{d}$ the dipole of the optical transition $e \rightarrow f$ and $\vec{E}$ the electric field at the QD position.

Each of the states $|e, 0>$ and $|f, 1>$ are also coupled to continua of states: continuum of the free-space optical mode, phonons of the semiconductor matrix, etc. [4]. Here, we consider only the coupling to the continuum of the free-space optical mode, related to the cavity losses, with a constant γ_c. When $g << \gamma_c$, the photon emitted by the recombination of an exciton efficiently escapes outside the cavity. The QD optical transition radiative recombination rate can be accelerated (Purcell effect) or inhibited. When $g >> \gamma_c$, the photon emitted by the QD stays long enough inside the cavity to be reabsorbed by the QD, re-emitted and so on. The spontaneous emission is reversible: the system is in the strong coupling regime. The one-by-one coupling of the two states $|e, 0\rangle$ and $|f, 1\rangle$ leads to mixed light–matter eigenstates with an energy splitting given by $2\hbar\sqrt{g^2 - \gamma_c^2}$ at resonance. To optimize the QD–cavity coupling, a large oscillator strength, a small cavity mode effective volume and low cavity losses are needed.

8.1.2 Motivation

The first application of QD in cavities is to fabricate highly efficient deterministic sources of quantum light. Today, most quantum communication or quantum computing protocols

Quantum Dots: Optics, Electron Transport and Future Applications, ed. Alexander Tartakovskii. Published by Cambridge University Press.

are realized with attenuated coherent sources. Since they present a Poisson statistic, these sources operate at a photon rate around 0.05 per pulse to limit the number of two photon events. In 2000, it was shown that QDs are deterministic source of single photons [30]. When several electron–hole pairs are created in a QD, the system returns to its ground state through the emission of a cascade of photons at different energies. Only one photon is emitted for each excitation pulse at a given wavelength. Yet the photons are emitted isotropically in high-refractive-index materials so that only 0.5% of the photons can be collected in a numerical aperture of 0.4. By coupling the QD to a cavity mode in the weak-coupling regime, it is possible to efficiently extract and collect the QD emission as first proposed in 1999 [13] and demonstrated in 2001 [40]. The QD emission rate in the cavity mode is F_P times larger than in any other modes so that a fraction proportional to $F_P/(F_P+1)$ is funneled into the cavity mode.

Other very promising applications rely on the spin of a QD resident carrier coupled to a cavity mode. Giant spin-dependent optical non-linearities are expected for a single spin strongly coupled to a cavity mode which can be used to achieve single shot readout of a spin, remote spin entanglement [19], delayed photon entanglers [20].

8.1.3 First demonstrations

The acceleration of spontaneous emission was first demonstrated in 1998 on QD ensembles in pillar cavities [14]. In 2001, it was evidenced for single QDs [23, 45, 32]. To reach the strong coupling regime, high-quality factor cavities and increased oscillator strength QDs were developed. In GaAs-based cavities, quality factors larger than 10^5 have been demonstrated [39, 7]. In 2004–2005, the strong-coupling regime was demonstrated for single QDs in cavities [52, 38, 35].

To control the spontaneous emission of a QD transition, the QD must be located at the maximum of the electric field of a cavity mode (spatial matching) and its emission energy must be resonant to the mode energy (spectral matching). Yet standard QD growth techniques produce QDs with random spatial distributions and wide inhomogeneous spectral features. As a result, one can easily estimate that the fabrication yields for QD–cavity devices around 10^{-3}–10^{-4}. For all the early demonstrations, many cavities were fabricated to find one with a QD spatially and spectrally matched to the cavity mode.

8.2 Determininistic coupling of a QD to a cavity

Over the past few years, much work has been devoted to the deterministic control of the coupling between a QD and a confined cavity mode [43, 44, 22, 51, 26, 15, 31, 17, 37, 27]. In this section, we try to give a brief overview of the techniques developed to obtain the spatial and spectral matching.

8.2.1 Toward deterministic coupling

Spatial matching To control the QD position within a cavity, two main approaches are used. The first one consists of measuring the position of a QD randomly located so as to build the cavity around it. Such a technique has been used in a pioneering piece of work [2]. Several QD layers were grown with a thin barrier between the QD layers so as to lead to the in-plane alignement of vertically stacked QDs. The position of the QD at the surface was measured with respect to metallic marks, using SEM imaging of the surface. E-beam lithography was then used to define a photonic crystal cavity around the QD, using the same metallic marks for alignment. The technique was improved in 2007 by using a single layer of QDs and measuring the QD position using AFM on the sample surface [18]. The technique we present in the next sections also relies on such a bottom-up approach: we use an optical technique to measure the QD position with nanometer-scale accuracy [8].

The second approach consists of developing techniques to control the QD position during the growth process. To favor the QD growth on a predefined pattern, several techniques are used. One consists of defining holes in the semiconductor by using lithography and etching: QD are then deposited by using conditions favoring growth in the holes [43, 44, 22]. QDs grown at the top of pyramids or inverted pyramids are another example of such techniques [6, 51, 26]. Other techniques consist of using a dielectric mask during the growth of the small-band-gap material defining the QDs. The mask is then removed before proceeding to the growth of the barrier to bury the QDs [15]. One of the challenges for all these methods is to obtain good optical properties for the QDs since re-growth tends to favor the insertion of non-radiative defects close to the QD [44]. Recently, great improvement has been made with QDs inserted in inverted pyramids, where linewidths of a few tens of microelectronvolts have been measured [31].

Deterministic spectral matching To achieve the QD–cavity spectral matching, most techniques consist of fabricating the device, measuring the detuning using micro-photoluminescence (μPL) and correcting the cavity geometry. In 2005, Hennessy and co-workers used a digital etching technique to change the resonance of a photonic crystal cavity mode: the cavity mode wavelength shifts by 3–4 nm by removing few atomic layers from the oxided surface of holes [17]. Another technique consists of laser annealing of microdisk cavities [37] by taking advantage of the poor heat dissipation of the structure. The disk size is reduced by thermal desorption. Another technique consists of reducing the diameter of micropillar cavities by using a focused ion beam [27].

8.2.2 In-situ lithography for deterministic cavity–QD coupling

In this section, we present an in-situ lithography technique which enables us to define a spectrally resonant micropillar around a chosen QD. We use a planar cavity sample consisting of two $Al_0.9Ga0.1As$/GaAs Bragg mirrors surrounding a λ-cavity embedding a dilute layer of QDs. A 1.7 μm silicon-nitride layer is deposited on the sample, which will serve as a hard mask for the pillar etching. A positive photo-resist is then spin-coated on the sample, and μPL is performed on the sample at 10 K. A red laser (wavelength between 700 nm and 900 nm) is used to create carriers in the QDs without exposing the resist. The microscope objective focusing the laser beam on the sample also allows us to collect the QD emission, which is sent to a spectrometer and CCD for spectral analysis. Piezo-actuators allow movement of the QDs sample with respect to the laser spot with an accuracy of a few nanometers.

Spatial matching By moving the sample with respect to the laser, one can find a QD emission signal. This signal is optimized by moving the QD on the tens of nanometer scale within the excitation spot. Carriers non-resonantly created by the red laser are more likely to be captured into the QD when the QD is at the center of the laser spot than on the edges. Figure 8.1a shows the PL emission intensity as a function of its position with respect to the objective optical axis: the signal mostly reproduces the gaussian profile of the excitation laser beam (Figure 8.1b). The width is around 1 μm as expected by the diffraction limit with an objective with numerical aperture of 0.55. Yet the maximum of this curve is pointed with an accuracy which is given by the signal to noise ratio, around 50 nm in the present measurement. Note that this procedure must be done for an excitation power well below

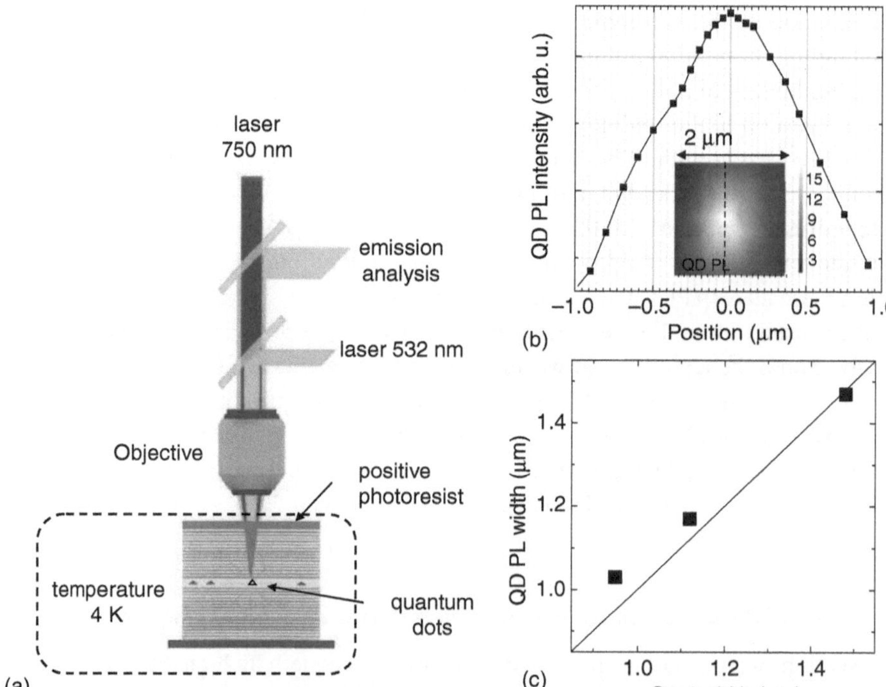

Figure 8.1 (a) Principle of the in-situ lithography technique. (b) Emission intensity of a QD line as a function of the distance to the objective axis. Insert: map of the QD emission in the plane. (c) Spatial width of the QD emission profile as a function of the spot size on the sample.

the saturation of the QD transition so that the emission line intensity is linear with the excitation power.

A green laser is overlapped with the red one with a 50 nm accuracy. When the QD is located at the center of the red laser beam, the green laser is used to expose the resist to define a disk centered on the QD. This disk will later be used as a direct mask for the pillar etching so that the QD will be located at the maximum of the electrical field of the pillar fundamental mode.

Spectral matching To ensure the spectral matching, one needs to control the diameter of the disk defined in the resist since the energy of the pillar mode increases when reducing the pillar diameter. The size of the disk defined in the resist depends on the spot size of the laser, the power used and the exposure time. Calibrations of the disk diameters for various exposing conditions are used during the in-situ lithography to define a disk diameter matched to the QD transition energy.

About the resist We use a positive photoresist optimized for the g-line of a mercury lamp to enable exposure with a 532 nm laser line. We found out that most resists designed to work at room temperature can also be used at cryogenic temperature. The only critical parameter seems to be the resist thickness, which should typically be below 500 nm.

8.2.3 On-demand Purcell effect

Figure 8.2 presents an example of a pillar fabricated by using the in-situ lithography technique. The bottom spectrum shows the PL measurement recorded during the in-situ lithography step. For this QD, the exciton line is at 1.3528 eV so that it was inserted in a pillar with a diameter of 1.7 μm. The middle spectrum is measured after pillar etching. The signal at the QD energy has strongly increased, as shown by the scales. The cavity mode can be seen as a sideband on the low-energy side of the exciton line. The detuning between the exciton and the cavity mode is close to zero at 10 K. By increasing the temperature, the exciton line is brought into perfect resonance with the cavity mode around 20 K and detuned for larger temperatures. The strong increase of signal at resonance with the cavity mode is a qualitative signature of the Purcell effect and efficient extraction efficiency [13, 11].

Several techniques can be used to measure the acceleration of the spontaneous emission [33]. An example is presented in Figure 8.2c for a pillar cavity, for which the biexciton line (XX) is resonant to the cavity mode at 4 K. Time-resolved measurements performed using a streak camera system show that the radiative lifetime of the XX line is 145 ps at resonance with the cavity mode. Out of resonance, at 45 K, the decay time of the line is around 1.1 ns. This measurement corresponds to a Purcell factor of $F_P^{measured} = 6 \pm 1$, very close to the expected value of the Purcell factor. It is important to note that large error bars remain on the precise measurement of the Purcell factor when using the temperature to control the detuning [33]. Indeed, when increasing temperature, diffusion of holes to the excited states can lead to an increase of the decay time and favor recapture processes [50].

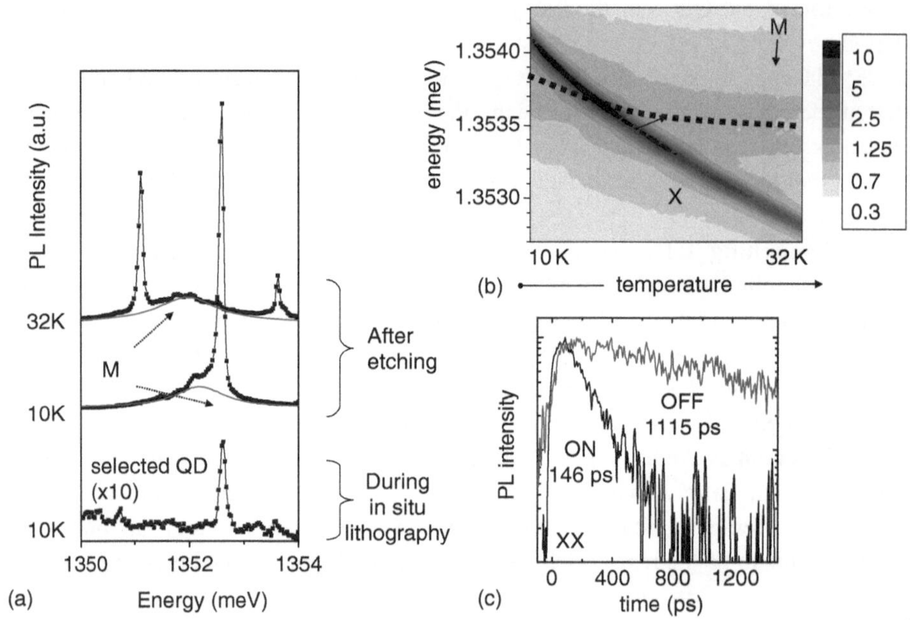

Figure 8.2 (a) PL spectrum of the QD during the in-situ lithography step and after the pillar etching. (b) Emission intensity of the device as a function of temperature and energy. (c) Time dependence of the emission measured on another device for the XX line in resonance with the mode at 5 K and spectrally detuned at 45 K.

Reproducibility

The number of the pillars one can make in a lithography step is only limited by the time devoted to the lithography. Figure 8.3 (left) presents an SEM image of a pillar field obtained using in-situ lithography. For each pillar, a QD line is spectrally very close to the cavity mode. A strong Purcell effect is observed for each device, showing good accuracy for the spatial matching. Figure 8.3 (right) represents the energy of the cavity mode as a function of the QD transition energy: the experimental points are on the diagonal with a standard deviation of only 0.45 meV.

Efficient single photon sources

Each pillar cavity fabricated by using an in-situ lithography step is a efficient single photon source. For each device, temperature is adjusted to finely tune a QD emission line in resonance to the pillar fundamental mode.

To estimate the brightness of the sources, the experimental setup is calibrated to obtain a correspondence between the number of photons collected in the first lens and a number of counts on the detector. Extraction efficiencies between 25% and 40% are measured. These values are consistent with the theoretically expected values given by $\eta = \frac{F_P}{F_P+1}\frac{Q}{Q_0}\alpha$, where

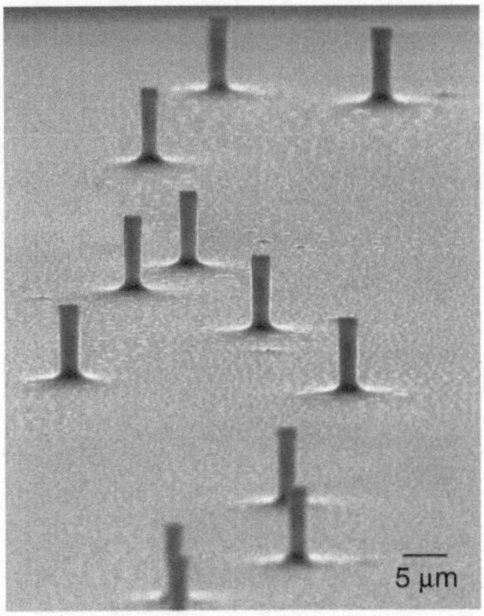

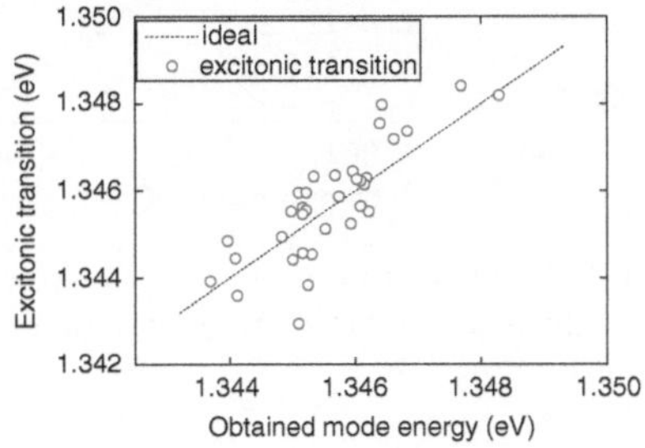

Figure 8.3 Left: scanning electron microscopy of micropillars determistically coupled to chosen single QDs. Right: energy of the excitonic transition as a function of the mode energy, measured on 33 pillars.

the first term is the fraction emission funneled into the cavity mode. The second term $\frac{Q}{Q_0}$ accounts for diffusions of photons at the pillar edges, with Q the quality factor of the pillar and Q_0 the planar cavity quality factor. Finally α is the fraction of emission in the upper space, for a emitter inside the planar cavity. In our case, $\alpha = 0.5$ since both Bragg mirrors have equal reflectivity, Q is close to Q_0 and F_P ranges from 5 to 10, depending on the pillar characteristics. To verify the single photon emission, photon correlation measurements can be performed using the standard Hanbury Twiss and Brown setup. The antibunching expected for single QD lines is observed.

8.2.4 Deterministic implementation of the strong-coupling regime

To demonstrate the strong-coupling regime, the same technique is used on a planar cavity presenting a higher quality factor. Quality factors around $Q = 140\,000$ are obtained for a cavity with 32/36 pairs in the top/bottom Bragg mirror. A dozen micropillars have been coupled to a single InGaAs-annealed QD using in-situ lithography. The strong-coupling regime is observed on each of the seven pillars for which resonance can be tuned using temperature. Figure 8.4 presents the energy of the lines observed on one of the pillars: three lines are observed, two corresponding to the pillar fundamental modes, one to a QD excitonic transition. The energy splitting of the fundamental mode results from some ellipticity in the lithography step, which amounted to $\approx$5% in this case [12]. Anticrossing between the exciton line and each of the pillar modes is observed. From the seven pillars showing the strong coupling regime, statistics are extracted: the average coupling constant amounts to $g = 44$ µeV for an average cavity mode linewidth $\gamma_C = 74$µeV. Figure 8.4

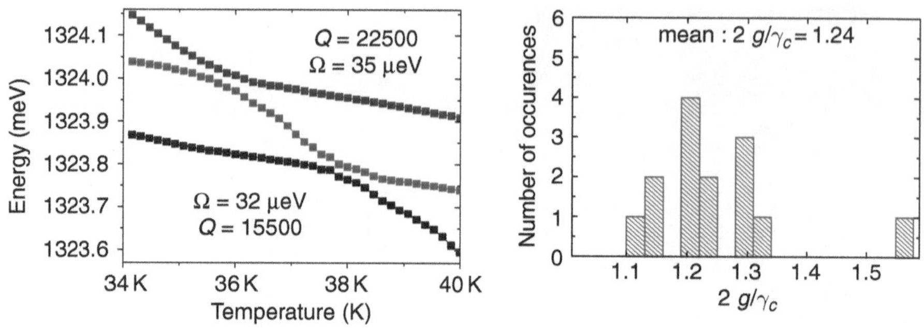

Figure 8.4 Left: energy of the emission lines measured on a pillar with a high quality factor. Right: statistics for the figure of merit $2g/\gamma_C$ of the strong coupling deduced from 14 anticrossings observed on seven pillars.

shows the statistics on the ratio $2g/\gamma_C$ with an average value around 1.24. This value is quite typical for pillar cavity systems [38]. We have recently developed annealed large InAs QDs, presenting oscillator strengths of around 40, for which we observe $2g/\gamma_C = 3$ [28].

8.3 An ultrabright source of entangled photon pairs

8.3.1 Extracting polarization entangled photon pairs

In 2000, Benson and coworkers theoretically proposed generating entangled photon pairs, using the radiative cascade of the last two photons emitted by a single QD [3]. The spin state of an XX state in a QD is undefined. Radiative recombination leads to a cascaded emission of two photons, with two possible intermediate states, depending on the X spin state. If both X states are degenerated (anisotropic exchange splitting, $S = 0$), the two recombination paths are indistinguishable. The two photons emitted at the X and XX energies are polarization entangled, with a high fidelity to the Bell state $\Psi_+ = \frac{1}{\sqrt{2}}(|HH\rangle + |VV\rangle)$. This property was experimentally demonstrated in 2006 [1, 53]. To be of any use for applications, these photon pairs needs to be efficiently collected. Extracting polarization-entangled photon pairs is more difficult than extracting single photons. Several conditions must be gathered so that the control of spontaneous emission is not detrimental to the entanglement [46]. First, the two photons are emitted at different energies. For the QDs we study, the XX line energy is 2–3 meV below the X line. Besides, the acceleration of spontaneous emission should be polarization independent. For instance, if the Purcell factor is larger for one linear polarization for both the X and XX lines, then the acceleration of spontaneous emission would lead to preferential recombination in one polarization. Finally, the Purcell effect ensures that the photons are mainly emitted into the cavity modes. If the radiation pattern of these mode is polarization dependent, one could know the recombination path followed by the system by knowing where the photons have been emitted. One should therefore use modes with polarization independent radiation patterns [29].

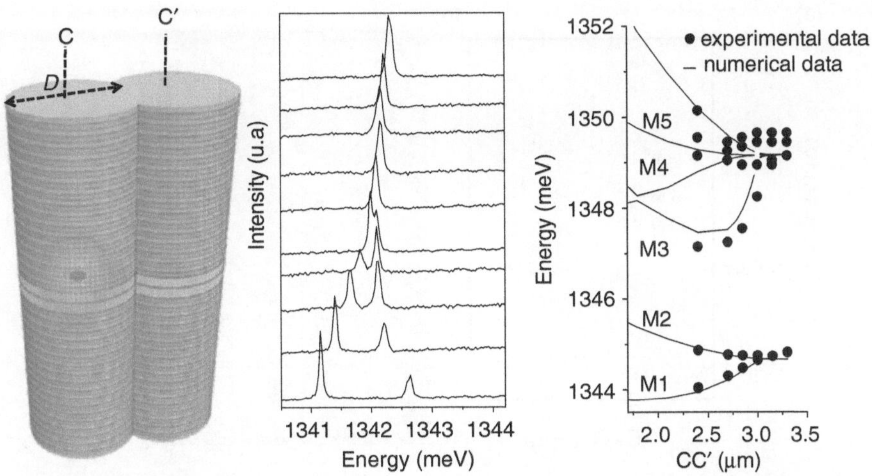

Figure 8.5 Left: schematic of the photonic molecule. Middle: emission spectra of a photonic molecule for $D = 3\,\mu\text{m}$ and decreasing CC′ from top to bottom. Right: energy of the modes of photonic molecules with diameter $D = 3\,\mu\text{m}$.

8.3.2 A cavity for extracting entangled photon pairs

We now present a structure which gathers all these requirements. It consists of a pillar microcavity embedding a single QD, coupled to a second identical but empty pillar (Figure 8.5). When reducing the center-to-center distance CC′ between the two pillars, coupling of the pillar modes lead to optical modes delocalized on the whole molecule. Bounding and antibounding modes are formed for each set of single pillar mode. Figure 8.5 presents the emission spectra and the energies of the mode of molecules with for $D = 3\,\mu\text{m}$ and various CC′. The diameter of the pillar D and the center-to-center distance CC′ can be used to adjust the molecule mode to single-QD X and XX lines.

Polarization properties For large CC′, the molecule present four polarization degenerated optical modes. Reducing CC′ leads to a splitting of the modes in two groups. Figure 8.6 presents the mode of molecules measured on a high quality factor sample ($Q = 60\,000$) with $D = 3\,\mu\text{m}$. Two linearly polarized modes are seen in each of the bounding and antibounding modes M1 and M2. This polarization splitting amounts to only 10–50 μeV, and is negligible if one uses a planar cavity with quality factor around $Q = 4000$ (mode linewidth=350 μeV). Finally, radiation patterns of the molecules have been measured for two linearly polarized orthogonal polarization. Radiation patterns in both polarization H (parallel to (CC′)) and V (perpendicular to (CC′)) present an overlap larger than 98% for modes M1 to M6 of the photonic molecule.

8.3.3 Fabrication of a bright source of entangled photon pairs

In most QDs, the exciton states are not polarization degenerate [41]: the anisotropic exchange splitting S can be as large as 200 μeV. This polarization splitting leads to

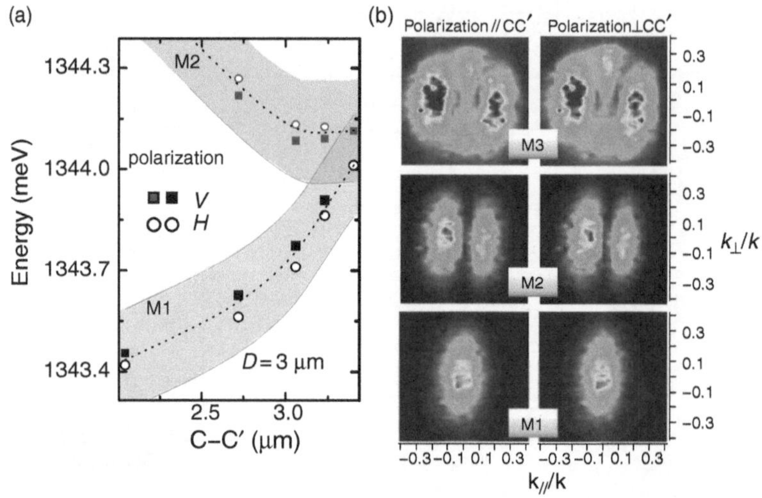

Figure 8.6 (a) Emission energy of the modes of photonic molecule with $D = 3.5\,\mu$m. The molecule have been fabricated from a sample with quality factor $Q = 60\,000$. The grey regions indicate the width of the modes we use to fabricate the sources of entangled photon pairs. (b) Radiation patterns measured for the first three modes of the molecule, for two linear polarizations: H, parallel to CC′, V, perpendicular to CC′.

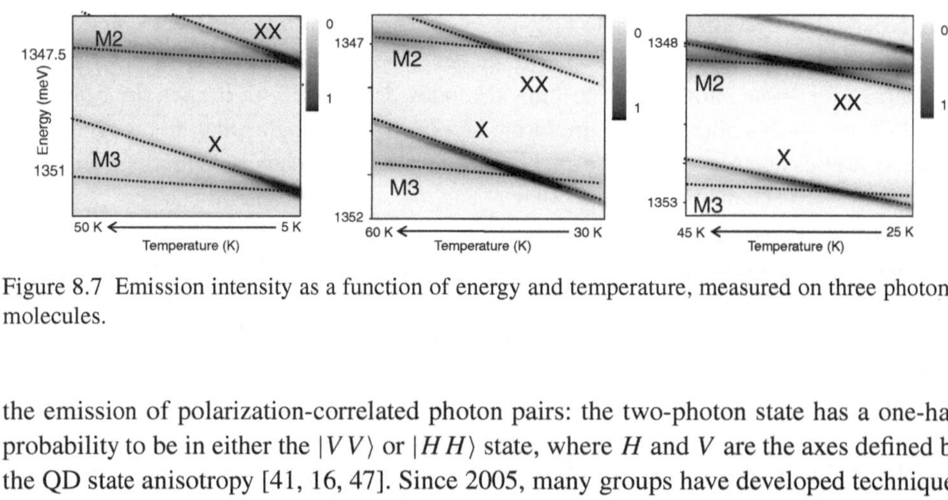

Figure 8.7 Emission intensity as a function of energy and temperature, measured on three photonic molecules.

the emission of polarization-correlated photon pairs: the two-photon state has a one-half probability to be in either the $|VV\rangle$ or $|HH\rangle$ state, where H and V are the axes defined by the QD state anisotropy [41, 16, 47]. Since 2005, many groups have developed techniques to control S [10, 48, 24]. In the present work, we have chosen to anneal the QDs at 867°C for 30 s, and we observe S of the order of 1–4 μeV depending on the QD. To fabricate the photonic molecule, we use in-situ lithography to define the first pillar centered on the QD. The sample is then moved by the chosen CC′ distance to define the second pillar. Dozens of molecules have been fabricated using this technique. Figure 8.7 presents the emission intensity as a function of energy and temperature for three devices. For each, we can find a temperature where both the X and XX line are matched to the modes M3 and M2 of the molecule. The strong increase of emission observed at resonance shows the simultaneous Purcell effect and efficient extraction of X and XX emission.

Entanglement and brightness We now present the results of photon correlation spectroscopy on the device matched at 5 K (left emission map in Figure 8.7). Time-resolved measurements show that each line experiences a Purcell factor of around 3–4 at 5 K. To measure the two-photon state and test the entanglement of the emitted pairs, the signal emitted by the molecule is sent to a non-polarizing beam-splitter and distributed into two arms where a set of quarter, half-wave plates and linear polarizers are used to analyze the correlation as a function of polarization. The signals on each arm are then sent to spectrometers coupled to avalanche photodiode with 300 ps temporal resolution. The spectrometers are set to the X and XX energy of the QD line. When $S = 0$, the two-photon state is the Bell state Ψ_+, which can be written in any linear or circular basis in the form:

$$\Psi_+ = \frac{1}{\sqrt{2}}(|HH\rangle + |VV\rangle) = \frac{1}{\sqrt{2}}(|DD\rangle + |DD'\rangle) = \frac{1}{\sqrt{2}}(|LR\rangle + |RL\rangle).$$

Measuring the polarization of the first photon fully determines the polarization of the second one. As a result, performing polarization-resolved photon correlation measurements in circular basis gives a good indication whether the two photon state presents a high fidelity to Ψ_+.

Figure 8.8a presents the measured $g_2^{X,XX}(\tau)$ function measured in a circular basis. Strong bunching is observed for the first photon in left circular polarization and the second photon in right circular polarization (R_X, L_{XX}), whereas no bunching is observed in the configuration (R_X, R_{XX}) showing that the emission of a left circularly polarized first photon is followed by a right polarized photon with a very high probability. The polarization of the correlation is measured in every basis (linear, diagonal, circular) and amounts to 45% to 65% in all bases. This observation is the first indication that the two-photon state is polarization entangled.

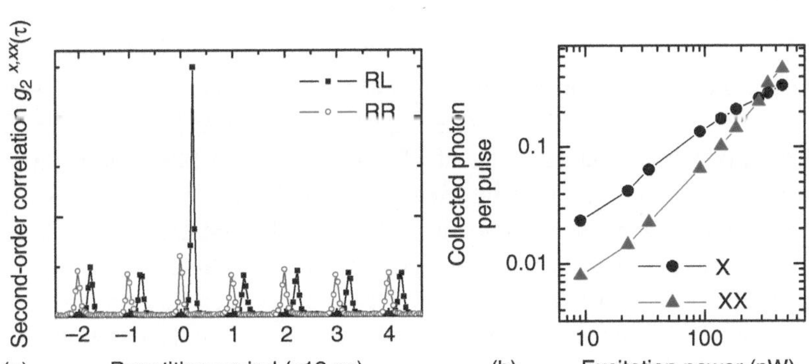

Figure 8.8 (a) $g_2^{X,XX}(\tau)$ measured in co-linear and perpendicular circular polarization. The curves have been horizontally shifted for clarity. (b) Number of photons per excitation pulse collected for each line in the first lens.

Table 8.1 *Common entanglement tests on the two-photon state. The last line gives the phase ϕ of the most probable state $\frac{1}{\sqrt{2}}(|HH\rangle + e^{i\phi}|VV\rangle)$ corresponding to the largest eigenvalue.*

	positive delays	bin 0	bin 1
Fraction of signal (delay > 0)	1	0.73	0.22
Peres criterion (< 0)	-0.16	-0.2	-0.3
Negativity (> 0)	0.33	0.39	0.6
Concurrence (> 0)	0.343	0.37	0.39
Fidelity $\psi+$	0.62	0.68	0.46
Largest eigenvalue	0.67	0.7	0.76
ϕ	0.17π	0.07π	0.41π

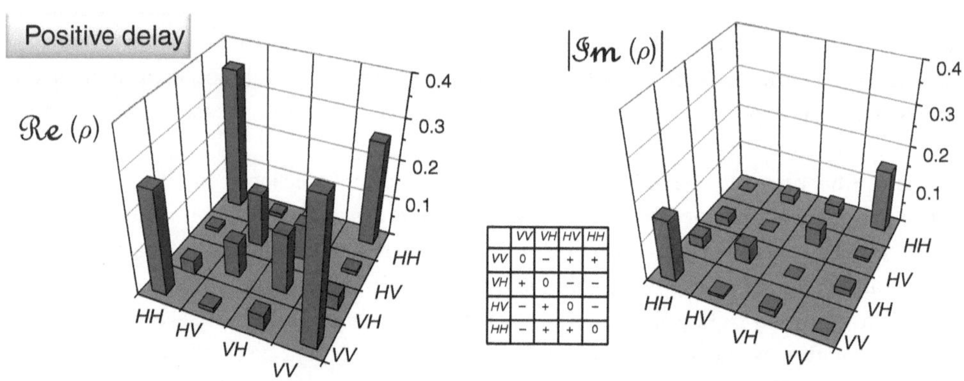

Figure 8.9 Real part and imaginary part of the density matrix of the two-photon state. For clarity, the signs of elements of the imaginary part are indicated in the table.

To measure the density matrix of the two-photon state, sixteen measurement are performed [21]. The density matrix, shown in Figure 8.9, presents large off-diagonal terms, expected for a two-photon entangled state. The fidelity to the Bell state is 0.62. Various criteria testing the entanglement are satisfied (see Table 8.1).

Figure 8.8b shows the number of photons collected in the first lens for each line. At saturation, 0.35 photons per excitation pulse are collected for each line. This value is consistent with the expected value, considering the Purcell factor for each line, the quality factor of the molecule which is close to the planar quality factor and the fact that one photon out of two is sent to the substrate side of the sample (Bragg mirror with equal reflectivity). At saturation, 0.12 entangled photon pairs are collected in the first lens within a numerical aperture of 0.4. This brightness exceeds by more than one order of magnitude that of any previous source of entangled photon pairs

8.3.4 Purcell effect and entanglement

In this section, we show how the implementation of the Purcell effect does not only enable high extraction efficiencies, but also increases the degree of entanglement of the source. Indeed, as proposed by Stace *et al.* in 2003 [46], reducing the radiative lifetime of the exciton state induces a broadening of the X lines which can become larger than the anisotropic exchange splitting of the state S. Another way to consider this effect is to consider the time evolution of the radiative cascade, as presented in reference [49]. After detection of the XX photon in a circularly polarized state, the system is in an entangled exciton–photon state $|H_{XX}X_H\rangle + |V_{XX}X_V\rangle$, where the photon at the XX energy is noted H_{XX} or V_{XX} depending on its polarization and the X state are $|X_H\rangle$, $|X_V\rangle$ are split by an energy S. After a delay τ, when the X recombines, a phase shift $\phi = S\tau/\hbar$ has developed between the two X states. The two-photon state is therefore in the maximally entangled state $|H_{XX}H_X\rangle + e^{iS\tau/\hbar}|V_{XX}V_X\rangle$. Yet the fidelity of this state to the Bell state depends on τ whenever $S \neq 0$. Implementing the Purcell effect on the X line allows us to reduce τ and minimize the effect of S.

The results presented in the previous section are obtained on a photonic molecule embedding a QD with $S = 2-3\,\mu$eV. Without the Purcell effect, one would expect this QD to emit polarization-correlated photon pairs, with a fidelity to the Bell state around $f = 0.4$. Table 8.1 presents the characteristics of the two-photon state for various delays τ. For $\tau < 500$ ps, the two-photon state presents a high-fidelity $f = 0.7$ to an entangled state with a small phase shift $\phi = 0.07\pi$. The fidelity to the Bell state is as high as $f = 0.68$. For the photons emitted with 500 ps $< \tau <$ 1 ns, the two-photon state presents a fidelity of $f = 0.76$ to the entangled state with $\phi = 0.41\pi$. This strong phase shift is due to the anisotropic exchange splitting $S = 2-3\mu$ eV. Yet, thanks to the Purcell effect, 73% of the X photons are emitted in the first 500 ps so that the fidelity to the Bell state when no positive delay photons are discarded amounts to $f = 0.62$. These measurements shows that the Purcell effect does indeed improve the fidelity of the source. In recent work, we also show that the use of Purcell effect allows quenching of the exciton spin relaxation during the radiative cascade and obtaining of a bright source of entangled photon pairs at 53 K [9].

8.4 Some perspectives

A large number of bright sources of single photons or entangled photon pairs can now be fabricated in a deterministic way. To enable the use of these sources in quantum information-processing protocols, several steps need to be achieved. Quasi-resonant excitation should be used to make the single photon or entangled photon pairs indistinguishable [42]. This resonant excitation may also improve the fidelity of the entangled photon-pair sources, which is currently limited by recapture processes. Another important step is to control the wavelength of two independent photon sources. To do so, electrical contact should be defined on the pillar to apply an electric field [25, 34, 5]. Finally, many theoretical proposals take advantage of the strong- or weak-coupling regime of a doped QD in a pillar cavity to achieve single-shot readout of a spin, remote spin entanglement [19], or delayed photon entanglers [20]. To do so, in-situ lithography would be performed on a cavity embedding a chemically doped QD layer.

Acknowledgements

The results presented in this paper would not have been possible without the hard work and expertise of my collaborators: A. Dousse, J. Suffczynski, L. Lanco, A. Lemaître, I. Sagnes, J. Bloch, O. Krebs, P. Voisin, A. Beveratos, E. Galopin, A. Miard. This work was partially supported by the French ANR P3N DELIGHT, ANR JCJC MICADOS.

References

[1] Akopian, N., Lindner, N. H., Poem, E. *et al.* 2006. Entangled photon pairs from semiconductor quantum dots. *Phys. Rev. Lett.*, **96**(13), 130501.
[2] Badolato, A., Hennessy, K., Atatüre, M. *et al.* 2005. Deterministic coupling of single quantum dots to single nanocavity modes. *Science*, **308**(5725), 1158–1161.
[3] Benson, O., Santori, C., Pelton, M. and Yamamoto, Y. 2000. Regulated and entangled photons from a single quantum Dot. *Phys. Rev. Lett.*, **84**(11), 2513–2516.
[4] Besombes, L., Kheng, K., Marsal, L. and Mariette, H. 2001. Acoustic phonon broadening mechanism in single quantum dot emission. *Phys. Rev. B*, **63**(15), 155307.
[5] Böckler, C., Reitzenstein, S., Kistner, C. *et al.* 2008. Electrically driven high-Q quantum dot-micropillar cavities. *Appl. Phys. Lett.*, **92**(9), 091107.
[6] Dalacu, D., Mnaymneh, K., Sazonova, V. *et al.* 2010. Deterministic emitter–cavity coupling using a single-site controlled quantum dot. *Phys. Rev. B*, **82**(3), 033301.
[7] Ding, L., Baker, C., Senellart, P. *et al.* 2010. High frequency GaAs nano-optomechanical disk resonator. *Phys. Rev. Lett.*, **105**(26), 263903.
[8] Dousse, A., Lanco, L., Suffczyński, J. *et al.* 2008. Controlled light–matter coupling for a single quantum dot embedded in a pillar microcavity using far-field optical lithography. *Phys. Rev. Lett.*, **101**(26), 267404.
[9] Dousse, A., Suffczyński, J., Krebs, O. *et al.* 2010. A quantum dot based bright source of entangled photon pairs operating at 53 K. *Appl. Phys. Lett.*, **97**(8), 081104.
[10] Ellis, D. J. P., Stevenson, R. M., Young, R. J. *et al.* 2007. Control of fine-structure splitting of individual InAs quantum dots by rapid thermal annealing. *Appl. Phys. Lett.*, **90**(1), 011907.
[11] Gayral, B. and Gérard, J. M. 2008. Photoluminescence experiment on quantum dots embedded in a large Purcell-factor microcavity. *Phys. Rev. B*, **78**(23), 235306.
[12] Gayral, B., Gérard, J. M., Legrand, B., Costard, E. and Thierry-Mieg, V. 1998. Optical study of GaAs/AlAs pillar microcavities with elliptical cross section. *Appl. Phys. Lett.*, **72**(12), 1421–1423.
[13] Gerard, J. -M. and Gayral, B. 1999. Strong Purcell effect for InAs quantum boxes in three-dimensional solid-state microcavities. *J. Lightwave Technol.*, **17**(11), 2089–2095.
[14] Gérard, J. M., Sermage, B., Gayral, B. *et al.* 1998. Enhanced spontaneous emission by quantum boxes in a monolithic optical microcavity. *Phys. Rev. Lett.*, **81**(5), 1110–1113.

[15] Gogneau, N., Gratiet, L. Le, Cambril, E. *et al.* 2008. One-step nano-selective area growth (nano-SAG) of localized InAs/InP quantum dots: First step towards single-photon source applications. *J. Crystal Growth*, **310**(15), 3413–3415.
[16] Hafenbrak, R., Ulrich, S. M., Michler, P. *et al.* 2007. Triggered polarization-entangled photon pairs from a single quantum dot up to 30 K. *New J. Phys.*, **9**(9), 315.
[17] Hennessy, K., Badolato, A., Tamboli, A. *et al.* 2005. Tuning photonic crystal nanocavity modes by wet chemical digital etching. *Appl. Phys. Lett.*, **87**(2), 021108.
[18] Hennessy, K., Badolato, A., Winger, M. *et al.* 2007. Quantum nature of a strongly coupled single quantum dot–cavity system. *Nature*, **445**(7130), 896–9.
[19] Hu, C. Y., Young, A., O'Brien, J. L., Munro, W. J. and Rarity, J. G. 2008. Giant optical Faraday rotation induced by a single-electron spin in a quantum dot: applications to entangling remote spins via a single photon. *Phys. Rev. B*, **78**(8), 085307.
[20] Hu, C. Y., Munro, W. J., O'Brien, J. L. and Rarity, J. G. 2009. Proposed entanglement beam splitter using a quantum-dot spin in a double-sided optical microcavity. *Phys. Rev. B*, **80**(20), 205326.
[21] James, D., Kwiat, P., Munro, W. and White, A. 2001. Measurement of qubits. *Physical Review A*, **64**(5), 1–15.
[22] Kiravittaya, S., Benyoucef, M., Zapf-Gottwick, R., Rastelli, A. and Schmidt, O. G. 2006. Ordered GaAs quantum dot arrays on GaAs(001): Single photon emission and fine structure splitting. *Appl. Phys. Lett.*, **89**(23), 233102.
[23] Kiraz, A., Michler, P., Becher, C. *et al.* 2001. Cavity-quantum electrodynamics using a single InAs quantum dot in a microdisk structure. *Appl. Phys. Lett.*, **78**(25), 3932–3934.
[24] Kowalik, K., Krebs, O., Golnik, A. *et al.* 2007. Manipulating the exciton fine structure of single CdTe/ZnTe quantum dots by an in-plane magnetic field. *Phys. Rev. B*, **75**(19), 195340.
[25] Laucht, A., Hofbauer, F., Hauke, N. *et al.* 2009. Electrical control of spontaneous emission and strong coupling for a single quantum dot. *New J. Phys.*, **11**(2), 023034.
[26] Leifer, K., Pelucchi, E., Watanabe, S. *et al.* 2007. Narrow ([approximate] 4 meV) inhomogeneous broadening and its correlation with confinement potential of pyramidal quantum dot arrays. *Appl. Phys. Lett.*, **91**(8), 081106.
[27] Lohmeyer, H., Kalden, J., Sebald, K. *et al.* 2008. Fine tuning of quantum-dot pillar microcavities by focused ion beam milling. *Appl. Phys. Lett.*, **92**(1), 011116.
[28] Loo, V., Lanco, L., Lemaître, A. *et al.* 2010. Quantum dot-cavity strong-coupling regime measured through coherent reflection spectroscopy in a very high-Q micropillar. *Appl. Phys. Lett.*, **97**(24), 241110.
[29] Larqué, M., Karle, T., Robert-Philip, I. and Beveratos, A. 2009. Optimizing H1 cavities for the generation of entangled photon pairs. *New J. Phys.*, **033022**(11).
[30] Michler, P., Kiraz, A., Becher, C. *et al.* 2000. A quantum dot single-photon turnstile device. *Science*, **290**(5500), 2282–2285.
[31] Mohan, A., Felici, M., Gallo, P. *et al.* 2010. Polarization-entangled photons produced with high-symmetry site-controlled quantum dots. *Nature Photonics*, **4**(March), 302.
[32] Moreau, E., Robert, I., Manin, L. *et al.* 2001. Quantum cascade of photons in semiconductor quantum dots. *Phys. Rev. Lett.*, **87**(18), 183601.
[33] Munsch, M., Mosset, A., Auffèves, A. *et al.* P. 2009. Continuous-wave versus time-resolved measurements of Purcell factors for quantum dots in semiconductor microcavities. *Phys. Rev. B*, **80**(11), 115312.
[34] Patel, R. B., Bennett, A. J., Farrer, I. *et al.* 2010. Two-photon interference of the emission from electrically tunable remote quantum dots. *Nature Photonics*, **4**, 632.

[35] Peter, E., Senellart, P., Martrou, D. *et al.* 2005. Exciton–photon strong-coupling regime for a single quantum dot embedded in a microcavity. *Phys. Rev. Lett.*, **95**(6), 067401.

[36] Purcell, E. M. 1946. *Phys. Rev.*, **69**(11–12), 674.

[37] Rastelli, A., Ulhaq, A., Kiravittaya, S. *et al.* 2007. In situ laser microprocessing of single self-assembled quantum dots and optical microcavities. *Appl. Phys. Lett.*, **90**(7), 073120.

[38] Reithmaier, J. P., Sek, G., Löffler, A. *et al.* 2004. Strong coupling in a single quantum dot–semiconductor microcavity system. *Nature*, **432**(7014), 197–200.

[39] Reitzenstein, S., Hofmann, C., Gorbunov, A. *et al.* 2007. AlAs/GaAs micropillar cavities with quality factors exceeding 150.000. *Appl. Phys. Lett.*, **90**(25), 251109.

[40] Santori, C., Pelton, M., Solomon, G., Dale, Y. and Yamamoto, Y. 2001. Triggered single photons from a quantum dot. *Phys. Rev. Lett.*, **86**(8), 1502–1505.

[41] Santori, C., Fattal, D., Vucković, J., Solomon, G. S., and Yamamoto, Y. 2002a. Indistinguishable photons from a single-photon device. *Nature*, **419**(6907), 594–597.

[42] Santori, C., Fattal, D., caronkovic, J. V., Solomon, G. S. and Yamamoto, Y. 2002b. Indistinguishable photons from a single-photon device. *Nature*, **419**(October), 594–597.

[43] Schneider, C., Strauss, M., Sunner, T. *et al.* 2008. Lithographic alignment to site-controlled quantum dots for device integration. *Appl. Phys. Lett.*, **92**(18), 183101.

[44] Schneider, C., Heindel, T., Huggenberger, A. *et al.* 2009. Single photon emission from a site-controlled quantum dot-micropillar cavity system. *Appl. Phys. Lett.*, **94**(11), 111111.

[45] Solomon, G. S., Pelton, M. and Yamamoto, Y. 2001. Single-mode spontaneous emission from a single quantum dot in a three-dimensional microcavity. *Phys. Rev. Lett.*, **86**(17), 3903–3906.

[46] Stace, T., Milburn, G. and Barnes, C. 2003. Entangled two-photon source using biexciton emission of an asymmetric quantum dot in a cavity. *Phys. Rev. B*, **67**(8), 1–15.

[47] Stevenson, R., Young, R., See, P. *et al.* 2006a. Magnetic-field-induced reduction of the exciton polarization splitting in InAs quantum dots. *Phys. Rev. B*, **73**(3), 1–4.

[48] Stevenson, R., Young, R. P., Gevaux, D. 2006b. Magnetic-field-induced reduction of the exciton polarization splitting in InAs quantum dots. *Phys. Rev. B*, **73**(3), 1–4.

[49] Stevenson, R., Hudson, A., Bennett, A. *et al.* 2008. Evolution of entanglement between distinguishable light states. *Phys. Rev. Lett.*, **101**(17), 1–4.

[50] Suffczyński, J., Dousse, A., Gauthron, K. *et al.* 2009. Origin of the optical emission within the cavity mode of coupled quantum dot-cavity systems. *Phys. Rev. Lett.*, **103**(2), 027401.

[51] Watanabe, S., Pelucchi, E., Leifer, K. *et al.* 2005. Patterning of confined-state energies in site-controlled semiconductor quantum dots. *Appl. Phys. Lett.*, **86**(24), 243105.

[52] Yoshie, T., Scherer, A., Hendrickson, J. *et al.* 2004. Vacuum Rabi splitting with a single quantum dot in a photonic crystal nanocavity. *Nature*, **432**(7014), 200–203.

[53] Young, R. J., Stevenson, R. M., Atkinson, P. *et al.* 2006. Improved fidelity of triggered entangled photons from single quantum dots. *New J. Phys.*, **8**(2), 29.

9

Quantum dots in photonic crystal cavities

A. Faraon, D. Englund, I. Fushman, A. Majumdar and J. Vučković

9.1 Introduction

During the past two decades, the development of micro- and nano-fabrication technologies has positively impacted multiple areas of science and engineering. In the photonics community, these technologies had numerous early adopters, which led to photonic devices that exhibit features at the nano-scale and operate at the most fundamental level of light–matter interaction [28, 39, 18, 29]. One of the leading platforms for these types of devices is based on gallium arsenide (GaAs) planar photonic crystals (PC) with embedded indium arsenide (InAs) quantum dots (QDs). The PC architecture is advantageous because it enables monolithic fabrication of photonic networks for efficient routing of light signals of the chip [26]. At the same time, PC devices have low loss and ultra-small optical mode volumes, which enable strong light–matter interactions. The InAs quantum dots are well suited for quantum photonic applications because they have excellent quantum efficiencies, large dipole moments, and a variety of quantum states that can be optically controlled [24, 3].

Currently, the development of these photonic technologies is geared mainly towards applications in quantum and classical information processing. The first proposals for quantum information processing using QDs in optical microresonators were developed more than a decade ago in the broader context of quantum information processing using quantum systems (such as atoms, ion, molecules) that can be optically controlled [23, 17]. Compared to other systems, the solid-state quantum photonic platform is attractive for quantum information applications because of its potential for large-scale integration [27]. For classical information processing these technologies are relevant because of the continuous drive to develop devices that have smaller footprint and operate at lower power [25]. In this respect, PCs with coupled QDs approach a fundamental limit, where light is confined in a region on the order of one cubic optical wavelength and the optically active material operates at the level of one quantum of excitation. This enables devices that can switch light at the single photon level and consume minute amounts of energy during operation.

Quantum Dots: Optics, Electron Transport and Future Applications, ed. Alexander Tartakovskii. Published by Cambridge University Press.

In this chapter we give an overview of our recent work on developing quantum photonic devices based on the QD–PC platform. Section 9.2 is a brief introduction to QDs, PCs, and the formalism to analyze coupled cavity/QD systems. In Section 9.3 we summarize the main experimental techniques utilized for fabrication, optical probing and tuning of these devices. In Section 9.4 we discuss how these techniques are used to probe strongly coupled QDs, and in Section 9.5 we focus on optical nonlinearities at the single photon level that are present in these systems. In the last section (Section 9.6) we comment on applications and future directions.

9.2 Quantum dots and photonic crystals

9.2.1 InAs quantum dots

Semiconductor QDs are nano-scale inclusions of a low-bandgap semiconductor inside a semiconductor with a larger bandgap. The bandgap difference acts as a potential barrier for carriers (electrons and holes), confining them inside the dot. Moreover, the dots are small enough (nanometer size) that the confined carriers can occupy only discrete energy levels, and the transitions between such levels in the conduction band and the valence band involve the absorption or emission of photons at optical frequencies. QDs can be formed spontaneously during epitaxial growth of lattice-mismatched materials, and such dots are called self-assembled [24]. For example, when InAs is deposited on GaAs, a strained planar layer, known as a wetting layer, initially forms. The strain energy that builds up in this layer is eventually partially relieved by the formation of nanometer-scale islands on the surface, which can subsequently be covered with a capping layer of GaAs.

InAs/GaAs QDs used in our experiments were self-assembled during molecular beam epitaxy (MBE), under conditions that give relatively sparse dots, with a surface density of $\sim 100\,\mu\text{m}^{-2}$. These QDs are engineered to emit at near infrared wavelengths ($\approx 930\,\text{nm}$) when operating at cryogenic temperatures (below $\approx 50\,\text{K}$) as discussed in Section 9.3.2. A typical spectrum of an ensemble of QDs is shown in Fig. 9.1a together with an atomic force microscope image (Fig. 9.1b) taken before the growth of the GaAs capping layer. These QDs self-assemble at random locations on the surface of the wafer.

9.2.2 Photonic crystals

Photonic crystals are formed by periodic arrangements of optical media with various refractive indexes [19]. The light in such structures is controlled via distributed Bragg reflection (DBR), that allows for the development of photonic bandgaps analogous to the electronic bandgaps in semiconductors. In such a material and at frequencies inside the photonic bandgap, the propagation of light is prohibited in all crystal directions. The 2D PCs drew most attention because they can be fabricated in a monolithic fashion using techniques already well developed in the semiconductor industry. The most common form for

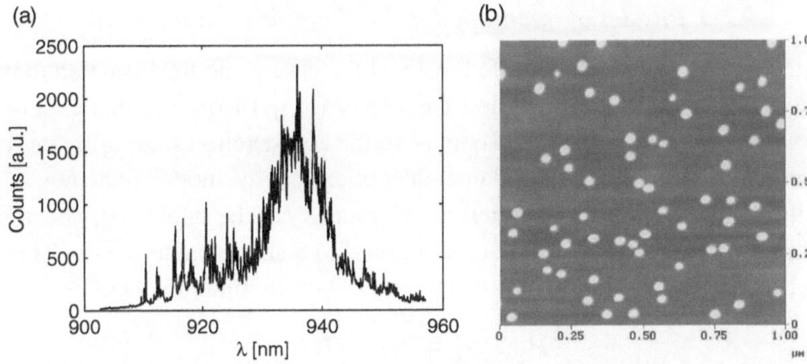

Figure 9.1 (a) Spectrum of an ensemble of QDs showing an inhomogeneous broadening of $\approx$ 20 nm. (b) Atomic force microscope image of QDs (un-capped) showing their random spatial distribution on the GaAs wafer (Image courtesy of Dr. Bingyang Zhang, Stanford University).

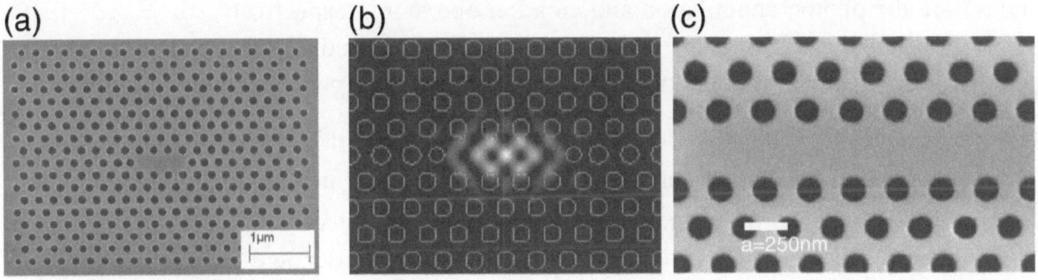

Figure 9.2 (a) Scanning electron microscope image of a linear three hole defect cavity fabricated in a 160 nm thick GaAs slab. (b) Simulated electric field intensity of the fundamental mode. (c) Fabricated W1 waveguide.

2D PCs is a periodic lattice (triangular of square) of holes patterned in a thin suspended membrane of high refractive-index material (n).

The confinement of light can be achieved by introducing perturbations into the periodicity of the photonic lattice, which leads to formation of PC cavities. The two main parameters used to characterize these cavities are the quality factor and the mode volume. The quality factor is defined as $Q = \lambda/\Delta\lambda$, with λ the wavelength in vacuum and $\Delta\lambda$ the cavity resonance linewidth, and V is defined in Section 9.2.3. Most of the results reported in this chapter were obtained using linear three-hole defect cavities (L3) [1]. A scanning electron microscope (SEM) image of a L3 cavity fabricated in GaAs is shown in Fig. 9.2, together with the profile (simulation) of the electric field intensity in the fundamental mode. The mode is TE (transversal electric)-like with the electric field polarized primarily in the PC plane. The design of the cavity can be engineered such that this mode can reach quality factors on the order of hundreds of thousands and optical mode volumes smaller than one cubic optical wavelength. The L3 cavity can be easily connected in a photonic network using PC waveguides. We used W1 waveguides created by omitting a row of holes in the triangular lattice (Fig. 9.2c).

9.2.3 Physics of quantum dots in photonic crystal cavities

Let us assume that a single QD is located inside a PC cavity and thus coupled to the cavity field. We also assume that the transition frequency (ω_{QD}) from the one-exciton (i.e. an electron–hole pair trapped in the QD) state $|e\rangle$ to the zero-exciton state $|g\rangle$ is on resonance or nearly on resonance with the fundamental optical cavity mode frequency ω_c. Under these conditions, the excitation of other cavity modes can be neglected, and the system can be modeled as a single two-level atom coupled to a single cavity mode. This coupled system can be described by the Jaynes–Cummings Hamiltonian [35, 41]:

$$H = H_E + H_F + H_I \,, \tag{9.1}$$

where $H_E = \hbar\omega_{QD}\sigma_z$, $H_F = \hbar\omega_c\big(a^\dagger a + \frac{1}{2}\big)$ and $H_I = i\hbar\big(g(\vec{r}_E)a^\dagger\sigma_- - g^*(\vec{r}_E)\sigma_+ a\big)$.

The three terms of the Jaynes–Cummings Hamiltonian are the excitonic Hamiltonian (H_E), the field Hamiltonian (H_F), and the exciton–field interaction Hamiltonian (H_I); a and $a^\dagger$ are the photon annihilation and creation operators, respectively, $\sigma_- = |g\rangle\langle e|$ and $\sigma_+ = |e\rangle\langle g|$ are the QD lowering and raising operators, respectively, while $\sigma_z = \frac{1}{2}\big(|e\rangle\langle e| - |g\rangle\langle g|\big)$ is the population operator. The coupling parameter $g(\vec{r}) = g_0\psi(\vec{r})\cos(\xi)$ is the product of the Rabi frequency $g_0 = \frac{\mu}{\hbar}\sqrt{\frac{\hbar\omega_c}{2\epsilon_M V}}$, a position-dependent part $\psi(\vec{r}) = \frac{E(\vec{r})}{|E(\vec{r}_M)|}$, and a polarization-dependent part $\cos(\xi) = \frac{\vec{\mu}\cdot\hat{e}}{\mu}$, where $\vec{r}_M$ denotes the point where the field intensity $\epsilon(\vec{r})|E(\vec{r})|^2$ is maximum and ϵ_M is the dielectric constant at this point ($\epsilon_M = \epsilon(\vec{r}_M)$). Electric field orientation at the location $\vec{r}$ is denoted as $\hat{e}$, $\vec{\mu}$ is the dipole moment matrix element between the states $|e\rangle$ and $|g\rangle$, and $\mu = |\vec{\mu}|$. Note that $\vec{\mu}$ is defined as $q\langle e|\vec{d}|g\rangle$, where $\vec{d}$ is the coordinate operator and q is a unit charge; $g(\vec{r}_E)$ denotes the value of the coupling parameter at the QD exciton location $\vec{r}_E$, and $|g(\vec{r}_E)|$ reaches its maximum value of $|g_0|$ when the exciton is located at the point $\vec{r}_M$ where the field intensity is maximum, and when its dipole moment is aligned with the electric field (i.e. when $\psi(\vec{r}_E) = 1$ and $\cos(\xi) = 1$). $E(\vec{r})$ is the electric field magnitude, and V is the cavity mode volume, defined as $V = \frac{\int\int\int \epsilon(\vec{r})|E(\vec{r})|^2 d^3\vec{r}}{\epsilon_M|E(\vec{r}_M)|^2}$. Note that H_I can be derived from the interaction Hamiltonian in the dipole approximation ($H_I = -\vec{\mu}\cdot\vec{E}$), after the expansion and quantization of electric field in terms of the cavity modes.

The losses of the system can be described in terms of the *cavity field decay rate* κ, equal to ($\omega_c/2Q$), and the *excitonic dipole decay rate* γ; κ is the decay rate of the resonant cavity mode, while γ includes losses to modes other than the cavity mode and to nonradiative decay routes. With losses taken into account, the eigenstates of the system Hamiltonian are:

$$\omega_{\pm} = \frac{\omega_c + \omega_{QD}}{2} - i\frac{\kappa + \gamma}{2} \pm \sqrt{g^2 + \frac{1}{4}(\delta - i(\kappa - \gamma))^2}, \tag{9.2}$$

where $\delta = \omega_{QD} - \omega_c$ is the QD/cavity detuning. Depending on the ratio of the coupling parameter $|g(\vec{r}_E)|$ to the decay rates κ and γ, we can distinguish two regimes of coupling between the exciton and the cavity field. For $\delta = 0$ and $\kappa >> \gamma$ (typical for our regime of operation) the system eigenstates are degenerate (i.e. real parts of the two solutions of

Eq. (9.2) are equal) for $g < \kappa/2$ (*weak coupling regime*) and non-degenerate with splitting $\sim 2g$ for $g > \kappa/2$ (*strong coupling regime*) [20]. In the strong-coupling case, the time scale of coherent coupling between the exciton and the cavity field is shorter than that of the irreversible decay into various radiative and nonradiative routes. *Rabi oscillation* occurs in this case, and (in the weak excitation limit) the time evolution of the system can be described by oscillations at frequency $2|g(\vec{r}_E)|$ between the states $|e, 0\rangle$ and $|g, 1\rangle$, where $|e, 0\rangle$ corresponds to one exciton in the QD and no photons in the cavity, and $|g, 1\rangle$ corresponds to zero excitons in the QD and one photon in the cavity. On the other hand, in the weak-coupling case, the irreversible decay rates dominate over the coherent coupling rate; in other words, the exciton–cavity field system does not have enough time to couple coherently before dissipation occurs.

Operation both in strong- and weak-coupling regimes has important consequences on the emission properties of the QD and the transmission properties of the coupled cavity/QD system. In the weak-coupling regime, the emission rate of the QD can be strongly enhanced via the Purcell effect [32] which is relevant for single-photon sources [33]. In this chapter we focus mainly on how the coupling of a single QD can modify the amplitude and the quantum statistics of coherent light transmitted through the cavity. In the weak-interaction limit excitation limit (when the average number of photons in the cavity is much smaller than one per cavity photon lifetime), the optical transmission function is given by the following formula:

$$T = \eta \left| \frac{\kappa}{i(\omega_c - \omega) + \kappa + \frac{g^2}{i(\omega_{QD} - \omega) + \gamma}} \right|^2, \tag{9.3}$$

where ω is the probe frequency, and η is a scaling factor that depends on the coupling efficiency of light into the cavity [6].

For typical parameters found experimentally in GaAs photonic crystal cavities coupled to InAs QDs ($\kappa/2\pi = 16GHz, \gamma/2\pi \sim 0.1GHz$), the normalized transmission function is shown in Fig. 9.3. What makes this system remarkable is that the presence of the coupled dipole can change the system from fully transparent to opaque even for modest values of the coupling rate g when the system is not in the strong-coupling regime [40]. This simple property is essential for implementing quantum repeaters for quantum information processing [40], or basic optoelectronic switches that operate at ultra-low energy levels. For switching applications the frequency of the quantum emitter can be controlled using external factors as local electric fields [12] or other optical fields [13].

9.3 Experimental techniques

9.3.1 Fabrication techniques

In our experiments, PCs are typically fabricated in 160 nm thick GaAs membranes grown by molecular beam epitaxy (MBE) on top of an $Al_xGa_{1-x}As$ sacrificial layer. The thickness is chosen to match half the QD emission wavelength ($\sim \lambda/2n$, where n is the GaAs

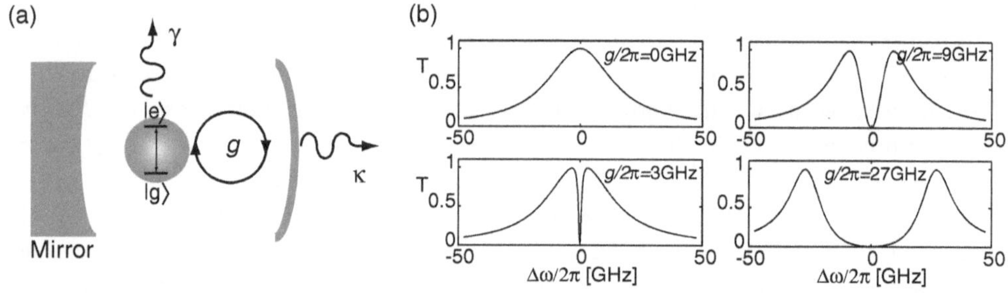

Figure 9.3 (a) Schematic representation of a one-sided optical cavity with a coupled optical dipole. The dipole couples to the cavity mode with rate g, the cavity field decay rate is κ and the spontaneous emission rate of the dipole is γ. (b) Theoretical transmission spectra of a coupled cavity/QD system ($\kappa/2\pi = 16\,GHz$, $\gamma/2\pi = 0.1\,GHz$) for different values of the coupling constant g. These spectra indicate that the transmission function of the cavity is significantly affected by the presence of the dipole even in the weak-coupling regime ($g < \kappa/2$)

refractive index). The membranes contain a QD layer in the middle. A distributed Bragg reflector (DBR) consisting of an AlAs/GaAs quarter wave stack was grown under the sacrificial layer. The DBR reflects all the light scattered from the QD into the GaAs substrate, thus improving collection efficiency in the out-of-plane direction.

The PCs are made using electron beam lithography and dry plasma etching. The pattern is first defined in the electron beam resist that is further used as an etch mask for the GaAs membrane. After patterning, the sacrificial layer is removed using wet etching, which results in suspended PC membranes. The concentration of Al in the sacrificial $Al_xGa_{1-x}As$ layer (x) can vary, depending on the desired wet etching rate (our best results were obtained with $x = 0.8$). Beside these basic semiconductor processing techniques, other steps may be required for more sophisticated devices involving electrical contacting or tuning.

9.3.2 Optical probing methods

During experiments, the sample is placed inside an optically accessible liquid-helium flow cryostat that maintains temperatures under 50 K. The optical measurement is done with a microscope setup. This type of setup, schematically shown in Fig. 9.4, uses the same lens both for focusing the probe light onto the sample and for collection of the output signal. In the case of resonant measurements, where the input and output signals have similar wavelengths, a cross-polarized configuration implemented using a polarizing beam splitter can be an efficient way to filter the output signal from the direct reflection of the input beam [6]. At the same time, wave-plates can be used to control the polarization of light incident on the sample and filters can be added in the output path to collect only part of the spectrum. The output signal can be sent to various measuring instruments (i.e. spectrometer, imaging camera, streak camera, photon counters) depending on the specific application.

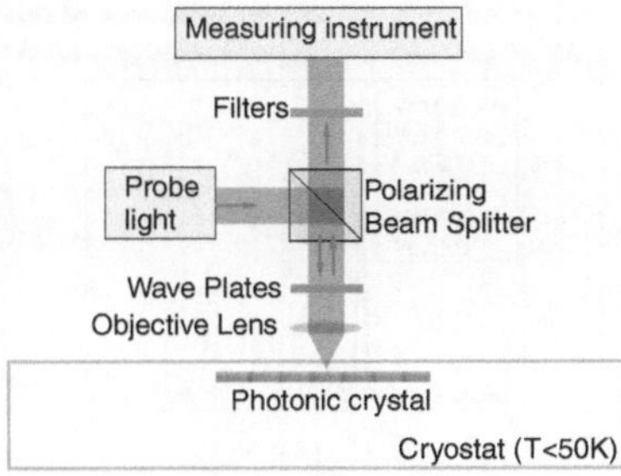

Figure 9.4 Schematic representation of an optical measurement setup used to probe photonic crystal cavities. Figure adapted from Ref. [12].

Depending on the wavelength of the probe light used to excite the QDs, there are two types of measurement: photoluminescence measurements and resonant measurements. In photoluminescence, the probe light excites free carriers that randomly recombine in the QD with the emission of photons that can be collected and measured. For this type of measurement, the probe is generally tuned above the GaAs bandgap. For resonant measurements, the probe is tuned on resonance with the cavity or the QD frequency.

9.3.3 Tuning and positioning techniques

As discussed in previous sections, InAs QDs have great advantages that make them excellent candidates for studying solid-state cavity quantum electrodynamics (CQED). However, one of the main drawbacks is their spectral inhomogeneous broadening and randomness in their spatial location (Fig. 9.1). This is caused by the limited amount of control over the self assembling process during the MBE growth. To overcome this difficulty, several frequency tuning and positioning techniques have been developed.

For frequency matching, the resonance of either the QD or the cavity can be controlled. One of the most common techniques to tune QDs and cavities into resonance is to change the temperature of the entire chip simply by controlling the temperature of the cryostat. The bandgap of both GaAs and InAs changes with temperature thus resulting in a shift of the QD emission lines. QDs can experience a red shift in their resonance of a couple of nanometers when temperature is changed from 4 K to about 50 K. At the same time, the cavity resonance is sensitive to temperature because of changes in the refractive index. Over the same temperature interval, the cavity resonance changes at a rate that is roughly three times slower than that of the QD.

Changing the temperature of the entire cryostat has some limitations: it is slow and all devices on the same chip are tuned at the same time. To overcome this difficulty, we developed local temperature tuning techniques that greatly improve the tuning speed and allow for simultaneous tuning of different cavities and QDs on the same chip [9]. Local temperature tuning was implemented by using the device shown in Fig. 9.5a whose temperature

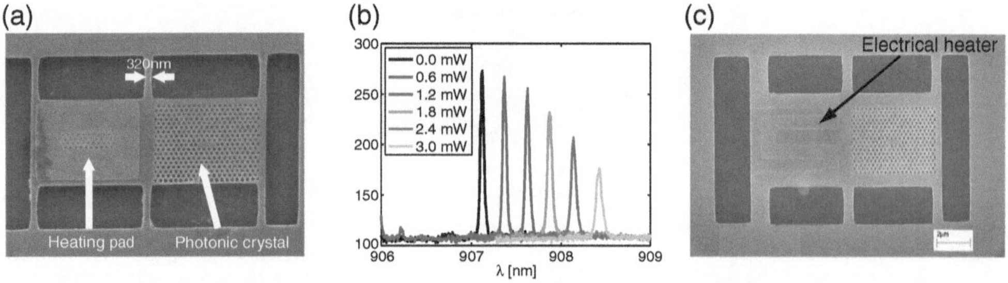

Figure 9.5 (a) Suspended structure with PC cavity and a metal-coated heating pad, used for local temperature tuning. (b) Local temperature tuning of a QD line while the power of the heating laser (measured before the focusing lens) was changed from 0 to 3 mW. (c) Scanning electron microscope of a PC structure equipped with a micron-scale electrical heater that can be used locally to control the temperature of the device. Figures adapted from Refs. [8, 9].

can be controlled via an external laser beam. The external laser was focused on a metal-coated heating pad located next to the cavity. The performance of the technique is shown in Fig. 9.5a, where a single QD line is tuned by up to 1.5 nm.

For larger-scale integration, where multiple components need to be controlled independently, it may be necessary for the temperature to be controlled via electrical heaters as demonstrated in Faraon and Vuckovic [8]. A prototype device that can be controlled in this way is shown in Fig. 9.5c. The temperature of devices of this type was controlled at speeds as high as 100 kHz, limited by the thermal relaxation time of the device.

Local temperature tuning provides fast and reliable control of the cavity and QD resonance. However, since both the resonator and the QD resonance depend on temperature, the technique does not provide enough control to tune into resonance several QDs and cavities on the same chip, as needed for future quantum networks. One solution may be a second local tuning technique that allows for fine control of the cavity resonance by using photorefractive materials [7]. The photorefractive material of choice was As_2S_3, a chalcogenide glass that changes its refractive index when excited with green light. Resonance shifts of up to 3 nm were observed for cavities resonant at 940 nm. During the tuning process the quality factor degraded from $Q \sim 7000$ to $Q \sim 4650$; however, the resulting Q is still sufficient to observe strong QD-cavity coupling.

The speed of the local temperature tuning techniques is limited by the thermal relaxation time of the PC device (typically below 1 MHz). For ultra-fast control, the QD can be modulated directly using electric fields [14] that modify the QD frequency via the quantum-confined Stark effect. We implemented [12] this type of control for QDs coupled to cavities (Fig. 9.6). A metallic electrode is placed next to the cavity such that it has a minimal overlap with the optical mode. When a bias voltage is applied, a depletion region and thus an electric field is created around the electrode. For voltages greater than $\sim 4\ V$ the depletion region reaches the QD in the cavity (Fig. 9.6b) and the QD experiences a frequency red shift. We were able to modulate the QD emission frequency up to speeds as

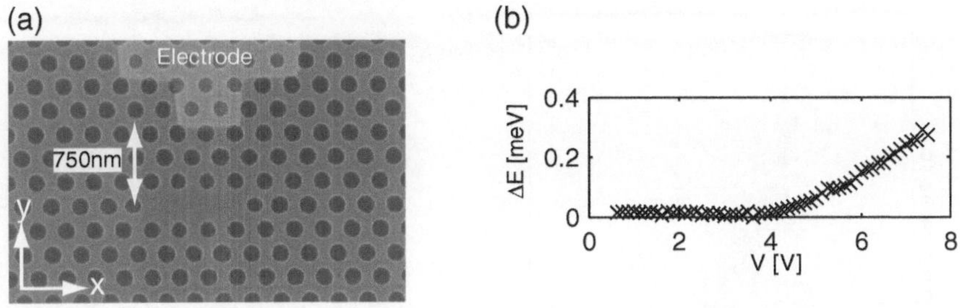

Figure 9.6 (a) Scanning electron microscope image of a PC cavity with a metallic electrode in close proximity. Electric fields can be created by applying a bias voltage on the electrode, thus tuning QDs coupled to the cavity. (b) Stark shift of a cavity-coupled QD as a function of the applied voltage. Figures adapted from Ref. [12].

high as 150 MHz (limited by the transmission line in the cryostat) but modulation speeds up to tens of GHz are possible using this method.

Beside the spectral alignment between the cavity and the QD, the second major challenge is the spatial alignment. During the MBE process, the QDs grow in random places on the wafer, which makes it difficult to fabricate the resonator such that it overlaps with the QD. For basic CQED experiments that employ a single cavity/QD system, it it possible to fabricate a large number of resonators and then search for one system where the cavity and the QD overlap [5, 42]. Another method developed by other groups is to precisely locate the position of the QD and then fabricate the resonator on top of it [15, 38]. In principle, this approach allows for more deterministic coupling of the QD to the cavity mode, and increases the likelihood that only a single QD is coupled to the cavity, but it is time consuming.

For more-complex devices that involve multiple QDs and cavities coupled in an on chip network, the alignment difficulty for all the techniques already described increases exponentially with the number of nodes in the network. Thus, to make this technology truly scalable, the QDs should be grown in predefined positions, which would solve the problem of spatial alignment. The spectral alignment could be done afterwards with local tuning techniques. There is an ongoing effort in growing QDs in predefined positions, but so far the radiative properties of these QDs did not match the ones of the self-assembled QDs [21, 34, 2, 43, 22].

9.4 Probing the strong-coupling regime

The strong-coupling regime occurs when the energy loss of the coupled cavity/QD system is low, such that a coherent back and forth energy transfer between the QD and the cavity mode can be sustained (see Section 9.2.3). Quantum mechanically, in this regime the states of the QD and the cavity become mixed into a new set of two eigenstates (with energies given by Eq. (9.2)) that have both cavity and QD character. One characteristic of systems in

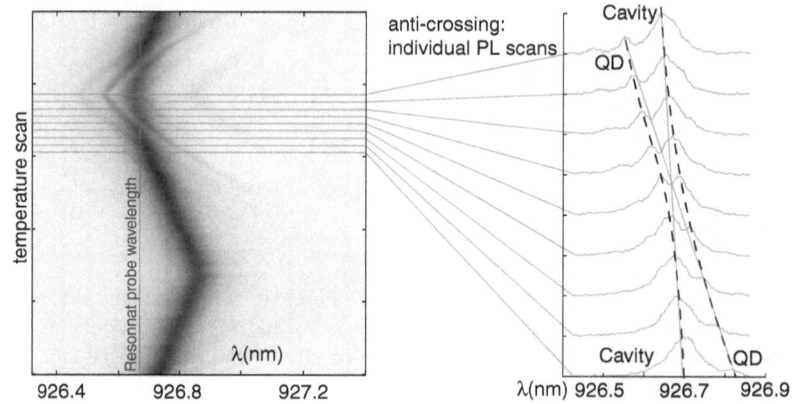

Figure 9.7 Cavity-QD anti-crossing in photoluminescence, a signature of the strong coupling regime, observed when tuning the QD across the cavity resonance using local temperature tuning. In the left panel the vertical axis represents the scan of the temperature with a triangular function. In the right panel, the dashed lines mark the wavelength of the polaritons, and the continuous lines mark the cavity and QD wavelength. Figures adapted from Refs. [6, 10].

the strong-coupling regime is that they should always display two spectral lines, regardless of the detuning between the cavity and the QD, a characteristic known as *anti-crossing* or *avoided crossing* of the spectral lines.

For QDs in micro-cavities, the anti-crossing can be easily observed in photoluminescence measurements as the QD is tuned on resonance with the cavity [42, 15, 30, 9]. The photoluminescence from a strongly coupled system is shown in Fig. 9.7 [6], where the tuning of the QD was done by local temperature tuning (see Section 9.3.3). The parameters of the system are: $\kappa/2\pi = 16$ GHz (linewidth 0.1 nm), Rabi frequency $g/2\pi = 8$ GHz (from Rabi splitting of $2g$ corresponding to 0.05nm), and $\gamma/2\pi \approx 0.1$ GHz. Since $g \approx \kappa/2$, the system operates on the onset of the strong-coupling regime where Rabi oscillations are only initiated, but a full oscillation is not completed.

Probing strong coupling via photoluminescence represents a very strong tool for characterization. However, for most applications the QDs-coupled cavities must be probed resonantly. Resonant probing enables the study of coherent interactions between the cavity, QD and laser fields. We reported resonant probing in Englund *et al.* [6], where the transmission function of the strongly coupled system shown in Fig. 9.7 was measured using a cross-polarized reflectivity measurement (see Section 9.3.2). For this measurement, the laser was kept at a constant resonant probe wavelength as marked in Fig. 9.7, while the cavity and QD were tuned using local temperature tuning. The expected intensity of the reflectivity signal is given by Eq. (9.3). When the QD is not coupled to the cavity (i.e. $g = 0$), the equation becomes the Lorentzian function that describes the transmission function of any resonator. A coupled QD ($g > 0$) will cause a drop in transmission (or reflectivity R) due to the coherent interaction with the laser and the cavity field. This

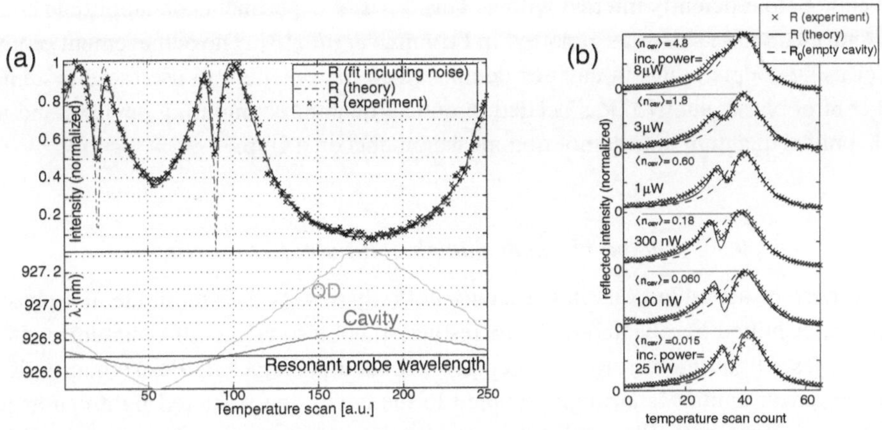

Figure 9.8 (a) Reflectivity spectrum as a function of the cavity/QD detuning. The bottom panel indicates the cavity, QD and probe laser wavelengths during the temperature tuning. (b) Dependence of the reflectivity spectrum with the incident power of the probe laser (measured before the objective lens). As the power is increased to levels above one photon per cavity lifetime ($\langle n_{cav} \rangle \sim 1$), the drop in reflectivity caused by the interaction with the QD vanishes. Figures adapted from Refs. [6, 10].

effect, shown in Fig. 9.8a, demonstrates that the transmission of a resonator can be controlled by using a single quantum emitter. This type of control is achieved for input optical powers corresponding to one photon per cavity lifetime (nanowatts). The system has ultra-high optical nonlinearities and can be controlled with minute power levels as discussed in the next sections.

9.5 Nonlinear optics at the single photon level

9.5.1 Amplitude and phase nonlinearities

Single QDs in PC cavities cause dramatic changes in the transmission function of the resonator, as previously discussed in Section 9.4. Because of the nonlinearity of the system, these changes are strongly dependent on the power of the probe laser as shown in Fig. 9.8b. The reflectivity is shown for various incident powers of the probe laser and the corresponding average photon number in the cavity (n_{cav}). The saturation behavior is modeled using a steady-state solution of the quantum master equation (solid line) as described in Tan [36]. The dashed curve shows the reflectivity ratio if no thermal fluctuations were present. At large power, both curves tend to unity as the QD/cavity spectrum approaches the Lorentzian shape of the empty cavity.

This optical nonlinearity can be exploited to control both the transmission and the phase of light interacting with the cavity, with control energies at the single photon level. Experiments can be designed where the intensity and the phase of a probe laser resonant with the QD can be controlled via a second laser beam that is detuned from the QD frequency but close enough to efficiently interact with it. This types of experiments on amplitude control and controlled phase shifts are reported in Fushman *et al.* [13]. The achievement of phase control using coupled QDs in cavities demonstrated the potential to use these systems to build control phase gates [37] as needed in optical quantum computing devices, and may enable optical quantum non-demolition measurements on a chip.

9.5.2 Photon blockade and photon-induced tunneling

The presence of a coupled QD in the cavity not only affects the amplitude and phase of the light transmitted or reflected from the resonator, but also affects its quantum statistics. While the laser light interacting with a empty resonator preserves its Poissonian statistics, when an optical emitter is strongly coupled to the cavity the scattered light can acquire non-classical character, either sub-Poissonian or super-Poissonian. This is the result of the interaction between the laser field and a quantum system that has eigenstates with anharmonic spacing.

In strong-coupling regimes, the energy eigenstates of the system are grouped in two-level manifolds with eigen-energies given by $n\omega_c \pm g\sqrt{n}$ (for $\omega_{QD} = \omega_c$), where n is the number of energy quanta in the cavity–QD system. The eigenstates can be written as:

$$|n, \pm\rangle = \frac{|g, n\rangle \pm |e, n-1\rangle}{\sqrt{2}}, \tag{9.4}$$

where $|g\rangle$ and $|e\rangle$ are the ground and excited states of the QD, and $|n\rangle$ is the photon number state of the optical mode. The energy splitting between the eigenstates in each manifold is nonlinear in n and given by $2g\sqrt{n}$ (Fig. 9.9). This anharmonicity in the splitting of the energy eigenstates gives rise to nonlinear optics phenomena at the single photon level, like photon blockade [16, 4] and photon-induced tunneling. In the case of photon blockade, the presence of one photon in the cavity blocks subsequent photons to enter the resonator. For example, let's consider that a coherent laser source is coupled to the cavity (Fig. 9.9a) and is resonant to one of the polaritons, say $|1, -\rangle$ (continuous arrow in Fig. 9.9b). Once a photon is coupled, the system is excited into the state $|1, -\rangle$, so the coupling of another photon with energy $\omega_c - g$ would require the system to transition to energy state $2(\omega_c - g)$. However, the system does not have an eigenstate at this energy, the closest being at $2\omega_c - g\sqrt{2}$, so the probability of coupling the second photon is reduced. Thus, it is preferential for only a single photon at at time to couple into the cavity and then to be emitted at the output port (sub-Poissonian statistics).

Another interesting regime of operation is photon-induced tunneling [10], when the laser is resonant with a n-photon transition to one of the levels in the nth order manifold ($\omega_c - g/\sqrt{n}$) (dashed arrow in Fig. 9.9b). In this case, coupling a single photon into the system has a low probability, but once a photon is coupled it becomes preferential for more photons to tunnel into the cavity to match the energy level in the nth order manifold. At the output port, it is more probable that photons exit in bunches (super-Poissonian statistics).

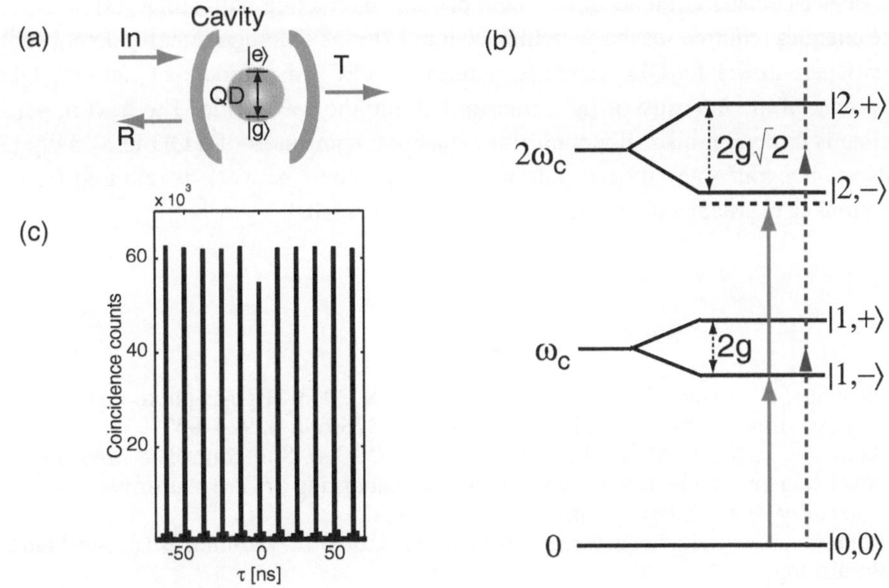

Figure 9.9 (a) Schematic representation of a single QD coupled to an optical resonator. (b) Anharmonic ladder of energy eigenstates for a strongly coupled cavity-QD system. The solid arrows indicate the one-photon transition to $|1, -\rangle$, while the dashed arrows indicate the n-photon transition (n large) to $|n, \pm\rangle$ (c) Measured second-order correlation data for laser pulses reflected from the cavity/QD system. The laser is tuned at the photon blockade frequency so photon anti-bunching is observed ($g^{(2)}(0) = 0.91$). Figures adapted from Refs. [6, 10].

These effects can be measured experimentally via second-order correlation measurements on the output field [10]. We probed photon blockade and photon-induced tunneling using laser pulses reflected from a strongly coupled cavity/QD system. The experimental data indicating photon anti-bunching (sub-Poissonian statistics) due to photon blockade are shown in Fig. 9.9c. The probing of photon blockade and photon-induced tunneling demonstrates optical nonlinearities at the single photon level in solid state systems. These types of nonlinearities open the possibility to develop deterministic sources of non-classical light and optical logic devices that operate at ultra-low power levels.

9.6 Applications and future directions

In the previous sections we discussed basic experiments that demonstrate how a QD strongly coupled to a cavity mode can be used to control both the amplitude and the quantum statistics of a light beam transmitted through the cavity. We also presented experimental techniques for local control of QDs and cavities on the chip. These experiments lay down the foundation for future quantum photonic devices. One example is the electro-optic switch based on a single QD that we recently developed [12]. The optical switching in this devices occurs at the level of single photons interacting with single excitons, with minute energies required for the switching action. In terms of integration, we demonstrated devices where the cavity/QD system is connected to PC waveguides [11] and the QD is used to control the intensity of light transmitted into the waveguide. The next important challenge is to deterministically control the quantum spin states of a QD in a cavity [31] and also to integrate these devices into a complex photonic network as required for most applications in classical and quantum information processing.

References

[1] Akahane, Y., Asano, T., Song, B.-S. and Noda, S. 2003. High-Q photonic nanocavity in a two-dimensional photonic crystal. *Nature*, **425**(6961), 944–947.

[2] Atkinson, P., Ward, M. B., Bremner, S. P. *et al.* 2006. Site-control of inAs quantum dots using *ex-situ* electron-beam lithographic patterning of GaAs substrates. *Japanese Journal of Applied Physics*, **45**(4A), 2519–2521.

[3] Benson, O. and Henneberger, F. 2009. *Semiconductor Quantum Bits*. Pan Stanford Publishing.

[4] Birnbaum, K. M., Boca, A., Miller, R. *et al.* 2005. Photon blockade in an optical cavity with one trapped atom. *Nature*, **436**, 87–90.

[5] Englund, D., Fattal, D., Waks, E. *et al.* 2005. Controlling the spontaneous emission eate of single quantum dots in a two-dimensional photonic crystal. *Phys. Rev. Lett.*, **95**(013904).

[6] Englund, D., Faraon, A., Fushman, I. *et al.* 2007. Controlling cavity reflectivity with a single quantum dot. *Nature (London)*, **450**(7171), 857–861.

[7] Faraon, A., Englund, D., Bulla, D. *et al.* 2008a. Local tuning of photonic crystal cavities using chalcogenide glasses. *Appl. Phys. Lett.*, **92**(043123).

[8] Faraon, A. and Vučković, J. 2009. Local temperature control of photonic crystal devices via micron-scale electrical heaters. *Appl. Phys. Lett.*, **95**(4), 043102.

[9] Faraon, A., Englund, D., Fushman, I. *et al.* 2007. Local quantum dot tuning on photonic crystal chips. *Appl. Phys. Lett.*, **90**(213110), 213110.

[10] Faraon, A., Fushman, I., Englund, D. *et al.* 2008b. Coherent generation of non-classical light on a chip via photon-induced tunnelling and blockade. *Nature Physics*, **4**(11), 859–863.

[11] Faraon, A., Fushman, I., Englund, D. *et al.* 2008c. Dipole induced transparency in waveguide coupled photonic crystal cavities. *Optics Express*, **16**(16), 12154.

[12] Faraon, A., Majumdar, A., Kim, H., Petroff, P. and Vučković, J. 2010. Fast electrical control of a quantum dot strongly coupled to a photonic-crystal cavity. *Phys. Rev. Lett.*, **104**(4), 047402.

[13] Fushman, I., Englund, D., Faraon, A. 2008. Controlled phase shifts with a single quantum dot. *Science*, **320**(5877), 769–772.

[14] Heller, W., Bockelmann, U. and Abstreiter, G. 1998. Electric-field effects on excitons in quantum dots. *Phys. Rev. B*, **57**(11), 6270–6273.

[15] Hennessy, K., Badolato, A., Winger, M. *et al.* 2007. Quantum nature of a strongly coupled single quantum dot-cavity system. *Nature (London)*, **445**(Feb), 896–899.

[16] Imamoğlu, A., Schmidt, H., Woods, G. and Deutsch, M. 1997. Strongly interacting photons in a nonlinear cavity. *Phys. Rev. Lett.*, **79**(8), 1467–1470.

[17] Imamoglu, A., Awschalom, D. D., Burkard, G. *et al.* 1999. Quantum information processing using quantum dot spins and cavity QED. *Phys. Rev. Lett.*, **83**(20), 4204–4207.

[18] Jalali, B. and Fathpour, S. 2006. Silicon photonics. *J. Lightwave Technol.*, **24**(12), 4600–4615.

[19] Joannopoulos, J. D. 1995. *Photonic Crystals: Molding the Flow of Light.* Princeton University Press.

[20] Kimble, H. J. 1994. *In Cavity Quantum Electrodynamics*, edited by P. Berman. San Diego: Academic Press.

[21] Kohmoto, S., Nakamura, H., Ishikawa, T. and Asakawa, K. 1999. Site-controlled self-organization of individual InAs quantum dots by scanning tunneling probe-assisted nanolithography. *Applied Physics Letters*, **75**(22), 3488–3490.

[22] Lee, J., Saucer, T. W., Martin, A. J. *et al.* 2011. Photoluminescence imaging of focused ion beam induced individual quantum dots. *Nano Letters*, **11**(3), 1040–1043.

[23] Loss, D. and DiVincenzo, D. P. 1998. Quantum computation with quantum dots. *Physical Review A*, **57**(1), 120–126.

[24] Michler, P. 2003. *Single Quantum Dots: Fundamentals, Applications, and New Concepts*. Springer-Verlag: Topics in Applied Physics.

[25] Miller, D. 2009. Device requirements for optical interconnects to silicon chips. *Proceedings of the IEEE*, **97**(7), 1166–1185.

[26] Noda, S. 2006. Recent progresses and future prospects of two- and three-dimensional photonic crystals. *J. Lightwave Tech.*, **24**(12), 4554–4567.

[27] Noda, S., Chutinan, A. and Imada, M. 2000. Trapping and emission of photons by a single defect in a photonic bandgap structure. *Nature*, **407**, 608–610.

[28] O'Brien, J. L., Furusawa, A. and Vučković, J. 2009. Photonic quantum technologies. *Nature Photonics*, **3**(12), 687–695.

[29] Ozbay, E. 2006. Plasmonics: merging photonics and electronics at nanoscale dimensions. *Science*, **311**(5758), 189–193.

[30] Press, D., Götzinger, S., Reitzenstein, S. *et al.* 2007. Photon antibunching from a single quantum-dot-microcavity system in the strong coupling regime. *Phys. Rev. Lett.*, **98**(11), 117402.

[31] Press, D., Ladd, T. D., Zhang, B. and Yamamoto, Y. 2008. Complete quantum control of a single quantum dot spin using ultrafast optical pulses. *Nature*, **456**(7219), 218–221.

[32] Purcell, E. M. 1946. Spontaneous emission probabilities at radio frequencies. *Physical Review*, **69**, 681.

[33] Santori, C., Fattal, D., and Yamamoto, Y. 2010. *Single-photon Devices and Applications*. Wiley-VCH.

[34] Schneider, C., Strauss, M., Sunner, T., *et al.* 2008. Lithographic alignment to site-controlled quantum dots for device integration. *Appl. Phys. Lett.*, **92**(18), 183101.

[35] Scully, M. O. and Zubairy, M. S. 1997. *Quantum Optics*. Cambridge: Cambridge University Press.

[36] Tan, S. M. 1999. A computational toolbox for quantum and atomic physics. *J. Opt. B*, **1**, 424–432.

[37] Thompson, R. J., Turchette, Q. A., Carnal, O. and Kimble, H. J. 1998. Nonlinear spectroscopy in the strong-coupling regime of cavity QED. *Phys. Rev. A*, **57**(4), 3084–3104.

[38] Thon, S. M., Rakher, M. T., Kim, H. *et al.* 2009. Strong coupling through optical positioning of a quantum dot in a photonic crystal cavity. *Appl. Phys. Lett.*, **94**(11), 111115.
[39] Vahala, K. J. 2003. Optical microcavities. *Nature*, **424**(6950), 839–846.
[40] Waks, E. and Vučković, J. 2006. Dipole induced transparency in drop-filter cavity-waveguide systems. *Phys. Rev. Lett.*, **96**(153601).
[41] Walls, D. F. and Milburn, G. J. 2008. *Quantum Optics*. Springer: Springer-Verlag Berlin Heidelberg.
[42] Yoshie, T., Scherer, A., Hendrickson, J. *et al.* 2004. Vacuum Rabi splitting with a single quantum dot in a photonic crystal nanocavity. *Nature*, **432**(Nov.), 200–203.
[43] Zwiller, V., Aichele, T., Hatami, F., Masselink, W. T. and Benson, O. 2005. Growth of single quantum dots on preprocessed structures: single photon emitters on a tip. *Appl. Phys. Lett.*, **86**(9), 091911.

10

Photon statistics in quantum dot micropillar emission

M. Aßmann and M. Bayer

10.1 Introduction

The ever-growing demand for fast optical data transmission calls for lasers offering high modulation rates and low energy consumption at the same time. Advances in growth and processing methods make quantum dot (QD) based lasers better candidates for this challenge than ever before. Placed in microresonators able to confine light in regions roughly the size of their wavelength, QDs pave the way to ultra-low threshold lasing. The most common resonator geometries aimed at three-dimensional light confinement are microdisks, photonic crystal membrane cavities and micropillars. The latter are especially good candidates for realizing microlasers suitable for applications as they offer directed emission and allow for parallel device processing. However, this increased efficiency also results in modified emission properties of QD lasers [8]. Semiconductor-specific processes like Pauli-blocking of states, the composite nature of the carriers involved and Coulomb interactions between carriers cause deviations from the standard atomistic laser picture. The main aim of our studies is to characterize microlaser emission in terms of photon statistics and coherence properties. Following Glauber, the most detailed description of a light field is given in a series of correlation functions describing coherence in different orders [10].

This chapter is organized as follows. Section 10.2 contains a brief review on the characteristic properties of micropillar lasers and discusses the emission properties of microlasers operated below and above threshold. Section 10.2.1 focuses on photon statistics and the classification of light fields. Section 10.3 presents different measurement techniques to determine second- and higher-order correlations. Section 10.4 gives an overview of time-integrated and time-resolved coherence properties of QD micropillar lasers of different sizes and material systems. Finally, Section 10.5 concludes this overview and gives an outlook on future challenges.

10.2 Theoretical background

Embedding QDs in a microcavity opens up the possibility to tailor their coupling to the electromagnetic field. We are focusing on the applications of microcavities in terms of

Quantum Dots: Optics, Electron Transport and Future Applications, ed. Alexander Tartakovskii. Published by Cambridge University Press.

lasers. For a comprehensive overview of other applications and the physics of QDs placed in microcavities in general, we refer the reader to [21].

Most interesting in terms of lasers is the so-called Purcell effect [20, 5, 16]. Cavities with high quality and small mode volume offer a modified local density of states compared to free space causing the spontaneous emission rate into a mode to be enhanced or suppressed. One of the important parameters characterizing a laser is the spontaneous emission coupling factor β, which is given by the ratio of the spontaneous emission directed into the lasing mode Γ_l and the total spontaneous emission rate into lasing and non-lasing modes $\Gamma_l + \Gamma_{nl}$:

$$\beta = \frac{\Gamma_l}{\Gamma_L + \Gamma_{nl}}. \tag{10.1}$$

High β-factors are indicators for low-threshold lasers. In small micropillar lasers, the high-quality modes are spectrally well separated, so the QDs couple predominantly to a single mode and the Purcell effect can be utilized to tailor the spontaneous emission into that mode. Other high-quality modes and leaky modes form the non-lasing dissipation channels. Advances in growth and design of QD-microcavity lasers made it possible to closely approach the optimal β-factor of unity [24, 19, 22, 25] which is characteristic for a so-called thresholdless laser. However, systems offering such high β-factors are difficult to characterize as the kink in the input–output curve, usually an unambiguous indicator for the onset of lasing, disappears gradually, as shown in Figure 10.1 [23]. To characterize such systems, it is necessary to consider the coherence properties of the emission as discussed in the following section.

10.2.1 Photon statistics

Besides obvious parameters like frequency, polarization or intensity, there are also further characteristics of light fields which manifest in their coherence properties, which can be described by a hierarchy of correlation functions stating with the field–field correlation function

$$g^{(1)}(\vec{r}_1, t_1, \vec{r}_2, t_2) = \frac{\langle E^-(\vec{r}_1, t_1) E^+(\vec{r}_2, t_2)\rangle}{\sqrt{\langle |E(\vec{r}_1, t_1)|^2\rangle \langle |E(\vec{r}_2, t_2)|^2\rangle}}. \tag{10.2}$$

E^- and E^+ denote the negative and positive frequency parts of a mode of the light field, respectively, and $g^{(1)}$ is a measure of phase correlations of a light field and reflects in the contrast of interference patterns of the em field. The two common quantities deduced from $g^{(1)}$ are the coherence time τ_{coh} and the correlation length l_{coh} which give the time and distance over which phase correlations are maintained, respectively. Still, a complete characterization of em fields which is also able to identify nonclassical states requires consideration of correlation functions of at least second order. Neglecting any spatial dependencies, the normal-ordered second-order photon number correlation function is given by $g^{(2)}$:

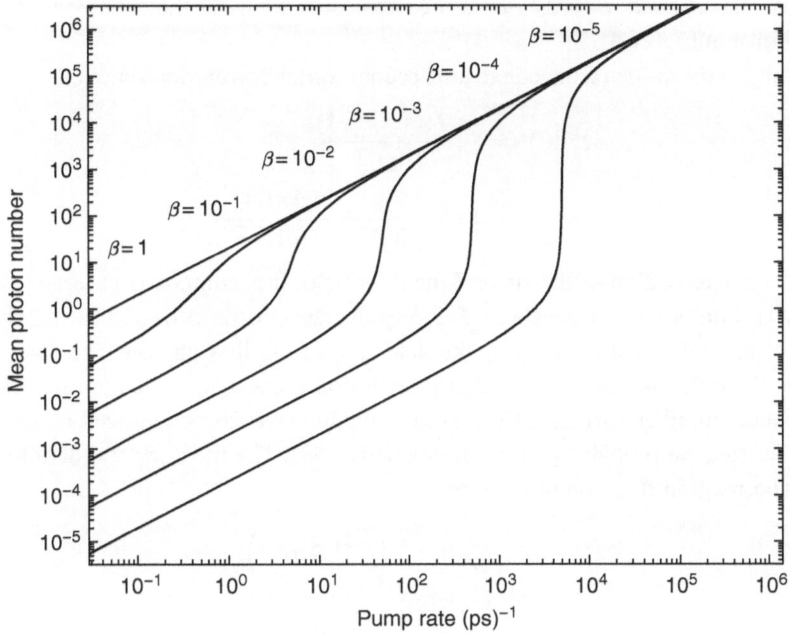

Figure 10.1 Input–output curves calculated from laser rate equations for β decreasing from 1 to 10^{-5} from top to bottom.

$$g^{(2)}(t_1, t_2) = \frac{\langle \hat{a}^\dagger(t_1)\hat{a}^\dagger(t_2)\hat{a}(t_1)\hat{a}(t_2)\rangle}{\langle \hat{a}^\dagger(t_1)\hat{a}(t_1)\rangle\langle \hat{a}^\dagger(t_2)\hat{a}(t_2)\rangle}, \tag{10.3}$$

where $\hat{a}^\dagger$ and $\hat{a}$ are the photon creation and destruction operators for the photon mode of interest, respectively. The normal ordering assures that the change of the state of the em field introduced by the detection of a photon is taken into account. For stationary light sources $g^{(2)}$ will depend only on the relative delay τ between two photon detections:

$$g^{(2)}(\tau) = \frac{\langle \hat{a}^\dagger(t)\hat{a}^\dagger(t+\tau)\hat{a}(t)\hat{a}(t+\tau)\rangle}{\langle \hat{n}(t)\rangle\langle \hat{n}(t+\tau)\rangle}, \tag{10.4}$$

where the averages are time averages and $\hat{n}$ denotes the photon number of the mode of interest. Here, $g^{(2)}$ can be considered as the conditional probability to detect a photon at a delay τ after the detection of a first photon, normalized to the probability of a second detection for photons which are emitted statistically independent of each other. For very large delays $\tau \to \infty$ the photon emission events are uncorrelated in any state of the light field, so $g^{(2)}(\tau \to \infty) = 1$. It is possible to distinguish three basic kinds of states of the light field, namely thermal, coherent and nonclassical light, by comparing $g^{(2)}(\tau = 0)$ to $g^{(2)}(\tau \to \infty)$ depending on whether the probability for simultaneous detection of two photons is increased, unaltered or decreased. The enhanced or decreased photon pair detection probability relaxes back towards 1 on a timescale depending on the coherence time of the light as shown in Figure 10.3.

Accordingly, the value of the equal-time second-order correlation function

$$g^{(2)}(0) = \frac{\langle n(t)(n(t)-1)\rangle}{\langle n(t)\rangle^2} \tag{10.5}$$

$$= 1 - \frac{1}{\langle n(t)\rangle} + \frac{\langle(\Delta n(t))^2\rangle}{\langle n(t)\rangle^2} \tag{10.6}$$

is a good characterization of the state of the light field. It is composed of three terms. The first term is a unity-valued constant. The negative second term describes the change of the state of the light field induced by the detection of the first photon. The positive third term takes the intrinsic noise of the photon-emission process into account in terms of the relative photon number variance. Correlation functions can be generalized up to arbitrary order to describe the probability of n-photon detections. The most general definition of an nth-order correlation function is given by

$$g^{(n)}(t_1 \ldots t_n) = \frac{\left\langle : \prod_{i=1}^{n} \hat{n}(t_i) : \right\rangle}{\prod_{i=1}^{n} \langle \hat{n}(t_i) \rangle}. \tag{10.7}$$

The double stops denote normal ordering of the underlying photon creation and annihilation operators. The statistical properties of coherent, thermal and nonclassical light will be discussed in more detail in the following sections.

10.2.2 Statistical properties of coherent, thermal and Fock states

This section discusses the statistical properties of different light fields. The expected outcomes in measurements of $g^{(2)}$ are discussed. Thermal light is emitted if the system under investigation can be considered as an ensemble of emitters in thermal equilibrium with a radiation field and showing the corresponding emission and absorption rates. Examples of this kind of emitters are incandescent light bulbs or even stars. The corresponding photon number distribution follows Bose–Einstein statistics. In the case of having a large number of states and photons, the probability of detecting n photons for a light field of mean photon count rate $\langle n\rangle$ can be simplified to a geometric distribution:

$$P_{th}(n, \langle n\rangle) = \frac{\langle n\rangle^n}{(1+\langle n\rangle)^{1+n}}. \tag{10.8}$$

One of the characteristic features of this distribution lies in the fact that the most probable photon number is always $n = 0$ regardless of the mean photon number as shown in Figure 10.2. As the distribution also decreases monotonically, it is obvious that the photon number variance $\langle(\Delta n(t))^2\rangle$ must be rather large for large mean photon numbers. In fact, the variance is given by $\langle(\Delta n(t))^2\rangle = \langle n\rangle^2 + \langle n\rangle$ [17]. As can be seen from Eq. (10.6),

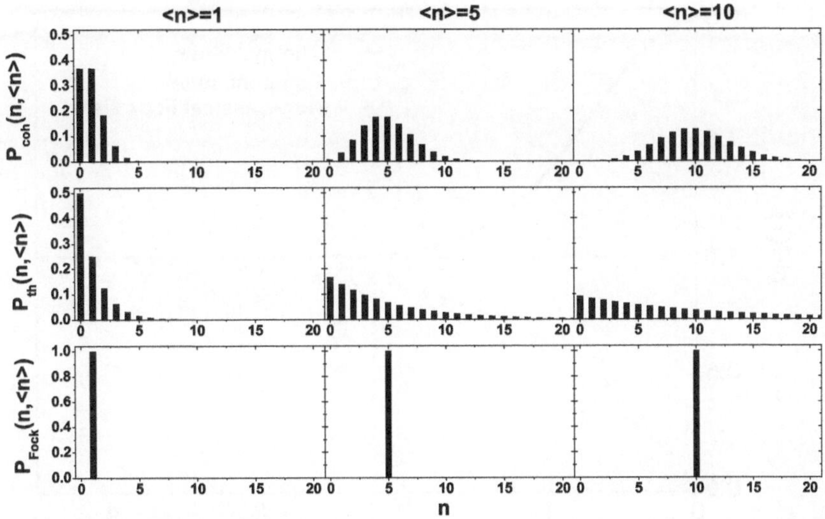

Figure 10.2 Photon number probability distributions for coherent (upper row), thermal (middle row) and Fock (bottom row) states of mean photon numbers $\langle n \rangle$ equal to 1, 5 and 10.

this results in $g^{(2)}(0) = 2$. This result also applies for pseudo-thermal light sources like Martienssen lamps [18]. Coherent states are minimum uncertainty states with uncertainty equally distributed between photon number and phase and also eigenstates of the photon annihilation operator. It can be shown that the latter requirement results in the photon number distribution being Poissonian [7]:

$$P_{coh}(n, \langle n \rangle) = e^{-\langle n \rangle} \frac{\langle n \rangle^n}{n!}. \tag{10.9}$$

As can be seen in Figure 10.2, this distribution is peaked at $\langle n \rangle$. The variance of this distribution is given by $\langle (\Delta n(t))^2 \rangle = \langle n \rangle$, resulting in $g^{(2)}(0) = 1$. This is a remarkable result as it reflects several properties of coherent states. First, it indicates that the joint detection probability factorizes into the product of the mean photon detection rates and therefore shows that the emitted photons are statistically independent. Second, it shows that the detection of one photon does not alter the state of the light field in this case, which demonstrates that coherent states are indeed eigenstates of the photon annihilation operator and also the closest analog to classical light fields there is.

Finally, Fock states are non-classical eigenstates of the photon number operator and have a simple photon number distribution:

$$P_{Fock}(n, \langle n \rangle) = \delta_{n,\langle n \rangle}. \tag{10.10}$$

This fact reflects in a vanishing variance $\langle (\Delta n(t))^2 \rangle = 0$ and a photon number dependent $g^{(2)}(0) = 1 - \frac{1}{\langle n \rangle}$.

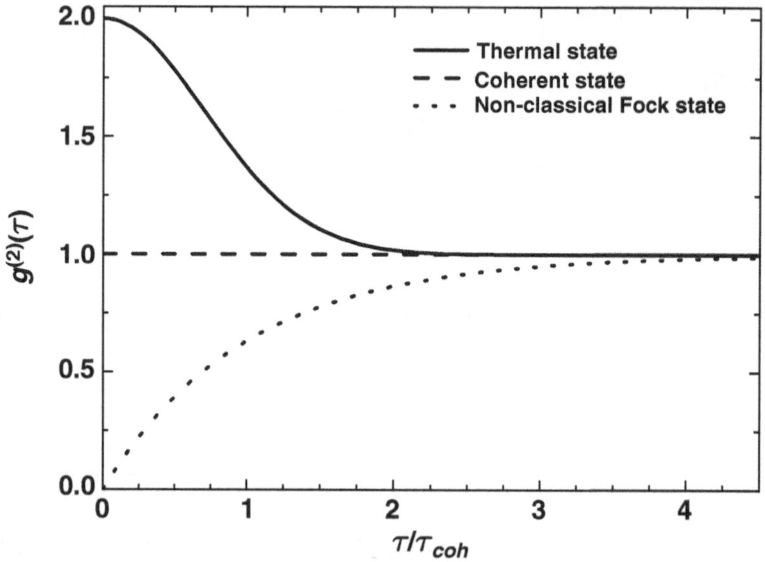

Figure 10.3 Correlation function $g^{(2)}(\tau)$ for stationary thermal (solid line), coherent (dashed line) and single photon (dotted line) states of the light field.

Regardless of the value of $g^{(2)}(0)$, $g^{(2)}(\tau)$ will return towards unity for delays with a decay constant on the order of the coherence time τ_{coh} for all stationary light fields as shown in Figure 10.3.

10.3 Experimental approaches

There is a wide variety of experimental techniques applied to measure $g^{(2)}(\tau)$. The most direct method consists of keeping track of the times of detection events of a single-photon detector followed by a straightforward calculation of $g^{(2)}(\tau)$ according to Eq. (10.4). However, directly determining photon statistics in this manner is a difficult task because such measurements are limited to timescales longer than the detector dead time. For detectors with single-photon sensitivity, such as single-photon avalanche photodiodes (SPADs), the dead time is typically on the order of several tens of nanoseconds. Several experimental approaches to overcome this limitation have been proposed and realized. This section gives a brief overview of the most common experimental techniques.

10.3.1 Hanbury Brown–Twiss setups

Hanbury Brown–Twiss type setups [11] overcome the dead-time limitation by using at least two independent detectors. The signal is split using a beam splitter and distributed evenly among the detectors. In such a scenario the second detector can still be active during the dead time of the first and a histogram of the delay times between consecutive photon detection events can be recorded by using the signals from the two diodes as start and stop events for a time-to-amplitude converter.

If the dark-count rate is negligible and the mean photon detection rate is much smaller than the inverse detector dead time, the resulting histogram is a measure of $G^{(2)}(\tau)$ for long integration times [26]. The upsides of this technique are the easy implementation, high detector efficiencies and the flexibility allowing to perform auto- or cross-correlation measurements with only minor changes to the setup. Drawbacks are the SPAD temporal resolution, which is usually on the order of a few hundred picoseconds for efficient SPADs and therefore limits the timescales that can be studied [15], the need to measure the normalization separately, and the growing complexity of the setup when trying to measure higher-order correlations as each additional order involves the usage of another detector.

10.3.2 Two-photon absorption setups

Two-photon absorption (TPA) in semiconductors is a well-known effect that can be visualized as an electronic transition from the valence band to the conduction band via an intermediate virtual state in the energy gap E_g. The TPA transition rates are proportional to $\langle \hat{a}^\dagger(t)\hat{a}^\dagger(t)\hat{a}(t)\hat{a}(t)\rangle$ and therefore a measure of $G^{(2)}(0)$. In order for such a transition to occur, two photons must be absorbed within a time interval given by the Heisenberg lifetime $\hbar/E_g$, which is on the order of a few femtoseconds for E_g corresponding to the visible light spectral range, allowing for excellent temporal resolution. TPA occurs for any photon energy between 0.5 E_g and E_g, usually resulting in a reasonable detection bandwidth [6]. To obtain the time-dependent $G^{(2)}(\tau)$, interferometric methods can be applied, e.g. by placing a Michelson interferometer in the beam path and scanning the relative delay between the two paths. In this case $G^{(2)}(\tau)$ can be determined by the following relationship:

$$\frac{I(\tau)}{I_1 + I_2} = 1 + G^{(2)}(\tau) + Re(F_2(\tau)e^{-2i\omega t}) + 4Re(F_1(\tau)e^{-i\omega t}), \qquad (10.11)$$

where $I(\tau)$ is the TPA signal at the detector (usually a photocathode operated in photon-counting mode) and I_1 and I_2 are the signals obtained using the two arms of the interferometer separately. The phase-interference terms proportional to F_1 and F_2 oscillate rapidly and can be filtered out at low frequency. The remaining term allows us to determine $G^{(2)}(\tau)$. Measurements of higher-order correlations are possible by interferometric means [12] or by using materials suitable for multi-photon absorption instead of TPA. Advantages of this technique are the unmatched temporal resolution and the efficiency in the infrared spectral region. Drawbacks lie in the complicated nature of extending this measurement scheme to non-stationary sources and the indistinguishablity between two photons and several photons being present as both cause a single detection event on a photocathode operated in photon counting mode.

10.3.3 Streak camera setups

Streak cameras convert the incoming light field to photoelectrons by using a photocathode, accelerate them and then deflect them by using electric fields synchronized with the excitation laser repetition rate applied orthogonally to the photoelectron direction of motion. They then hit a phosphor screen on a position depending on the relative time of arrival of the initial photon. The resulting afterglow is the signal that can be recorded using a CCD camera and gives the time dependence of the light field with a temporal resolution of down to 2 ps. Using a streak camera for correlation measurements requires further customization. Usually the streak image is recorded by integrating over many repeated streak cycles. However, information about correlations is present only in single pictures and gets washed out by this integration. It is therefore necessary to record images of single signal pulses. This prerequisite significantly reduces the possible data acquisition rate due to the limited readout rates achievable for CCDs. State-of-the-art CCDs with sufficient quantum efficiency for recording single streak pictures can usually be operated at 100 Hz at best. However, it is possible to add a second slow deflection voltage orthogonally to the other one in order to record several pulses on one screen. Doing so allows us to record up to 40 pulses on one screen and allows for an effective repetition rate of 4 kHz. From those pictures it is then possible to calculate $G^{(2)}(\tau)$ by counting all the photon pairs in these pulses and creating a histogram [27]. The normalization constant can easily be determined by averaging over all the single pulses, which yields the average count rates at the respective times [2]. Advantages of this technique lie in the possibility to determine correlations between all photons inside one pulse – not just consecutive ones – with picosecond temporal resolution and the ability to measure time-resolved correlations even for non-stationary fields. Also, it is possible to extract second- and higher-order correlations from a single measurement run [1] . Drawbacks are the low quantum efficiency of the approach and the need to reconstruct the single photon detections from the pattern recorded by the CCD in real time.

10.4 Experimental results

In this section we investigate the coherence properties of QD micropillar lasers by using the streak camera technique. Three different QD micropillar lasers with different characteristics were studied. All of them have nominally cylindrical shape. One sample based on a II–VI material system and two III–V material-system-based samples were used. The II–VI-based sample was grown by molecular beam epitaxy. The distributed Bragg reflectors consist of 15 upper layers and 18 bottom layers in which $ZnS_{0.06}Se_{0.94}$ (48 nm) layers were used as high-index material and a 25.5-period MgS (1.7 nm)/ZnCdSe (0.6 nm) superlattice was used as low-index material. The central λ cavity contains a single sheet of CdSe/ZnSe quantum dots with an approximate density of $\sim 5 \times 10^{10}\,\mathrm{cm}^{-2}$. A pillar with 1.5 μm diameter was used for the measurements. The cavity quality factor was estimated to be $\sim$1850.

The III–V micropillar samples were grown by molecular beam epitaxy on a GaAs substrate. The distributed Bragg reflectors consist of 20 upper and 23 lower alternating layers

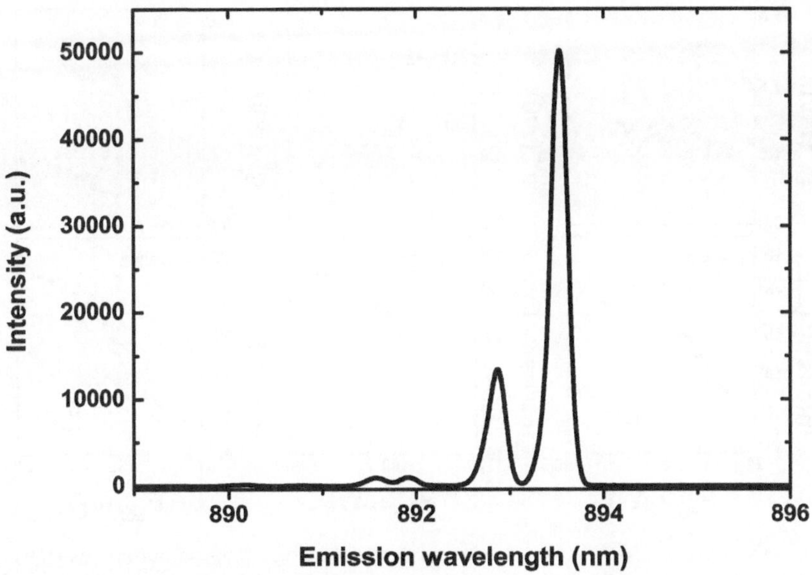

Figure 10.4 Longitudinal emission mode spectrum of the low-Q micropillar laser. The fundamental mode is twofold degenerate and shows emission at 893.53 nm. The polarization splitting cannot be resolved. The first four excited modes can also be seen.

of AlAs (79 nm)/GaAs (67 nm) $\lambda/4$ pairs for the low-Q micropillar and 26 upper and 33 lower alternating layers of AlAs (74 nm)/GaAs (68 nm) $\lambda/4$ pairs for the high-Q micropillar. The central λ cavity contains one layer of self-assembled InGaAs quantum dots with a density of $\sim 3 \times 10^{10}\,\mathrm{cm}^{-2}$ in the low-Q case and one layer of self-assembled AlGaInAs quantum dots with a density of $\sim 6 \times 10^{9}\,\mathrm{cm}^{-2}$ in the high-Q case, from which cavities with diameters of several micrometres were fabricated by means of high-resolution electron beam lithography and plasma-induced reactive ion etching. Micropillars with diameters of 5 μm (low-Q) and 8 μm (high-Q) were used for the experiments. The Q-factors are deduced from the linewidths of the fundamental modes seen in the modal spectra. Taking the finite resolution of the spectrometer into account, they are estimated as 9000 and 19000, respectively. A typical mode spectrum taken at high excitation power under non-resonant, pulsed optical excitation of 1 mW for the low-Q III–V micropillar is shown in Figure 10.4. The fundamental mode is located at 893.53 nm.

The first excited mode can be seen at an emission wavelength of 892.85 nm. This difference is large enough to single out the fundamental mode using an interference filter with 1 nm spectral width.

10.4.1 Time-integrated measurements

The main aim of the measurements was to investigate the basic emission properties of the three considered samples under varying excitation power. At low excitation densities, a

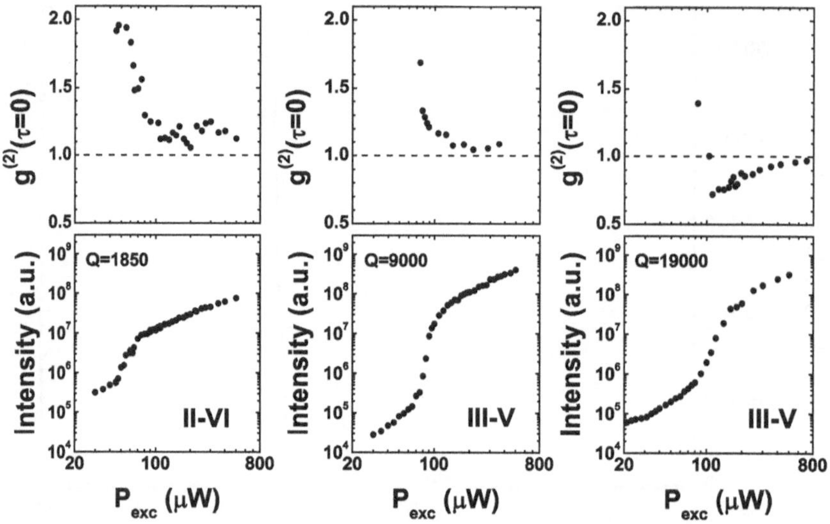

Figure 10.5 Values of $g^{(2)}(0)$ (upper panel) and corresponding input–output curves (lower panel) for three different QD lasers. The left column shows results for a 1.5 μm diameter II–VI cavity. The other columns show results for III–V cavities of diameters 5 μm (middle column) and 8 μm (right column). Dotted lines in the upper panel denote the coherent limit.

broad emission peak from the QD ensemble is seen, superimposed by a series of narrow high intensity peaks marking the microcavity modes. With increasing excitation density, the integrated fundamental mode intensities of all pillars show a characteristic slope change in double-logarithmic plots as shown in the lower panel of Figure 10.5. This nonlinearity marks the onset of stimulated emission in the microlaser structures. The nonlinear region is apparently broadened over a range of excitation densities. For all three samples the width in excitation powers of this broadened region roughly equals the excitation power at its onset which complicates the definition of a well defined lasing threshold. Determining the β-factor is another nontrivial task as the samples operate in a regime where the kink in the IO curve does not scale with β^{-1}. Theoretical analysis reveals β-factors on the order of ~0.1 for the III–V cavities and a slightly higher β-factor on the order of ~0.13 for the II–VI cavity [27]. Complementary measurements of the equal-time second-order correlation function using the experimental setup presented in Section 10.3.3 allow for a more detailed analysis of the properties of the emitted light. Results are shown in the upper panel of Figure 10.5. At high excitation densities far above the lasing threshold all samples show lasing emission identified by values of $g^{(2)}(0)$ of approximately 1 which are clear evidence for the Poissonian nature of the underlying photon statistics. The low-Q III–V and the II–VI sample are still subject to some small excess fluctuations which manifest in values of $g^{(2)}(0) = 1.1$ and 1.2, respectively. The origin of this small amount of excess noise is not completely clear. Possible reasons include relevant contributions of spontaneous emission from early and late times in the pulse and efficient cavity feeding effects [28, 4, 14]. Further studies on the influence of spontaneous emission at different emission times inside the pulse will be described in Section 10.4.2. Below threshold, the behavior is rather different for the three samples. None of the samples shows a saturation of $g^{(2)}(0)$ at a value of 2 as would be expected for a classical low-β laser. For the II–VI cavity $g^{(2)}(0)$ saturates for low excitation powers at a value of $\approx 1.9 - 1.95$. The small difference from the expected value for a low-β laser is a manifestation of a limited number of emitters. A detailed treatment shows that thermal radiation emitted by a system consisting of a fixed number N of emitters shows an emitter-number-dependent second-order correlation function [13] already without considering further cavity-QED effects:

$$g_{th}^{(2)}(0) = 2\left(1 - \frac{1}{N}\right). \tag{10.12}$$

Values between 1.9 and 1.95 are therefore expected if 20–40 QDs contribute to the emission in the II–VI cavity. Considering the QD density of this sample this number corresponds to roughly 8 % – 17 % of the QDs inside the micropillar coupling to the the cavity mode. Considering the spectral overlap between the distribution of the QD emission energies and the cavity mode and possible cavity feeding effects, this value is reasonable. Interpretation of the results below threshold is more complicated for the III–V samples. For these structures there is no resolvable saturation of $g^{(2)}(0)$ below threshold. The quantum efficiency of the S-20 photocathode inside the streak camera is about one to two orders of magnitude worse for the wavelength range around 900 nm where the III–V samples operate compared to the 500 nm range where the II–VI sample operates. Correspondingly background noise contributes more strongly and the necessary measurement duration becomes much longer already in the threshold region for the III–V samples. Under these circumstances, it becomes impossible to monitor the photon statistics below threshold. The highest measured values of $g^{(2)}(0)$ are about 1.7 for the low-Q sample and approximately 1.4 for the high-Q sample.

10.4.2 Time-resolved measurements

For large τ, $g^{(2)}(\tau)$ must necessarily return to 1 as photons emitted with a large delay are statistically independent. For thermal light, straight application of classical coherence theory predicts a decay from the value at $\tau = 0$ towards a value of unity on a timescale of the order of the coherence time of the light. The dynamical evolution of the second-order correlations of the III–V cavities, shown for selected excitation powers in Figure 10.6, differs from this prediction. For the low-Q cavity (left panel) and excitation below the threshold region, $g^{(2)}(\tau)$ drops smoothly from a value slightly below two towards unity within the first picoseconds. However, for delays of around 40 ps, values below unity appear. For the high-Q cavity (right panel) and excitation in the threshold region, pronounced long-lasting oscillations of $g^{(2)}(\tau)$ are visible. The oscillations become obviously damped with increasing excitation power. Above threshold, we find that $g^{(2)}(\tau) \approx 1$ in both the low-Q case and the high-Q case. The observed oscillations in the photon correlations are a result

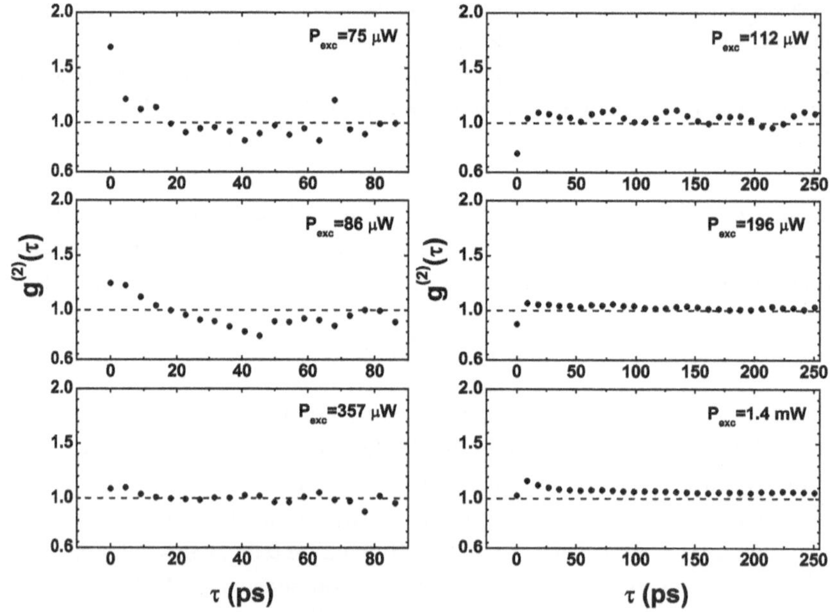

Figure 10.6 Temporal evolution of $g^{(2)}$ for selected pump powers. Results are shown for the low-Q (left panel) and high-Q (right panel) III–V cavities of Figure 10.5. An unexpected outcome is the appearance of dynamical antibunching $g^{(2)}(\tau) < 1$ for low and intermediate excitation densities at finite τ for the low-Q cavity. The intensity correlation function does not merely drop to a value of unity, but takes on smaller values with subsequent oscillations. This is also particularly apparent for the high-Q cavity at $P_{exc} = 112$ mW. In the threshold region the high-Q cavity shows antibunching caused by the small number of quantum dots coupling to the cavity.

of the dynamical coupling between photons and carriers and can be qualitatively understood as follows. For a microcavity operated at steady state in the spontaneous-emission regime, $g^{(2)}(\tau)$ decays from nearly two to one on the timescale of the coherence time. In the regime of dominating stimulated emission, $g^{(2)}(\tau)$ equals one independently of time delay. The oscillations are observed in the regime of transition from spontaneous to stimulated emission in a system that, under these conditions, contains only very few photons emitted by very few quantum dots. This is the transition regime of cavity QED lasers discussed in [23]. Unlike in the situation in the lasing regime, here the loss of a photon from the cavity represents a severe perturbation of the system, which strongly influences the coupled carrier–photon dynamics. Systems of emitters coupled to a cavity mode are known to exhibit different kinds of oscillations of the emission intensity. Relaxation oscillations can occur close to the threshold region when the laser is switched on or perturbed, and Rabi oscillations can occur in the regime in which the dissipation is small in comparison with the light–matter coupling strength. The dynamics of the correlation function $g^{(2)}(\tau)$ can be traced back to this behavior. Theory predicts that both kinds of oscillations can be triggered by photon emission events. In reaction to the perturbation, the system tries to re-establish equilibrium, and, in doing so, undergoes quantum oscillations. In both cases, the origin of these oscillations is the feedback due to the cavity, which can lead to out-of-phase oscillations of photon number and lasing medium. These oscillations become damped as the pump rate increases and a regime of stimulated emission is reached in which the photon number is high enough that single photon losses no longer affect the system considerably. In the case discussed before, the perturbation of the few-emitter system can become so prominent that, for example, the subsequent emission of a photon is suppressed. This leads to the dynamic antibunching both for zero delay and for times after enhanced photon pair emission during the oscillations. These oscillations potentially also appear in intensity measurements. However, as the moment of photon emission is stochastic, any time averaging blurs the oscillations. Nevertheless, the oscillations also carry over to the correlation functions similar to $g^{(2)}(\tau)$. There they survive the averaging, as although the moment of arrival of the first photon is still stochastic, a second photon is picked whose delay, τ, relative to the first is fixed for all detected photon pairs. For increasing Q, the cavity feedback is enhanced, causing the quantum fluctuations to become more pronounced as reflected by the oscillations of $g^{(2)}(\tau)$.

It is also worthwhile to study the dynamics during the build-up and break-down of coherence during a pulse. The starting point of such a study is again the time-resolved recording of the individual photon emission events in the output pulse which provides a complete mapping of the second-order correlation function of the emitted light in t and τ. Instead of determining the τ-dependence by calculating the intensity-weighted average of $g^{(2)}(t, \tau)$ over a time interval t during the emission pulse, as done in the earlier discussions, now the t-dependence at a vanishing photon pair delay time of $\tau = 0$ is the central quantity to be studied [3]. While the former treatment gives a good characterization of the photon statistics and coherence time of the emitted pulse as a whole, the latter is a good measure of the time-resolved second-order coherence properties and therefore also the build-up and break-down of coherence during a pulse. For these measurements a micropillar with 6 μm diameter was used. All other characteristics of this pillar are comparable to the low-Q III–V micropillar discussed before. The transition region to lasing was identified by measuring the input–output curve shown in the inset of Figure 10.7. A nonlinear behavior between excitation powers of 60 and 150 μW marks this region. At lower excitation powers, $g^{(2)}(t, 0)$ takes on the expected static value of 2 without showing any dynamics. More interesting are the excitation powers (filled squares in Figure 10.7) in and above the threshold region for which the intensity-dependent second-order coherence properties of the emission were determined. Figure 10.7 shows $g^{(2)}(t, 0)$ for these excitation powers alongside the temporal emission-intensity profiles. The light exhibits thermal behavior at the very beginning and at the very end of the emission pulse. After the generation of carriers in the barrier states by the pump pulse, these carriers rapidly relax into the QD states [9]. As long as a small number of carriers is present in the QD states, spontaneous recombination processes determine the output. When the population becomes sufficiently strong, the system is driven into the regime of coherent emission, characterized by a decrease in the second-order correlation function toward the value of 1, and by a faster decay of the

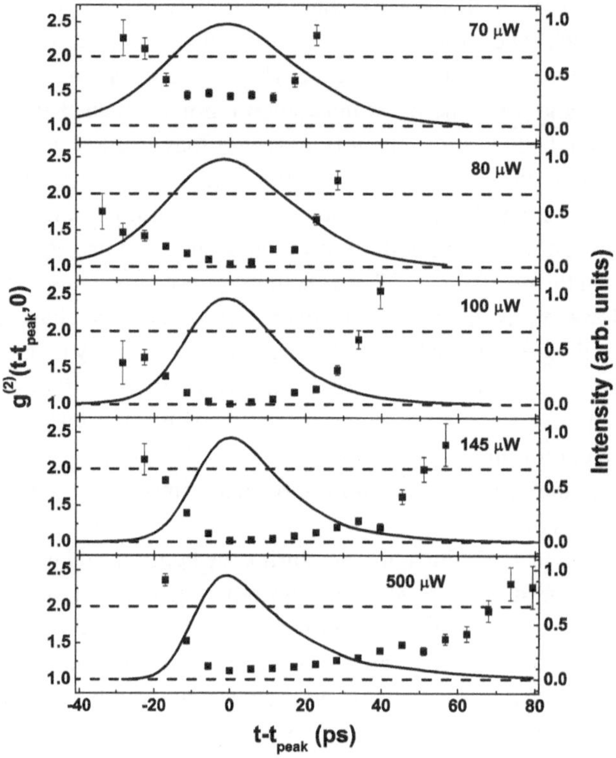

Figure 10.7 Time evolution of the second-order photon correlation function $g^{(2)}(t, 0)$ (symbols) compared to the normalized output intensity (solid lines) for the micropillar fundamental mode of a pillar nominally identical to the one shown in the middle column of Figure 10.5. Dashed lines denote the limiting cases of $g^{(2)}(t, 0)$ equal to unity or two for coherent and thermal light, respectively. The power density for pulsed excitation increases from top to bottom, and t_{peak} corresponds to the time of the maximum of the emission intensity for each pump power.

emission intensity due to the stimulated processes, visible in an apparent temporal narrowing of the emission peak. The decrease toward $g^{(2)}(t, 0) = 1$ becomes more pronounced for higher excitation densities, finally leading to a broadening of the dip, as more carriers are excited in the system and stimulated emission can be maintained for a longer time. While the transition from thermal to coherent emission in the beginning of the pulsed emission can take 40 ps or even more in the threshold region, it happens on a timescale on the order of 10–15 ps far above threshold.

10.5 Summary and outlook

During the past few years the interest in experimental techniques for characterization of light fields on ultrashort timescales has increased significantly. In this chapter, we have addressed the time-resolved coherence properties of the emission from QD micropillar lasers by using a streak camera measurement scheme. While it could be shown that structures with low β-factors and many QDs coupling to the fundamental mode behave as expected, intriguing effects like antibunching on short timescales and behavior similar to relaxation oscillations have been found for high-quality resonators with only few QDs coupling to the fundamental mode. Despite these first insights, there are still many open questions and challenges concerning the physics of efficient QD micropillar lasers. First of all, the simplifying picture of having a fixed number of QDs coupling to the cavity mode is not necessarily justified under all circumstances. It has been shown that even QDs that are strongly detuned from the cavity mode can couple to the cavity by means of non-resonant QD–cavity coupling, for example via phonons or the build-up of a continuum of multi-exciton transitions for strongly excited systems. In systems where these effects become important, the number of QDs coupling to the cavity might depend strongly on the experimental parameters and might even become time-dependent. As the distribution of the QD exciton ground-state energies is always broad for self-assembled QDs, the efficiency of QD based lasers might be significantly improved by harvesting such mechanisms and having initially off-resonant QDs couple to the cavity mode in a controlled manner. Characterizing the micropillar laser emission on ultrafast timescales provides a first step towards understanding those phenomena.

References

[1] Aßmann, M., Veit, F., Bayer, M., van der Poel, M. and Hvam, J. M. 2009. Higher-order photon bunching in a semiconductor microcavity. *Science*, **325**(5938), 297.

[2] Aßmann, M., Veit, F., Tempel, J.-S. *et al.* 2010a. Measuring the dynamics of second-order correlation functions inside a pulse with picosecond time resolution. *Opt. Express*, **18**, 20229.

[3] Aßmann, M., Veit, F., Bayer, M. *et al.* 2010b. Ultrafast tracking of second-order photon correlations in the emission of quantum-dot microresonator lasers. *Phys. Rev. B*, **81**(16), 165314.

[4] Ates, S., Ulrich, S. M., Ulhaq, A. *et al.* 2009. Non-resonant dot–cavity coupling and its potential for resonant single-quantum-dot spectroscopy. *Nature Photonics*, **3**, 724–728.

[5] Bayer, M., Reinecke, T. L., Weidner, F. *et al.* 2001. Inhibition and enhancement of the spontaneous emission of quantum dots in structured microresonators. *Phys. Rev. Lett.*, **86**(14), 3168–3171.

[6] Boitier, F., Godard, A., Rosencher, E. and Fabre, C. 2009. Measuring photon bunching at ultrashort timescale by two-photon absorption in semiconductors. *Nat. Phys.*, **5**, 267–270.

[7] Gerry, C. C. and Knight, P. L. 2005. *Introductory Quantum Optics*. First edition. Cambridge University Press.

[8] Gies, C., Wiersig, J., Lorke, M. and Jahnke, F. 2007. Semiconductor model for quantum-dot-based microcavity lasers. *Phys. Rev. A*, **75**(1), 013803.

[9] Gies, C., Wiersig, J. and Jahnke, F. 2008. Output characteristics of pulsed and continuous-wave-excited quantum-dot microcavity lasers. *Phys. Rev. Lett.*, **101**(6), 067401.

[10] Glauber, R. J. 1963. Coherent and incoherent states of the radiation field. *Phys. Rev.*, **131**(6), 2766–2788.

[11] Hanbury Brown, R. and Twiss, R. Q. 1956. A test of a new type of stellar interferometer on Sirius. *Nature*, **178**, 1046–1048.

[12] Hayat, A., Nevet, A. and Orenstein, M. 2010. Ultrafast partial measurement of fourth-order coherence by HBT interferometry of upconversion-based autocorrelation. *Opt. Lett.*, **35**(5), 793–795.

[13] Hennrich, M. 2003. *Kontrollierte Erzeugung Einzelner Photonen in Einem Optischen Resonator Hoher Finesse*. Ph.D. thesis, TU München.

[14] Laucht, A., Kaniber, M., Mohtashami, A. *et al.* 2010. Temporal monitoring of non-resonant feeding of semiconductor nanocavity modes by quantum dot multiexciton transitions. *Phys. Rev. B*, **81**(24), 241302.

[15] Li, G., Zhang, T. C., Li, Y. and Wang, J. M. 2005. Photon statistics of light fields based on single-photon-counting modules. *Phys. Rev. A*, **71**(2), 023807.

[16] Lodahl, P., Floris van Driel, A., Nikolaev, I. S. *et al.* 2004. Controlling the dynamics of spontaneous emission from quantum dots by photonic crystals. *Nature*, **430**, 654–657.

[17] Loudon, R. 2000. *The Quantum Theory of Light*. Vol. 3. Oxford Science Publications.

[18] Martienssen, W. and Spiller, E. 1964. Coherence and fluctuations in light beams. *American Journal of Physics*, **32**(12), 919–926.

[19] Park, H.-G., Kim, S.-H., Kwon, S.-H. *et al.* 2004. Electrically driven single-cell photonic crystal laser. *Science*, **305**(5689), 1444–1447.

[20] Purcell, E. M., Torrey, H. C. and Pound, R. V. 1946. Resonance absorption by nuclear magnetic moments in a solid. *Phys. Rev.*, **69**(1-2), 37–38.

[21] Reitzenstein, S. and Forchel, A. 2010. Quantum dot micropillars. *J. Phys. D: Applied Physics*, **43**(3), 033001.

[22] Reitzenstein, S., Bazhenov, A., Gorbunov, A. *et al.* 2006. Lasing in high-Q quantum-dot micropillar cavities. *Appl. Phys. Lett.*, **89**(5), 051107.

[23] Rice, P. R. and Carmichael, H. J. 1994. Photon statistics of a cavity-QED laser: a comment on the laser–phase-transition analogy. *Phys. Rev. A*, **50**(5), 4318–4329.

[24] Solomon, G. S., Pelton, M. and Yamamoto, Y. 2001. Single-mode spontaneous emission from a single quantum dot in a three-dimensional microcavity. *Phys. Rev. Lett.*, **86**(17), 3903–3906.

[25] Strauf, S., Hennessy, K., Rakher, M. T. *et al.* 2006. Self-tuned quantum dot gain in photonic crystal lasers. *Phys. Rev. Lett.*, **96**(12), 127404.

[26] Ulrich, S. M., Gies, C., Ates, S. *et al.* 2007. Photon statistics of semiconductor microcavity lasers. *Phys. Rev. Lett.*, **98**(4), 043906.

[27] Wiersig, J., Gies, C., Jahnke, F. *et al.* 2009. Direct observation of correlations between individual photon emission events of a microcavity laser. *Nature*, **460**, 245–249.

[28] Winger, M., Volz, T., Tarel, G. *et al.* 2009. Explanation of photon correlations in the far-off-resonance optical emission from a quantum-dot–cavity system. *Phys. Rev. Lett.*, **103**(20), 207403.

11

Nanoplasmonics with colloidal quantum dots

V. V. Temnov and U. Woggon

11.1 Introduction

Colloidal semiconductor nanocrystals were the first model systems to evidence radius-dependent energy shifts of excitonic states caused by three-dimensional quantum confinement (for reviews see e.g. [1, 2, 3, 4, 5, 6, 7, 8, 9, 10] and references therein). Besides the large tunability of optical emission wavelength, such nanocrystal quantum dots exhibit high quantum efficiency q and can nowadays replace organic dyes in various applications, e.g. act as biomarkers or active laser medium. In the broad field of nanoplasmonics, combinations of metallic nanostructures with semiconductor nanocrystals have potential applications in optoelectronics. The enhancement of spontaneous emission rate near metallic surfaces makes nanocrystals attractive candidates as probes of electromagnetic field distribution and nano-antenna effects. Charge separation at the metal–semiconductor interface can be applied in photocatalytic processes as was recently shown for production of hydrogen in multicomponent metal–semiconductor nanocrystal structures [11, 12]. The modification of optical and electronic properties of colloidal nanocrystals close to a metallic surface is a longstanding issue of research and will be reviewed in this chapter, in particular with a focus on the latest developments. We will start with an overview of fundamental properties of colloidal quantum dots along with a presentation of the most recent results in the field of functionalized colloidal nanostructures. We consider semiconductor and metallic nanostructures separately and describe their properties as individual building blocks for future complex nanosystems. In the following sections we deal with coupling schemes of quantum dots to metal surfaces, discuss practical applications in all-optical plasmonic devices and outline perspectives in quantum optics with surface plasmons.

11.2 Optical and electronic properties of colloidal semiconductor quantum dots

After a period of tremendous progress in physical understanding and application of size-dependent properties, nanocrystal quantum dots have gained renewed attention. New material combinations, geometries, morphologies and specific functionalities have been presented in recent literature (for reviews see e.g. [13, 14]). Of particular interest are effects of hybridization, a term which is used in different meanings to describe the coupling

Quantum Dots: Optics, Electron Transport and Future Applications, ed. Alexander Tartakovskii. Published by Cambridge University Press.

of quantum states. Reports are published about hybridization of two nanocrystal wave functions to form quantum dot molecules, hybridization of cavity photons and quantum confined nanocrystal excitons, hybridization between exciton states of nanocrystals of different sizes mediated by the interaction with the evanescent electromagnetic field of a surface plasmon in a metallic nanostructure, hybrid nanoparticles and nanomaterials created by arranging nanosized domains of metallic, semiconducting, and magnetic materials into a single nanostructure, or hybridization of 0D-quantum states with 2D-quantum states in structures of mixed dimensionality. Such artificially designed nanostructures allow us to engineer nanomaterials with multiple functionalities for novel device architectures.

While the combination of various materials that may have very different structural and physical properties is a fascinating task, the knowledge and understanding of individual components and their optical, structural and electronic properties are the foundations of any engineering. Thus we will start with an overview about basic properties of colloidal semiconductor nanocrystals such as anisotropy and exchange splitting, morphology and shape control, to provide later an overview about coupled metal–semiconductor nanostructures.

The traditional core/shell semiconductor nanocrystal is a nanoparticle containing a semiconductor core with sizes in the range of 2–20 nm surrounded by a wide-gap semiconductor shell with a thickness of a few nanometers. The most prominent system is the CdSe/ZnS core/shell nanocrystal system which was first introduced to enhance the quantum efficiency of CdSe core nanocrystals by an additional confining semiconductor shell. A great advantage of colloidal nanocrystal quantum dots, besides the wide tunability of optical transition energies, is the large cross-section for light absorption in the UV–VIS spectral range (large optical dipole moment), high photostability and quantum yield. Since the field of functionalization of colloidal nanocrystal quantum dots is a vividly developing and very interdisciplinary one, a long list of widely discussed and very important questions exists. Examples are the control of electron and hole localization via suited energy gap and band offset combinations between the core and the shell, in particular for narrow band-gap semiconductors; theoretical calculations which allow to predict localization and transfer mechanisms in novel structures; the role of shells in the process of forming nanocrystal complexes, aggregates, molecules and chains; the dynamics of energy and charge transfer in hybrid structures containing core/shell nanocrystals and their engineering by varying the properties of the shell. Great progress has been made in the accurate theoretical description of the optical properties of core/shell nanocrystals and their aggregates, in particular if the size falls into the range between atomistic models and the solid state effective mass approximation (for a recent review see [15]).

There has been much work on synthetic procedures of semiconductor nanocrystals to enable shape control giving rise to various architectures such as elongated quantum dots (QDs), tetrapods, nanowires and quantum rods (QRs) (see e.g. [16, 17, 18, 19, 20] and references therein). The energy band gap of various quantum rods has been determined by optical and STM measurements [21, 22, 23]. The results confirm that the band gap of CdSe quantum rods (QRs) depends mainly on the width (the dimension of the electronic confinement) and only slightly on the length. The homogeneous linewidth and coupling to phonons have been compared in CdSe dots and rods by spectral hole burning [24, 25] and high-resolution spectroscopy [26] at low temperature. Rods have a larger homogeneous linewidth than nanodots, ~ 20 μeV vs. 6 μeV, respectively.

In comparative studies of nanocrystal rods and dots it has been shown that nanorods exhibit nearly one order of magnitude larger absorption cross-section (scaled to their volume) [20]. In part this is caused by different local field factors in rods and dots, however, in CdSe nanorods the optical selection rules depend also on the nanocrystal shape [27, 28, 29, 30, 31, 32, 33]. Very large anisotropic exchange interactions result in significant mixing between the bright and dark exciton states. In single CdSe/ZnS core–shell nanorods (NR) a fine structure splitting between 1 and 2 meV is caused by exciton exchange interaction. A shape-dependent change in the symmetry of the 1D-exciton ground state transforms the NR exciton optical transition from a dark state into a bright state at radii below a critical radius of $R = 3.7$ nm. This hypothesis is supported by the measured emission dynamics of NR ensembles as a function of temperature and its comparison with spherical NCs of similar radius $R = 3$ nm (see Fig. 11.1). While the NC data can be fitted well by the standard model of thermally activated population of the bright state from the dark state which gives the radiative decay times of 15 ns and 142 ns respectively, an exchange splitting energy of 1.4 meV and a transfer time between these two states (requiring a spin flip) of 1.4 ns. The NR decay time is shorter for similar radii and rises from 3.6 ns at $T = 5$ K with increasing

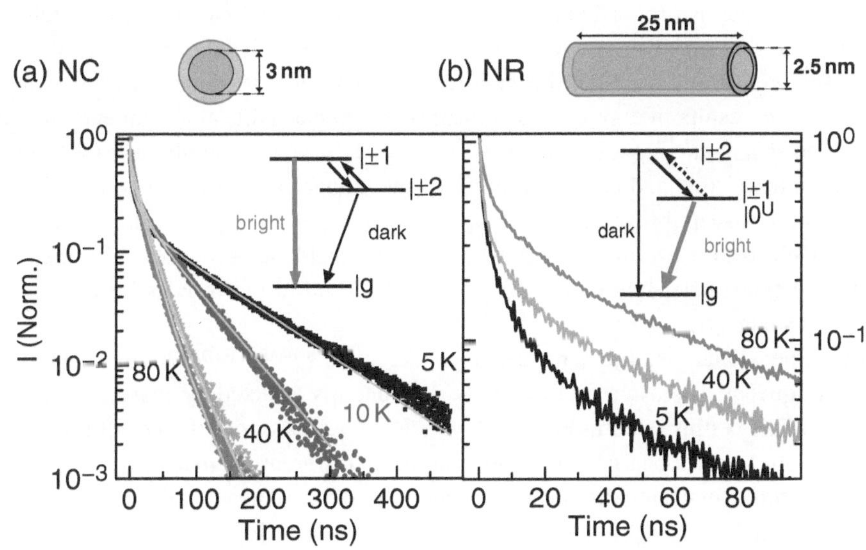

Figure 11.1 Ensemble photoluminescence dynamics in (a) spherical CdSe/ZnS core/shell quantum dots (core radius $R = 3$ nm) and (b) CdSe/ZnS nanorods ($R = 2.5$ nm, $L = 25$ nm). Temperature dependence of ensemble photoluminescence decay times is different for spherical nanocrystals and nanorods, see Ref. [30].

temperature according to a $1.2\sqrt{T}$[ns$\sqrt{\text{K}}$] law. The relaxation processes occurring within the states of the exciton fine structure in QDs are reviewed in [34, 35].

In recent years significant advances in the understanding of how to control nanocrystal morphology have been made. The morphology of nanocrystals, i.e. its domain structure, determines how nanocrystal complexes and different shapes can be formed. In addition, it determines conditions of anchoring metallic particles or organic ligands to nanocrystals, creating nanocrystal heterostructures or seeded growth of nanostructures (see e.g. [36, 37] and references therein). The control of reaction chemistry by using nanocrystal catalysts to promote anisotropic growth or through the directed attachment of preformed nanocrystals is standard nowadays. Surface coatings and ligand design can significantly help to stabilize nanoparticles or to alter their optical properties in a reproducible way. The spatial variation of the electronic band gap at the nanoscale, i.e. the build up of direct or indirect electronic transitions, has a strong impact on electron–hole transfer processes across a heterostructure. An overview about some of the most recent developments can be found in [13, 14, 38, 39, 40, 41] and references therein.

11.3 Surface plasmons in low-dimensional metallic nanostructures

When the electrons in bulk metal are displaced with respect to the ions, the charge separation induces an electric field, which tends to restore the electrical neutrality. This relaxation process is governed by oscillations of the electrons at the so-called plasma frequency $\omega_{\rm p} = \sqrt{\frac{n_e e^2}{\epsilon_0 m}}$, which is proportional to the electron density n_e. Since in most metals there is at least one free electron per ion, the resulting electron plasma density is very high ($n_e \sim 10^{22}$ cm^{-3}) and $\omega_{\rm p}$ is in the near-ultraviolet frequency range. High plasma frequency $\omega_{\rm p} > \omega$ results in exponential screening of optical fields at a frequency ω within the so-called *skin depth*, $\delta_{\rm skin}$. For the majority of metals the skin depth lies in the range of 10–20 nm in the visible and near-infrared frequency range. A simple Drude model for the linear dielectric susceptibility of a metal $\epsilon_m(\omega) = 1 - \omega_{\rm p}^2/\omega(\omega + i/\tau_c)$, where τ_c stands for an effective electron scattering time, represents a reasonable approximation for the optical properties of free-carrier-like metals (Ag,Au,Cu,Al) used in plasmonics because of their small losses (long τ_c).

In low-dimensional metallic nanostructures like metal surfaces (2D), wires (1D) or dots (0D), the frequency of plasma oscillations is substantially affected by charge separation at the surface. This is illustrated in Fig. 11.2, which shows major differences of these mixed electromagnetic/surface charge density excitations in different geometries.

In case of zero-dimensionality (0D), the electron gas oscillates at a fixed resonance frequency and may lead to significant enhancement of optical fields close to the surface of the nanoparticle due to the nanoantenna effect. This field enhancement is of key importance for understanding the interaction mechanisms between plasmonic nanoparticles and nanoscale light emitters. The resonance frequency for metal nanoparticles with the diameter $d \ll \delta_{\rm skin}$ equals $\omega_{\rm p}/\sqrt{3}$ and is red shifted due to retardation effects as the diameter

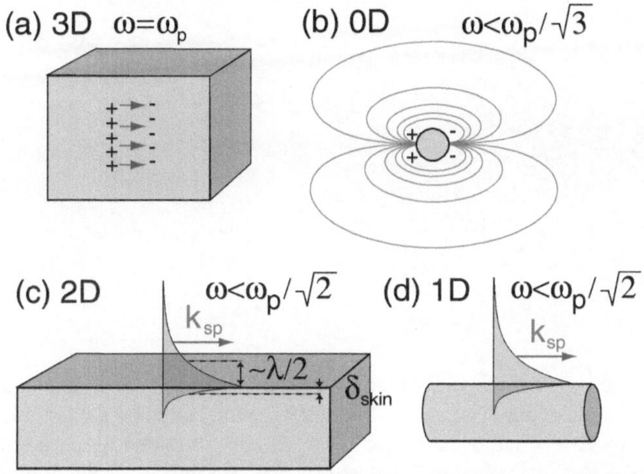

Figure 11.2 Oscillations of electron gas in bulk metal (3D) are modified by surface effects in metallic nanoparticles (0D), nanowires (1D) and surfaces (2D). Whereas metallic nanoparticles possess a specific surface plasmon resonance frequency determined by their shape and size, surface plasmon polaritons in 1D and 2D form a broad frequency band $\omega(k_{\mathrm{sp}})$ and may propagate over macroscopic distances.

increases [42]. Therefore by varying the diameter of 0D metallic nanoparticles between a few nanometers and a few tens of nanometers, their plasmon frequency can be tuned over the entire visible spectral range. A systematic tuning of the plasmonic resonances in 0D metallic nanoparticles can also be achieved by variation of their shape or internal structure [43, 44, 45, 46, 47, 48]. For example, two spherical nanoparticles separated by sub-wavelength distance couple via their near-field dipole–dipole interaction resulting in hybridization of their surface plasmon resonances. Metallic nanoshells consisting of a dielectric core surrounded by a thin (10–20 nm) shell of noble metal are characterized by a lower energy (red-shifted) bonding plasmon mode and a higher energy (blue-shifted) antibonding plasmon mode [44, 45, 49]. Thus, plasmon coupling can be viewed here as an analogous process to molecular hybridization.

In case of metallic surfaces (2D) and wires (1D) there also exist surface charge density waves propagating along the metal surface, often referred to as surface plasmon polaritons. Those coupled electromagnetic – surface charge density waves on flat metal surfaces and wires can propagate over macroscopic distances largely exceeding the optical wavelength λ. In the direction normal to the surface the electromagnetic fields are exponentially bound and strongly confined. On the vacuum side the typical exponential decay length is a fraction of the optical wavelength in vacuum λ. Inside the metal the surface plasmon intensity decays within the skin depth. Surface plasmon polaritons exist in a broad frequency band below the cut-off frequency $\omega_{\mathrm{p}}/\sqrt{2}$ and are characterized by the dispersion relation $\omega(k_{\mathrm{sp}})$, where k_{sp} is the wave

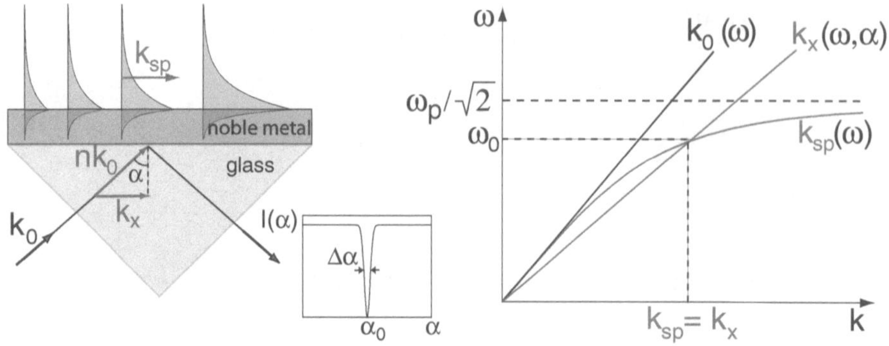

Figure 11.3 Illumination of a thin metal film through the glass prism (Kretschmann geometry) leads to excitation of surface plasmons at the metal–air interface manifested by a sharp minimum of angular reflectivity $I(\alpha)$ at specific angle of incidence α_0. It can be explained by phase-matching between incident light and the surface plasmon visualized by the crossing of dispersion curves $k_{\mathrm{sp}}(\omega_0) = k_x(\omega_0, \alpha_0)$ and leading to strong electric field enhancement at the surface.

vector of the surface plasmon. This dispersion relation is identical for 1D and 2D cases as long as the nanowire radius significantly exceeds the skin depth [50] and is given by $k_{\mathrm{sp}}(\omega) = k_0\sqrt{\frac{\epsilon_m(\omega)}{\epsilon_m(\omega)+1}}$, where $k_0 = \omega/c$ is the wavevector of light in vacuum.

Excitation of surface plasmons in low-dimensional structures by optical fields is an important issue. It is clear that 0D metallic nanoparticles represent an optical dipole and thus can be excited by plane electromagnetic waves. The situation is entirely different for surface plasmon polaritons in 1D and 2D geometries. Since their dispersion curve $\omega(k_{\mathrm{sp}})$ never crosses the light line in a vacuum ($k_{\mathrm{sp}} > k_0$ for all frequencies), they cannot be excited by plane electromagnetic waves impinging on a metal–air interface. An elegant and most widely used way to excite surface plasmons is provided by the so-called Kretschmann geometry (see Fig. 11.3), where a thin metal film deposited on top of a dielectric prism is illuminated by a collimated light beam through the prism with refractive index $n > 1$. For a particular combination of the incident angle α_0 and light frequency ω_0, the in-plane component of the wave vector of incident light $k_x(\omega, \alpha) = (n\omega/c)\sin\alpha$ matches the wave vector of a surface plasmon polariton $k_{\mathrm{sp}}(\omega_0) = k_x(\omega_0, \alpha_0)$. In case of a monochromatic laser source the angular dependence of reflectivity $I(\alpha)$ shows a sharp dip at $\alpha = \alpha_0$ indicating that a significant fraction of incident energy is converted into a surface plasmon propagating along the metal–air interface. The metal layer should be relatively thin (typically $\sim$ 50 nm) in order to allow for efficient coupling of incident light from the glass–metal to the metal–air interface. Owing to a phase-matched excitation, the intensity of a surface plasmon grows upon propagation along the surface and can get up to 300 times larger as compared to the incident wave, depending on energy dissipation in the metal [51]. In a realistic situation this field enhancement is lower due to other effects reducing the surface plasmon propagation length such as, for example, surface roughness [52].

11.4 Coupling of quantum dots to metal surfaces

In order to understand the interaction mechanism of nanoscale light emitters and surface plasmons in low-dimensional metallic nanostructures, we consider the simplest case of a point-like optical dipole source interacting with a flat metal surface. Keeping in mind realistic solid-state-based experimental geometries we assume that a dielectric spacer layer between the metal surface and the colloidal quantum dots provides a well-defined thickness d.

The total energy relaxation rate of an optical dipole in proximity of the metal surface can be represented as a sum of several contributions [53]:

$$\gamma_{\text{total}}(d) = \gamma_{\text{nr}} + \gamma_{\text{freespace}}(d) + \gamma_{\text{d-d}}(d) + \gamma_{\text{sp}}(d). \tag{11.1}$$

Except for the intrinsic non-radiative recombination rate γ_{nr}, other terms are sensitive to dipole orientation (parallel versus perpendicular to the surface, see Fig. 11.4) and the distance d.

For large $d \sim \lambda$ the dominant energy relaxation mechanism is photoluminescence in free space given by $\gamma_{\text{freespace}}(d)$. A fraction of light emitted by the dipole is reflected by the metallic surface. When averaged over all possible emission directions this effect results into weak oscillations of the spontaneous emission rate with a periodicity $\cos(k_0 d)$. This behavior was first discovered by Drexhage in 1970 [54]. The dipole located very close to the metal surface can be described by a short-range dipole–dipole interaction with a mirror dipole providing the third term $\gamma_{\text{d-d}}(d) \sim 1/d^3$ in Eq. (11.1). It is somewhat analogous to the electrostatic Coulomb interaction of a static dipole with a metal surface. However, in the case of a dipole oscillating at optical frequencies, the mirror dipole has different dipole moment and phase, and is strongly damped. The damping originates from the Joule heating in a metal caused by electron–phonon collisions. The majority of experimental data indicate that this non-radiative energy relaxation channel dominates for $d < 10$ nm and is

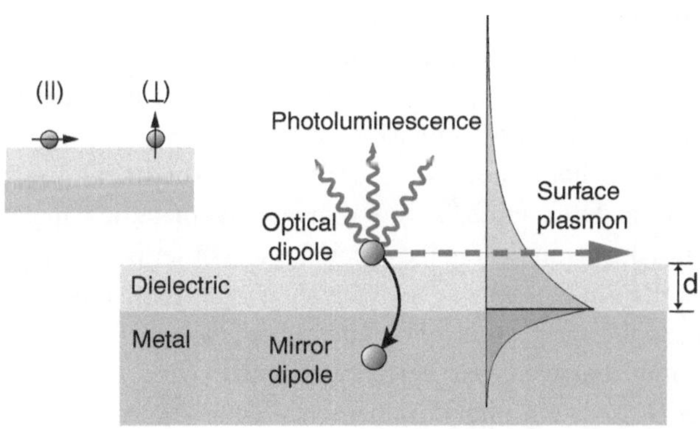

Figure 11.4 Energy relaxation pathways for a nanoscale light emitter in the vicinity of a flat metal surface strongly depend on distance d and dipole polarization (parallel versus perpendicular to the surface).

often associated with the phenomenon of photoluminescence quenching for molecules and quantum dots deposited on metal surfaces. For intermediate distances $d \sim 20$ nm the dipole emission into surface plasmon polaritons (the last term $\gamma_{\mathrm{sp}}(d)$ in Eq. (11.1)) is expected to dominate [51]. This interaction channel lies in the heart of active plasmonics. It allows transfer of the energy from an optically active medium into surface plasmon polaritons, which may transport electromagnetic energy over macroscopic distances (typically tens of microns) limited by surface plasmon propagation length.

The rigorous mathematical treatment [55] allows us to represent the total energy relaxation rate in integral form, different for dipole orientations parallel and perpendicular to the metal surface:

$$\frac{\gamma^{\perp}(d)}{\gamma_0} = 1 + \frac{3q}{2}\mathrm{Im}\left[\int_0^{\infty} r_p(u)\frac{\exp(-4\pi d\sqrt{u^2-1}/\lambda)u^3}{\sqrt{u^2-1}}du\right] \tag{11.2}$$

$$\frac{\gamma^{||}(d)}{\gamma_0} = 1 + \frac{3q}{4}\mathrm{Im}\left[\int_0^{\infty} [(u^2-1)r_p(u)+r_s(u)]\frac{\exp(-4\pi d\sqrt{u^2-1}/\lambda)u}{\sqrt{u^2-1}}du\right]. \tag{11.3}$$

The decay rates $\gamma^{\perp}(d) = \gamma^{\perp}_{\mathrm{total}}(d)$ and $\gamma^{||}(d) = \gamma^{||}_{\mathrm{total}}(d)$ are normalized to the dipole emission rate in the absence of metal structure $\gamma_0 = \gamma^{\perp,||}(d \to \infty) = \gamma_{\mathrm{nr}} + \gamma_{\mathrm{freespace}}$ $(d \to \infty)$, $q = 1 - \gamma_{\mathrm{nr}}/\gamma_0$ stands for the quantum efficiency of photoluminescence, and $r_s(u)$ and $r_p(u)$ denote Fresnel reflection coefficients as defined in Ref. [55]. The integral is taken over the dimensionless parameter $u = k_x/k_0$, where k_x is the in-plane component of emitted radiation. The range $0 < u < 1$ corresponds to the radiative decay in free space $\gamma_{\mathrm{freespace}}(d)$. Since the wave vector of surface plasmons obeys the inequality $k_{\mathrm{sp}} > k_0$, the range $1 < u < 2$ adequately describes the surface plasmon emission rate $\gamma_{\mathrm{sp}}(d)$. The near-field dipole–dipole interaction $\gamma_{\mathrm{d-d}}(d)$ is described by the remaining part of the integral: $2 < u < \infty$ (see Ref. [56] for details).

Here, surface plasmon polaritons can be treated as ordinary electromagnetic waves similar to the waveguide modes in a dielectric waveguide. The role of surface plasmons is to increase the local density of states for spontaneous emission of radiation. On flat metal–dielectric interfaces the density of states due to surface plasmons is roughly two times higher than for free space radiation, i.e. we do not expect any dramatic Purcell factors for the enhancement of the spontaneous emission rate.

The above discussion should help us to understand experimental results obtained for colloidal CdSe quantum dots coupled to low-dimensional metallic nanostructures. We will introduce and discuss the results not in the chronological order in which they were obtained and published, but rather in logical way to explain the underlying physics and to isolate different contributions to the energy relaxation rate in Eq. (11.1). To avoid possible confusions it is worth mentioning that whatever detection pathway (free space radiation or surface plasmon emission) is used, the photoluminescence decay time τ_{PL} always measures the inverse total decay rate given by Eq. (11.1).

Ueda and co-workers have fabricated and investigated a hybrid multilayer system [57] almost identical to our idealized model in Fig. 11.4. An optically thick gold film was sputtered on optical quality glass substrate. A thin polymer layer with variable thickness $d = 5-43$ nm was spin-coated on a gold film and then covered with a monolayer of closely packed core shell CdSe/ZnS quantum dots with radius $R = 5$ nm. The CdSe nanocrystals were impulsively excited by ultrashort laser pulses centered at 2.75 eV, which lead to nearly instantaneous excitation of NC ground state excitonic transition at 2 eV. The time-resolved photoluminescence technique with a streak camera was used to measure the photoluminescence decay time $\tau_{PL} = \gamma_{total}^{-1}$. The results of this investigation showed a pronounced dependence of the non-radiative dipole–dipole relaxation rate scaling as $\gamma_{nr} \sim 1/d^{3.1}$ for short distances. However, from these measurements it was not possible to identify the contribution of surface plasmons.

Excess to the surface plasmon emission rate γ_{sp} is provided by the possibility to decouple them into free space radiation at defects of the metal surface. The most obvious pathway is surface plasmon scattering on surface roughness, which, as we will see below, plays a crucial role in the majority of experimental studies but is very hard to quantify. Another option is to engineer the deterministic outcoupling of surface plasmon polaritons on metallic nanostructures. Suitable metal nanostructures are provided by colloidal chemistry allowing the growth of high-aspect-ratio single-crystal silver nanowires with a typical diameter of 100–200 nm and lengths of the order of a few micrometers. These silver nanowires act as surface plasmon resonators: 1D surface plasmon polaritons form Fabry–Perot cavity modes by reflection at the nanowire tips [58]. The details of growth and optical characterization of these low-Q plasmonic cavities are given elsewhere [59].

Applying electro-chemical layer-by-layer deposition, silver nanowires can be covered by a fused silica spacer layer of desired thickness followed by a monolayer of CdSe quantum dots [60]. Figure 11.5a shows an example of the hybrid coupled silver nanowire/CdSe quantum dot system, which allows us to study the emission process of surface plasmon polaritons in silver nanowires [61]. The major difference to the experimental scheme in Fig. 11.4 is that surface plasmons can be efficiently scattered into free space radiation at both nanowire tips. This results in bright emission from the nanowire tips measured on single nanowires by using microphotoluminescence spectroscopy. The intensity of emission from the tips is proportional to the intensity of surface plasmon polaritons excited in silver nanowires. The dependence of tip intensity on the spacer thickness d in Fig. 11.5b represents the key result of our experimental investigation. It shows a pronounced maximum at $d \simeq 15$ nm, which is a clear qualitative signature of efficient surface plasmon excitation at intermediate distances as predicted by theory. Microphotoluminescence spectroscopy allows us to measure the emission intensity not only from the nanowire tips but also from any position along the nanowire. A remarkable result is that the emission intensity from the central part of the nanowire shows a qualitatively similar behavior on thickness with a maximum at $d \simeq 15$ nm, although the peak intensity is significantly lower. Surface plasmons scatter into free space radiation on surface defects at any position along the nanowire (see SEM images in the inset of Fig. 11.5b) but with much lower scattering efficiency compared with the nanowire tips. The increasing surface roughness for a larger spacer thickness leads to the strong suppression of surface plasmon contribution for $d > 30$ nm.

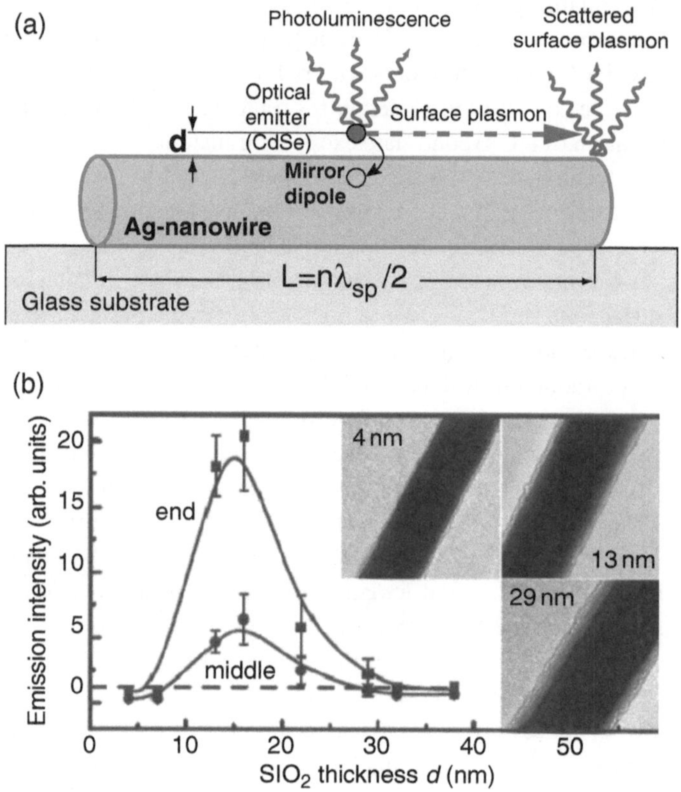

Figure 11.5 (a) CdSe quantum dots coupled to single crystal silver nanowires. Metal–QD distance is controlled by a thickness d of a fused silica shell grown by electro-chemical layer-by-layer deposition. (b) Photoluminescence emission intensity from the nanowire tips and from the central part of the nanowire shows a pronounced maximum at $d \simeq 15$ nm, which represents a fingerprint of surface plasmon excitation. Surface plasmons are emitted either at the nanowire tips or by scattering on the residual surface roughness along the wire (SEM images in the inset show the increase of surface roughness for larger shell thickness $d = 4, 13, 29$ nm).

The enhancement of the photoluminescence decay rate of CdSe quantum dots interacting with rough metal surfaces and gold colloids has been discovered. Kulakovich *et al.* [62] observed a three-fold enhancement of the CdSe NC photoluminescence decay rate from a layer of CdSe quantum dots separated by a 15 nm polymer layer from a monolayer of monodisperse gold colloids with 12–15 nm diameter. Okamoto and co-workers [63] reported a 23-fold enhancement of luminescence from CdSe nanocrystals deposited right on top of a 50 nm thin gold layer produced by thermal evaporation. In the same study CdS/ZnSe core/shell nanocrystals demonstrated nearly no PL enhancement compared with quantum dots on a dielectric substrate. Ito *et al.* [64] investigated the interaction of CdSe/ZnS core/shell nanocrystals with gold surfaces characterized by different surface roughness: PL quenching on smooth surfaces contrasts PL enhancement on rough surfaces. The observed maximum of the PL enhancement factor at optical wavelength of 540 nm on rough metal surfaces [64] resembles the behavior on gold colloids [62] and is assisted by localized 0D surface plasmon resonances. Indeed, on rough surfaces the local density of states of electromagnetic radiation can be enhanced due to the nano-antenna effect [65] strongly increasing both the excitation and emission rate of QDs [66]. Similar conclusions were drawn by Zhai *et al.* [67], who report on the correlation between the PL enhancement of CdSe/ZnS quantum dots and the percolation threshold in semi-continuous gold films.

In the majority of aforementioned investigations the analysis of the temporal evolution of photoluminescence traces in coupled metal–QD systems appeared to be useful. As discussed in Section 11.2 the dynamics of photoluminescence of non-resonantly pumped quantum dots is often masked by complex intradot relaxation dynamics resulting in blinking behavior. In this context the quantitative analysis of photoluminescence decay dynamics for nanocrystal ensembles coupled to rough metal surfaces, which are usually obtained with different fabrication methods and under different excitation conditions, is extremely challenging. However, the experimental observations by Okamoto *et al.* [63] and Fedutik and co-workers [61] provide a clear indication that the timescale of the fast component in PL decay traces become about two times shorter for CdSe quantum dots in close proximity to the metal surfaces.

Sophisticated statistical analysis of photoluminescence traces of single quantum dots coupled to smooth metal surfaces provides further insight into the underlying physics [68, 69]. An elegant statistical analysis of photoluminescence decay times obtained during bright emission periods of the blinking process developed by Pelton and co-workers [68] yields clear monoexponential decay rates of quantum dots in the vicinity of smooth metal surfaces. These decay rates appeared to be statistically distributed due to different orientations of optical transition dipole moments relative to the metal surface. The orientational distribution of decay rates was obtained experimentally and found in quantitative agreement with the theoretical treatment by Eqs. (11.2) and (11.3) [68]. The intrinsic randomness of optical dipole orientations of spherical quantum dots can be avoided by using nanorods, where the dipole moment points along the nanorod axes. Because of a high aspect ratio the nanorods deposited on a metal surface lie in the surface plane. For 1D metal or dielectric nanowires the nanorods show deterministic alignment along the nanowire axes [60].

So far we have discussed the interactions of quantum dots with radiation in a weak-coupling regime relying on the modification of spontaneous emission rate of non-resonantly excited quantum dots. Another important area of quantum optics deals with the strong-coupling regime, where energy can be periodically transferred between the QD-exciton and the mode of the electromagnetic field, resulting into Rabi-splitting in the optical spectra. Such a regime of strong exciton–plasmon interaction was first realized by Pockland and co-workers [70] with dye molecules and only recently extended by Gomez *et al.* [71] to colloidal CdSe quantum dots interacting with surface plasmons in a thin silver film. Using Kretschmann geometry the researchers have varied the angle of incidence α

(see Fig. 11.3) of a collimated white light beam and observed a clear Rabi-splitting in the line shape of the reflected light $I(\alpha, \omega)$ at conditions of resonant surface plasmon excitation $\alpha = \alpha_0$. The demonstration of room-temperature strong coupling between surface plasmons and colloidal QDs has important practical implications. High photostability and ease of fabrication of colloidal quantum dots provide a fundamental framework for engineering all-optical nonlinear devices that rely on strong light–matter interactions at the nanoscale (see e.g. [72]).

11.5 Practical application: QD-based all-optical plasmonic modulator

Efficient interaction of CdSe quantum dots with surface plasmon polaritons was exploited as a basic operation principle for an all-optical plasmonic modulator [73] shown in Fig. 11.6. It consists of a plasmonic slit–groove microinterferometer covered with a densely packed layer of colloidal CdSe nanocrystals.

The plasmonic microinterferometer represents a sub-wavelength slit–groove pair milled in a silver film by a focussed ion beam (FIB). If the entire area of the microinterferometer is illuminated by a collimated spatially coherent laser beam, surface plasmons are excited at the groove and propagate towards the slit, where they are re-scattered into free space radiation and interfere with the directly transmitted light. The amplitude of the plasmonic transmission through the slit signal can be changed by modifying the absorption for surface plasmons. Quantum dots on top of the metal surface introduce absorption, which is particularly strong if the plasmon frequency is resonant to the QD ground-state excitonic transition. In a plasmonic modulator the control surface plasmons are used to non-resonantly excite QDs and thus change the absorption for the signal surface plasmons. The intensity of light transmitted through the slit can be thus changed by $\sim 15\%$, suggesting that this device can serve as an efficient all-optical switch.

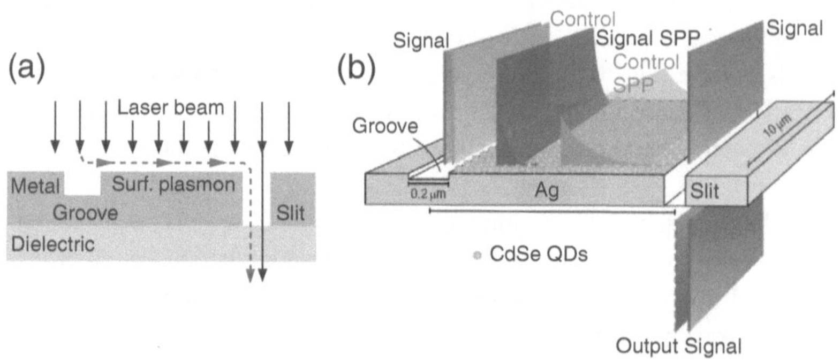

Figure 11.6 (a) In the plasmonic slit–groove microinterferometer surface plasmons are excited at the groove, propagate towards the slit, where they are converted to free-space radiation interfering with directly transmitted light. (b) CdSe quantum dots cover the entire interferometer area. Non-resonant excitation by control SPP is used to change the absorption for signal SPP and thus modulate the optical transmission through the slit at signal wavelength.

11.6 Perspective: quantum optics with surface plasmons

Interest in experimental studies reviewed above was largely motivated by the perspectives of using coupled QD/plasmon systems for quantum optical applications. Back in 1976, Bonifacio and Mocawitz realized that the properties of cooperative spontaneous emission can be significantly modified for a monolayer of optically excited atoms or molecules interacting with surface plasmon polaritons [74]. Tremendous progress in fabrication capabilities of coupled hybrid QD–metal nanostructures stimulated further theoretical concepts like surface plasmon laser (SPASER) [75], superradiance in plasmonic nanoshells coupled to quantum dots [76] and strong coupling between ultrathin metal nanowire plasmons and QD-excitons [77]. Technological limitations in fabrication of semiconductor quantum dots and metallic nanostructures still make the implementation of these concepts very challenging for experimentalists. For example, the size dispersion of semiconductor quantum dots causes a strong inhomogeneous broadening of excitonic transition energies and thus strongly alters the expected performance of superradiance [76]. The theoretically suggested mechanism of strong coupling between surface plasmon polaritons in nanowires and quantum dots relies on the geometric singularity of the SP electromagnetic field in metal nanowires as their diameter decreases significantly below 100 nm. So far it has been possible to observe only a slight decrease in the QD radiative lifetime for ~ 100 nm thin nanowires [78]. Further progress in this direction is limited by the increasing requirements for the surface quality of metallic nanowires and dielectric nanoshells. Recent progress in fabrication techniques of atomically smooth metal nanostructures for plasmonics should help to solve at least some of the outlined technological problems [52].

Some recent progress in active nanoplasmonics has come from replacing semiconductor quantum dots by different active media such as dye molecules or NV-centers in diamond. An optically pumped coupled dye gold nanoparticle composite revealed clear signatures of a SPASER [79]. Hybrid NV-center/nanowire structures represent a robust arrangement for the generation and transport of surface plasmons over macroscopic distances at single-photon (single-plasmon) level [80]. These promising examples show the tremendous potential but also the challenges in the field of nanoplasmonics with colloidal nanocrystals which are worth studying for both fundamentals and applications.

Acknowledgement

Financial support by the EU-network of excellence PHOREMOST, Deutsche Forschungsgemeinschaft and Région Pays de la Loire is gratefully acknowledged.

References

[1] Efros, Al. L. and Efros, A. L. 1982. Interband absorption of light in a semiconductor sphere. *Sov. Phys. Semicond.*, **16**, 772.

[2] Brus, L. E. 1984. Electron–electron and electron–hole interactions in small semiconductor crystallites: the size dependence of the lowest excited electronic state. *J. Chem. Phys.*, **80**, 4403.
[3] Alivisatos, A. P. 1996. Semiconductor clusters, nanocrystals, and quantum dots. *Science*, **271**, 933.
[4] Woggon, U. 1996. Optical properties of semiconductor quantum dots, *Springer Tracts in Modern Physics*, **136**.
[5] Gaponenko, S. V. and Woggon, U., *Landolt-Börnstein New Series III/34C2 Optical Properties of Semiconductor Nanostructures*, ed. C. Klingshirn, Springer-Verlag, 2004, Chap. 5.5. II–VI Semiconductor Quantum Dots – Nanocrystals, pp. 284–347.
[6] Wise, F. W. 2000. Lead salt quantum dots: the limit of strong quantum confinement. *Accounts Chem. Res.*, **33**, 773.
[7] Eychmuller, A. Mews, A. and Weller, H. 1993. A quantum-dot quantum well – CdS/HgS/CdS. *Chem. Phys. Lett.*, **208**, 59.
[8] Dabbousi, B. O., RodriguezViejo, J., Mikulec, F. V. *et al.* 1997. (CdSe)ZnS core-shell quantum dots: Synthesis and characterization of a size series of highly luminescent nanocrystallites. *J. Phys. Chem. B*, **101**, 9463.
[9] Hines, M. A. and Guyot-Sionnest, P. 1996. Synthesis and characterization of strongly luminescing ZnS-capped CdSe nanocrystals. *J. Phys. Chem.*, **100**, 468.
[10] Murray, C. B., Kagan, C. R. and Bawendi, M. G. 2000. Synthesis and characterization of monodisperse nanocrystals and close-packed nanocrystal assemblies. *Ann. Rev. Mater. Sci.*, **30**, 545.
[11] Amirav, L. and Alivisatos, A. P. 2010. Photocatalytic hydrogen production with tunable nanorod heterostructures. *J. Phys. Chem. Lett.*, **1**, 1051–1054.
[12] Berr, M., Vaneski, A., Susha, A. S. *et al.* 2010. Colloidal CdS nanorods decorated with subnanometer sized Pt clusters for photocatalytic hydrogen generation. *Appl. Phys. Lett.*, **97**, 093108-1–093108-3.
[13] Costi, R., Saunders, A. E. and Banin, U. 2010. Colloidal hybrid nanostructures: a new type of functional materials, *Angew. Chem. Int. Ed.*, **49**, 4878–4897.
[14] Choi, C. L. and Alivisatos, A. P. 2010. From artificial atoms to nanocrystal molecules: preparation and properties of more complex nanostructures. *Ann. Rev. Phys. Chem.*, **61**, 369–389.
[15] Wang, L. W. 2010. Novel computational methods for nanostructure electronic structure calculations. *Ann. Rev. Phys. Chem.*, **61**, 19–39.
[16] Peng, X. G., Manna, L., Yang, W. D. *et al.* 2000. Shape control of CdSe nanocrystals. *Nature* **404**, 59-61.
[17] Talapin, D. V., Koeppe, R., Gotzinger, S. 2003. Highly emissive colloidal CdSe/CdS heterostructures of mixed dimensionality. *Nano Lett.*, **3**, 1677–1681.
[18] Saraidarova, T., Reisfeld, R., Sashchiuk, A. and Lifshitz, E. 2007. Synthesis and characterization of PbS nanorods and nanowires. *Physica E*, **37**, 173–177.
[19] Mulvihill, M. J., Habas, S. E., Jen-La Plante, I., Jiamin W., and Mokari, T. 2010. Influence of size, shape, and surface coating on the stability of aqueous suspensions of CdSe nanoparticles. *Chem. Mater.*, **22**, 5251–5257.
[20] Giblin, J. and Kuno, M. 2010. Nanostructure absorption: a comparative study of nanowire and colloidal quantum dot absorption cross sections. *J. Phys. Chem. Lett.*, **1**, 3340–3348.
[21] Katz, D., Wizansky, T., Millo, O. *et al.* 2002. Size-dependent tunneling and optical spectroscopy of CdSe quantum rods. *Phys. Rev. Lett.*, **89**, 086801-1–086801-4.

[22] Talaat, H., Abdallah, T., Mohamed, M. B., Negmc, S. and El-Sayed, M. A. 2009. The sensitivity of the energy band gap to changes in the dimensions of the CdSe quantum rods at room temperature: STM and theoretical studies. *Chem. Phys. Lett.*, **473**, 288–292.

[23] Bartnik, A. C., Efros, Al. L., Koh, W.-K., Murray, C. B. and Wise, F. W. 2010. Electronic states and optical properties of PbSe nanorods and nanowires *Phys. Rev. B*, **82** 195313-1–195313-16.

[24] Tavenner-Kruger, S., Young-Shin Park, Lonergan, M., Woggon, U. and Wang, H. 2006. Zero-phonon linewidth in CdSe/ZnS core/shell nanorods. *Nano Lett.*, **6**, 2154–2157

[25] Palinginis, P., Tavenner, S., Lonergan, M. and Wang, H. 2003. Spectral hole burning and zero phonon linewidth in semiconductor nanocrystals. *Phys. Rev. B*, **67**, 201307-1–201307-4.

[26] Littleton, B. N., Fernee, M. J., Gomez, D. E., Mulvaney, P. and Rubinsztein-Dunlop, H. 2009. High-resolution line width measurement of single CdSe nanocrystals at long time scales. *J. Phys. Chem. Lett.*, **113**, 5345–5348.

[27] Chamarro, M., Gourdon, C., Lavallard, P., Lublinskaya, O. and Ekimov, A. I. 1996. Enhancement of electron–hole exchange interaction in CdSe nanocrystals: A quantum confinement effect. *Phys. Rev. B*, **53**, 1336–1342.

[28] Efros, A. L., Rosen, M., Kuno, M. *et al.* 1996. Band-edge exciton in quantum dots of semiconductors with a degenerate valence band: dark and bright exciton states. *Phys. Rev. B*, **54**, 4843–4856.

[29] Li, J and Wang, L. W. 2003. Shape effects on electronic states of nanocrystals. *Nano Lett.*, **3** 1357–1363.

[30] LeThomas, N., Herz, E., Schops, O. and Woggon, U. 2005. Exciton fine structure in single CdSe nanorods. *Phys. Rev. Lett.*, **94**, 016803-1–016803-4.

[31] Shabaev, A. and Efros, Al. L. 2004. 1D exciton spectroscopy of semiconductor nanorods. *Nano Lett.*, **4**, 1821–1025.

[32] Protasenko, V., Bacinello, D. and Kuno, M. 2006. Experimental determination of the absorption cross-section and molar extinction coefficient of CdSe and CdTe nanowires. *J. Phys. Chem. B*, **110**, 25322–25331.

[33] Shaviv, E., Salant, A. and Banin, U. 2009. Size dependence of molar absorption coefficients of CdSe semiconductor quantum rods. *ChemPhysChem*, **10**, 1028–1031.

[34] Kim, J., Wong, C. Y. and Scholes, G. D. 2009. Exciton fine structure and spin relaxation in semiconductor colloidal quantum dots. *Acc. Chem. Res.*, **42**, 1037–1046.

[35] Scholes, G. D., Kim, J., Wong, C. Y. *et al.* 2006. Nanocrystal shape and the mechanism of exciton spin relaxation. *Nano Lett.*, **6**, 1765–1771.

[36] Mokari, T., Rothenberg, E., Popov, I., Costi, R. and Banin, U. 2004. Selective growth of metal tips onto semiconductor quantum rods and tetrapods. *Science*, **304**, 1787–1790.

[37] Saunders, A. E., Popov, I. and Banin, U. 2006. Synthesis of hybrid CdS-Au colloidal nanostructures. *J. Phys. Chem. B*, **110**, 25421–25429.

[38] Borys, N. J.,Walter, M. J., Huang, J., Talapin, D.V. and Lupton, J. M. 2010. The role of particle morphology in interfacial energy transfer in CdSe/CdS heterostructure nanocrystals. *Science*, **330**, 1371–1374.

[39] Panzer, M. J., Aidala, K. E., Anikeeva, P. O. *et al.* 2010. Nanoscale morphology revealed at the interface between colloidal quantum dots and organic semiconductor films. *Nano Lett.*, **10**, 2421–2426.

[40] Mauser, C., Da Como, E., Baldauf, J. *et al.* 2010. Spatio-temporal dynamics of coupled electrons and holes in nanosize CdSe-CdS semiconductor tetrapods. *Phys. Rev. B*, **82**, 081306-1–081306-4.

[41] Lutich, A. A., Mauser, C., Da Como, E. *et al.* 2010. Multiexcitonic dual emission in CdSe/CdS tetrapods and nanorods. *Nano Lett.*, **10**, 4646–4650.

[42] Pelton, M., Aizpurua, J. and Bryant, G. 2008. Metal–nanoparticle plasmonics. *Laser and Photon. Rev.*, **2**, 136–159.

[43] Kelly, K. L., Coronado, E., Zhao, L. L. and Schatz, G. C. 2003. The optical properties of metal nanoparticles: the influence of size, shape, and dielectric environment. *J. Phys. Chem. B*, **107**, 668–677.

[44] Prodan, E., Radloff, C., Halas, N. J. and Nordlander, P. 2003. A hybridization model for the plasmon response of complex nanostructures. *Science*, **302**, 419–422.

[45] Nordlander, P., Oubre, C., Prodan, E., Li, K. and Stockman, M. I. 2004. Plasmon hybridizaton in nanoparticle dimers. *Nano Lett.*, **4**, 899–903.

[46] Jain, P. K., Eustis, S. and El-Sayed, M. A. 2006. Plasmon coupling in nanorod assemblies: optical absorption, discrete dipole approximation simulation, and exciton-coupling model. *J. Phys. Chem. B*, **110**, 18243–18253.

[47] Wang, H., Brandl, D. W., Nordlander, P. Halas, N. J. 2007. Plasmonic nanostructures: artificial molecules. *Acc. Chem. Res.*, **40**, 53–62.

[48] Jain, P. K. and El-Sayed, A. M. 2010. Plasmonic coupling in noble metal nanostructures. *Chem. Phys. Lett.*, **487**, 153–164.

[49] Sheikholeslami, S., Young-wook Jun, Jain, P. K. and Alivisatos, A. P. 2010. Coupling of optical resonances in a compositionally asymmetric plasmonic nanoparticle dimer. *Nano Lett.*, **10** 2655–2660.

[50] Wang, K. and Mittleman, D. M. 2006. Dispersion of surface plasmon polaritons on metal wires in the terahertz frequency range. *Phys. Rev. Lett.*, **96**, 157401.

[51] Weber, W. H., and Ford, G. W. 1981. Optical electric-field enhancement at a metal surface arising from surface-plasmon excitation. *Opt. Lett.*, **6**, 122–124.

[52] Nagpal, P., Lindquist, N., Oh, S. and Norris, D. J. 2009. Ultrasmooth patterned metals for plasmonics and metamaterials. *Science*, **325**, 594–597

[53] Ford, G. W. and Weber, W. H. 1984. Electromagnetic interactions of molecules with metals surfaces. *Phys. Reports*, **113**, 195–278.

[54] Drexhage, K. H. 1970. Influence of a dielectric interface on fluorescence decay time. *J. Luminescence*, **1-2**, 693–701.

[55] Chance, R. R. Prock, A. and Silbey, R. 1978. In: Prigogine, I. and Rice, S. A. (eds.), *Advances in Chemical Physics XXXVII*. NewYork: Wiley.

[56] Weber, W. H. and Eagen, C. F. 1979. Energy tranfer from an excited duy molecule to the surface plasmons of an adjacent metal. *Opt. Lett.*, **4**, 236–238.

[57] Ueda, A., Tayagaki, T. and Kanemitsu, Y. 2008. Energy transfer from semiconductor nanocrystal monolayers to metals surfaces revealed by time-resolved photoluminescence spectroscopy. *Appl. Phys. Lett.*, **92**, 133118.

[58] Ditlbacher, H., Hohenau, A., Wagner, D. *et al.* 2005. Silver nanowires as surface plasmon resonators. *Phys. Rev. Lett.*, **95**, 257403.

[59] Allione, M., Temnov, V. V., Fedutik, Y., Woggon, U. and Artemyev, M. V. 2008. Surface plasmon mediated interference phenomena in low-Q silver nanowire cavities. *Nano. Lett.*, **8**, 31.

[60] Fedutik, Y., Temnov, V. V., Woggon, U., Ustinovich, E. and Artemyev, M. V. 2007a. Exciton–plasmon interaction in a composite metal–insulator-semiconductor nanowire system. *J. Am. Chem. Soc.*, **129**, 14939.

[61] Fedutik, Y., Temnov, V. V., Schöps, O., Woggon, U., and Artemyev, M. V. 2007b. Exciton–plasmon-photon conversion in plasmonic nanostructures. *Phys. Rev. Lett.*, **99**, 136802.

[62] Kulakovich, O., Strekal, N., Yaroshevich, A. *et al.* 2002. Enhanced luminescence of CdSe quantum dots on gold colloids. *Nano. Lett.*, **2**, 1449–1452.

[63] Okamoto, K., Vyawahare, S. and Scherer, A. 2006. Surface-plasmon enhanced bright emission from CdSe quantum-dot nanocrystals. *J. Opt. Soc. Am. B*, **23**, 1674–1678.

[64] Ito, Y., Matsuda, K. and Kanemitsu, Y. 2007. Mechanism of photoluminescence enhancement in single semiconductor nanocrystals on metal surfaces. *Phys. Rev. B*, **75**, 033309.

[65] Anger, P., Bharadwaj, P. and Novotny, L. 2006. Enhancement and quenching of single-molecule fluorescence. *Phys. Rev. Lett.*, **96**, 113002.

[66] Matsuda, K., Ito, Y. and Kanemitsu, Y. 2008. Photoluminescence enhancement and quenching of single CdSe/ZnS nanocrystals on metal surfaces dominated by plasmon resonant energy transfer. *Appl. Phys. Lett.*, **92**, 211911.

[67] Zhai, Y. Y., Song, H., Zhou, Z. K., Li, M. and Hao, Z. H. 2008. Enhanced radiative emission rate of CdSe/ZnS quantum dots on semi-continuous gold films. *Eur. Phys. J. Appl. Phys.*, **42**, 109–112.

[68] Wu, X., Sun, Y. and Pelton, M. 2009. Recombination rates for single colloidal quantum dots near a smooth metal film. *Phys. Chem. Chem. Phys.*, **11**, 5867–5870.

[69] Ma, X., Tan, H., Kipp, T. and Mews, A. 2010. Fluorescence enhancement, blinking supression, and gray states of individual semiconductor nanocrystals close to gold nanoparticles. *Nano. Lett.*, **10**, 4166–4174.

[70] Pockland, I., Brillante, A. and Mobius, D. 1982. Surface-plasmon enhanced bright emission from CdSe quantum-dot nanocrystals. *J. Chem. Phys.*, **77**, 6289–6295.

[71] Gomez, D. E., Vernon, K. C., Mulvaney, P. and Davis, T. J. 2010b. Surface plasmon mediated strong exciton-photon coupling in semiconductor nanocrystals. *Nano. Lett.*, **10**, 4166–4174.

[72] Gomez, D. E., Vernon, K. C., Mulvaney, P. and Davis, T. J. 2010a. Coherent superposition of exciton states in quantum dots induced by surface plasmons. *Appl. Phys. Lett.*, **96**, 073108.

[73] Pacifici, D., Lezec, H. J. and Atwater, H. A. 2007. All-optical modulation by plasmonic excitation of CdSe quantum dots. *Nature Phot.*, **1**, 402–406.

[74] Bonifacio, R. and Morawitz, H. 1976. Cooperative emission of an excited molecular monolayer into surface plasmons. *Phys. Rev. Lett.*, **36**, 1559–1562.

[75] Bergman, D. J. and Stockman, M. I. 2003. Surface plasmon amplification by stimulated emission of radiation: quantum generation of coherent surface plasmons in nanosystems. *Phys. Rev. Lett.*, **90**, 027402.

[76] Temnov, V. V. and Woggon, U. 2005. Superradiance and subradiance in an inhomogeneously broadened ensemble of two-level systems coupled to a low-Q cavity. *Phys. Rev. Lett.*, **95**, 243602.

[77] Chang, D. E., Sorensen, A. S., Hemmer, P. R. and Lukin, M. D. 2006. Quantum optics with surface plasmons. *Phys. Rev. Lett.*, **97**, 053002.

[78] Akimov, A. V., Mukherjee, A., Yu, C. L. *et al.* 2007. Generation of single optical plasmons in metallic nanowires coupled to quantum dots. *Nature*, **450**, 402–406.

[79] Noginov, M. A., Zhu, G., Belrgave, A. M. *et al.* 2009. Demonstration of a spaser-based nanolaser. *Nature*, **460**, 1110–1113.

[80] Kolesov, R., Grotz, B., Balasubramanian, G. *et al.* 2009. Wave–particle duality of single surface plasmon polaritons. *Nature Phys.*, **5**, 470–474.

Part IV

Quantum dot nano-laboratory: magnetic ions and nuclear spins in a dot

12

Dynamics and optical control of an individual Mn spin in a quantum dot

L. Besombes, C. Le Gall, H. Boukari and H. Mariette

We show in this review that the spin state of a single magnetic atom embedded in an individual semiconductor quantum dot can be optically probed. A high degree of spin polarization can be achieved for an individual Mn atom using quasi-resonant or fully resonant optical excitation of the quantum dot at zero magnetic field. Under quasi-resonant excitation, optically created spin-polarized carriers generate an energy splitting of the Mn spin and enable magnetic moment orientation controlled by the photon helicity and energy. Monitoring the time dependence of the intensity of the fluorescence during a resonant optical pumping process allows us to directly probe the dynamics of the initialization of the Mn spin. The dynamics and the magnetic field dependence of the optical-pumping mechanism shows that the spin lifetime of an isolated Mn atom at zero magnetic field is controlled by a magnetic anisotropy induced by the built-in strain in the quantum dots. The Mn spin state prepared by optical pumping is fully conserved for a few microseconds. These experiments open the way to full optical control of the spin state of an individual magnetic atom in a solid state environment.

12.1 Introduction

The ability to control spins in semiconductor nanostructures is an important issue for spintronics and quantum information processing. Single-spin detection and control is a key but very challenging step for any spin-based solid-state quantum computing device. In the past few years, efficient optical techniques have been developed to control the spin of individual carriers [34] or ensemble of nuclei [22] in semiconductor quantum dots (QDs). Thanks to their expected long spin coherence time, magnetic atoms in a semiconductor host could be alternative media to store quantum information in the solid state. However, as these localized spins interact weakly with their environment, they can hardly be controlled by electrical or optical methods. Recently, QDs containing individual Mn atoms have been observed in both II–VI [4] and III–V [21] compounds. In these systems, since the confined carriers and Mn spin functions become strongly mixed, the optical excitation of the QD strongly affects the spin state of the Mn atom through the exchange interaction offering a possibility of full optical control [36].

Quantum Dots: Optics, Electron Transport and Future Applications, ed. Alexander Tartakovskii. Published by Cambridge University Press.

When a single Mn atom is included in a II–VI semiconductor QD (CdTe in ZnTe), the spin of the optically created electron–hole pair (exciton) interacts with the $5d$ electrons of the Mn (total spin $S = 5/2$). This leads to a splitting of the once simple photoluminescence (PL) spectrum of an individual QD into six $(2S + 1)$ components. This splitting results from the spin structure of the confined heavy holes which are quantized along the QDs' growth axis with their spin component taking only the values $J_z = \pm 3/2$. In the first approximation, the hole–Mn exchange interaction reduces to an Ising term $J_z.S_z$ and shifts the emission energy of the QD, depending on the relative projection of the Mn and hole spins [26]. As the spin state of the Mn atom fluctuates during the optical measurements, the six lines are observed simultaneously in time-average PL spectra. The intensities of the lines reflect the probability for the Mn to be in one of its six spin components when the exciton recombines [5].

In this review, we show that one can exploit the injection of spin polarized carriers in an individual II–VI QD to optically initialize the spin state of an embedded Mn atom. We use quasi-resonant (on a QD excited state) or resonant optical excitation of one of the six exciton levels of an Mn-doped QD to prepare by optical pumping the spin state of the magnetic atom. Under resonant excitation conditions, scattered photons coming from spin-flips of the exciton without change of the Mn spin are observed. This fluorescence signal is used to probe the dynamics of the initialization of the Mn spin during the optical pumping process. A pumping efficiency of about 75% is obtained with an initialization time in the range of tens nanoseconds. We show that the efficiency of the optical pumping is affected by a magnetic field of a few tens of milliteslas applied in the Voigt or Faraday configurations. This reveals the influence of the Mn fine structure on the spin dynamics: the magnetic anisotropy of the Mn spin induced by biaxial strains slows down the spin relaxation at zero magnetic field. Partial relaxation of the strain or an anisotropic in-plane strain distribution at the Mn spin location can restore coherent precession of the Mn spin away from the optical axis, decreasing the efficiency of the optical pumping.

12.2 II–VI diluted magnetic semiconductor QDs

Carrier–Mn coupling was mostly studied in bulk diluted magnetic semiconductors (DMS) made of II–VI semiconductors in which Mn impurities was introduced (see the review papers [13, 14]). In these materials, Mn impurities have the d^5 electronic configuration and substitute the cations from column II (Zn, Cd or Hg) up to 100%. An important point is that Mn is an isoelectronic impurity in II–VIs – by contrast to the acceptor character observed in GaAs and similar III–Vs [38, 27]. The Mn ground state is 6S (or 6A1 in cubic or hexagonal symmetry), introducing localized, isotropic spins with $S = 5/2$. If not interacting, these localized spins follow Maxwell–Boltzmann statistics, resulting in a magnetization induced by an applied field given by a Brillouin function. Optical spectroscopy around the bandgap reveals the so-called "giant Zeeman effect", with a spin splitting proportional to the Mn magnetization [15]. Several studies have demonstrated this proportionality and measured the strength of the coupling [41]. Magneto-optical spectroscopy is now a very sensitive method for measuring locally the magnetization of the Mn system [7]. Altogether, this excellent knowledge of II–VI DMS, and of the coupling between the localized spins and carriers, constitutes a very firm basis for the further studies described below.

In a magnetic QD, the sp–d interaction takes place with a single carrier or a single electron–hole pair. However, besides effects related to the carriers–Mn exchange interaction, such as giant Faraday rotation and giant Zeeman shift, it was found that even a small content of Mn introduced in a II–VI semiconductor material can strongly suppress PL if the energy gap E_g exceeds the energy of the Mn internal transition. This strongly limits the study of individual DMS QDs [8]. The first studies of individual QDs doped with Mn atoms were reported by Maksimov *et al.* [31]. They studied CdMnTe QDs inserted in CdMgTe barriers in which the optical transition energies are lower than the energy of the internal transition of the Mn atom. This suppresses the non-radiative losses due to the transfer of confined carriers to the Mn internal levels. This system allowed observation of the formation of quasi zero-dimensional magnetic polaron.

Another way of reducing the non-radiative losses was to introduce the magnetic atoms in to the QDs barriers. This was achieved for self-assembled CdSe dots embedded in ZnMnSe barriers by Seufert *et al.* [39]. In this system, the interaction between the confined exciton and the magnetic atoms is due to the spread of the wave function in the barriers. In these DMS structures, the formation of a ferromagnetically aligned spin complex was demonstrated to be surprisingly stable as compared to bulk magnetic polaron [28, 42] even at elevated temperature and high magnetic fields. The PL of a single electron–hole pair confined in one magnetic QD, which sensitively depends on the alignment of the magnetic atoms spins, allowed the measurement of the statistical fluctuation of the magnetization on the nanometer scale. Quantitative access to statistical magnetic fluctuations was obtained by analyzing the linewidth broadening of the single dot emission. This all optical technique allowed to address a magnetic moment of about 100 μ_B and to resolve changes in the order of a few μ_B [2, 10, 20].

A large effort has also been made to incorporate magnetic ions in chemically synthesized II–VI nanocrystals [33]. The incorporation of the magnetic atoms is strongly dependent on the growth conditions and is controlled by the adsorption of atoms on the nanocrystal surface during growth [12]. The doping of nanocrystals with magnetic impurities also leads to interesting magneto-optical properties [6] but once again, in these highly confined systems, the transfer of confined carriers to the Mn internal levels strongly reduces their quantum efficiency and prevents the optical study of individual Mn-doped nanocrystals. However, by looking to magnetic circular dichroism absorption spectra, it is possible to observe a giant excitonic Zeeman splitting and to deduce directly the sp–d exchange interaction [1]. Recently, a very robust light-induced spontaneous spin polarization [3] was obtained in large CdSe nanocrystals containing a few percent of Mn and emitting at lower energy than the internal transition of the Mn.

CdTe/ZnTe self-assembled QDs usually present an emission energy below the internal transition of the Mn atom. The incorporation of magnetic atoms is then possible without losing the good optical properties of these QDs. Most of the experimental studies on these QDs were focused on the interaction of a single carrier spin with its paramagnetic environment (large number of magnetic atoms) [29, 9]. However, CdTe/ZnTe QDs structures doped with a low density of Mn allow us to access optically the spin state of a single magnetic atom interacting with a single electron–hole pair or a single carrier [25].

The CdTe/ZnTe QDs samples used in this study are all grown on ZnTe substrates. A 6.5 monolayer thick CdTe layer is deposited at 280°C by atomic layer epitaxy on a ZnTe barrier grown by molecular beam epitaxy at 360°C. The dots are formed by the well-established tellurium deposition/desorption process [40] and protected by a 100 nm thick ZnTe top barrier. Single Mn atoms are introduced in CdTe/ZnTe QDs during their growth adjusting the density of Mn atoms to be roughly equal to the density of QDs [30]. The height of the QDs' core is a few nanometers and their diameter is in the 10–20 nm range.

12.3 Optical Mn spin orientation

To optically pump the Mn spin, Mn-doped QDs were quasi-resonantly excited with a tunable continuous wave (CW) dye laser. In order to record the dynamics of the preparation of the Mn spin (spin transients), the linear polarization of the excitation laser was modulated between two orthogonal states by switching an electro-optic modulator, and converted to circular polarization with a quarter-wave plate. Trains of resonant light with variable duration were generated from the CW laser by using acousto-optical modulators with a switching time of 10 ns. The circularly polarized collected light was dispersed by a 1 m double monochromator before being detected by a fast avalanche photodiode in conjunction with a time-correlated photon counting unit.

Figures 12.1 and 12.2 summarize the main features of the time-resolved optical orientation experiment. The PL of the exciton–Mn (X–Mn) complex was excited about 20 meV above the PL (top of Fig. 12.1a), of an excited state of the X–Mn complex [17]; the PL intensity was detected in circular polarization (e.g., σ^-, corresponding to the recombination of the -1 exciton). The relative intensities of the six lines dramatically depend on the excitation energy (bottom of Fig. 12.1a): as each line corresponds to one value of the Mn spin projection S_z, the whole process creates a non-equilibrium occupation probability of the Mn spin states. Under these conditions, switching the circular polarization of the excitation produces a change of the σ^- PL intensity (Fig. 12.1b) with two transients: first an abrupt one with the same sign for all six lines, reflecting the population change of the spin-polarized excitons; then a slower transient reflecting an increase or a decrease of the occupation of the detected Mn spin state. This slow transient has an opposite sign for the two extreme PL lines (i.e., when monitoring the Mn spin states $S_z = \pm + 5/2$, Fig. 12.2b, and a characteristic time that is inversely proportional to the pump intensity (Fig. 12.2a). This is the signature of an optical pumping process which realizes a spin orientation of the Mn atom. We first discuss the details of this process, then use it to study the spin dynamics of the single Mn in the QD.

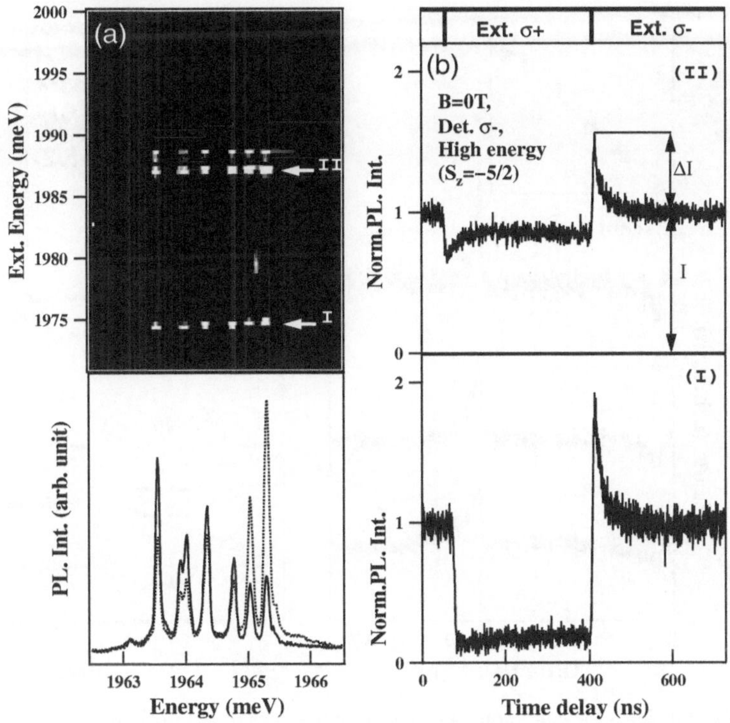

Figure 12.1 (a) PL and PLE of a Mn-doped QD (QD1) at $B = 0$ T and $T = 5$ K. The PL is detected in circular polarization under an alternate σ^-/σ^+ excitation at two different wavelengths: 1987.0 meV (solid line) and 1987.4 meV (dotted line). (b) PL transient under polarization switching at $B = 0$ T. The PL is detected on the high-energy line of X–Mn in σ^- polarization (Mn spin $S_z = -5/2$). Transient (I) (resp. (II)) was observed under resonant excitation at 1975 meV (resp. 1987 meV). Figures adapted from Ref. [23].

The relevant sub-levels of X–Mn and Mn are schematized in Fig. 12.2c. For the sake of simplicity, we omit the dark exciton states which should be included for a quantitative analysis and consider that the dynamics can be described by two spin relaxation rates, one for the Mn alone $\Gamma_{Mn,G}$ and one within the X–Mn complex Γ_{X-Mn} [19, 5]. As spin orientation results from a cumulative effect of relaxation in presence of the exciton, it can be performed only if Γ_{X-Mn} is faster than $\Gamma_{Mn,G}$. When exciting one of the low energy excited states of the QD, two mechanisms are expected to contribute to the observed spin orientation: the excitation of the QD can show a dependence on the Mn spin state [17], and the relaxation of the Mn spin within the X–Mn system is driven by the interaction with the spin-polarized carriers which have been injected.

Under spin-selective excitation, the spin relaxation of X–Mn tends to empty the spin state of the Mn that is most absorbant [19]. Under injection of spin polarized carriers, relaxation processes tend to anti-align the Mn spin with the X exchange field to reach a

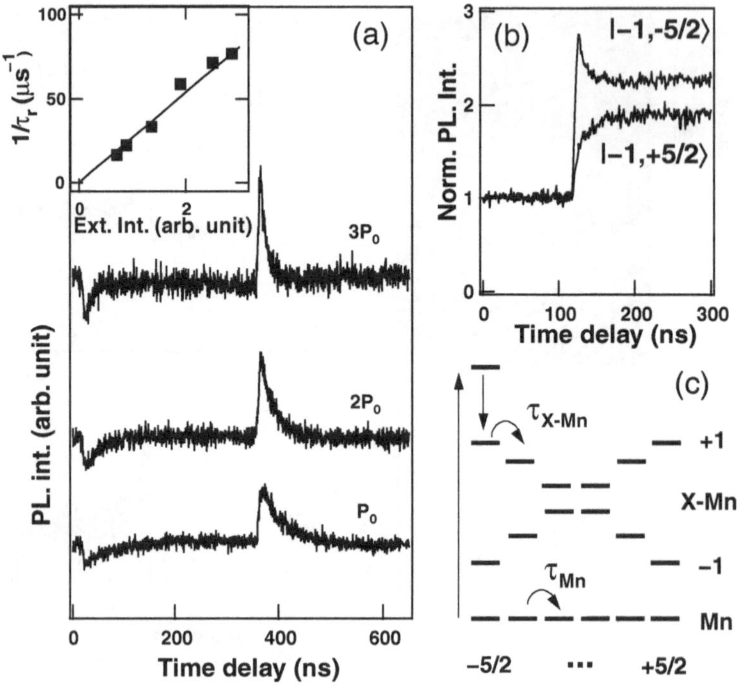

Figure 12.2 PL transients measured on QD1 at different values of the excitation power. Inset: power dependence of the inverse response time τ_r, taken at the $1/e$ point of the spin-related transient. (b) PL transients recorded in σ^- polarization on the high ($S_z = -5/2$) and low (S_z=+5/2) energy line of the X–Mn complex. (c) Simplified level diagram of a Mn-doped QD, as a function of Mn spin (X–Mn: bright exciton–Mn). Figures adapted from Ref. [23].

thermal equilibrium on the X–Mn levels [19]. Hence, optical pumping with σ^- photons, for instance, tends to decrease the population of the spin state $S_z = -5/2$ and increase that of $S_z = +5/2$, as observed in Fig. 12.2b. Both mechanisms, absorption selectivity and spin injection, depend on the structure of the excited states, resulting in a pumping signal which depends on the excitation energy (Fig. 12.1). An efficient pumping of the Mn spin can be performed within a few tens of ns, showing that at $B = 0$ T the spin relaxation time of the Mn alone is long enough compared to the X–Mn dynamics.

Having established a method to prepare the Mn spin, we can perform pump-probe experiments to observe how long the Mn polarization can be conserved (Fig. 12.3). We prepare a non-equilibrium distribution of the Mn spin with a $\sigma^\pm$ pump pulse. The pump laser is then switched off, and switched on again after a dark time τ_{dark}. The amplitude of the pumping transient after τ_{dark} depends on the Mn spin relaxation in the dark. For QD1 (Fig. 12.3a) no transients are observed after a dark time of 3.5 μs. This demonstrates that in the absence of charges fluctuations (i.e. neutral QD (see inset of Fig. 12.3a)) the prepared

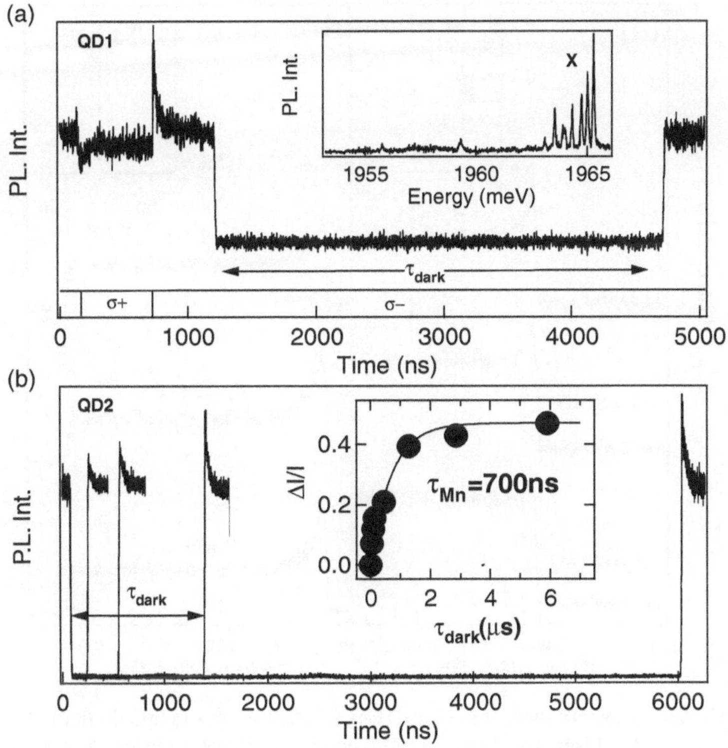

Figure 12.3 (a) PL transients recorded on QD1 (corresponding PL in the inset) under the optical polarization sequence displayed at the bottom of the plot. The spin distribution prepared by optical pumping is conserved during $\tau_{dark} = 3.5\,\mu$s. (b) PL transients recorded on QD2. The amplitude of the pumping signal is restored after $\tau_{dark} \sim 3\,\mu$s. From the delay dependence of this amplitude we deduce a Mn relaxation time of ~ 700 ns.

Mn spin is conserved over microseconds. For QD2 (Fig. 12.3b) the amplitude of the pumping signal is restored after $\tau_{dark} \sim 3\,\mu$s. From the delay dependence of this amplitude we deduce a Mn relaxation time of ~700 ns. These two examples show that the measured spin relaxation is not intrinsic to the Mn spin but depends on its local environment (strain, presence of carriers, etc.).

More information on the Mn local environment can be obtained from the magnetic-field dependence of the optical pumping signal. For an isotropic Mn spin, the decoherence of the precessing spin in a transverse field gives rise to the standard Hanle depolarization curve with a Lorentzian shape and a width proportional to $1/T_2$ [32]. In the present case, a magnetic field in the Faraday configuration (B_z) does not change significantly the PL transients (Fig. 12.4b): a weak increase of the spin orientation efficiency is observed as soon as a field of a few milliteslas is applied. By contrast, an in-plane field (B_x) induces a coherent precession of the Mn spin away from the optical axis (= QDs' growth axis),

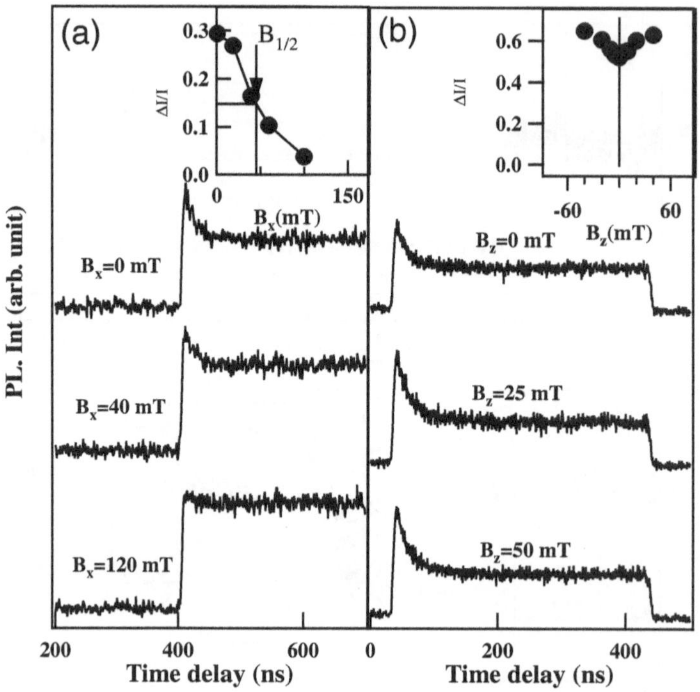

Figure 12.4 Mn spin transient measured on QD1 as a function of a magnetic field applied in-plane (a) and out-of-plane (b). Inset: transverse field dependence of the transient amplitude $\Delta I/I$ (see Fig. 12.1). $B_{1/2}$ is the half width at half maximum.

so that the average spin polarization, and therefore the amplitude of the optical pumping signal, decays (Fig. 12.4a).

It is known from electron paramagnetic resonance spectroscopy that the ground state of Mn^{2+} presents a fine structure [35]. In a cubic crystal, it results from a hyperfine coupling with the Mn nuclear spin, $A\,\mathbf{I}\cdot\mathbf{S}$ (with $I = 5/2$ and $A \approx 0.7\mu eV$), and the crystal field. In addition, in epitaxial structures, built-in strains due to the lattice mismatch induce a magnetic anisotropy with an easy axis along the QD's axis. It scales as $D_0 S_z^2$, with D_0 proportional to the tetragonal strain [35]. The total Hamiltonian of an Mn spin in a cubic crystal under biaxial strain and magnetic field reads:

$$\mathcal{H}_{Mn} = g\mu_B \vec{H}.\vec{S} + \mathcal{A}\vec{I}.\vec{S} + \mathcal{D}_0\left[S_z^2 - \frac{1}{3}S(S+1)\right] + \frac{1}{6}a\left[S_x^4 + S_y^4 + S_z^4 - \frac{1}{5}S(S+1)(3S^2+3S-1)\right]. \quad (12.1)$$

The resulting fine structure under a magnetic field applied in-plane or out-of-plane is shown in Fig. 12.5a. At zero field, the Mn electronic spin is quantized along the growth axis and the different electronic spin doublets are separated by an energy proportional to D_0. Each level is further split into six lines by the hyperfine coupling.

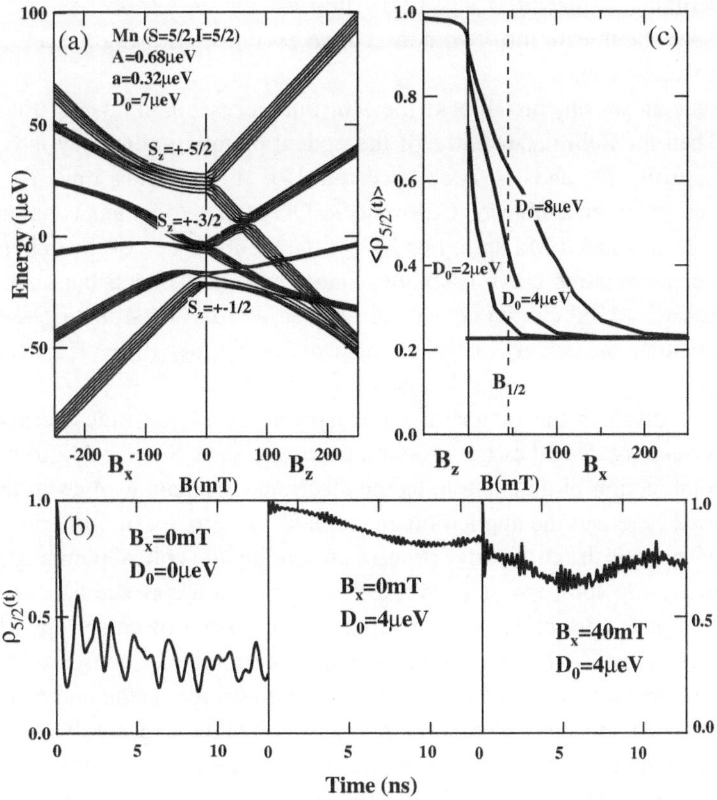

Figure 12.5 (a) Magnetic field dependence of the fine structure of the Mn spin with out-of-plane (right) and in-plane (left) field, calculated with $A = 0.68$ μeV, $D_0 = 7$ μeV and a cristal field parameter $a = 0.32$ μeV. (b) Time evolution of the density matrix element $\rho_{+5/2}(t)$ calculated for different values of D_0 and of the transverse magnetic field B_x ($\rho_{+5/2}(0) = 1$). (c) Time-average value of $\langle \rho_{+5/2}(t) \rangle$ versus B_x for different values of D_0 ($T_2 = \infty$). The $B_{1/2}$ found experimentally on QD1 is indicated with a dotted line. Figures adapted from Ref. [23].

This fine structure controls the Mn spin dynamics at zero or weak magnetic field. At zero field, in the absence of anisotropy, the precession of the electronic spin of the Mn in its own hyperfine field should erase any information stored on the electronic spin [18]. In the presence of magnetic anisotropy, the precession of the Mn spin in the nuclear field is blocked even at $B = 0\,T$. This is shown in Fig. 12.5b, where the evolution of the density matrix element $\rho_{5/2}(t)$ is calculated for the electronic + nuclear spins system of a single Mn atom using the master equation $d\rho_S(t)/dt = -i/\hbar[\mathcal{H}_{Mn}, \rho_S(t)]$ (no relaxation) with the initial condition $\rho_{5/2}(0) = 1$ ($S_z = +5/2$ and random nuclear spin I_z). With $D_0 = 0$, the coherent evolution of the electronic spin in the nuclear field erases the orientation of the electronic spin in a few hundreds of picoseconds. With a D_0 of a few μeV, this free precession is blocked and the Mn spin state can be conserved for a long time, in agreement with the experiment displayed in Fig. 12.3a. In particular, it is conserved during the time interval between the injection of two consecutive excitons, allowing the optical pumping mechanism to take place.

The magnetic anisotropy also blocks the Mn spin precession in a weak transverse magnetic field. Then the field dependence of the optical pumping efficiency is controlled by both the anisotropy D_0 and the coherence time T_2. A dephasing time $T_2^* \simeq 1$ ns has been measured in an ensemble of $CdMn_{0.02}Se$ QDs [37]. A longer coherence time T_2 is expected for an isolated Mn spin. For $T_2 \geq 1$ ns, the influence of the coherence on the width of the depolarization curve is smaller than 4 mT and its contribution to the experimental curve (Fig. 12.4a) can be neglected. Hence, in order to estimate D_0, we can take $T_2 = \infty$ to calculate the density matrix element $\rho_{5/2}(t)$ for $\rho_{5/2}(0) = 1$, assuming various values of D_0.

Figure 12.5c displays the magnetic field dependence of the time-averaged value of $\rho_{5/2}(t)$. This quantity describes the probability for the state $S_z = +5/2$ to be conserved after the recombination of an exciton, as the electronic Mn spin evolves in the hyperfine field, the crystal field and the applied magnetic field. A decrease in this spin conservation progressively destroys the cumulative process controlling the optical pumping mechanism. For a free precessing spin, $\langle \rho_{5/2}(t) \rangle \approx 0.24$ as soon as a transverse field is applied and the depolarization is controlled by T_2 [32]. In the presence of anisotropy, the Mn spin does not precess at weak field and the Mn spin state is partially conserved (Fig. 12.5b). When the transverse magnetic field is strong enough to overcome the magnetic anisotropy ($g_{Mn}\mu_B B_x \gg D_0$), the time average of $\rho_{5/2}(t)$ reaches the expected value for a coherently precessing spin ($\approx$ 0.24). This progressive decrease of $\langle \rho_{5/2}(t) \rangle$ gives rise to a field-induced depolarization curve which depends on the value of D_0. A half-height field $B_{1/2} \simeq 45$ mT, as observed experimentally in Fig. 12.4b, is obtained for $D_0 \approx 6$ μeV. As a value $D_0 \approx 12$ μeV is expected for CdTe coherently grown on ZnTe [35], the measured value is consistent with a Cd and Zn interdiffusion or a partial relaxation of the mismatch strain at the Mn location.

A magnetic anisotropy large enough to block most of the Mn spin relaxation also explains the very weak influence of an out-of-plane field. However, the Zeeman splitting cancels the residual non-diagonal coupling induced by crystal field (equation (12.1)) and slightly improves the Mn spin conservation (left part in Fig. 12.5c), thus accounting for the small increase of the optical pumping efficiency experimentally observed in Fig. 12.4b.

12.4 Resonant optical pumping of a single Mn spin

As proposed by Govorov *et al.* [19], the direct resonant excitation of one optical transition of the ground X–Mn complex can also be used to perform a direct and more efficient optical pumping of the Mn spin. In this optical-pumping process, a laser resonantly drives one of the exciton–Mn transition ($|-1, S_z = -5/2\rangle$ in Fig. 12.6) with a Rabi frequency $\Omega = \mathcal{P}\mathcal{E}/\hbar$, where $\mathcal{P}$ is the dipolar moment of the QD transition and $\mathcal{E}$ the amplitude

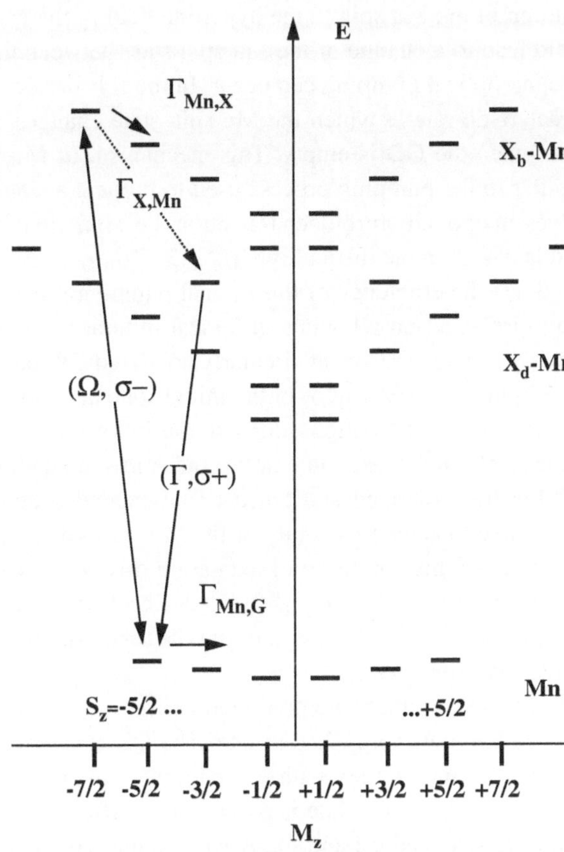

Figure 12.6 Energy levels of an Mn-doped QD involved in the resonant optical pumping mechanism described in the text (black: bright excitons (X_b)); grey: dark excitons (X_d)). The states are displayed as a function of their total angular momentum M_z and energy E. For the resonant optical pumping, the QD is resonantly driven on the state $S_z = -5/2$ with a $\sigma-$ laser pulse with a Rabi frequency Ω. The scattered photons obtained after a spin-flip of the exciton (rate $\Gamma_{X,Mn}$) are recorded in $\sigma+$ polarization. Figure adapted from Ref. [24].

of the electric field of the resonant laser. Photon absorption occurs only if the Mn spin in the QD is in the $S_z = -5/2$ spin state. The resultant exciton can radiatively recombine via the same channel, or a spin-flip process can project the X–Mn complex in a state with $S_z \neq -5/2$. After a few cycles of absorption–emission, the probability of detecting the Mn in the $S_z = -5/2$ state decreases. In this mechanism, we have assumed that the Mn spin was conserved once the exciton has recombined. The conservation of the Mn spin between the recombination of an exciton and the absorption of a photon can be altered in two ways: either by a relaxation process involving an exchange of energy, or by a coherent evolution [23, 18]. A coherent evolution can be neglected if the fine structure of the Mn atom is dominated by a magnetic anisotropy along the growth axis [23]. Otherwise, processes such as the coherent evolution of the Mn spin in the hyperfine field of the Mn nucleus or in the tetragonal crystal field lead to a change of the Mn spin state between the injection of two excitons. In that case, no optical pumping can occur. In the following, we will use $\Gamma_{Mn,G}$ to describe the characteristic rate at which the Mn spin state changes due to coherent or incoherent processes, when the QD is empty. This mechanism of Mn spin manipulation discussed here is similar to the pumping process used to prepare a single carrier spin in a QD [11, 16]. It involves an optically forbidden transition (i.e. spin flip of the Mn interacting with the exciton) and is based on the inequality $\Gamma_{Mn,X} > \Gamma_{Mn,G}$.

To demonstrate and test the efficiency of this optical pumping process, we developed a two-wavelength pump-probe set-up allowing an optical initialization and read-out of the Mn spin [24]. In this experiment, a resonant circularly polarized CW laser (*resonant pump*) tuned on a X–Mn level pumps the Mn spin. In the initial state at $t = 0$, the Mn atom is in thermal equilibrium. The resonant creation of an exciton followed by a spin relaxation of the Mn in the exchange field of the exciton empties the spin state under excitation. Then, a second laser train, linearly polarized and tuned on an excited state of the QD (*quasi-resonant probe*), injects excitons independently of the Mn spin state S_z. A spin relaxation of the X–Mn complex under this condition of excitation drives the Mn atom back to an equilibrium where all spin states are equally populated. Recording one of the six PL lines under this periodic sequence of excitation, we monitor the time evolution of the probability of occupation of a given Mn spin state.

The main features of this experiment are presented in Fig. 12.7. In this example, the σ^+ PL signal is recorded on the low energy X–Mn line. The QD is resonantly excited on the high energy state of the X–Mn complex with σ^- photons. This excitation can create an exciton in the dot only if the Mn spin state is $S_z = -5/2$. After this pumping sequence, the resonant pump laser is switched off and followed by a linearly polarized excitation on an excited state (*quasi-resonant probe*). The amplitude of this quasi-resonant PL depends on the population of $S_z = -5/2$ and, at the beginning of the probe pulse, is a probe of the resonant pumping efficiency reached at the end of the pump pulse.

This is illustrated in Fig. 12.7(i), which presents the difference in the two PL signals produced by the probe when the resonant pump laser was OFF or ON in the pump-probe sequence presented underneath the curve Fig. 12.7(ii). The difference of the two PL signals reflects the population difference between a sequence with optical pumping and a sequence where $S_z = -5/2$ is evenly populated. The height of the difference signal at the beginning of the probe pulse, which reaches 75%, gives a direct measurement of the efficiency of the spin optical pumping. The PL transients observed during the probe pulse correspond to the progressive destruction of the non-equilibrium distribution prepared by the pump. This reset process is produced by the injection of unpolarized excitons and its rapidity depends on the intensity of the probe laser.

A more direct way to probe the optical pumping process is to monitor the time evolution of the fluorescence signal observed during the resonant excitation. Excitation transfer can occur within the X–Mn complex during the lifetime of the exciton and gives rise to a weak PL on all the QDs levels. Whatever the spin relaxation processes involved in this excitation

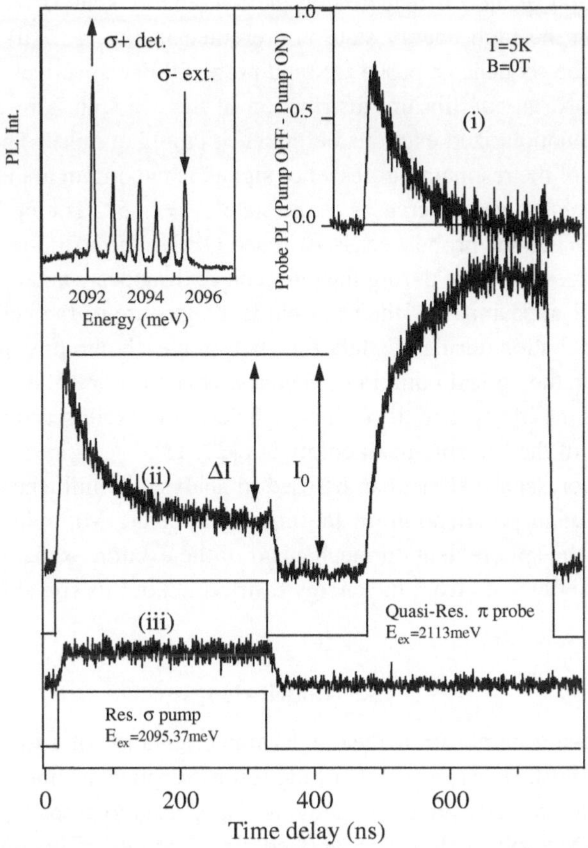

Figure 12.7 PL transients recorded on the low energy line of an Mn-doped QD under the quasi-resonant (QD excited state: *probe*) and resonant (QD ground state: *pump*) optical excitation sequence displayed at the bottom. The inset presents the QD PL under non-resonant excitation and the configuration of the resonant excitation and detection. (i) Difference between the PL produced by the probe, when the pump is OFF and when the pump is ON, (ii) PL from the pump and the probe, and (iii) resonant PL produced by the pump alone. Because of the Mn spin memory in the absence of injected carriers, no signature of pumping is observed when the linearly non-resonant probe is OFF (iii). The optical pumping process is directly observed on the resonant fluorescence produced by the pump and latter on the PL from the probe laser. I_0 is the amplitude of the fluorescence at the beginning of the pump pulse and ΔI the amplitude of the transient. $\Delta I/I_0$ is the efficiency of the spin optical pumping. Figures adapted from Ref. [24].

transfer, this signal depends on the absorption of the pump laser which is controlled by the occupation of $S_z = -5/2$: it monitors the spin selective absorption of the QD and is then a direct probe of the pumping efficiency of the Mn spin. The pumping efficiency is then given by $\Delta I/I_0 \approx 75\%$ (see Fig. 12.7), in agreement with the pumping efficiency measured on the probe sequence.

The time evolution of the PL detected on the low-energy state of X–Mn under a resonant excitation on the high energy state is presented in Fig. 12.7(ii) and (iii) for two different pump-probe sequences: probe ON and probe OFF, respectively. When the probe laser is switched ON, an equilibrium distribution of the Mn spin is restored by the non-resonantly injected unpolarized excitons before each pumping pulse. The absorption, and then the amplitude of the resonant fluorescence signal, is maximum at the beginning of the pump pulse and progressively decreases as the state $S_z = -5/2$ is emptied by the optical pumping process. When the probe laser is switched OFF in the pump-probe sequence, the resonant fluorescence transients during the pump pulse disappears and a weak constant PL is observed. This disappearance of the transient is a signature of the perfect conservation of the Mn spin distribution during the dark time between each pumping pulse. The steady-state PL depends on the optical pumping efficiency, which is controlled by the ratio of the relaxation rates for the Mn spin in the exchange field of the exciton and the relaxation and coherent evolution of the Mn spin in an empty dot [23, 18].

The resonant fluorescence signal can be used to analyse the influence of the excitation power, wavelength and polarization on the efficiency of the Mn spin optical pumping. Information about the spin relaxation mechanism of the exciton exchange coupled with a Mn spin can also be obtained from the energy distribution of this signal [24].

12.5 Conclusion

To conclude, our results demonstrate the optical spin orientation of a single magnetic atom in a semiconductor host. Quasi-resonant or fully resonant optical excitation of an individual Mn-doped QD with circularly polarized photons can be used to prepare a non-equilibrium distribution of the Mn spin without any applied magnetic field. This distribution is fully conserved for a few microseconds. We show that spin-flip scattered photons can be used to probe the dynamics of the initialization of the Mn spin during the resonant optical pumping process. With this new preparation and readout scheme it should be possible to initialize a Mn spin, manipulate it on a microseconds timescale with resonant microwave excitation, and reliably read the final state.

The observation and the optical addressing of an individual atomic spin in a QD open new possibilities in information storage. The device presented here is a first step towards the development of new memories in which an information digit could be stored on the spin state of an individual atom.

References

[1] Archer, P. I., Santangelo, S. A. and Gamelin, D. R. 2007. Direct observation of sp–d exchange interactions in colloidal Mn^{2+} and Co^{2+} doped CdSe quantum dots. *Nanoletters*, **7**, 1037.

[2] Bacher, G., Maksimov, A. A., Schömig, H. *et al.* 2002. Monitoring statistical magnetic fluctuations on the nanometer scale. *Phys. Rev. Lett.*, **89**(12), 127201.

[3] Beaulac, R., Schneider, L., Archer, P. I., Bacher, G. and Gamelin, D. R. 2009. Light-induced spontaneous magnetization in doped colloidal quantum dots. *Science*, **325**(5943), 973–976.

[4] Besombes, L., Léger, Y., Maingault, L. *et al.* 2004. Probing the spin state of a single magnetic ion in an individual quantum dot. *Phys. Rev. Lett.*, **93**, 207403.

[5] Besombes, L., Léger, Y., Bernos, J. *et al.* 2008. Optical probing of spin fluctuations of a single paramagnetic Mn atom in a semiconductor quantum dot. *Phys. Rev. B*, **78**, 125324.

[6] Bhattacharjee, A. K. and Pérez-Conde, J. 2003. Optical properties of paramagnetic ion-doped semiconductor nanocrystals. *Phys. Rev. B*, **68**(4), 045303.

[7] Boukari, H., Kossacki, P., Bertolini, M. *et al.* 2002. Light and electric field control of ferromagnetism in magnetic quantum structures. *Phys. Rev. Lett.*, **88**(20), 207204.

[8] Chernenko, A. V., Dorozhkin, P. S., Kulakovskii, V. D. *et al.* 2005. Auger recombination of excitons in semimagnetic quantum dot structure in a magnetic field. *Phys. Rev. B*, **72**(4), 045302.

[9] Clément, T., Ferrand, D., Besombes, L., Boukari, H. and Mariette, H. 2010. Dynamical equilibrium between magnetic ions and photocarriers in low Mn-doped single quantum dots. *Phys. Rev. B*, **81**(15), 155328.

[10] Dorozhkin, P. S., Chernenko, A. V., Kulakovskii, V. D. *et al.* 2003. Longitudinal and transverse fluctuations of magnetization of the excitonic magnetic polaron in a semimagnetic single quantum dot. *Phys. Rev. B*, **68**(19), 195313.

[11] Dreiser, J., Atatüre, M., Galland, C. *et al.* 2008. Optical investigations of quantum dot spin dynamics as a function of external electric and magnetic fields. *Phys. Rev. B*, **77**(7), 075317.

[12] Erwin, S. C., Zu, L., Haftel, M. I. *et al.* 2005. Doping semiconductor nanocrystals. *Nature*, **436**, 91–94.

[13] Furdyna, J. K. 1998. Diluted magnetic semiconductors. *J. Appl. Phys.*, **64**, R29.

[14] Furdyna, J. K. and Kossut, J. (eds.) 1988. *Diluted Magnetic Semiconductors*. Semiconductors and Semimetals, vol. 25. Academic Press, Boston.

[15] Gaj, J. A., Planel, R. and Fishman, G. 1979. *Solid State Commun.*, **29**, 435.

[16] Gerardot, B. D., Brunner, D., Dalgarno, P. A. *et al.* 2008. Optical pumping of a single hole spin in a quantum dot. *Nature*, **451**(7177), 441–444.

[17] Glazov, M. M., Ivchenko, E. L., Besombes, L. *et al.* 2007. Fine structure of exciton excited levels in a quantum dot with a magnetic ion. *Phys. Rev. B*, **75**, 205313.

[18] Goryca, M., Ferrand, D., Kossacki, P. *et al.* 2009. Magnetization dynamics down to a zero field in dilute (Cd,Mn)Te quantum wells. *Phys. Rev. Lett.*, **102**, 046408.

[19] Govorov, A. O. and Kalameitsev, A. V. 2005. Optical properties of a semiconductor quantum dot with a single magnetic impurity: photoinduced spin orientation. *Phys. Rev. B*, **71**, 035338.

[20] Hundt, A., Puls, J. and Henneberger, F. 2004. Spin properties of self-organized diluted magnetic $Cd_{1-x}Mn_xSe$ quantum dots. *Phys. Rev. B*, **69**(12), 121309.

[21] Kudelski, A., Lemaitre, A., Miard, A. *et al.* 2007. Optically probing the fine structure of a single Mn atom in an InAs quantum dot. *Phys. Rev. Lett.*, **99**, 247209.

[22] Latta, C., Hogele, A., Zhao, Y. *et al.* 2009. Confluence of resonant laser excitation and bidirectional quantum-dot nuclear-spin polarization. *Nature Physics*, **5**(10), 758–763.

[23] Le Gall, C., Besombes, L., Boukari, H. *et al.* 2009. Optical spin orientation of a single manganese atom in a semiconductor quantum dot using quasiresonant photoexcitation. *Phys. Rev. Lett.*, **102**, 127402.

[24] Le Gall, C., Kolodka, R. S., Cao, C. L. *et al.* 2010. Optical initialization, readout, and dynamics of a Mn spin in a quantum dot. *Phys. Rev. B*, **81**, 245315.

[25] Léger, Y., Besombes, L., Fernández-Rossier, J., Maingault, L. and Mariette, H. 2006. Electrical control of a single Mn atom in a quantum dot. *Phys. Rev. Lett.*, **97**(10), 107401.

[26] Léger, Y., Besombes, L., Maingault, L. and Mariette, H. 2007. Valence-band mixing in neutral, charged, and Mn-doped self-assembled quantum dots. *Phys. Rev. B*, **76**, 045331.

[27] Linnarsson, M., Janzén, E., Monemar, B., Kleverman, M. and Thilderkvist, A. 1997. Electronic structure of the GaAs:Mn_{Ga}scenter. *Phys. Rev. B*, **55**(11), 6938–6944.

[28] Mackh, G., Ossau, W., Yakovlev, D. R. *et al.* 1994. Localized exciton magnetic polarons in $Cd_{1-x}Mn_xTe$. *Phys. Rev. B*, **49**(15), 10248–10258.

[29] Mackowski, S., Gurung, T., Jackson, H. E., Smith, L. M. and Karczewski, G. 2005. Exciton-controlled magnetization in single magnetic quantum dots. *Appl. Phys. Lett.*, **87**, 072502.

[30] Maingault, L., Besombes, L., Léger, Y., Bougerol, C. and Mariette, H. 2006. Inserting one single Mn ion into a quantum dot. *Appl. Phys. Lett.*, **89**, 193109.

[31] Maksimov, A. A., Bacher, G., McDonald, A. *et al.* 2000. Magnetic polarons in a single diluted magnetic semiconductor quantum dot. *Phys. Rev. B*, **62**(12), R7767–R7770.

[32] Myers, R. C., Mikkelsen, M. H., Tang, J.-M. *et al.* 2008. Zero-field optical manipulation of magnetic ions in semiconductors. *Nature materials*, **7**, 203.

[33] Norris, D. J., Yao, N., Charnock, F. T. and Kennedy, T. A. 2001. High-quality manganese-doped ZnSe nanocrystals. *Nanoletters*, **1**, 3.

[34] Press, D., Ladd, T. D., Zhang, B. and Yamamoto, Y. 2008. Complete quantum control of a single quantum dot spin using ultrafast optical pulses. *Nature*, **456**(7219), 218–221.

[35] Qazzaz, M., Yang, G., Xin, S. H. *et al.* 1995. Electron paramagnetic resonance of Mn2+ in strained layer semiconductor superlattices. *Solid State Commun.*, **96**, 405.

[36] Reiter, D. E., Kuhn, T. and Axt, V. M. 2009. All-optical spin manipulation of a single manganese atom in a quantum dot. *Phys. Rev. Lett.*, **102**, 177403.

[37] Scheibner, M., Kennedy, T. A., Worschech, L. *et al.* 2006. Coherent dynamics of locally interacting spins in self-assembled $Cd_{1-x}Mn_xSe$/ZnSe quantum dots. *Phys. Rev. B*, **73**(8), 081308.

[38] Schneider, J., Kaufmann, U., Wilkening, W., Baeumler, M. and Köhl, F. 1987. Electronic structure of the neutral manganese acceptor in gallium arsenide. *Phys. Rev. Lett.*, **59**(2), 240–243.

[39] Seufert, J., Bacher, G., Scheibner, M. *et al.* 2001. Dynamical spin response in semimagnetic quantum dots. *Phys. Rev. Lett.*, **88**(2), 027402.

[40] Tinjod, F., Gilles, B., Moehl, S., Kheng, K. and Mariette, H. 2003. II–VI quantum dot formation induced by surface energy change of a strained layer. *Appl. Phys. Lett.*, **82**, 4340.

[41] Twardowski, A., Swiderski, P., von Ortenberg, M. and Pauthenet, R. 1984. *Solid State Commun.*, **50**, 509.

[42] Yakovlev, D. R., Kavokin, K. V., Merkulov, I. A. *et al.* 1997. Picosecond dynamics of magnetic polarons governed by energy transfer to the Zeeman reservoir. *Phys. Rev. B*, **56**(15), 9782–9788.

13

Optical spectroscopy of InAs/GaAs quantum dots doped with a single Mn atom

O. Krebs and A. Lemaître

13.1 Introduction

Most of the current semiconductor devices rely on intentional, density- and spatially controlled doping with impurities. Dopants of donor and acceptor type enable both to change locally the electronic properties (conductivity, chemical potential, built-in electric field, etc.) and to tune these properties by metallic gates. Using such doping modulation has been shown to be very fruitful in the past two decades to fabricate and investigate semiconductor quantum dots (QDs) in the Coulomb blockade regime where the number of resident charges can be deterministically tuned one by one. In parallel, incorporating magnetic dopants in a semiconductor matrix has long been motivated by the possibility of inducing new properties and developing new functionalities. Observation of ferromagnetism in diluted magnetic semiconductors like $Ga_{1-x}Mn_xAs$ (with x in the range of a few percent) by the end of the 1990s has more specifically stimulated a lot of work [14]. Even though the Curie temperature of GaMnAs below ~ 200 K is likely to limit its potential use for applications, this compound still behaves like an ideal system to investigate the setup and control of Zener-type ferromagnetism in semiconductors, where the Mn atoms incorporated in the GaAs matrix provide both localized magnetic moments and free carriers. Combining the properties of quantum dots with those induced by magnetic doping is naturally an attractive track to explore, both to tailor new spin-based quantum properties and to investigate the fundamental interactions between carriers and magnetic impurities at the microscopic level. In this perspective, the limiting case of a single magnetic atom in a single quantum dot is obviously the elementary system of highest interest. Furthermore, the spin degree of freedom of a magnetic impurity, which is attached to the lattice at a fixed position, could be considered as an interesting candidate in the current quest for an ideal quantum bit in condensed matter [3]. Embedded in quantum dots, such isolated spins could benefit from their well-established properties, and advanced optical and electrical methods for control of individual dots. With self-assembled quantum dots, there is a clear difficulty in fabricating singly doped QDs because of the random nature of dopant incorporation during the semiconductor growth. Actually, this technique turned out to be quite successful thanks to the possibility provided by micro-photoluminescence spectroscopy to scan a large number of dots rapidly (typically one thousand per hour) and thus to identify and select those presenting the signature of a single magnetic impurity. The first demonstration was achieved for

Quantum Dots: Optics, Electron Transport and Future Applications, ed. Alexander Tartakovskii. Published by Cambridge University Press.

CdTe quantum dots doped by a single Mn atom in 2004 [4], and this pioneering work was followed by extensive investigations showing many remarkable new features. The present chapter is devoted to the case of InAs/GaAs self-assembled QDs doped by a single Mn dopant which strongly differs from the case of II–VI based QDs due to the acceptor nature of Mn impurities in III–V semiconductors.

13.2 The Mn acceptor impurity A^0

13.2.1 Electronic configuration

In the high-dilution regime (below 10^{18} cm^{-3}) Mn atoms are incorporated in GaAs predominantly at substitutional positions on Ga sites and present acceptor character. Indeed, as a transition metal, the Mn atom ($4s^2\,3d^5$) provides only two electrons to form the sp^3 valence bonds with its nearest neighbors. A third electron could be taken from the d levels, leading in such a case to a neutral impurity in $3d^4$ configuration, as observed in GaP [16]. However, investigations of the electron paramagnetic resonance in GaAs:Mn have established the preferred $A^0 = 3d^5 + \text{hole}$ configuration [21]. The sp^3 bonds are thus formed by taking the missing electron from the lattice, which in turn creates a negative center denoted A^-. The associated Coulombic potential localizes a hole (h) from the valence band which leads to the neutral acceptor state A^0. This Mn-bound hole is characterized by a Bohr radius of around 1 nm and a rather large binding energy of around -110 meV. At high concentration, when the Coulombic potential of A^- is screened, most of the Mn impurities are ionized, which provides the free carriers crucial for the ferromagnetism of GaMnAs. The Mn impurity in InAs has been investigated much less. Apart from a likely weaker binding energy (so far, rather poorly documented) and larger Bohr radius, the same configuration $3d^5 + h$ is observed, as clearly indicated by the direct imaging of the bound hole wavefunction by scanning tunneling microscope (STM). The images of a single Mn impurity show essentially the same typical bow-tie shapes that prove the existence of similar A^0 states in InAs [20] and GaAs [26].

13.2.2 Effective spin $J = 1$

Let us now consider the magnetic properties of A^0 resulting from its half-filled d shell. Following Hund's rule, the five electrons in the $3d^5$ configuration have parallel spins, resulting in a total spin $S = 5/2$, and occupy 5 different d-orbitals so that their total orbital angular momentum vanishes. As other spin configurations are found at much higher energy, they can be neglected and the $3d^5$ electrons can be described by their total spin $S = 5/2$ only. The strong interest in Mn-doped systems comes from the sp–d exchange interaction of the d electrons with conduction and valence electrons. Within the A^0 complex, the hole–Mn p–d exchange which reads $\varepsilon\, \boldsymbol{J}_{\mathbf{h}} \cdot \boldsymbol{S}$ in the spherical approximation, where $\boldsymbol{J}_{\mathbf{h}}$ is the total angular momentum of the hole ($J_h = 3/2$), is anti-ferromagnetic ($\varepsilon > 0$). The ground state of A^0 is given by the anti-ferromagnetic configuration of total angular momentum $J = 1$, where $\boldsymbol{J} = \boldsymbol{J}_{\mathbf{h}} + \boldsymbol{S}$ [23, 21, 6, 12]. This level is well separated from the above $J = 2$ configuration by $2\varepsilon \sim 10$ meV, a splitting significantly larger than $k_B T$ for temperatures below 50 K. Under such experimental conditions, higher levels ($J = 2, 3, 4$) have vanishing occupations, so that in the neutral A^0 form, a Mn impurity should behave essentially as an effective spin 1. Yet the Mn-bound hole experiences the potential perturbations induced by strain or composition inhomogeneity over its spatial extension around the Mn site. Such a property is highly susceptible to lifting the threefold degeneracy of the $J = 1$ effective spin. In InAs/GaAs quantum dots, a zero-field splitting will indeed occur and therefore will drastically affect the optical signature of A^0 as compared to that of a single Mn atom in a II–VI quantum dot.

13.2.3 Fabrication of Mn-doped quantum dot samples

The samples of self-assembled Mn-doped QDs that have been studied so far [17, 15], are grown by molecular beam epitaxy on a [001] GaAs substrate. The QD layer forms spontaneously by depositing an InAs film close to the critical thickness of 1.7 monolayer in 5 s, while the substrate rotation is interrupted to provide a certain thickness gradient. One thus obtains a narrow transition region, a few millimeters wide, at the onset of the 3D growth mode, where the low QD density, of the order of 10^8 cm^{-2}, is optimal for single-dot spectroscopy. The Mn doping is achieved by opening the Mn cell shutter during the InAs deposition. As the substrate temperature ($\sim$ 500 °C) is chosen to obtain QDs with the highest radiative yield, the Mn segregation is very strong. Mn atoms tend to accumulate on the surface and only a very low fraction remains in the QD layer. As confirmed by micro-photoluminescence measurements, the efficiency of actual QD doping is rather poor, with at most 1% of the QDs doped with a Mn atom, whereas the nominal flux was adjusted to reach one Mn per QD on average. In the present study, two types of structure were designed. The first one consists of a Mn-doped QD layer inserted in between an electron reservoir and a Schottky gate to provide charge-tuneable QDs [24, 25]. The second type is simply an Mn-doped QD layer embedded at the center of a thick GaAs layer, and surrounded by GaAlAs barriers. In this sample, the QDs turn out to be positively charged, probably owing to a residual p-type doping.

13.3 Micro-photoluminescence setup in magnetic field

To investigate the Mn-doped QDs that are sparsely distributed in the QD layer we need to scan a large number of individual dots in a reasonable time. Therefore a micro-photoluminescence (μ-PL) experiment at low temperature appears to be quite suitable. The application of a magnetic field of a few Tesla also turns out to be necessary to identify the "magnetic" dots and to study accurately their spin-dependent interactions. In the experiments presented below, the sample is attached to a vertical plate facing a 2 mm focal length aspheric lens (N.A. 0.5), which can be moved by piezo-motors to provide relative positioning in all three directions. This compact microscope, fixed at the end of the sample rod, is put inside a split-coil magneto-optic cryostat and cooled down to 1.5 K in pumped liquid helium. With this design the microscope lens is rigidly connected to the sample mount, which ensures very high stability against temperature drift or magnetic-field-induced stresses. A specific dot can be investigated for several hours or even days without the need for re-positioning. To find QDs exhibiting sets of spatially and spectrally correlated lines, the primary characteristic feature of Mn doping, the sample has to be scanned over a large area, consisting of continuous courses of $\sim$ 200 μm, while visually analyzing the PL signal. For this reason, the samples are not covered with a shadow mask as frequently done in μ-PL experiments. The collected PL is dispersed by a 0.6 m focal length double spectrometer and detected by a nitrogen-cooled CCD array camera. Thanks to the good signal-to-noise ratio the scan speed amounts to 1 μm per 5 s, yet limited to a 12 meV wide spectral range. Interestingly, the microscope, which integrates both the sample and the optical lens, can be rotated as a whole about the vertical axis of the cryostat to change the magnetic field direction with respect to the optical axis from parallel (Faraday configuration) to perpendicular (Voigt configuration). One can afterwards refocus the laser spot on the position of a given QD thanks to sample surface imaging.

13.4 Zero-field signature of Mn doping

We first present the specific signature of a single Mn atom observed in a QD PL spectrum under non-resonant excitation and in zero magnetic field, see Fig. 13.1a. It consists of two doublets of lines with the same splitting δ in the 100 μeV range and separated from each other by a larger splitting Δ in the millielectronvolt range. A much weaker line, labeled O in the figure, is also generally seen between the doublets. At first sight, such a pair of doublets looks very similar to that formed by an exciton–biexciton couple as the anisotropic electron–hole exchange usually splits the bright exciton levels by a few tens of microelectronvolts. The confusion can be easily discarded by two rapid checks. First, the relative intensity of the doublets does not depend on the excitation power for the Mn-doped QD. Second, the individual lines are not linearly polarized, in contrast to those of excitons in undoped dots. Furthermore, in Mn-doped QDs the exciton (X^0), biexciton ($2X^0$) and the singly charged excitons (X^+ or X^-) all exhibit the same characteristic spectrum of Fig. 13.1a in zero field with only small differences that will be discussed in next sections.

The PL spectrum shown in Fig. 13.1a actually corresponds to an X^+ trion. In this excited state, the QD contains one electron in its lowest conduction state, while two holes occupy the highest level in the valence band in a singlet spin configuration, as illustrated in Fig. 13.1b. From the point of view of spin-dependent interactions this state amounts to a single electron spin. The X^+ trion relaxes via spontaneous emission, a process where the electron of spin $\uparrow$ ($\downarrow$) recombines with a $\Downarrow$ ($\Uparrow$) hole to emit a σ^- (σ^+) photon. The final state of X^+ relaxation consists of a single hole. The exchange interaction with the magnetic impurity in the A^0 configuration therefore involves a single electron in the excited state and

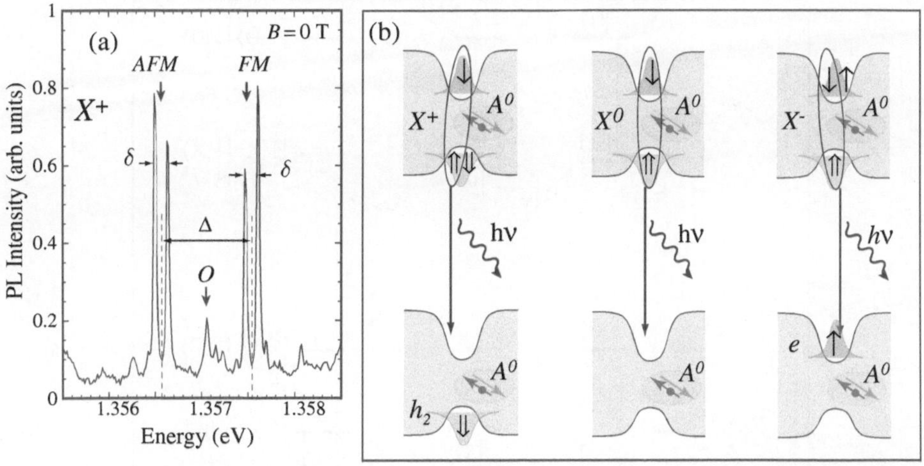

Figure 13.1 (a) PL spectrum of a charged exciton X^+ from a quantum dot doped by a single Mn atom exhibiting two characteristic equidistant doublets in zero magnetic field. A non-resonant excitation is provided by a He–Ne laser line. (b) Schematics of the optical transitions for trions $X^{\pm}$ and excitons X^0 in presence of a spectator A^0 impurity.

a single hole in the final state. Obviously, for a negative trion X^- a similar situation holds with only the interchange of the electron and the hole, whereas for an exciton the exchange with A^0 takes place simultaneously in the same level for both electron and hole.

13.5 Energy levels for trions in zero field

Figure 13.2 depicts the evolution of the levels involved in the X^+ spectrum, first owing to the QD perturbation (strain and confinement) on the $J = 1$ spin of A^0, then to the effective exchange with either the electron or the hole confined in the dot. This approach assumes that the Mn-bound hole is only weakly perturbed by the dot potential as originally suggested by A. Govorov [12]. This provides a very good framework for the interpretation of the experimental results. Yet it is worth mentioning that calculations carried out in the envelope function approximation predict in contrast that the Mn-bound hole should delocalize in the whole QD volume [7]. Leaving behind this open issue, let us follow the perturbative description. The primary effect of the QD potential is to split the A^0 ground state level, shifting the $|J_z = 0\rangle$ state above the $|J_z = \pm 1\rangle$ level, as a consequence of the strong anisotropy of the lens-shaped QDs and their related ϵ_{zz} uniaxial strain. This effect, which results from the higher light-hole character of $|0\rangle$ as compared to $|\pm 1\rangle$ states [21], is analogous to the shift of the light-hole states with respect to the heavy-hole states in the QD. Next, since the bound hole of A^0 inevitably experiences some in-plane anisotropy, e.g. owing to the proximity of a QD edge or a local shear strain, the $|\pm 1\rangle$ states are also split by an energy $\sim\delta$, which leads to the states denoted $|x\rangle$ and $|y\rangle$ in Fig. 13.2. The

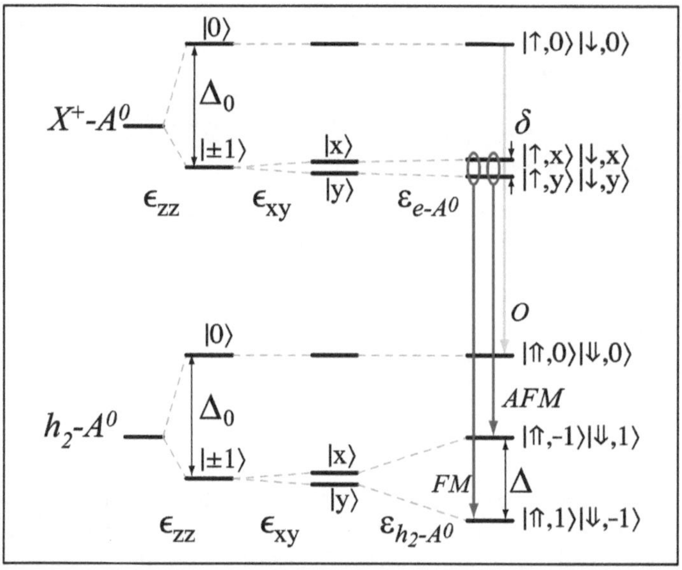

Figure 13.2 Diagram of energy levels involved in X^+ PL lines as a function of the main perturbations acting on A^0 spin, i.e. ϵ_{zz} uniaxial strain, ϵ_{xy} shear strain, and effective exchange with the unpaired electron spin in X^+ state or with the QD hole h_2 in the ground state.

initial threefold degeneracy of the bulk A^0 ground state turns out to be completely lifted in a QD. The exchange interaction with the unpaired electron in the X^+ state does not significantly alter the A^0 levels because of its relatively small intensity. This will be discussed in more detail below. To the first order, the levels are unchanged and the $X^+ - A^0$ eigenstates simply read as $|\uparrow \text{ or } \downarrow, x\rangle$, $|\uparrow \text{ or } \downarrow, y\rangle$, and $|\uparrow \text{ or } \downarrow, 0\rangle$. Conversely, the $h_2 - A^0$ exchange interaction in the final state drastically changes the levels. The QD hole spin acts like a longitudinal magnetic field along z or $-z$ depending on the spin orientation $\Uparrow$ or $\Downarrow$, because of its heavy character ($\langle\Uparrow |J_{h_2,x}| \Downarrow\rangle \sim 0$ and $\langle\Uparrow |J_{h_2,y}| \Downarrow\rangle \sim 0$). The exchange interaction $\varepsilon_{h_2-A^0}$ is, to the first order, equivalent to an Ising coupling $\propto J_{h_2,z}J_z$ between the two spins. The upper level corresponding to the $|\Uparrow \text{ or } \Downarrow, 0\rangle$ states is unchanged, while the two lower levels are now split by the exchange energy $|\Delta| \gg \delta$, which restores for the most part the circular symmetry of the eigenstates. In this approximation, one ends up with two levels of ferromagnetic (FM) and anti-ferromagnetic (AFM) spin configuration corresponding to the eigenstates $(|\Uparrow, +1\rangle, |\Downarrow, -1\rangle)$ and $(|\Uparrow, -1\rangle, |\Downarrow, +1\rangle)$ respectively. Their relative position, namely the sign of Δ, will be determined from the evolution in magnetic field.

The zero-field spectrum of Fig. 13.1 is deduced straightforwardly from this diagram of levels. Since $|\langle x| \pm 1\rangle|^2 = |\langle y| \pm 1\rangle|^2 = 1/2$, the states belonging to the δ-split doublet of $X^+ - A^0$ are optically coupled both to the FM and AFM levels of $h_2 - A^0$ following usual selection rules for trions. This gives rise to the equidistant doublets split by Δ in Fig. 13.1a.

The central line 0 corresponds to the transitions $|\uparrow, 0\rangle \rightarrow |\Uparrow, 0\rangle$ and $|\downarrow, 0\rangle \rightarrow |\Downarrow, 0\rangle$. Its intensity is much weaker because of the very low population of A^0 in $|0\rangle$ state at thermal equilibrium since the vertical shift Δ_0, around 2 meV in most investigated QDs, greatly exceeds $k_B T \sim 150\,\mu$eV at 1.8 K.

13.6 Photoluminescence in a longitudinal magnetic field

In a longitudinal magnetic field B_z, assumed here to be positive, the levels are split due to the Zeeman effect experienced by their constitutive particles, see Fig. 13.3a, and consecutively the PL lines split into σ^+ and σ^- polarized components. The dominant Zeeman term is governed by A^0 owing to its effective g-factor $g_{A^0} \sim 3$ as compared to $g_h \sim 0.8$ and $g_e \sim -0.5$ for hole and electron respectively. However, as the A^0 complex is present in both the initial and final states of the transitions and does not couple itself to light, its Zeeman shift is not expected to manifest strongly in the PL evolution. This is rather well-verified experimentally except for two specific features clearly visible in Fig. 13.3b. First, there is a very distinctive cross pattern in the density plot of the PL intensity, consisting of two rapidly shifting lines which anti-cross with the normal X^+ lines, of effective g-factor given by the sum $|g_e| + |g_h|$, both in FM and AFM configuration. Obviously, the anti-crossing observed near the zero field corresponds to the anisotropic coupling between $|\pm 1\rangle$ states responsible for the doublets in Fig. 13.1. The lines forming the cross pattern

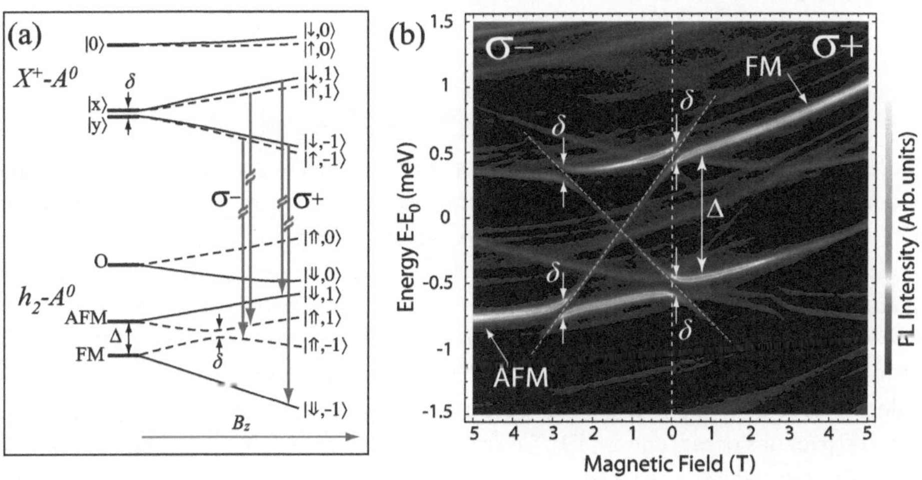

Figure 13.3 (a) Evolution of the levels involved in X^+ PL spectrum as a function of magnetic field. The anisotropic coupling $\delta/2$ between $|+1\rangle$ and $|-1\rangle$ A^0 states gives rise to an anti-crossing of the $|\Uparrow, 1\rangle$ and $|\Uparrow, -1\rangle$ states when the applied field compensates their initial splitting Δ. (b) Density plot of PL spectra of QD1 around $E_0 = 1.357$ meV as a function of the longitudinal magnetic field and measured separately in σ^+ and σ^- polarizations. A linear interpolation between spectra recorded every 50 mT is used to improve the graphical rendering. The cross pattern due to the "forbidden" transitions $|\pm 1\rangle \rightarrow |\mp 1\rangle$ is the hallmark of Mn doping.

correspond thus to the normally forbidden transitions $|\uparrow,\pm1\rangle \to |\Uparrow,\mp1\rangle$. They become weakly permitted thanks to the $\delta/2$ coupling taking place both in the initial state $X^+ - A^0$ and final state $h_2 - A^0$ of the transition. In the latter case, it produces an anti-crossing of the $|\Uparrow,+1\rangle$ and $|\Uparrow,-1\rangle$ levels at the magnetic field which compensates their initial exchange splitting Δ as shown in Fig. 13.3a and confirmed by the anti-crossing with both FM and AFM lines at the finite field ~ 3 T in Fig. 13.3b. Quite interestingly, the distance between both lines forming the cross provides a direct measurement of the effective A^0 g-factor since it grows as $4g_{A^0}\mu_B B_z$, where μ_B is the Bohr magneton. In Fig. 13.3b one finds $g_{A^0} = 3.13$, which is slightly larger than the value of 2.77 found in bulk [21]. This is mainly caused by a small modification of the eigenstates in the $J = 1$ ground state induced by the QD uniaxial strain. The second manifestation of the strong Zeeman effect on A^0 is related to its "orientation" by the magnetic field. In a positive field the population of A^0 tends to relax to the spin state $|-1\rangle$. This effect is clearly visible in the PL evolution, which shows the intensity enhancement (vanishing) of the FM (AFM) branch in σ^+ polarization due to the thermal relaxation $|\Downarrow,+1\rangle \to |\Downarrow,-1\rangle$, and symmetrically the intensity enhancement (vanishing) of the AFM (FM) branch in σ^- polarization. Actually, this strong polarization of A^0 in the field turns out to be very useful to identify assuredly which line corresponds to the FM and AFM configurations and subsequently to establish the FM character of the effective exchange between the QD hole and A^0. One can also notice that the position of the cross, mostly on the σ^- side, provides the same information. Besides, the small shift of the anti-crossing from a true zero field to $B_z \approx 120$ mT on the σ^+ side indicates a finite AFM $e - A^0$ exchange splitting of $\sim 40\,\mu$eV.

13.7 Excitons versus trions

A specific feature of Mn doping in InAs/GaAs quantum dots in comparison to CdTe quantum dots is the very similar zero field signatures observed for excitons (X^0), biexcitons ($2X^0$), and positive or negative trions [17]. A typical example is shown in Fig. 13.4a for a single quantum dot in a charge tunable structure. The gate voltage was set to favor the formation of the X^- trion, but non-resonant excitation provided by a 633 nm HeNe laser line still allowed us to observe the neutral exciton and biexciton states. In all three cases, the Mn doping produces two equidistant doublets with lines of similar intensity, while in Mn-doped CdTe quantum dots the PL spectrum of trions comprises 11 lines against 6 lines for neutral excitons or biexcitons [19]. The replication of this spectral signature observed in InAs dots is due to the strong anisotropic splitting Δ_0 of A^0 which is only weakly perturbed by the presence of a single electron in the dot, see Fig. 13.2. The intrinsic fine structure of A^0, in particular the $|x\rangle$ and $|y\rangle$ δ-split levels, is involved in all transitions either in the initial or final states, which gives rise to the pair of doublets constantly observed in zero field. One can notice that the four lines forming the zero-field spectral signature of the different excitonic states do not have the same intensity. For X^0 and $2X^0$ the inner lines are more intense than the outer lines. This property is due to the anisotropic $\delta/2$ coupling,

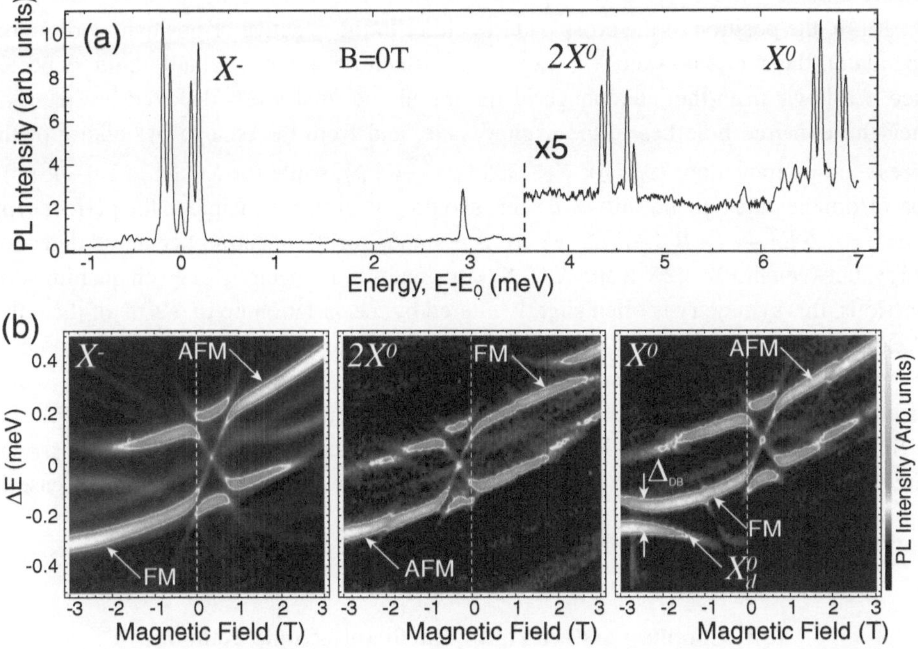

Figure 13.4 (a) Zero-field PL spectrum of an individual QD doped by a single Mn atom in a charge tunable sample around E_0 = 1.2535 eV. Three groups of correlated lines are identified as exciton (X^0), biexciton ($2X^0$) and negatively charged exciton (X^-). (b) Evolution of these spectral lines measured in σ^+ circular polarization as a function of a longitudinal magnetic field.

which to the first order in δ/Δ leads to X^0 eigenstates in the form $|\downarrow\Uparrow, +1\rangle \pm \frac{\delta}{\Delta}|\downarrow\Uparrow, -1\rangle$ or $|\uparrow\Downarrow, -1\rangle \pm \frac{\delta}{\Delta}|\uparrow\Downarrow, +1\rangle$. The AFM and FM X^0 states keep respectively a $|x\rangle$ or $|y\rangle$ character, which favors the transition to the level of A^0 with the same character. For trions, the opposite behavior is generally observed, although the AFM and FM levels still have in principle a $|x\rangle$ or $|y\rangle$ character. The inner lines of the doublets turn out to be less intense than the outer ones. This feature of both X^+ and X^- trions is the direct consequence of the finite AFM electron–A^0 coupling which, as discussed above, shifts the anti-crossing to a small but non-zero magnetic field.

The observation of Mn doping as a neutral A^0 acceptor in an n-Schottky structure is somewhat puzzling as one should expect the recombination of the Mn-bound hole with the free electrons to form negative centers A^- of spin 5/2. This point is notably intriguing at the gate voltages where the quantum dots are charged by an excess electron and lead to the formation of X^- states. The apparent robustness of A^0 against ionization is likely to be the result of a very slow ionization process in competition with the recapture of a photo-created hole. So far, there is no direct evidence of this scenario but it is partly supported by the long PL decay time of $D^0 - A^0$ observed in GaAs:Mn [1]. Probing resonantly the X^- with a narrow laser line could confirm this interpretation in the near future.

More drastic differences show up when a magnetic field is applied. As shown in Fig. 13.4b, the position of the cross pattern, either in the negative or positive field region, depends on the excitonic state X^0, $2X^0$ or X^-. Such behavior is actually quite expected since it follows from the interchange of the initial and final levels depicted in Fig. 13.2, when the unpaired hole lies in the excited state, and from the change of helicity of the corresponding transitions (e.g. for $X^+ \mid\Uparrow\Downarrow\uparrow\rangle \xrightarrow{\sigma-} \mid\Uparrow\rangle$, while for $X^- \mid\uparrow\downarrow\Uparrow\rangle \xrightarrow{\sigma+} \mid\uparrow\rangle$). Apart from the diamagnetic shift $\propto B_z^2$ this should give rise in principle to the perfect symmetry $(B_z, \Delta E) \leftrightarrow (-B_z, -\Delta E)$, where ΔE stands for the spectrally centered detection energy, between the X^+ ($2X^0$) and X^- (X^0) magneto-optic spectra of a given quantum dot. For trions, this symmetry is often slightly altered by the variation up to ~20% of the hole–A^0 exchange interaction in the presence (X^-) or absence (X^+) of a singlet electron pair. Conversely, this symmetry is well respected for exciton and biexciton transitions which indeed share the same excitonic levels and therefore exhibit the same exchange splitting Δ. Small variations of the A^0 level splitting δ, within the $2X^0$ state and the QD ground state $|0\rangle$, can take place but do not significantly perturb the symmetry as is the case in Fig. 13.4b.

13.8 Coupling to dark states due to anisotropic exchange

The magnetic field also reveals a distinctive feature of neutral excitons with respect to trions. The FM branch of X^0 in Fig. 13.4 undergoes an anti-crossing Δ_{DB} with an extra line at a finite field of −2.5 T. This extra line whose intensity vanishes far from the anti-crossing region and which moves in a direction opposite to that of the σ^+-polarized $\mid\downarrow\Uparrow\rangle$ exciton corresponds to the dark exciton X_d^0 in the $\mid\downarrow\Downarrow\rangle$ state. The redshift of ~0.5 meV in the zero field of the dark exciton state with respect to the bright exciton state (see Fig. 13.5b) is due to the electron–hole exchange interaction. In the case of trions, which always comprise a spin singlet of holes or electrons, the electron–hole exchange cancels out and there is no dark state. However, there are still optically forbidden transitions corresponding to a change of angular momentum projection $\Delta m = \pm 2$. They are quite equivalent to the dark exciton optical recombination, and for this reason they also appear in the magneto-PL image of trions due to a similar coupling to the allowed transitions. This gives rise to anti-crossings with the FM and AFM lines around ±2.5 T in Fig. 13.4b.

Interestingly, the strength of this anti-crossing Δ_{DB}, which is truly specific to Mn-doped dots, does not scale with the dominant exchange energy Δ, but rather expresses the anisotropy of the Mn position in the quantum dot [15]. Figure 13.5a shows magneto-PL images of a quantum dot for which it is particularly pronounced, with a ratio $\Delta_{\mathrm{DB}}/\Delta$ close to 1, while it usually falls in the 0%–50% range. As a result, the transitions involving the dark exciton levels can be observed relatively far from the anti-crossing field for both the exciton and biexciton. In the zero magnetic field, replicas of the δ-split doublets due to X_d^0 are clearly observed and one can even distinguish the associated cross patterns. This provides a means to measure the X^0 (X_d^0) exchange interactions Δ (Δ') with A^0,

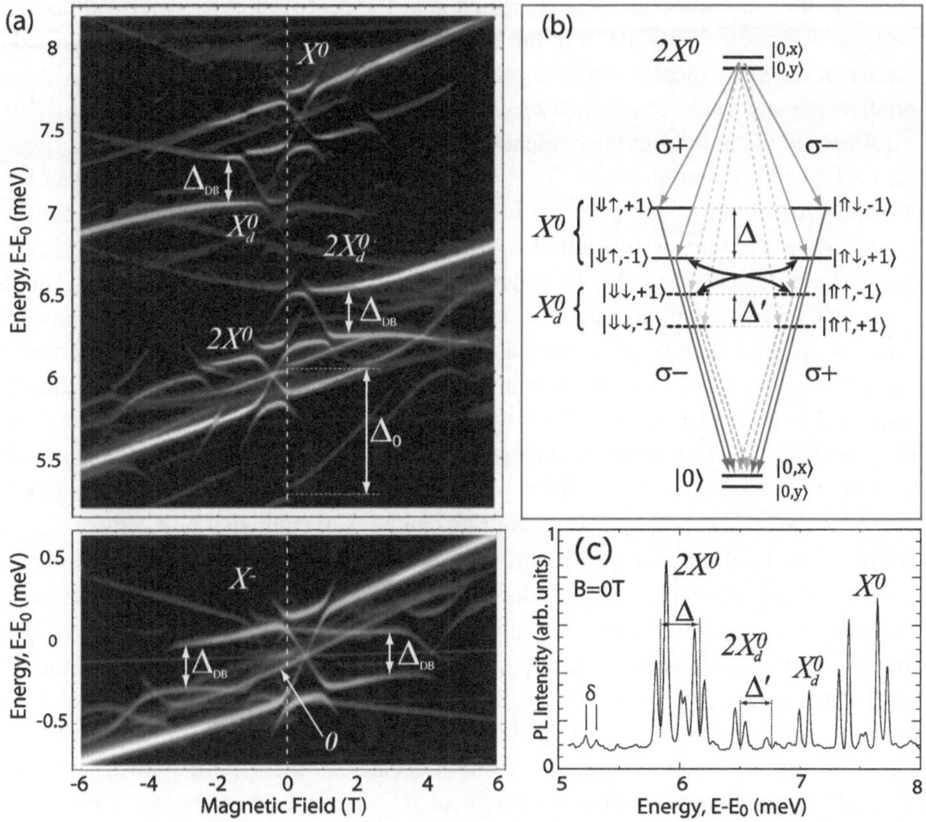

Figure 13.5 (a) Density plots of PL spectra from a Mn-doped QD measured in σ^+ circular polarization as a function of a longitudinal magnetic field. Zoom over the X^0 and $2X^0$ lines (top) and X^- (bottom) state around $E_0 = 1.2552$ meV. The anisotropy-induced magnetic coupling between hole spins $\Uparrow$ and $\Downarrow$ induces a large anti-crossing Δ_{DB} between bright (allowed) and dark (forbidden) transitions. The diamagnetic shift has been subtracted to make the symmetry between X^0 and $2X^0$ more obvious. (b) Main levels involved in the $2X^0 \to X^0 \to |0\rangle$ cascade at the origin of the symmetry between X^0 and $2X^0$. The weak splitting due to anisotropic electron–hole exchange has been neglected. (c) Zero-field spectrum zoomed over the region of X^0 and $2X^0$ lines.

see Fig. 13.5b, c, so that the individual electron–A^0 and hole–A^0 exchange energies can be determined separately. In the present case, they amount respectively to -30 μeV and 300 μeV. There is also an extra δ-split doublet below the $2X^0$ doublets that usually shows up for large Δ_{DB}. This doublet is not related to the dark states, but results from a finite mixing of the $J_z = \pm 1$ and $J_z = 0$ states of A^0 making the transition $|2X^0, x \text{ or } y\rangle \to |\Downarrow\uparrow, 0\rangle$ possible. Its distance to the central O line of $2X^0$ provides straightforwardly the anisotropic splitting Δ_0 of A^0 which amounts here to 0.65 meV.

For X^-, the strong anisotropic coupling Δ_{DB} gives rise to a perceptible replica of the cross pattern in the negative field side. Note that this is not a symmetric replica of the main cross by the transformation $B \to -B$, which could appear if circular polarizations were not respected, but it truly results from the forbidden transitions that are partly permitted in σ^+ polarization. Since these transitions persist up to high fields, one can measure with precision their Zeeman shift $(3g_h + g_e)/2\,\mu_B B$ and therefore determine the g-factors of both the electron and the hole from the comparison with the shift of allowed transitions $(3g_h - g_e)/2\,\mu_B B$. For the quantum dot of Fig. 13.5 one obtains $g_e = -0.60$ and $g_h = 0.87$.

The anti-crossing Δ_{DB} of the X^0 and $2X^0$ transitions results from couplings between dark and bright excitons as illustrated by the double arrows in Fig. 13.5b. They correspond to the hole spin-flip $|\Uparrow\rangle \leftrightarrow |\Downarrow\rangle$ while other spins remain unchanged. This is, however, not a direct coupling since it would violate the Kramers degeneracy theorem for half-integer spins. In that respect, one can indeed note that it vanishes for the central O line of X^- which exhibits no splitting in the zero field. There is a splitting of the central line of $2X^0$ (not visible for X^0 because of low signal-to-noise ratio), but this is the result of another effect, namely the anisotropic electron–hole exchange which yields linearly polarized eigenstates of the form $|\Downarrow\uparrow \pm \Uparrow\downarrow, 0\rangle$ split as usual by a few tens of microelectronvolts. The hole spin-flip coupling can be well simulated by a term $\propto J^3_{h_2,x} J_z$ within the $J = 1$ subspace of A^0. Its origin can be confidently ascribed to the combination of two anisotropic terms [15]: (i) the local anisotropy experienced by the Mn-bound hole which confers a non-zero in-plane angular momentum to the $|\pm 1\rangle$ states, and (ii) the anisotropic exchange interaction due to the out-of-center Mn position in the quantum dot volume which induces a coupling $\propto J^3_{h_2,x} S_x$ [5]. The latter can be seen as a consequence of the light–hole part in the valence ground state due to the lateral confinement [2]. This part does not participate in the optical transition because of the symmetry of its envelope, but can significantly contribute to the very local $p-d$ exchange interaction with the Mn spin. This effect can alternatively be taken into account by assuming the mixing of the heavy-hole state with an optically inactive light hole [17, 18, 11].

13.9 Theoretical simulation of PL spectra

The above qualitative description of the spin interactions in Mn-doped InAs quantum dots is supported by a model based on a spin Hamiltonian [15]. To reproduce the PL spectra in magnetic field as shown in Fig. 13.6 for a negative trion, four distinct spins are taken into account, namely $S_{Mn} = 5/2$ for the Mn atom, $J_{h_1} = 3/2$ for the Mn-bound hole, and $S_e = 1/2$ and $J_{h_2} = 3/2$ respectively for the electron and hole confined in the quantum dot. All spin configurations of A^0 $(J = 1, 2, 3, 4)$ are thus naturally taken into account. Even though a model of effective spin $J = 1$ would turn out to be sufficient in many cases, it is sometimes necessary to include explicitly the coupling between the $J = 1$ and $J = 2$ levels. Besides, such an approach allows us in principle to derive from the PL spectra the actual exchange interactions between the individual spins. The particles corresponding to the four spins are assumed to occupy the ground state of their respective confinement potential.

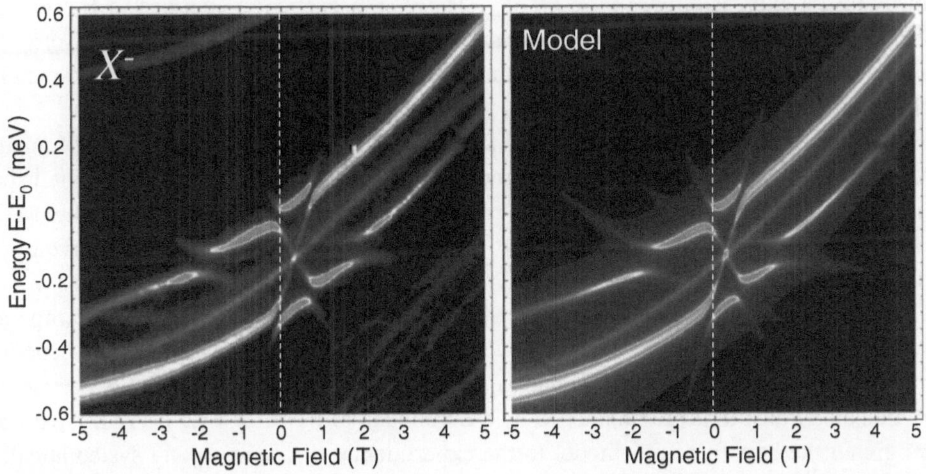

Figure 13.6 Density plots of PL spectra from a Mn-doped QD measured in σ^+ circular polarization as a function of a longitudinal magnetic field. Zoom over the X^- state around $E_0 = 1.273$ meV. Corresponding simulation performed with a spin model using the parameters : $g_e = -0.5$, $g_{h_1} = g_{h2} = 0.8$, $g_{Mn} = 2$, $\varepsilon_{Mn-h_1} = 3.75$ meV, $\varepsilon_{Mn-h_2} = 0.25$ meV, $\varepsilon_{h_1-h_2} = 0.425$ meV, $\rho = 0.35$, $T = 10$ K.

The spin Hamiltonian has to include both truly spin-dependent interactions and those due to the quantum dot which act on the angular momentum of the states. The potential anisotropy experienced by h_1 and responsible for the A^0 splitting in Fig. 13.2 is described by a general anisotropic strain tensor [27]. It consists of a dominant uniaxial compressive strain $\epsilon_\parallel$ and a shear strain $\epsilon_\perp$ in its perpendicular plane. For the h_2 hole a $\Delta_{HL} \sim 30$ meV splitting between the heavy-hole ($J_{h_2,z} = \pm 3/2$) and light-hole ($J_{h_2,z} = \pm 1/2$) states accounts for the strong QD confinement along z. The magnetic field induces the Zeeman splitting of each spin and the diamagnetic shift of the electron and h_2 hole energies independently of the spin state. The exchange interactions between the different spins are taken in the form of Heisenberg Hamiltonians [6, 13]. For both holes h_1 and h_2, this amounts to treating the issue of their indiscernibility in the Heitler–London scheme [10]. A specific anisotropic part $\mathcal{H}_a$ is added to the p–d exchange between the Mn spin and h_2 in order to reproduce the Δ_{DB} anti-crossing. The exchange experienced by the unpaired electron with A^0 is experimentally very small, in good agreement with theoretical calculations showing a mutual cancellation of both e–Mn and e–h_1 exchanges [22]. As long as only the $J = 1$ levels of A^0 are concerned, these exchanges can thus be neglected. Eventually, the relevant exchange Hamiltonian for trions reads as follows:

$$\hat{\mathcal{H}}_X = \varepsilon_{Mn-h_1}\hat{\mathbf{S}}_{Mn}\cdot\hat{\mathbf{J}}_{h_1} + \varepsilon_{Mn-h_2}\hat{\mathbf{S}}_{Mn}\cdot\hat{\mathbf{J}}_{h_2} + \varepsilon_{h_1-h_2}\hat{\mathbf{J}}_{h_1}\cdot\hat{\mathbf{J}}_{h_2} + \hat{\mathcal{H}}_a. \tag{13.1}$$

The anisotropic part $\mathcal{H}_a$ was derived in Ref. [5] in the case of a spherical quantum dot where it depends linearly on a parameter ρ representing roughly the relative distance of the Mn atom to the QD center. Here, because of the large Δ_{HL} splitting, it essentially reduces to the expression $\rho\varepsilon_{Mn-h_2} J^3_{h_2,x} S_x$. The 2-spin exchange energies in Eq. (13.1) which depend on the actual overlap between the particles, the strength ρ of the anisotropic exchange, and the components of the strain tensor acting on h_1, are all fitting parameters of the model.

To calculate the theoretical PL emission spectra as shown in Fig. 13.6, the Hamiltonians of the initial and final states of the trion transition are first diagonalized. The PL intensity emitted in the transition from $|i\rangle$ to $|f\rangle$ is calculated by taking into account both the oscillator strength $\propto |\langle f|\hat{P}_{\sigma^+}|i\rangle|^2$ (where $\hat{P}_{\sigma^+}$ is the dipolar operator for a σ^+ polarization) and the population of the initial state $|i\rangle$. They are determined from the populations in the different states of A^0 assumed to obey a Boltzmann distribution at an effective temperature T ~10 K. The latter is slightly above the sample temperature to take into account some heating due to the optical excitation. Eventually, to compare with the experiments the calculated transitions are broadened by a Lorentzian of FWHM = 25 μeV. The parameters giving the best fit of our model to the experimental data correspond systematically to an AFM h_1–h_2 exchange interaction which overcomes the AFM h_2–Mn interaction. This results in the observed FM exchange between h_2 and A^0. However, the determination of both these parameters is not very constrained. Since the experiments show essentially the interaction with A^0 in state $J = 1$, only the combination of both these parameters that provides the correct AFM-FM splitting Δ really matters. For example, even by taking a zero h_1–h_2 exchange it is possible to generate a good fit of the experimental plot in Fig. 13.6. In this case the h_2–Mn interaction has to be taken FM. This sounds to be in contradiction with observations in GaMnAs, but actually such sign reversal of the exchange interaction has been recently predicted when the Mn impurity still accepts a localized state [9]. In that respect it is however noteworthy that the strength of the sole $p-d$ exchange $xN_0\beta/3J_hS_z$ [8], estimated for $Ga_{1-x}Mn_xAs$ at the x concentration corresponding to one Mn per QD, turns out to be one order of magnitude smaller than the effective exchange $\Delta/2$. The origin of this strong exchange in quantum dots, is obviously still an open issue which requires further investigations of the Mn-doped quantum dots.

13.10 Conclusion

The optical spectroscopy of Mn-doped InAs/GaAs quantum dots reveals a rich physical system made of a few spins in interaction within a nanometer-size volume of semiconductor. The quantum-dot potential and its associated strain field act perturbatively on the acceptor state of the Mn impurity, preserving the neutral $A^0 = 3d^5 +$ hole configuration, as in GaMnAs at very low Mn concentration. The effective $J = 1$ spin of the A^0 ground state, substantially modified by the anisotropic QD potential, is clearly demonstrated by the spectral PL lines of exciton or trions and their evolution in a magnetic field. One of the most salient observations is likely to be the strong ferromagnetic effective exchange of A^0 with a hole in the quantum dot, which differs in sign and amplitude from the p–d exchange in GaMnAs and for which there is to date no microscopic theory. Interestingly, the ferromagnetic hole–A^0 complex with total angular momentum projection $\sim \pm 5/2$ and strong anisotropy composes a potential nanomagnet, that could be switched with a resonant optical excitation. Further investigations are required to measure the lifetime of its orientation and the feasibility of its optical pumping.

This work has been supported by French´ANR-P3N contract QUAMOS and the région Ile-de-France.

References

[1] Akimov, I. A., Dzhioev, R. I., Korenev, V. L. *et al.* 2009. Electron-spin dynamics in Mn-doped GaAs using time-resolved magneto-optical techniques. *Phys. Rev. B*, **80**, 081203.

[2] Baldereschi, A. and Lipari, N. O. 1973. Spherical model of shallow acceptor states in semiconductor. *Phys. Rev. B*, **8**, 2697.

[3] Baudin, E., Benjamin, E., Lemaître, A. and Krebs, O. 2011. Optical pumping and nondestructive readout of a single magnetic impurity spin in an InAs/GaAS quantum dot. *Phys. Rev. Lett.*, **107**, 197402.

[4] Besombes, L., Leger, Y., Maingault, L. *et al.* 2004. Probing the spin state of a single magnetic ion in an individual quantum dot. *Phys. Rev. Lett.*, **93**(20), 207403.

[5] Bhattacharjee, A. K. 2007. Exciton fine structure and coherent spin precession in transition-metal-doped semiconductor quantum dots. *Phys. Rev. B*, **76**, 75305.

[6] Bhattacharjee, A. K. and Benoit à la Guillaume, C. 1999. Model for the Mn acceptor in GaAs. *Solid State Comm.*, **113**, 17.

[7] Chutia, S. and Bhattacharjee, A. K. 2008. Electronic structure of Mn-doped III-V semiconductor quantum dots. *Phys. Rev. B*, **78**, 195311.

[8] Dietl, T., Ohno, H. and Matsukura, F. 2001. Hole-mediated ferromagnetism in tetrahedrally coordinated semiconductors. *Phys. Rev. B*, **63**, 195205.

[9] Dietl, T. 2008. Hole states in wide band-gap diluted magnetic semiconductors and oxides. *Phys. Rev. B*, **77**, 085208.

[10] Fazekas, P. 1999. *Lecture Notes on Electron Correlation and Magnetism.* Series in Modern Condensed Matter Physics – Vol. 5. Singapore: World Scientific Publishing.

[11] Goryca, M., Plochocka, P., Kazimierczuk, T. *et al.* 2010. Brightening of dark excitons in a single CdTe quantum dot containing a single Mn^{2+} ion. *Phys. Rev. B*, **82**(16), 165323.

[12] Govorov, A. O. 2004. Optical probing of the spin state of a single magnetic impurity in a self-assembled quantum dot. *Phys. Rev. B*, **70**(3), 035321.

[13] Govorov, A. O. 2008. Optical and electronic properties of quantum dots with magnetic impurities. *C.R. Physique*, **9**, 857.

[14] Jungwirth, T., Sinova, J., Masek, J., Kuèera, J. and MacDonald, A. H. 2006. Theory of ferromagnetic III, MnV semiconductors. *Review Of Modern Physics*, **78**, 809.

[15] Krebs, O., Benjamin, E. and Lemaître, A. 2009. Magnetic anisotropy of singly Mn-doped InAs/GaAs quantum dots. *Phys. Rev. B*, **80**, 165315.

[16] Kreissl, J., Ulrici, W., El-Metoui, M. *et al.* 1996. Neutral manganese acceptor in GaP: an electron-paramagnetic-resonance study. *Phys. Rev. B*, **54**, 10508.

[17] Kudelski, A., Lemaître, A., Miard, A. *et al.* 2007. Fine structure of a single Mn acceptor level in an InAs/GaAs quantum dot. *Phys. Rev. Lett.*, **99**, 247209.

[18] Léger, Y., Besombes, L., Maingault, L., Ferrand, D. and Mariette, H. 2005. Geometrical effects on the optical properties of quantum dots doped with a single magnetic atom. *Phys. Rev. Lett.*, **95**(4), 047403.
[19] Léger, Y., Besombes, L., Fernandez-Rossier, J., Maingault, L. and Mariette, H. 2006. Electrical control of a single Mn atom in a quantum dot. *Phys. Rev. Lett.*, **97**(10), 107401.
[20] Marczinowski, F., Wiebe, J., Tang, J.-M. *et al.* 2007. Local electronic structure near Mn acceptors in InAs: surface-induced symmetry breaking and coupling to host states. *Phys. Rev. Lett.*, **99**, 157202.
[21] Schneider, J., Kaufmann, U., Wilkening, W., Baeumler, M. and Köhl, F. 1987. Electronic structure of the neutral manganese acceptor in gallium arsenide. *Phys. Rev. Lett.*, **59**, 240.
[22] Sliwa, C. and Dietl, T. 2008. Electron–hole contribution to the apparent s–d exchange interaction in III–V dilute magnetic semiconductors. *Phys. Rev. B*, **78**, 165205.
[23] Szczytko, J., Mac, W., Stachow, A. *et al.* 1996. The s, p-d exchange interaction in GaAs heavily doped with Mn. *Solid State Comm.*, **99**, 927.
[24] Warburton, R. J., Dürr, C. S., Karrai, K. *et al.* 1997. Charged excitons in self-assembled semiconductor quantum dots. *Phys. Rev. Lett.*, **79**, 5282.
[25] Warburton, R. J., Schaflein, C., Haft, D. *et al.* 2000. Optical emission from a charge-tunable quantum ring. *Nature*, **405**, 926–929.
[26] Yakunin, A. M., Silov, A. Yu., Koenraad, P. M. *et al.* 2004. Spatial structure of an individual Mn acceptor in GaAs. *Phys. Rev. Lett.*, **92**, 216806.
[27] Yakunin A. M. *et al.* 2007. Warping a single Mn acceptor wavefunction by straining the GaAs host. *Nat. Mat.*, **6**, 512.

14

Nuclear spin effects in quantum dot optics

B. Urbaszek, B. Eble, T. Amand and X. Marie

14.1 Introduction

This chapter is devoted to the description of the interaction of polarized light with carrier spins and nuclear spins in semiconductor quantum dots. A historical starting point of these original experiments is the close analogy between quantum dot physics and atomic physics. In 1952, Brossel Kastler and Winter investigated mercury atoms in a weak magnetic field which splits the electron Zeeman levels. By irradiation of the atoms with circularly polarized light the authors could selectively populate one of the electron Zeeman levels [9]. This process has since been referred to as *optical pumping*. Soon afterwards the first optical pumping of carrier spins in a semiconductor was reported [31]. The initial pumping of spin-orientated conduction electrons in silicon induced by polarized light led to polarization of the nuclear spins of the atoms of the silicon lattice via the hyperfine interaction. This dependence of the nuclear magnetization on the polarization of the absorbed light is at the heart of the experiments described in this chapter. A review of the nuclear spin effects in bulk semiconductors can be found in [37]. The hyperfine interaction between carrier and nuclear spins gives even more spectacular results in quantum dots as shown in pioneering work on optically detected nuclear magnetic resonance ODNMR [23] and orientation of one spin species will have a strong influence on the other [25, 7]. Below we detail a selection of the most remarkable consequences of nuclear spin physics on the optical properties of quantum dots. Section 14.1 contains a brief review of the underlying concepts, Section 14.2 shows how arbitrarily aligned nuclear spins are the main source for carrier spin decoherence, Section 14.3 is devoted to dynamical nuclear polarization under non-resonant and resonant light excitation and Section 14.4 summarizes further experiments with strong nuclear effects.

14.1.1 Samples and charge states

One of the great attractions of quantum dot physics is the precise charge control on a single carrier level in charge tunable or chemically doped samples. Here we focus on the three most relevant scenarios: a conduction electron–valence hole pair (for convenience named exciton) with two optically active "bright" states ($\Uparrow\downarrow$ or $\Downarrow\uparrow$) and two "dark" states

Quantum Dots: Optics, Electron Transport and Future Applications, ed. Alexander Tartakovskii. Published by Cambridge University Press.

($\Uparrow\uparrow$ or $\Downarrow\downarrow$), the negatively charged exciton (trion) X^- ($\Uparrow\uparrow\downarrow$ or $\Downarrow\uparrow\downarrow$) and the positively charged exciton X^+ ($\Uparrow\Downarrow\uparrow$ or $\Uparrow\Downarrow\downarrow$). Here $\uparrow$ ($\downarrow$) and $\Uparrow$ ($\Downarrow$) represent the conduction electron spin and valence hole pseudo-spins. Owing to strong localization of the carrier wave function, direct and exchange Coulomb effects are very strong in dots. For the trions the direct Coulomb interaction leads to a renormalization of the transitions energies in the millielectron volt range but exchange terms cancel [7]. For neutral excitons (in zero magnetic field and in the absence of strong nuclear polarization) selection rules are determined by the short-range and long-range electron–hole Coulomb exchange interaction, the latter including an anisotropic contribution [2] sensitive to the dot shape, strain and composition.

14.1.2 Optical carrier spin orientation

The interaction of an electromagnetic wave with carriers in a semiconductor is governed by strict optical selection rules [37]. Energy and angular momentum are conserved for transitions between the valence and conduction band of a typical cubic semiconductor like GaAs. The conduction states are *s*-like, so the electron angular momentum is simply $m_s = \pm 1/2$ ($\uparrow$ or $\downarrow$). The *p*-like valence states are determined by spin–orbit coupling and we consider here only the states with total angular momentum of $J^h = 3/2$. The states with a projection of $J_z^h = \pm 3/2$ ($\Uparrow$ or $\Downarrow$) are called heavy holes, $J_z^h = \pm 1/2$ are called light holes. The quantization axis z is chosen perpendicular to the quantum dot plane and in most experiments z is also parallel to the excitation light propagation direction. Absorption of a photon can increase the valence electron angular momentum $J_z^{ve} = -J_z^h$ by $+1$ for a σ^+ polarized photon or lower it by -1 for a σ^- polarized photon; see Figure 14.1b for all possible transitions between $J^h = 3/2$ valence and conduction states. The selection rules for photon absorption and emission are identical. The heavy-hole and light-hole valence states are separated by an energy Δ_{HL} due to quantum confinement and strain. For certain experiments the light-hole states can safely be ignored and optical spin state preparation is straightforward. In practice, strain, alloy intermixing and shape anisotropy can introduce heavy hole to light hole coupling and all transitions indicated in Figure 14.1b, with very different probabilities, are possible between the mixed states.

14.1.3 High-resolution optical spectroscopy in the spectral and temporal domains

Optical spectroscopy of individual quantum dot spin states is often performed at low temperatures (4 K) using a confocal microscope, where the detection area on the sample has a typical diameter of 1 μm and is imaged onto a single mode fibre which serves as a pin-hole. An example of the emission of a single InAs dot in the X^+ state is shown in Figure 14.1c in the absence of any external magnetic fields [4]. The lower panel shows photoluminescence (PL) emission following linearly polarized laser excitation, and as expected many σ^+ as σ^- polarized photons are detected. When changing to σ^+ polarized excitation in the upper panel, the dot shows as expected a stronger σ^+ polarized emission. But, in addition, the σ^+

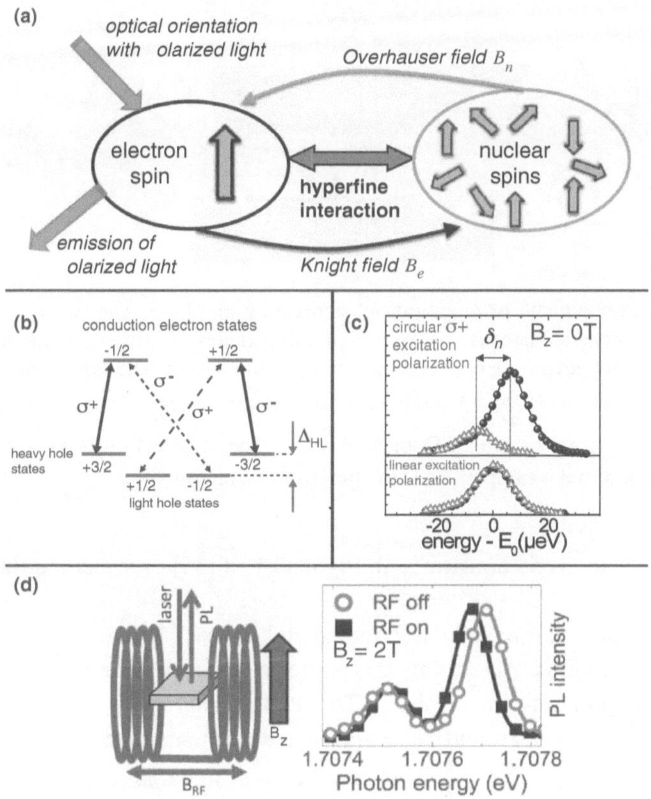

Figure 14.1 (a) Interplay between photon, electron and nuclear spin polarization. (b) Optical selection rules. (c) High-resolution PL spectrum of a trion X^+ in a single InAs dot at $B_z = 0$ using a Fabry–Perot interferometer following σ^+ excitation in the wetting layer [4]. E_0 is the trion emission energy. (d) Direct demonstration of nuclear spin effects in ODNMR in GaAs interface fluctuation dots, after [35].

and σ^- polarized photons do not have the same emission energy [30]. As described below, this considerable energy splitting is caused by the creation of an internal magnetic field due to the preferential alignment of the nuclear spins parallel to the injected spin-down electron. Two physical quantities can be measured: (i) the average photon emission polarization degree P_c is given by the difference of the integrated intensities of the σ^+ and σ^- polarized emission (divided by the sum) and gives directly in the case of the X^+ emission the mean electron polarization as $\langle \hat{S}_z^e \rangle = -P_c/2$; (ii) the splitting in energy δ_n is proportional to the nuclear spin polarization. The interdependence of the electron polarization $\Leftrightarrow$ nuclear polarization can be monitored in one experiment. Application of a small oscillating (radiofrequency RF domain) transverse magnetic field in resonance with the *nuclear* Zeeman splitting can randomize the nuclear spin orientation. As nuclear spin polarization modifies the *electron* Zeeman splitting, a change in this splitting can be seen when turning

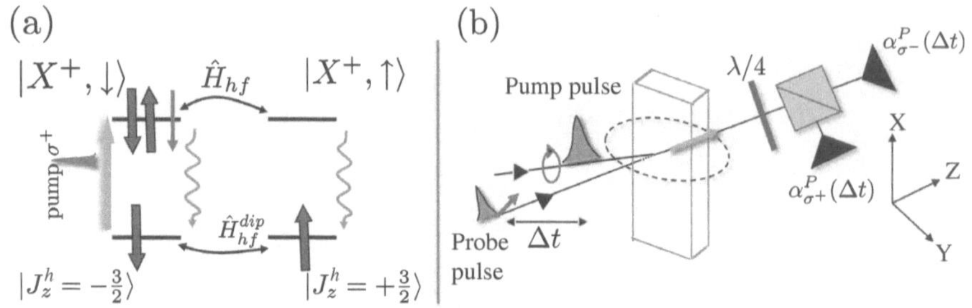

Figure 14.2 (a) Level scheme of quantum dot containing one hole. The two excited and the two ground states are degenerate because of the two possible values of spin projection along the growth axis. The vertical wavy arrows denote spontaneous recombination. The hole spin can relax via the dipolar hyperfine interaction $\left(\hat{H}^{dip}_{hf}\right)$ and the electron contact hyperfine interaction couples the trion states $(\hat{H}_{hf})$. (b) Principle of the PCD measurement. The Z direction is parallel to the growth direction of the quantum dot sample; Δt is the pump-probe delay.

on an off the RF field, as demonstrated in Figure 14.1d [35] or sweeping the field through resonance [23].

To probe the state of the hole left in the dot after X^+ radiative recombination, a complementary technique, measuring the photo-induced circular dichroism (PCD) in a pump/probe configuration can be used. The PCD signal gives the difference of spin polarization carried in the ground state and in the trion state, see Figure 14.2a, namely $\left(\rho^h_{\Uparrow}(t)-\rho^h_{\Downarrow}(t)\right)-\left(\rho^{X+}_{\uparrow}(t)-\rho^{X+}_{\downarrow}(t)\right)$. As the PCD signal is proportional to the imaginary part of the complex refractive index of the material, it is a complementary technique to Faraday and Kerr rotation measurements that probe the real part of the refractive index [24].

To perform PCD measurements, a picosecond Ti:sapphire laser tuned to the X^+ trion transition of the p-doped dots is split into pump and probe beams, which are quasi colinearly focused onto the sample placed inside a cryostat (Figure 14.2b). The pump beam is circularly polarized to generate selectively spin-polarized trions, whereas the probe beam is linearly polarized. The delay between the pump and probe pulses is controlled with a retardation line. After transmission through the sample, the probe beam is spatially resolved into its two circular components ; the difference of their linear absorption through the sample, $PCD(t)=\alpha^P_{\sigma+}(t)-\alpha^P_{\sigma-}(t)$, is measured in a balanced optical bridge as a function of the pump-probe delay. The absorption of a σ^+ (resp. σ^-) probe component is linked to the pump-induced state occupancy $\rho^h_{\Downarrow}(t)-\rho^{X^+}_{\downarrow}(t)$ $\left(\text{resp. } \rho^h_{\Uparrow}(t)-\rho^{X^+}_{\uparrow}(t)\right)$.

14.1.4 Hyperfine interaction between carriers and nuclear spins

The strength of the hyperfine interaction in quantum dots is increased compared to semiconductor bulk or quantum well structures due to the strong localization of the electron wavefunction over typically only 10 000 lattice sites. This number is too small for efficient cancellation of the total nuclear spin [11, 38], yet too large to address each nuclear spin state individually. In III–V quantum dots like GaAs and InAs, 100% of the lattice sites have a non-zero nuclear spin and these materials are taken here as model systems. Even for solids with very few isotopes carrying a nuclear spin like carbon [14], ZnO [33] or CdSe [21] it is still the hyperfine effects that play a key role in the carrier spin-state evolution. There are two main types of hyperfine interaction [1]. First we consider the Fermi contact interaction which relies on a physical overlap of the carrier wavefunction with the lattice site (nuclear spin). This type of interaction is efficient for s-type wavefunctions (periodic part of the Bloch function) of conduction electrons and negligible for p-type valence electrons (holes).

The Fermi contact hyperfine interaction between an electron of spin $\hat{S}^e = \frac{1}{2}\hat{\sigma}^e$ confined to a quantum dot and N nuclei is ([16] and references therein):

$$\hat{H}_{hf} = \frac{\nu_0}{2} \sum_j A^j |\psi(\bar{r}_j)|^2 \left(2\hat{I}_z^j \hat{S}_z^e + [\hat{I}_+^j \hat{S}_-^e + \hat{I}_-^j \hat{S}_+^e]\right), \tag{14.1}$$

where ν_0 is the two atom unit cell volume, $\bar{r}_j$ is the position of the nuclei j with spin $\hat{I}^j$, the different nuclear species j. The nuclear spin is 3/2 for Ga and As, 5/2 for Al, 9/2 for In. A^j is the constant of the hyperfine interaction with the electron in the order or 50 μeV for In, Ga and As; $\psi(\bar{r})$ is the electron envelope function.

The first part of equation (14.1) shows that, in the presence of a mean lattice nuclei polarization, the electron "feels" an effective magnetic (Overhauser) field B_n. Polarized electrons in turn act like an effective magnetic (Knight) field B_e on the nuclear spins, as indicated in Figure 14.1a. The maximum Overhauser field in GaAs can reach several tesla, the Knight field is in the millitesla range [16].

The second term in equation (14.1) allows for spin transfer via simultaneous spin flips (flip-flops) of a carrier and nuclear spin. Depending on the exact experimental conditions, the competition between spin generation and decay determines the net result of these spin flips, which can lead to electron spin dephasing [7], nuclear spin dephasing [1, 38] or dynamic nuclear polarization [23]. In the latter case, $\delta_n = 2\tilde{A}\langle\hat{I}_z\rangle/N = g_e\mu_B B_n$ relates the Overhauser shift δ_n to the average nuclear polarization, where N is the number of nuclei in the dot.

Another type of hyperfine interaction is the dipole–dipole interaction. The p-symmetry of the periodic part of the valence Bloch wavefunction results in negligible overlap with the nuclear spins yielding negligible Fermi contact interaction. Therefore the dipole–dipole interaction between carrier and nuclear spins, dominant for example in carbon [14], at least one order of magnitude weaker than the Fermi contact interaction, will dominate for valence states [18]. Derivation of the hole–nuclear spin Hamiltonian is not straightforward [41], since the dipolar hyperfine interaction depends on the valence band mixing $\beta = |\beta|e^{i\theta}$, which quantifies the deviation of the confined hole from the ideal, pure heavy-hole case. The valence band mixing β, arising from anisotropic strain field within the quantum dot and/or shape anisotropy [3], makes it necessary to consider modified hole states by adding a light-hole component:

$$|\widetilde{\pm 3/2}\rangle = \frac{1}{\sqrt{1+|\beta|^2}}\left(|\pm 3/2\rangle + \beta|\mp 1/2\rangle\right). \tag{14.2}$$

As a direct consequence of the energy splitting Δ_{HL} between heavy-hole and light-hole states (tens of millielectronvolts) in InAs/GaAs quantum dots $|\beta|$ should be much smaller than unity. In this particular hole basis set, the dipolar Hamiltonian takes the following form for a given nucleus:

$$\hat{H}_{hf}^{dip} = \frac{2\tilde{C}}{N}\left(\hat{I}_z\hat{S}_z^h + \alpha\left[\frac{\hat{I}_+\hat{S}_-^h e^{i\theta} + \hat{I}_-\hat{S}_+^h e^{-i\theta}}{2}\right]\right), \tag{14.3}$$

where $\theta = \arg(\beta)$ plays no physical role and can be omitted in the following, $\hat{S}_\pm^h$ denote the hole rising/lowering pseudo-spin $1/2$ operators, and finally $\alpha = \frac{2|\beta|}{\sqrt{3}}$ is the anisotropy factor, which tends to zero for pure heavy holes [22, 41].

$\hat{H}_{hf}$ and $\hat{H}_{hf}^{dip}$ (see equations (14.1) and (14.3)) have a similar form, but there are important differences: (i) the ratio $\tilde{C}/\tilde{A}$ is about $\simeq 0.1$ [41, 20, 13], where $\tilde{A}$ is the average constant of the hyperfine interaction with the electron, as demonstrated directly in Figures 14.3e and 14.4d, and (ii) another significant difference is the relative strength of the flip-flop term, which is at the origin of building-up of nuclear spin polarization $\langle \hat{I}_z\rangle \neq 0$, and also a potential source of a carrier spin decoherence, see Section 14.2.2.

14.2 Carrier spin decoherence

14.2.1 Electron spin dephasing due to nuclear field fluctuations

A mean nuclear spin polarization can be associated with an Overhauser field B_n acting on the electron spin. For a nuclear spin polarization clearly below 100%, the nuclear field fluctuates around a mean value. The fluctuations can be written as an effective field δB_n. Several theoretical studies have predicted that the dominant mechanism of electron spin polarization decay in quantum dots at low temperature in the absence of external magnetic fields is due to the interaction with these nuclear field fluctuations [11, 38]. The origin of these fluctuations lies in the finite number of nuclei within the dot. Since the nuclear spin operators $\hat{I}_x, \hat{I}_y, \hat{I}_z$ do not commute, it is impossible to determine the x, y and z components of the nuclear spin system with equal precision. For the experiments in this section the mean nuclear spin polarization is zero, i.e. repeated measurements of the expectation value of B_n at time intervals longer than the nuclear spin correlation time of the order of 10^{-4} s give an average of zero. But an electron spin will interact during its lifetime with a non-zero δB_n of fixed but arbitrary orientation (during about 10^{-4} s, i.e. frozen fluctuation model as in [38]).

The dispersion of the nuclear hyperfine field B_n in the absence of dynamic nuclear polarization is described here by a Gaussian distribution $W(B_n) \propto exp(-B_n^2/\delta B_n^2)$ characterized by its width δB_n. Here $\delta B_n^2 = 2\langle B_n^2\rangle/3$. $W(B_n)$ has a spherical symmetry indicating

B_n has no preferred spatial orientation. To estimate the magnitude of δB_n, consider a quantum dot made of N identical nuclear spins I. The average amplitude of the fluctuating hyperfine field is $\delta B_n = \frac{1}{g_e \mu_B} \frac{2\tilde{A}}{\sqrt{N}} \sqrt{I(I+1)}$ and the maximum Overhauser field, corresponding to a 100% nuclear spin polarization, is $B_n^{max} = \frac{1}{g_e \mu_B} 2\tilde{A} I$ in the order of several tesla in InAs and GaAs. As a result $\delta B_n \sim B_n^{max}/\sqrt{N}$, which corresponds to an effective field in InAs dots of typically 30 mT which arbitrarily changes orientation on a time scale of 10^{-4} s [38], orders of magnitude longer than electron spin dynamics. The electron spin can thus coherently precess around δB_n. The amplitude and direction of the effective nuclear field vary strongly from dot to dot. The average electron spin $\langle S(t) \rangle$ in an ensemble of dots will thus decay as a consequence of the random distribution of the local nuclear effective field. For repeated measurements on a single quantum dot the hyperfine interaction has the same effect as for an ensemble of dots: the signal integration times are commonly much longer than 10^{-4} s, so δB_n changes orientation several times during one measurement and the average is taken over a large number of nuclear spin configurations. For the sake of simplicity this spin dephasing mechanism is often termed "spin relaxation".

The positively charged excitons X^+ is the ideal configuration to probe the electron spin relaxation mediated by nuclei in quantum dots with optical experiments. The analysis of the circular polarization of the X^+ luminescence in p-doped quantum dots following a circularly polarized laser excitation probes the spin polarization of the electron directly as $\langle \hat{S}_z^e \rangle = -P_c/2$.

Figure 14.3a displays the circular polarization dynamics of the photoluminescence from an ensemble of p-modulation doped InAs/GaAs quantum dots [7]. The inset presents the time evolution of the polarized luminescence intensity components. The circular polarization dynamics in figure 14.3a presents two regimes. The polarization decays within the first 800 ps down to about 1/3 of its initial value; then it remains stable with no measurable decay on the radiative life-time scale. These results are in very good agreement with the predicted electron spin relaxation due to the hyperfine interaction with the nuclei [38].

A magnetic field B_z applied along the z growth axis (Faraday configuration) can screen δB_n and stabilize the electron spin. B_z must be larger than δB_n, to ensure that the Zeeman interaction of the electron spin with the magnetic field is stronger than the interaction with the nuclei [7, 15]. The effect of the random nuclear field fluctuations on the electron spin coherence can also be directly measured by recording the electron spin dynamics in a transverse magnetic field [24, 34].

14.2.2 Hole spin dephasing

As the hyperfine interaction with the fluctuating nuclear field is severely limiting electron spin coherence, the hole spin with negligible Fermi contact hyperfine interaction seems a more promising candidate for achieving long spin coherence times. With this motivation in mind, detailed investigations led to promising reports [10] under certain experimental conditions (as detailed in Chapter 4), but also surprisingly short hole spin dephasing times were

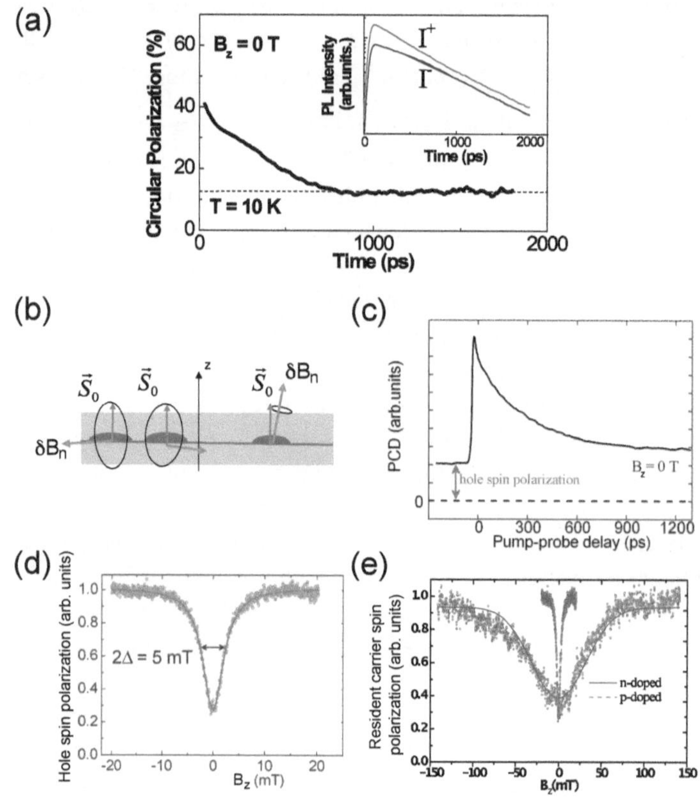

Figure 14.3 (a) Time evolution of the circular polarization of the X^+ emission [7] in an ensemble of InAs dots. (b) Scheme of the dephasing of electrons and holes spins due to nuclear field fluctuations δB_n. (c) PCD signal as a function of pump-probe delay, when the p-doped quantum dots are excited by a periodic train of ultra-short pulses. (d) PCD normalized amplitude at negative pump-probe delay $t = -130$ ps (i.e. reflecting the hole spin polarization induced by the previous pump pulse) versus the applied longitudinal magnetic field B_z, performed on p-doped quantum dot sample. The HWHM equals 2.5 mT. The solid line is a fit using the model developed in [19]. (e) PCD normalized amplitude at negative pump-probe delay $t \sim 13$ ns (i.e. the resident carrier spin polarization) versus the applied longitudinal magnetic field, performed on n-doped quantum dots. The HWHM equals 47 mT. Similar measurement on p-doped quantum dots is added on the same figure to compare the efficiency of hyperfine-induced dephasing for electrons and holes.

measured [18]. Below we give an overview of hole spin dynamics in the model system of p-doped InAs quantum dots, revealing that spin decoherence of the resident hole, at cryogenic temperatures, is limited by the dipole–dipole hyperfine interaction with randomly oriented nuclear spins.

Figure14.3c shows the PCD signals versus pump-probe delay, obtained when no magnetic field is applied. Then, we observe a non-zero PCD signal at negative pump-probe delays, indicating that the spin polarization is not fully relaxed within the $T_L = 13$ ns repetition period of the laser pulses. This long-living component of the PCD signal is unambiguously associated with the net spin polarization of the *resident* holes, the only species present in the sample after the radiative recombination of X^+ trions ($\tau_{rad} \sim$ 800 ps). From this experimental observation, one can understand how the spin polarization of the resident holes in the dots is built up by a resonant pulsed excitation, in the following sequential way (see Figure 14.2a). (i) Before optical excitation, there is no net hole spin polarization, so $\rho^h_{\Uparrow} = \rho^h_{\Downarrow}$. (ii) The circularly polarized pump beam σ^+, photo-creates spin oriented trions X^+ with spin $\downarrow$ electrons. (iii) During the lifetime of these excited states, the efficient electron–nuclear hyperfine interaction leads to a coherent coupling of their spin projections along the light propagation direction. (iv) And finally, the spontaneous decay of the trion states by emission of polarized photons then leads to an unbalanced hole spin population with $\rho^h_{\Uparrow} > \rho^h_{\Downarrow}$ (see also Section 4.5.1).

Figure14.3d shows the experimental data for the PCD signal at negative delays, $PCD(0^-)$, i.e. the hole spin polarization at 13 ns, as a function of the applied magnetic field. Note that the application of a small magnetic field B_z in the millitesla range has a dramatic impact on the hole spin polarization. The Lorentzian-like dependence with a half width at half maximum (HWHM) of only 2.5 mT is interpreted as the progressive magnetic field quenching of the hyperfine-induced hole spin relaxation [18].

Also, PCD measurements allow us to revisit electron spin dynamics. Figure 14.3e presents the PCD evolution versus longitudinal magnetic field performed on an n-doped quantum dots ensemble. The electron–nuclei interaction takes place in the ground state of the n-type quantum dots after X^- recombination (Figure 14.3e), whereas in p-type dots the electron in the X^+ interacts with nuclei before recombination. The external magnetic field can efficiently cancel the effect of the hyperfine interaction on the carrier-spin dephasing time, if its magnitude becomes larger than the corresponding effective nuclear field fluctuation. This field is of the order of a few milliteslas for holes, instead of several tens of milliteslas for electrons, as can be seen in the comparison of PCD measurments shown in figure14.3b [15] for electrons and holes. The relative strength of the hole-to-electron hyperfine coupling is deduced from comparison of both HWHMs (i.e. of the external field strength necessary to screen the hyperfine effects), and is in the order of 10%.

14.2.3 Towards long carrier spin coherence and relaxation times

For both electron and hole spins the hyperfine interaction with randomly orientated nuclear spins severely limits the spin *coherence* and *relaxation* times. The latter can be prolonged by simply applying an external magnetic field $B_z \gg \delta B_n$, but there is no such simple measure for the coherence time [34]. A promising route is to narrow down the nuclear spin distribution through dynamic nuclear polarization [5]. To progress from the regime of electron spin dephasing due to the nuclear spin bath to transferring a well-defined electron spin polarization state in optical experiments, several experimental parameters have to be optimised. For example, increasing the pump power will induce more spin flip-flops between carrier and nuclear spins per unit time, enabling dynamic nuclear polarization [28] as will be detailed in the following section.

14.3 Dynamic nuclear polarization

14.3.1 Optical pumping of nuclear spins under non-resonant excitation

Illumination of a quantum dot sample with circularly polarized laser light results in the optical creation of carriers with well-defined spin. In this section we consider non-resonant excitation, where the carriers are optically created in an excited state of the quantum dot or in the wetting layer surrounding the dot. The created carriers will subsequently relax in energy and are trapped by the quantum dots. For holes, the energy relaxation from the wetting layer is accompanied by a spin relaxation, the captured hole spin state is arbitrary. So for optical hole spin orientation excited states *inside* the dot have to be pumped. The electron spin is more robust and is partially conserved when relaxing from higher energy states (barrier/wetting layer) to the dot ground state [7]. It is important to note, for all that follows, that the hole and electron spin states in the dot ground state do not relax during their radiative lifetime.

Favourable experimental conditions for achieving strong dynamic polarization of nuclear spins are: injection of well-defined carrier spins at a high rate (high excitation power) and stabilization of carrier and nuclear spins by an external magnetic field B_z. It has been shown that both electron and hole spins are sensitive to nuclear effects, but so far dynamic nuclear polarization has been clearly demonstrated only as originating from electron spin pumping.

In experiments in longitudinal magnetic fields $|B_z| > 0$ following, for example, non-resonant excitation with linearly polarized light (equal injection probability for spin-up and spin-down carriers, so $\langle \hat{S}_z^e \rangle = 0$) first the Zeeman splitting in the absence of nuclear fields is measured and $\delta_z^e = \delta_z$. Then, the excitation light polarization is switched to circular ($\langle \hat{S}_z^e \rangle \neq 0$), and measurable changes δ_n (Overhauser shift) in the Zeeman splitting are now related to the additional field B_n as $\delta_z^e = \delta_z + \delta_n$.

The model developed in reference [17] gives an implicit expression for δ_n as a function of B_z, optical pump power, nuclear spin relaxation time T_d, the hyperfine interaction correlation time τ_c:

$$\delta_n = 2\tilde{A}\langle I_z \rangle = -\frac{2\tilde{A}\tilde{Q}\langle \hat{S}_z^e \rangle}{1 + \frac{T_{e1}(\delta_n)}{T_d}}, \tag{14.4}$$

where $\tilde{Q} = \sum_j x_j \frac{I^j(I^j+1)}{S(S+1)}$ (in $In_{1-x}Ga_xAs$ for example j = In, Ga, As) and $\frac{1}{T_{1e}} = \left(\frac{\tilde{A}}{N\hbar}\right)^2 \frac{2 f_e \tau_c}{1+(\frac{\delta_z+\delta_n}{\hbar}\tau_c)^2}$. The average fraction of time the quantum dot contains an electron is f_e and the electron Zeeman splitting due to the external magnetic field B_z is δ_z, where $\delta_z^e = \delta_z + \delta_n$.

Equation (14.4) has only one real solution when $\delta_z = g_e \mu_B B_z$ and δ_n have the same sign, but may have up to three solutions when the signs are opposite, depending on the experimental conditions that determine τ_c (and f_e), resulting in pronounced bistabilities of the nuclear spin polarization, see Figure 14.4.

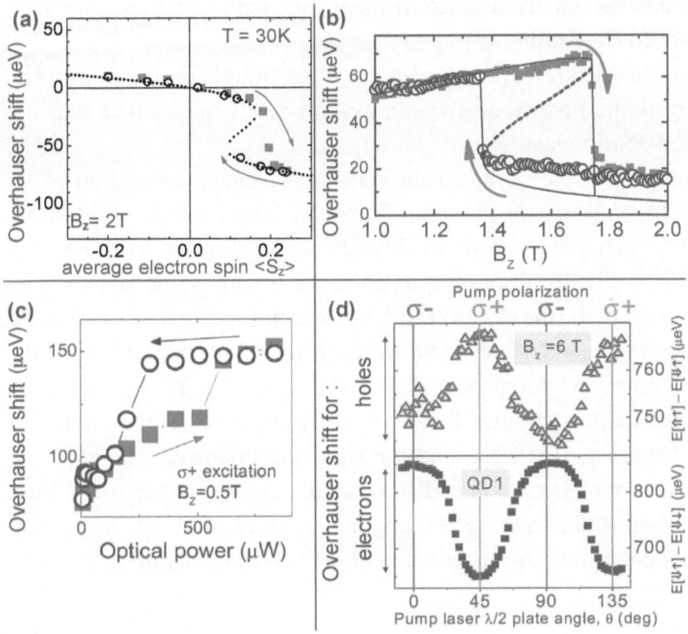

Figure 14.4 (a) Overhauser shift as a function of average electron spin of an X^+ at 30 K after [42] in InAs dots. (b) Bistability of the nuclear polarization in InAs dots as a function of the magnetic field sweep direction [36]. Circles are a fit with the model from [17]. (c) Dependence of the Overhauser shift on excitation power for InP/GaInP dots [39]. (d) Measurement of the electron and hole nuclear spin interaction in a neutral InP dot at $B_z = 6T$. The angle of a $\lambda/2$ plate is varied to change the polarization of the pump laser resulting in a change of nuclear spin polarization [13].

The physical origin for these predicted and widely observed bistabilities and strong non-linearities is feedback, see Figure 14.4 , as the electron Zeeman splitting δ_z and δ_n are of comparable magnitude. So to increase the nuclear polarization (i.e. δ_n) an electron has to flip its spin and change Zeeman level. Owing to the strong difference in order of magnitude between the electron and nuclear gyromagnetic factors, the flip-flop process between electron and nuclear spins is in principle forbidden due to energy mismatch. Thus the electron spin flip will be more likely if the electron Zeeman splitting is small. So positive feedback is possible, if every electron–nuclear spin flip-flop decreases the electron Zeeman splitting slightly, making the next flip-flop already a bit more likely. This applies to the case of B_z and B_n having opposite signs, i.e. (partially) cancelling each other. Negative feedback is the result of B_z and B_n having the same sign, so that every spin flip increases the Zeeman splitting, hence decreasing the flip-flop probability. This strong feedback explains the bistable nuclear and electron polarization observed in experiments [36, 28, 40, 39, 8]. The correlation time τ_c of the electron–nuclear spin interaction, (limited by the carrier capture time, the radiative lifetime or tunnelling events), corresponds to a broadening of the

Zeeman levels and eventually overlap of the levels, permitting electron spin flips without violating energy conservation [42].

To determine the electron spin and nuclear spin polarization for a given set of experimental parameters the history of the spin system needs to be taken into account, i.e. the behaviour is non-Markovian.

The dynamic polarization created through non-resonant excitation of an InP dot was monitored via the emission of the neutral exciton bright states $|\Uparrow\downarrow\rangle$ and $|\Downarrow\uparrow\rangle$ and the dark states $|\Uparrow\uparrow\rangle$ and $|\Downarrow\downarrow\rangle$ (here weakly allowed again due to heavy hole – light hole mixing) [13]. It is therefore possible to compare for the same dot the effect of a finite nuclear polarization on (i) the electron spin by measuring the energy difference between the $|\Downarrow\uparrow\rangle$ and $|\Downarrow\downarrow\rangle$ state and (ii) very interestingly, the effect on the hole spin by measuring the energy difference between the $|\Uparrow\uparrow\rangle$ and $|\Downarrow\uparrow\rangle$. Figure 14.4d shows directly, that the hyperfine constant is about a factor of 10 stronger for electrons as for holes (same conclusion as from Figure 14.3e), and the signs are opposite. The exact origin of heavy hole–light hole coupling is crucial at different stages of this experiment. First, it allows the observation of dark states as there is a slight admixture of bright states. Second, the hole mixing will also determine the exact value of the Overhauser effect on the holes.

14.3.2 Novel nuclear spin physics under resonant optical pumping

There are several scenarios that result in the construction of nuclear polarization following optical pumping: (i) an electron with a well-defined spin is repeatedly injected into the dot or (ii) spin $\uparrow$ and $\downarrow$ electrons are injected with equal probability, but, for reasons of energy conservation, the spin transfer (via a spin flip-flop) to a nuclear spin does not have the same probability for the spin $\uparrow$ and $\downarrow$ electron. The latter mechanism is based on the so called optical solid effect [1, 27] with very encouraging experimental results reported by [12].

In the case of non-resonant optical excitation, it is difficult to pin down the exact moment when the simultaneous carrier and nuclear spin flips occur as several energy relaxation steps occur between absorption and emission. But it is safe to assume that nuclear effects have only very little effect on the absorption process under non-resonant excitation.

Under resonant excitation the spin flip-flop occurs either during photon absorption or emission, so the hyperfine interaction determines in this case which photon polarization leads to absorption and emission. In addition, by flipping nuclear spins the quantum dot ground-state resonance shifts in energy, possibly away from or towards the laser energy. Construction of nuclear polarization in resonant experiments is possibly assisted by mechanisms beyond the simple flip-flop term of equation (14.1) and the exact mechanisms are still under debate for different experimental conditions [32, 43, 12, 26].

The absorption of a spectrally narrow laser by a single quantum dot transition (X^-) has been measured [32]. At zero applied field, the laser is scanned across the resonance energy and as expected a Lorentzian shaped absorption profile is obtained, see Figure 14.5a. The same experiments at a field of 4.5 T, show strong absorption but, as the laser is scanned

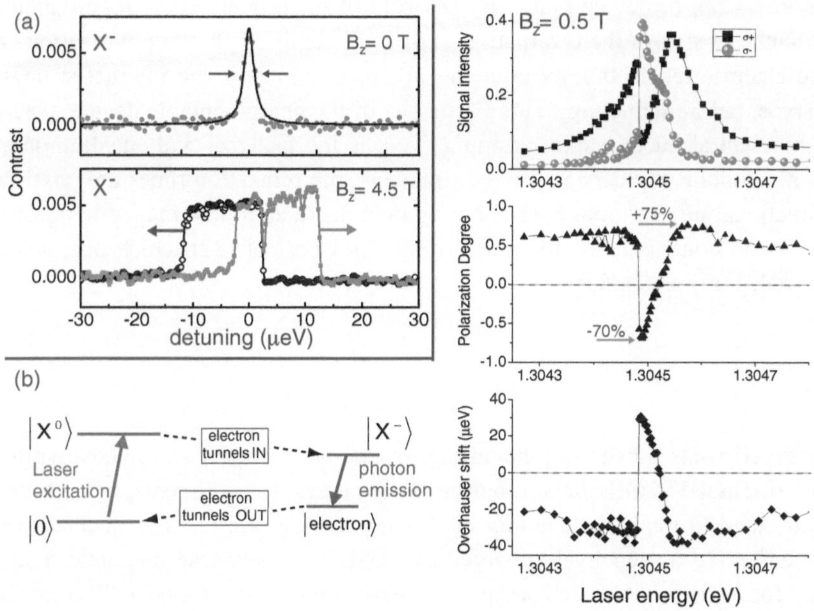

Figure 14.5 (a) Trion absorption at zero magnetic field (Lorentzian fit with linewidth 2 μeV) and strong deviation from Lorentzian lineshape at 4.5 T. (b) Left: energy level diagram with resonant neutral exciton X^0 excitation and charged exciton X^- emission in the tunnelling regime ([26]). Right: experimentally measured signal intensity, polarization degree, and Overhauser shift versus laser energy (laser detuning) for a σ^+ polarized pump in an external field of $B_z = 0.5\,T$ at fixed bias in the center of the hybridization region (see text).

across the resonance, the resonant energy is shifted and remains locked to the laser energy for a certain energy range. The exact shape of this particular absorption curve depends on the direction of the laser scan. In stark contrast to earlier experiments, this nuclear-spin polarization is bidirectional, allowing the combined electron nuclear spin system to track the changes in laser frequency dynamically on both sides of the resonance. This behaviour is referred to as "dragging" and is observed to different extents in a variety of experiments [43, 12, 20]. In a simple picture, the quantum dot uses the nuclear spins as an adjustable parameter to stay in resonance with the excitation laser and for a wide range of experimental parameters it is impossible to isolate elementary quantum dot excitations from the strong influence of nuclear spins.

Using a charge tunable structure in the tunnelling regime [26] allows us to explore an original control scheme over the nuclear spin–electron spin interaction on driving an optical transition resonantly. The neutral exciton X^0 transition is pumped and emission from the trion X^- is monitored. Following resonant σ^+ excitation, the spontaneous emission either preserves the circular polarization of the excitation source or inverts it, see Figure 14.5b (right panel). The photon emission polarization can be switched from predominantly σ^+ emission to σ^- emission either with a small change in pump wavelength, as in Figure 14.5b, or device bias. The key role of nuclear effects is highlighted by the sudden change of sign of the Overhauser shift as the emitted photon polarization changes sign. The electron left in the dot after the X^- recombination can interact strongly with nuclear spins, before tunneling to the Fermi sea of the charge tunable device. One advantage of this natural nuclear spin pumping cycle is the fact that without illumination the dot does not contain any carriers, so long nuclear spin relaxation times are possible [26]. The bidirectional nuclear polarization tuning demonstrated at low magnetic fields (0.5 T) in Figure 14.5b complements the optical dragging effect of [32] which operates at high magnetic fields (Figure 14.5a).

14.4 Perspectives

Selected experiments illustrating the strong link between carrier and nuclear spin effects have been discussed in this chapter, but numerous elegant experiments, not detailed here, have contributed to the understanding of this fascinating physics. For instance, manipulating an ensemble of negatively charged excitons in a transverse magnetic field allows frequency focussing of the electron spin precession into certain modes. The information about electron spin precession is imprinted in the nuclei and thereby can be stored for tens of minutes in darkness [24], confirming the long nuclear spin memory. Any attempt to prolong carrier spin coherence is of importance for using quantum dots in future quantum state (qubit) manipulation schemes. The state of the art of coherent electron spin manipulation in transport experiments in gate-defined quantum dots is also addressing the beneficial and detrimental aspects of the coupling to nuclear spins [25]. One experimental target is to polarize as many nuclear spins as possible to finally achieve a net reduction in the nuclear spin fluctuations δB_n once the nuclear polarization is approaching 100%. So far the maximum nuclear polarization achieved in different material systems is about 60% [6, 42, 12]. Future experiments will clarify if this presents some fundamental limit or if almost complete nuclear spin polarization is accessible in experiments in dots where each lattice nucleus has non-zero nuclear spins [21].

We have so far neglected the fact that the nuclear spin system with $I > 1/2$ in quantum dots with strong strain effects and alloy intermixing experiences strong nuclear quadrupolar effects. These effects due to electric field gradients in the lattice introduce mixing between nuclear spin levels and contribute to the suppression of nuclear spin diffusion [1]. This can help to explain the build up of fairly robust nuclear spin polarization in the absence of magnetic fields [30, 4, 12] in addition to the stabilization due to the Knight field. In transverse magnetic fields very surprising electron spin depolarization curves, termed the anomalous Hanle effect, exhibit strong non-linearities and bistabilities [29], most likely also connected to strong quadrupolar effects.

We thank Patrick Maletinsky, Christoph Kloeffel and Christian Latta for providing the original figures and Olivier Krebs, Atac Imamoglu and Alexander Tartakovskii for fruitful discussions and ANR QUAMOS and ITN Spinoptronics for financial support.

References

[1] Abragam, A. 1961. *Principles of Nuclear Magnetism*. Oxford University Press.

[2] Bayer, M. *et al.* 2002. Fine structure of neutral and charged excitons in self-assembled In(Ga)As/(Al)GaAs quantum dots. *Phys. Rev. B*, **65**(19), 195315.

[3] Belhadj, T., Amand, T., Kunold, A. *et al.* 2010. Impact of heavy hole–light hole coupling on optical selection rules in GaAs quantum dots. *Appl. Phys. Lett.*, **97**(5), 051111.

[4] Belhadj, T., Simon, C.-M., Amand, T. *et al.* 2009. Controlling the polarization eigenstate of a quantum dot exciton with light. *Phys. Rev. Lett.*, **103**(8), 086601.

[5] Bluhm, H. *et al.* 2010. Enhancing the coherence of a spin qubit by operating it as a feedback loop that controls its nuclear spin bath. *Phys. Rev. Lett.*, **105**(21), 216803.

[6] Bracker, A. S., *et al.* 2005. Optical pumping of the electronic and nuclear spin of single charge-tunable quantum dots. *Phys. Rev. Lett.*, **94**(4), 047402.

[7] Braun, P.-F., Marie, X., Lombez, L. *et al.* 2005. Direct observation of the electron spin relaxation induced by nuclei in quantum dots. *Phys. Rev. Lett.*, **94**(11), 116601.

[8] Braun, P.-F., Urbaszek, B., Amand, T. *et al.* 2006. Bistability of the nuclear polarization created through optical pumping in In_{1-x} Ga_x As quantum dots. *Phys. Rev. B*, **74**(24), 245306.

[9] Brossel, J., Kastler, A. and Winter, J. 1952. *J. de Physique et le Radium*, **13**, 668.

[10] Brunner, D., Gerardot, B. D., Dalgarno, P. A. *et al.* 2009. A coherent single-hole spin in a semiconductor. *Science*, **325**(5936), 70–72.

[11] Burkard, G., Loss, D. and DiVincenzo, D. P. 1999. Coupled quantum dots as quantum gates. *Phys. Rev. B*, **59**(3), 2070–2078.

[12] Chekhovich, E. A., Makhonin, M. N., Kavokin, K. V. *et al.* 2010. Pumping of nuclear spins by optical excitation of spin-forbidden transitions in a quantum dot. *Phys. Rev. Lett.*, **104**(6), 066804.

[13] Chekhovich, E. A., Krysa, A. B., Skolnick, M. S., and Tartakovskii, A. I. 2011. Direct measurement of the hole–nuclear spin interaction in single InP/GaInP quantum dots using photoluminescence spectroscopy. *Phys. Rev. Lett.*, **106**(2), 027402.

[14] Childress, L., Gurudev Dutt, M. V., Taylor, J. M. *et al.* 2006. Coherent dynamics of coupled electron and nuclear spin qubits in diamond. *Science*, **314**(5797), 281–285.

[15] Desfonds, P., Eble, B., Fras, F. *et al.* 2010. Electron and hole spin cooling efficiency in InAs quantum dots: the role of nuclear field. *Appl. Phys. Lett.*, **96**(17).

[16] Dyakonov, M. I. 2008. *Spin Physics in Semiconductors*. Springer Series in Solid State Physics, **157**.

[17] Eble, B., Krebs, O., Lemaître, A. *et al.* 2006. Dynamic nuclear polarization of a single charge-tunable InAs/GaAs quantum dot. *Phys. Rev. B*, **74**(8), 081306.

[18] Eble, B., Testelin, C., Desfonds, P. *et al.* 2009. Hole–nuclear spin interaction in quantum dots. *Phys. Rev. Lett.*, **102**(14), 146601.

[19] Eble, B., Desfonds, P., Bernardot, F., Testelin, C. and Chamarro, M. 2010. Hole and trion spin dynamics in quantum dot under excitation by a train of circulary polarized pulses. *Phys. Rev. B*, **81**(045322), 8.

[20] Fallahi, P., Yilmaz, S. T. and Imamoğlu, A. 2010. Measurement of a heavy-hole hyperfine interaction in InGaAs quantum dots using resonance fluorescence. *Phys. Rev. Lett.*, **105**(25), 257402.

[21] Feng, D. H., Akimov, I. A. and Henneberger, F. 2007. Nonequilibrium nuclear-electron spin dynamics in semiconductor quantum dots. *Phys. Rev. Lett.*, **99**(3), 036604.

[22] Fischer, J., Coish, W. A., Bulaev, D. V. and Loss, D. 2008. Spin decoherence of a heavy hole coupled to nuclear spins in a quantum dot. *Phys. Rev. B*, **78**(15), 155329.

[23] Gammon, D., Brown, S. W., Snow, E. S. *et al.* 1997. Nuclear spectroscopy in single quantum dots: nanoscopic Raman scattering and nuclear magnetic resonance. *Science*, **277**, 85–88.

[24] Greilich, A., Shabaev, A., Yakovlev, D. R. *et al.* 2007. Nuclei-induced frequency focusing of electron spin coherence. *Science*, **317**(5846), 1896–1899.

[25] Hanson, R., Kouwenhoven, L. P., Petta, J. R., Tarucha, S. and Vandersypen, L. M. K. 2007. Spins in few-electron quantum dots. *Rev. Mod. Phys.*, **79**(4), 1217–1265.

[26] Kloeffel, C., Dalgarno, P. A., Urbaszek, B. *et al.* 2011. Controlling the interaction of electron and nuclear spins in a tunnel-coupled quantum dot. *Phys. Rev. Lett.*, **106**(4), 046802.

[27] Korenev, V. L. 2007. Nuclear spin nanomagnet in an optically excited quantum dot. *Phys. Rev. Lett.*, **99**(25), 256405.

[28] Krebs, O., Eble, B., Lemaître, A. *et al.* 2008. Hyperfine interaction in InAs/GaAs self-assembled quantum dots: dynamical nuclear polarization versus spin relaxation. *C.R. Physique*, **9**, 874.

[29] Krebs, O., Maletinsky, P., Amand, T. *et al.* 2010. Anomalous Hanle effect due to optically created transverse Overhauser field in single InAs/GaAs quantum dots. *Phys. Rev. Lett.*, **104**(5), 056603.

[30] Lai, C. W., Maletinsky, P., Badolato, A. and Imamoğlu, A. 2006. Knight-field-enabled nuclear spin polarization in single quantum dots. *Phys. Rev. Lett.*, **96**(16), 167403.

[31] Lampel, G. 1968. Nuclear dynamic polarization by optical electronic saturation and optical pumping in semiconductors. *Phys. Rev. Lett.*, **20**(10), 491–493.

[32] Latta, C., Hogele, A., Zhao, Y. *et al.* 2009. Confluence of resonant laser excitation and bidirectional quantum-dot nuclear-spin polarization. *Nature Phys.*, **5**(8), 758.

[33] Liu, W. K., Whitaker, K. M., Smith, A. L. *et al.* 2007. Room-temperature electron spin dynamics in free-standing ZnO quantum dots. *Phys. Rev. Lett.*, **98**(18), 186804.

[34] Lombez, L., Braun, P.-F., Marie, X. *et al.* 2007. Electron spin quantum beats in positively charged quantum dots: nuclear field effects. *Phys. Rev. B*, **75**(19), 195314.

[35] Makhonin, M. N., Chekhovich, E. A., Senellart, P. *et al.* 2010. Optically tunable nuclear magnetic resonance in a single quantum dot. *Phys. Rev. B*, **82**(16), 161309.

[36] Maletinsky, P., Lai, C. W., Badolato, A. and Imamoğlu, A. 2007. Nonlinear dynamics of quantum dot nuclear spins. *Phys. Rev. B*, **75**(3), 035409.

[37] Meier, F. and Zakharchenya, B. 1984. Optical orientation. Modern Problems in Condensed Matter Sciences (North-Holland, Amsterdam), **8**.

[38] Merkulov, I. A., Efros, A. L. and Rosen, M. 2002. Electron spin relaxation by nuclei in semiconductor quantum dots. *Phys. Rev. B*, **65**(20), 205309.

[39] Skiba-Szymanska, J. *et al.* 2008. Overhauser effect in individual InP/GaInP dots. *Phys. Rev. B*, **77**(16), 165338.

[40] Tartakovskii, A. I. *et al.* 2007. Nuclear spin switch in semiconductor quantum dots. *Phys. Rev. Lett.*, **98**(2), 026806.

[41] Testelin, C., Bernardot, F., Eble, B. and Chamarro, M. 2009. Hole–spin dephasing time associated with hyperfine interaction in quantum dots. *Phys. Rev. B*, **79**(19), 195440.

[42] Urbaszek, B. *et al.* 2007. Efficient dynamical nuclear polarization in quantum dots: temperature dependence. *Phys. Rev. B*, **76**(20), 201301.

[43] Xu, X. *et al.* 2009. Optically controlled locking of the nuclear field via coherent dark-state spectroscopy. *Nature (London)*, **459**, 1105.

Part V

Electron transport in quantum dots fabricated by lithographic techniques from III–V semiconductors and graphene

15

Electrically controlling single spin coherence in semiconductor nanostructures

Y. Dovzhenko, K. Wang, M. D. Schroer and J. R. Petta

15.1 Introduction

In 1998, Daniel Loss and David DiVincenzo published a seminal paper describing how semiconductor quantum dots could be used to create spin qubits for quantum information processing [28]. They recognized that a single spin in a magnetic field forms a natural two-level system that can serve as a quantum bit. Moreover, owing to the weak magnetic moment of the electron, the spin is relatively well isolated from the environment leading to long coherence times. To confine single spins, Loss and DiVincenzo envisioned the quantum dot architecture shown in Fig. 15.1. A GaAs/AlGaAs heterostructure confines electrons to a two-dimensional electron gas (2DEG). Depletion gates are fabricated on top of the structure to provide a tunable confinement potential, trapping a single electron in each quantum dot. Neighboring quantum dots are tunnel coupled, with the coupling strength controlled by the electrostatic potential. The orientation of a single spin can be controlled by using electron spin resonance (ESR), while nearest-neighbor coupling is mediated by the depletion gate tunable exchange interaction.

It is fair to say that in 1998 many of the requirements of the Loss–DiVincenzo proposal had not been implemented, starting with the most basic necessity of a single electron lateral quantum dot [8]. The purpose of this chapter is to describe several experiments inspired by the Loss–DiVincenzo proposal. Many powerful experiments have been performed since 1998 and, given the space constraints here, we cannot give each experiment the attention it deserves. Instead of focusing on the chronological development of the field, we choose to describe single spin preparation, coherence, control, and readout in double quantum dots (DQD). There are a number of review articles that detail experimental and theoretical work in this area. Several early publications describe the physics and transport properties of quantum dots [1, 23]. Transport in DQDs is nicely covered by van der Wiel *et al.* [41]. A thorough review article by Hanson *et al.* presents theory and experiment related to spins in DQDs [16]. Coupling of electron and nuclear spin systems is reviewed by Coish and Baugh [9]. The theory of quantum control of the singlet–triplet qubit is outlined in papers by Levy [27] and Taylor *et al.* [40] and in Chapter 16 of the present book, by Ribeiro and Burkard.

Quantum Dots: Optics, Electron Transport and Future Applications, ed. Alexander Tartakovskii. Published by Cambridge University Press.

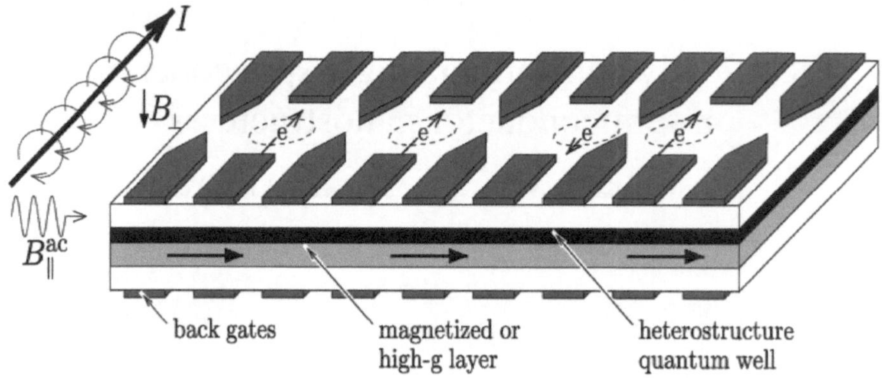

Figure 15.1 Quantum dot architecture proposed by Loss and DiVincenzo.

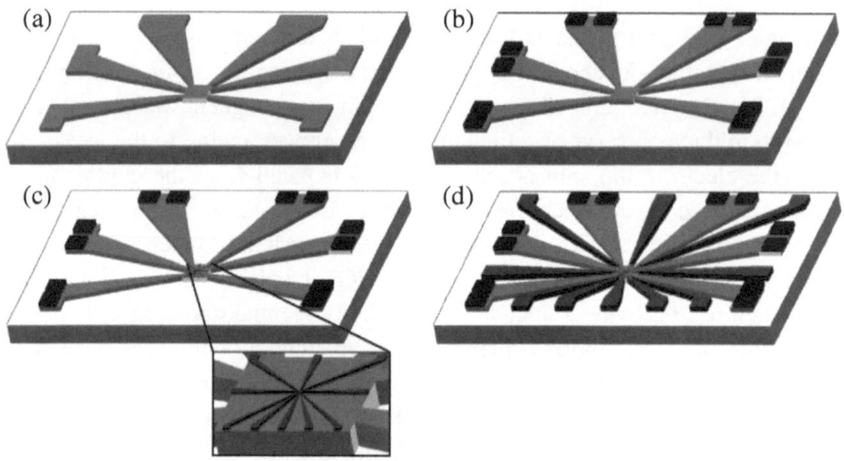

Figure 15.2 Sample fabrication consists of four main steps: (a) mesa etch, (b) ohmic contact deposition and annealing, (c) e-beam lithography to define the depletion gate pattern, and (d) a final photolithography step to connect the depletion gates to wirebonding pads.

15.2 Sample fabrication

In this section we describe how a working quantum dot device is made. The starting point for sample fabrication is the growth of the GaAs/AlGaAs heterostructure that supports the 2DEG [14]. Once the wafer has been grown, the fabrication of a device can be broken down into the four steps illustrated in Fig. 15.2: (a) mesa etch, (b) ohmic contact deposition and annealing, (c) electron beam (e-beam) lithography, and (d) final photolithography.

15.2.1 Heterostructure

The GaAs/AlGaAs heterostructure is grown using molecular beam epitaxy. Figure 15.3 shows the growth profile for a typical heterostructure with the 2DEG located 110 nm below

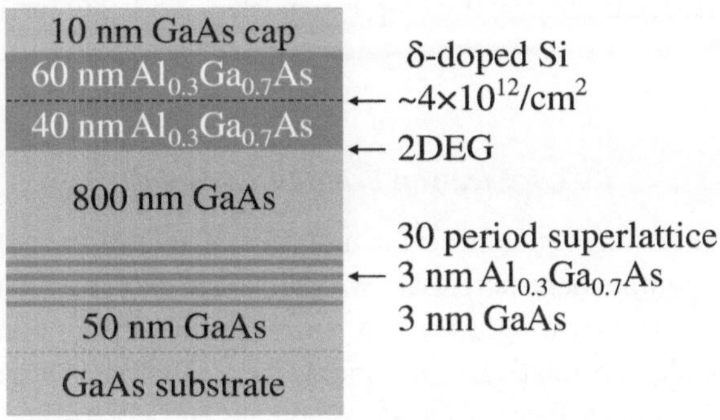

Figure 15.3 Growth profile of a typical GaAs/AlGaAs heterostructure (not to scale).

the surface of the wafer. The 2DEG charge densities are on the order of 2–3 $\times$ $10^{11}/\text{cm}^2$ and the carrier mobility is $\sim 200\,000\,\text{cm}^2/(\text{Vs})$.

15.2.2 Mesa etch

A sample usually starts with a piece of a GaAs/AlGaAs heterostructure wafer 6 mm $\times$ 6 mm in size, which is small enough to fit in most commercially available chip carriers and large enough to support 4–6 quantum dot devices. It is crucial to eliminate unnecessary regions of 2DEG on the wafer by performing a mesa etch, thereby reducing potential leakage current from the gates to the underlying 2DEG. A mesa etch also allows different quantum dot devices fabricated on the same chip to be electrically isolated. Overlap between the 2DEG and the depletion gates is limited to the e-beam lithography defined fine gate pattern, while the wirebonding pads are located on regions of the chip where the 2DEG has been etched away. The mask for etching the mesa is defined by positive-resist photolithography. A UV ozone clean is performed after development in order to clean any resist residue. The mesa is subsequently wet etched by a mixture of H_2SO_4 : H_2O_2 : H_2O with the ratio of 1: 8: 240. We choose an etch depth that exceeds the 2DEG depth to ensure electrical isolation of the etched area at low temperatures.

15.2.3 Ohmic contacts

Ohmic contacts are required to make electrical contact to the 2DEG. It is helpful to have two ohmic pads on each arm of the mesa in order to allow four-probe electrical measurements. Ohmic contacts should also overlap with the edge of the mesa, otherwise edge-state formation at large perpendicular magnetic fields will lead to poor electrical contact to the 2DEG. We use a bilayer lithography step for easy liftoff of thick metal films. After the resist is developed, the sample is UV ozone cleaned, etched in NH_4OH, rinsed in deionized water, blown dry with nitrogen gas and immediately placed in the evaporator. Ohmic contacts are made by evaporating a stack of 62 Å of Pt, 2000 Å of Au, 1000 Å of Ge, 660 Å of Pt, and 630 Å of Au. After liftoff, the chip is annealed at 525 °C for 100 s in a forming gas environment.

15.2.4 Electron beam lithography

Fabrication of the depletion gate pattern is the most crucial part of the process as the depletion gates create the confinement potential. We define the depletion gate pattern using a single layer of 950 k molecular weight PMMA and a 30 keV e-beam lithography tool. The fine gate structure is aligned relative to the mesa using alignment mark crosses that are included in the ohmic contact layer. After development and UV ozone clean, 30 Å of Ti and 170 Å of Au are evaporated.

15.2.5 Final photolithography

High-resolution e-beam lithography requires relatively thin metal layers in order to keep the depletion gate aspect ratio (height/width) reasonable and to promote good liftoff. These thin metal layers cannot run over the edge of the mesa and maintain electrical continuity owing to the large mesa height of ~ 150 nm. As a result, a final photolithography/liftoff step is required to complete the device. The key design requirement here is that the metal thickness be greater than the mesa height, to ensure that the metal will form a continuous path up and over the mesa edge. We define the final metal layer using a bilayer photolithography process identical to the ohmic contact layer resist recipe. After development, UV ozone clean, and a NH_4OH etch, 50 Å of Ti and 2000 Å of Au are deposited via thermal evaporation. The sample is ready for measurement after liftoff.

15.3 Measurement technology

In this section we describe several electrical measurement techniques used to probe and control a gate-defined multiple quantum dot. Before presenting experimental details, it is important to gain some appreciation of the energy scales associated with the quantum dot device [16]. Starting from the large energy scales, the electrostatic charging energy is on the order of 4 meV. This is followed by the orbital level spacing $\sim$1–2 meV. For two electrons confined in a single dot, the two-electron singlet–triplet splitting is of the order of 500 μeV in GaAs and can reach 5 meV energies in InAs nanowire quantum dots [25, 3, 30]. For spatially separated electrons, the singlet–triplet energy splitting is $4t^2/U \sim 30$ μeV, where $t \sim 5$–20 μeV is the interdot tunnel coupling and U is the single dot charging energy [28, 33]. In GaAs, $|g|$=0.44, resulting in a typical Zeeman splitting of E_z=$|g|\mu_B B \sim 130$ μeV at a magnetic field of B = 5 T [15]. Owing to the small energy scales associated with the device (micro- to millielectronvolt), it is necessary to work at millikelvin temperatures, where the thermal energy scale is smaller than most of the energy scales listed above. Substantial electrical filtering is required in order to achieve low electron temperatures.

15.3.1 Electrical filtering

Both a.c. and d.c. wiring must be carefully filtered to eliminate unwanted noise sources. The d.c. lines connecting the BNC breakout box to the cryostat are filtered using D-sub capacitive filters. The d.c. wiring loom in the cryostat consists of twisted pairs. At the sample holder, the d.c. lines pass through metal film resistors that are thermally anchored to the mixing chamber. The tail piece on the sample holder contains copper powder, which provides an additional level of filtering.

Electrical pulses and continuous wave (CW) microwave excitation can be applied to the depletion gates to modulate the quantum dot confinement potential or drive spin rotations. High-frequency signals are transmitted to the sample by using semi-rigid coax cables. Figure 15.4 illustrates the dilution refrigerator microwave wiring configuration. UT-85-BeCu-SS coax is used between the room temperature hermetic bulkhead connector and the 1 K pot plate. UT-85-SS-SS coax connects the 1 K pot plate to the still plate and the still plate to the mixing chamber plate. Anritsu bias-tees are mounted at the mixing chamber plate, allowing the high-frequency signal to be added to the quasi-static d.c. gate voltages. Copper UT-85 semi-rigid coax connects the bias tee and the sample holder. Microwave attenuators are placed in-line at the 1 K pot plate, still plate, and mixing chamber plate, for

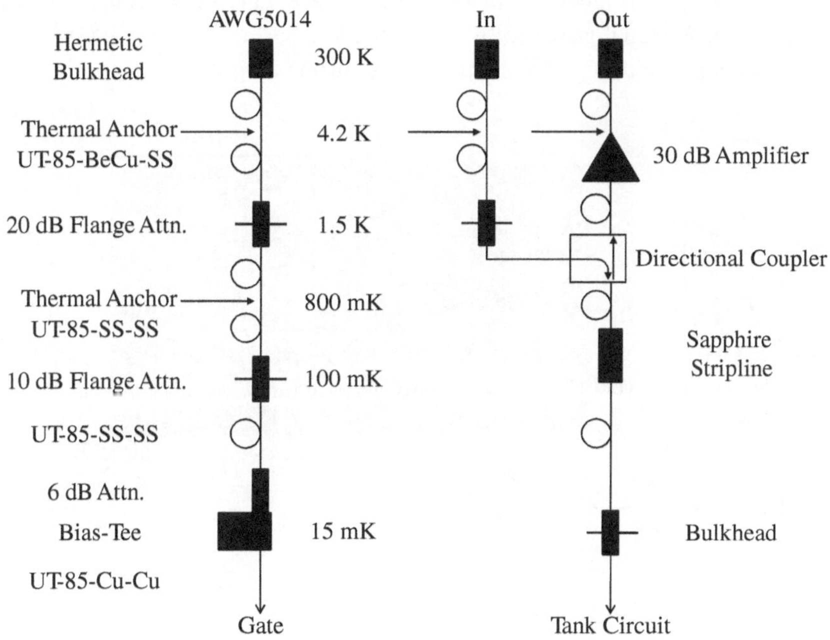

Figure 15.4 Cryostat wiring for the pulsed gate lines and rf-reflectometry circuit. The coaxes are looped to allow for thermal contraction.

a total of 25–40 dB of attenuation. The attenuators provide filtering and thermally anchor the center conductor of the coax cables.

15.3.2 Sample holder design

Early DQD experiments were performed by using a commercially available 32-pin LCC chip carrier [33]. Copper UT-34 semi-rigid coax cables were simply soldered to several of the contacts on the chip carrier and then connected to the coax lines installed in the cryostat. Given the gains it allows in the data-acquisition rate, rf-reflectometry has recently become a standard measurement technique. The rf-reflectometry circuit requires surface mount components that cannot be easily located in the 32-pin chip carrier. As a result, most groups have switched to custom-made planar sample holders. An early version is described by Reilly *et al.* [36] and a more recent design can be found in Edward Laird's PhD thesis [26].

15.3.3 Basic measurement circuit

Figure 15.5 shows the circuit diagram used to measure a triple quantum dot (TQD) sample. The device can be probed by measuring the conductance through the TQD, g_D, or the conductance of the quantum point contact (QPC) charge sensor, g_Q. For example, the conductance of the left QPC is measured using a simple current biased measurement. The sine wave output of a CW signal generator is connected to a current limiting 100 MΩ resistor, which is in turn connected to ohmic contact #2. A current amplifier is connected to ohmic

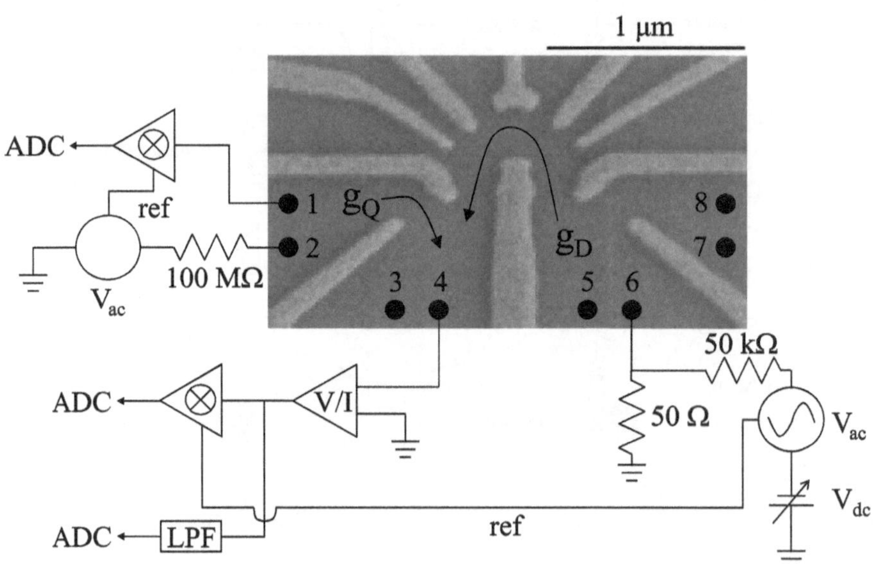

Figure 15.5 Circuit diagram used to measure the TQD conductance, g_D, and the QPC conductance, g_Q. Ohmic contacts are numbered.

contact #4, providing a virtual ground. The a.c. voltage drop across the QPC is measured by connecting ohmic contact #1 to a lock-in amplifier. The output of the lock-in amplifier is connected to an analog-to-digital converter (ADC). Similarly, the conductance through the TQD is measured using a voltage biased circuit. The sine wave output of a CW signal generator is sent to a 1000 : 1 voltage divider, with the output of the divider connected to ohmic contact #6. The TQD conductance and d.c. current can be simultaneously measured by splitting the output of the current amplifier. One port is connected to a low-pass filter for d.c. current measurements, while the other port is connected to a lock-in amplifier for a.c. conductance measurements.

Coulomb blockade spectroscopy

Samples are initially tested by measuring the conductance through the quantum dot, g_D, as a function of gate voltage, V_G, as illustrated in Fig. 15.6. Coulomb blockade peaks are observed when an energy level is swept through the Fermi level of the source and drain contacts. The width of the Coulomb blockade peak is proportional to the electron temperature, allowing for convenient electron thermometry [23]. In a DQD, the conditions for a finite conductance are restricted [41]. At zero bias, the energy level of the left and right dots must be aligned with the Fermi level of the source and drain electrodes. The addition of a third dot further restricts the conditions for a non-zero conductance to be measured. Charge sensing is convenient since it can probe the device in configurations where the dot conductance is undetectable [38].

Charge sensing

In 1993 Field *et al.* demonstrated charge sensing using a QPC charge detector [12]. The main idea is to bias the QPC at one of the steps in conductance, where a small change in gate voltage leads to a large change in conductance. Addition of an electron to the quantum dot effectively reduces the QPC channel width and lowers the conductance by a few percent. Charge sensing has been applied to a variety of devices.

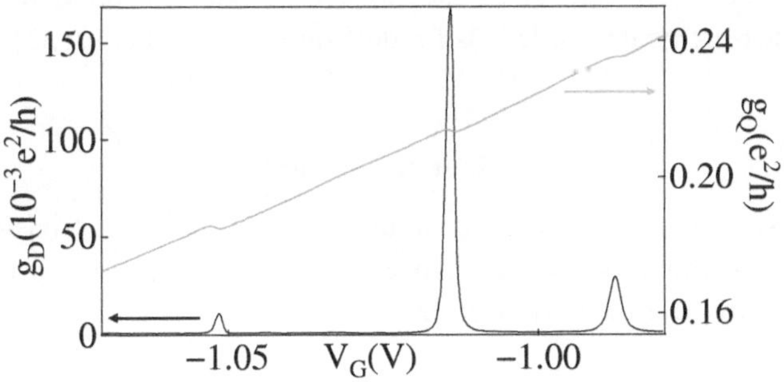

Figure 15.6 Simultaneous measurement of conductance through the dot, g_D, and QPC charge detector, g_Q, as a function of gate voltage, V_G. The charge sensor conductance increases by a few percent each time an electron is removed from the dot.

The current biased measurement circuit illustrated in Fig. 15.5 is used to probe the QPC conductance. Figure 15.6 shows a plot of the conductance through the QPC, g_Q, as a function of gate voltage, V_G. The overall slope is due to direct capacitive coupling between the gate and the QPC. However, when an electron is ejected from the quantum dot, we observe a small increase in the QPC conductance. These features in the QPC conductance data are correlated with the Coulomb blockade peaks observed in transport. Recently, it has been shown that replacing the QPC with a sensor quantum dot results in large gains in the sensitivity [5].

In 1998, Schoelkopf *et al.* developed the rf single electron transistor (rf-SET), which allows measurements to be performed on much faster timescales [37]. The main idea is to embed the device (SET, QPC, or quantum dot) in a LC circuit in such a way that the impedance of the device acts to damp the LC circuit. By driving the rf tank circuit and measuring the amplitude of the reflected signal, it is possible to perform measurements on timescales that are much faster than the intrinsic RC time constants associated with the d.c. wiring (typically limited to a few kilohertz). The cryostat wiring required to perform rf-reflectometry is illustrated in Fig. 15.4. An LC tank circuit located adjacent to the device is coupled to the source of the SET, QPC or sensor quantum dot, creating an effective RLC circuit. The amplitude of the reflected signal is sensitive to the resistance of the sensor, and therefore the state of the quantum dot. This voltage is amplified by a cryo-amplifier and demodulated at room temperature [36]. The main advantage is that rf-reflectometry measurements can be performed on 100 ns timescales, allowing for single shot measurements of the electron spin [5].

15.3.4 Pulsed-gate methods

The quantum dot confinement potential can be modulated by applying voltage pulses to the depletion gates. The voltage pulses are added to the dc gate voltage using Anritsu bias-tees, as shown in Fig. 15.4. Pulsed-gate methods were used in early "pump-probe" experiments on vertical quantum dots to measure the singlet–triplet relaxation time and the orbital relaxation time [13]. The techniques were then extended to lateral single dots for spin relaxation measurements and DQDs for quantum control [11, 17, 33].

15.4 Quantum control

In the previous sections, we covered quantum dot fabrication and basic measurement techniques. We now move on to quantum control of the qubit, in particular the most fundamental operations for the two-electron singlet–triplet qubit.

15.4.1 Double quantum dot energy level diagram

The focus of our discussion is the DQD, the platform in which most recent experiments have been performed [16]. The DQD forms a simple double well confinement potential for

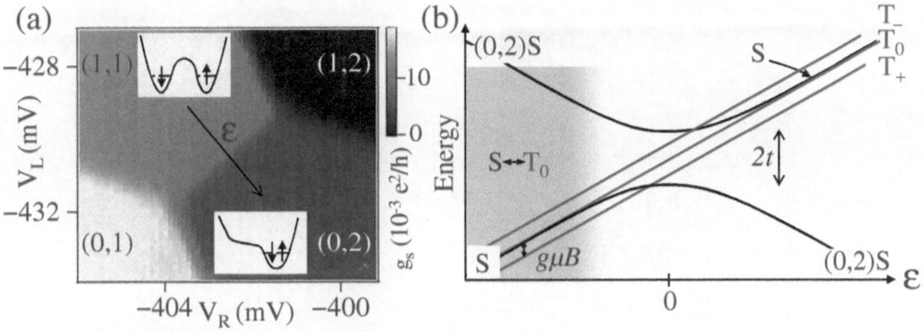

Figure 15.7 (a) Charge stability diagram measured using the QPC charge sensor. A single electron can be transferred from the left dot to the right dot by simply increasing the detuning, ϵ. (b) Energy level diagram in the two-electron regime (see text).

the electrons. The first step in the experiment is to determine the occupancy of the DQD using charge sensing. Figure 15.7a shows the charge stability diagram of the DQD, which is extracted by measuring the QPC conductance as a function of the DQD gate voltages, V_L and V_R. We focus on the region of parameter space near the (1,1)–(0,2) charge transition, where (N_L, N_R) denotes the number of electrons in left and right dots. The asymmetry of the double well potential is adjusted by varying the detuning parameter, ϵ. The ground state at positive detuning is the (0,2) charge state, where both electrons reside in the right dot. By sweeping to negative detuning, a single electron can be transferred from right to left, forming a (1,1) charge state. Zero detuning corresponds to a situation where the (1,1) and (0,2) charge states are degenerate. Here the interdot tunnel coupling, t, lifts the degeneracy, resulting in an anti-crossing with magnitude $2t$.

We now consider the spin degree of freedom, which results in the level diagram shown in Fig. 15.7b. The four possible two-electron spin states are the singlet (total S = 0) and three triplets (total S = 1). The (0,2) charge state is characterized by a large singlet–triplet splitting $J_{(0,2)} \sim 400$ μeV. As a result, we can neglect the (0,2) triplet states. Deep in the (1,1) charge state, the singlet–triplet splitting is on the order of 20 neV and all four spin states are accessible. Therefore, at positive detuning the ground state is the two-electron singlet (0,2)S and at negative detuning (and zero field) the (1,1) triplets and (1,1) singlet have nearly the same energy. Application of an external magnetic field lifts the triplet-state degeneracy, resulting in a two-level system consisting of the (1,1) singlet state (denoted S) and the m_s=0 (1,1) triplet state, denoted T_0. Owing to the hybridization of the charge states, the energy difference $J(\epsilon)$ between S and T_0 is a function of detuning. In the basis of T_0 and S, the resulting Hamiltonian is:

$$H_{m_s=0} = \begin{pmatrix} 0 & g\mu_B \delta B_{nuc} \\ g\mu_B \delta B_{nuc} & -J(\epsilon) \end{pmatrix}. \tag{15.1}$$

Here the off-diagonal terms are set by the z-component of the hyperfine field gradients, δB_{nuc}. When $J = 0$, the hyperfine fields drive a random σ_x rotation. When $|J| \gg |B_{nuc}|$,

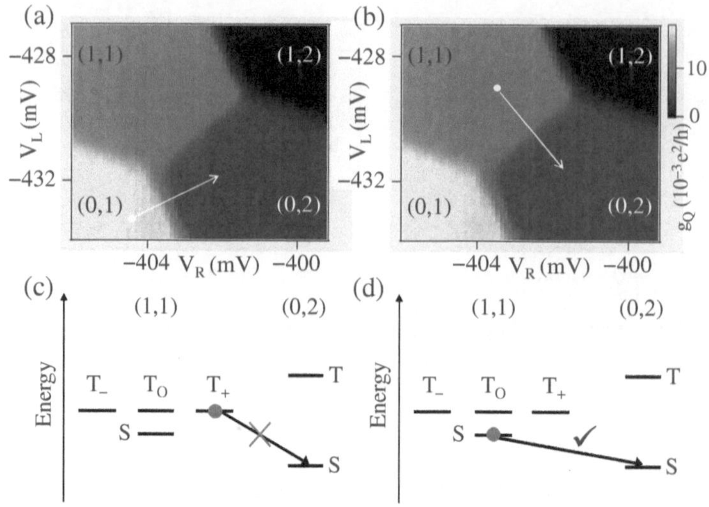

Figure 15.8 (a) A spin singlet state, (0,2)S, is prepared by pulsing from the (0,1) to (0,2) region of the charge stability diagram. (b) Spin-to-charge conversion is achieved by pulsing from the (1,1) to (0,2) region of the charge stability diagram. (c) Due to Pauli blockade a (1,1)T_+ state cannot tunnel to (0,2)S. (d) Spin selection rules allow tunneling from (1,1)S to (0,2)S.

the exchange term dominates, resulting in a σ_z rotation. Electrical control of the exchange interaction is used to prepare, manipulate, and read out spin states, as described next.

15.4.2 Spin preparation

We initialize the DQD in a spin-singlet state using pulsed-gate techniques. Referring to the charge stability diagram shown in Fig. 15.8a, we start with the gate voltages residing in the (0,1) region of the charge stability diagram, where one electron occupies the right quantum dot. A voltage pulse is then applied which shifts the system into the (0,2) region of the charge stability diagram, adding a second electron to the right dot. Owing to the large singlet–triplet splitting, $J_{(0,2)}$, the electron loads into the (0,2)S state. Experiments based on optical control of single spins have demonstrated an initialization efficiency of 99.8% [4].

15.4.3 Spin-to-charge conversion

Spin detection is performed using a method called spin-to-charge conversion, which is illustrated in Fig. 15.8b. Assume that at the end of the experiment the qubit is in the (1,1) region of the charge stability diagram with an unknown spin state. In order to deduce the spin state, we pulse the detuning to a point in the (0,2) region of the charge stability diagram. The detuning is adjusted so that the energy of the (1,1) charge state lies above the energy of the (0,2)S state, but below the energy of the (0,2)T states. If the (1,1) charge state was a triplet, the interdot charge transition will be forbidden due to the Pauli exclusion principle. However, if the (1,1) state was a singlet, the transition to (0,2)S is allowed. The spin-dependent charge transition is detected using the QPC charge sensor, allowing an indirect measurement of the spin state.

15.4.4 Quantum errors: relaxation and decoherence

The main source of decoherence in GaAs quantum dots is the contact hyperfine interaction [19, 18]. In a typical quantum dot the electron wavefunction overlaps with roughly 10^6 lattice nuclei, each of which carries spin I = 3/2. As a result, the electron experiences an effective Overhauser field with magnitude $B_{nuc} \sim$ 1–3 mT. Temporal and spatial variations in the Overhauser field lead to decoherence. Spin–orbit interactions are also important since they couple the spin degree of freedom to the orbital degree of freedom. When combined with the electron–phonon interaction, spin–orbit and hyperfine terms can lead to spin relaxation [10, 42]. The effects of environmental coupling have been extensively investigated [16].

There is an important distinction between spin-relaxation and spin dephasing or decoherence. Spin relaxation involves a transition between two different spin states separated by an energy gap, for example the Zeeman splitting or singlet–triplet energy. For spin relaxation to occur, the electron spin must flip and the electron must also lose energy. Spin relaxation is characterized by a time T_1, which exhibits a strong magnetic field dependence in quantum dots [24]. Spin dephasing generally involves a phase randomization process and is characterized by two timescales, T_2^* and T_2. T_2^* is an ensemble averaged spin dephasing time. The decoherence time, T_2, describes the decay of quantum coherence and is typically measured using spin-echo. We now describe how T_1, T_2^*, and T_2 are measured in a two-electron DQD.

15.4.5 Spin relaxation – T_1

Pump-probe pulsed gate techniques were first used to measure the orbital lifetime and singlet–triplet relaxation time in vertical quantum dots [13]. The method was extended to lateral single quantum dots by Elzerman *et al.* and lateral DQDs by Johnson *et al.* and Petta *et al.* [11, 17, 34]. A widely used technique for extracting T_1 is to initialize the qubit in an excited spin state, wait for a variable time t, then measure the excited state occupation using spin-to-charge conversion. The probability of remaining in the excited state decays with a characteristic relaxation time T_1.

Focusing on the two-electron DQD, the experimental cycle begins by initializing the sample in the (0,1) region of the charge stability diagram. A voltage pulse shifts the device to the (1,1) region of the charge stability diagram and the DQD is then initialized in the (1,1)S state or one of the (1,1) triplet states. The potential well is then tilted so that the

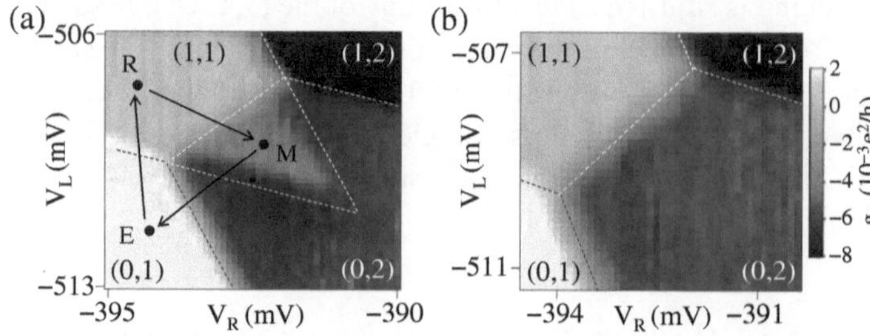

Figure 15.9 (a) Charge stability diagram measured with voltage pulses applied to the depletion gates. The pulse sequence consists of three steps: (E)mpty into (0,1), (R)eset in (1,1), and (M)easure in (0,2). A representative pulse sequence is overlayed on the data. Data are acquired at an external field of 100 mT and with a dwell time at point M of 8 μs. (b) Similar data, taken at an external field of 0 mT and a dwell time at point M of 80 μs. Hyperfine fields drive fast spin rotations when $B \leq B_{nuc}$, leading to rapid spin relaxation. In this case, the pulse triangle is completely absent.

(0,2) charge state becomes energetically favored. If the initial state was a singlet, the left electron will tunnel into the right dot on a timescale set by the interdot tunnel coupling. If the initial state was one of the triplets, then the interdot charge transition is forbidden until the (1,1) triplet mixes into the (1,1)S.

Figure 15.9a shows the charge stability diagram with the pulse sequence superimposed for clarity. Technical details are described by Petta *et al.* [34]. With pulses applied, a triangular shaped feature appears in the (1,1) region of the charge stability diagram. Spin-blocked interdot charge transitions result in some (1,1) charge state occupation after the pulse from (1,1) to (0,2). The decay of this signal gives T_1. For a spin to relax, the spin must flip and energy must be transferred to the environment in the form of phonon emission. As a result, the spin relaxation time is strongly dependent on the magnetic field and detuning [17]. At high magnetic fields, where $B \gg B_{nuc}$, T_1 approaches 1 ms. When $B \leq B_{nuc}$, hyperfine fields can drive rapid rotations from triplet to singlet, lifting spin-blockade, resulting in short a T_1 of 2 μs. Recent measurements in GaAs have observed T_1 as long as 1 s [2]. In Si, a T_1 of 6 s has been reported [29].

15.4.6 Inhomogeneous dephasing – T_2^*

In the limit of $J(\epsilon) < B_{nuc}$ hyperfine fields dominate the Hamiltonian in Eqn (15.1) and lead to a random σ_x rotation. An initially prepared spin singlet state will dephase under the influence of the hyperfine fields [33]. Repeated measurements of the singlet decay, averaging over a distribution of hyperfine fields, lead to a measurement of the ensemble averaged spin dephasing time, T_2^*. The pulse sequence used to measure T_2^* is shown in Fig. 15.10. The system is initialized in the (0,2)S state at point P in the (0,2) region of the charge stability diagram. The singlet is then separated by moving to point S in the (1,1) region of

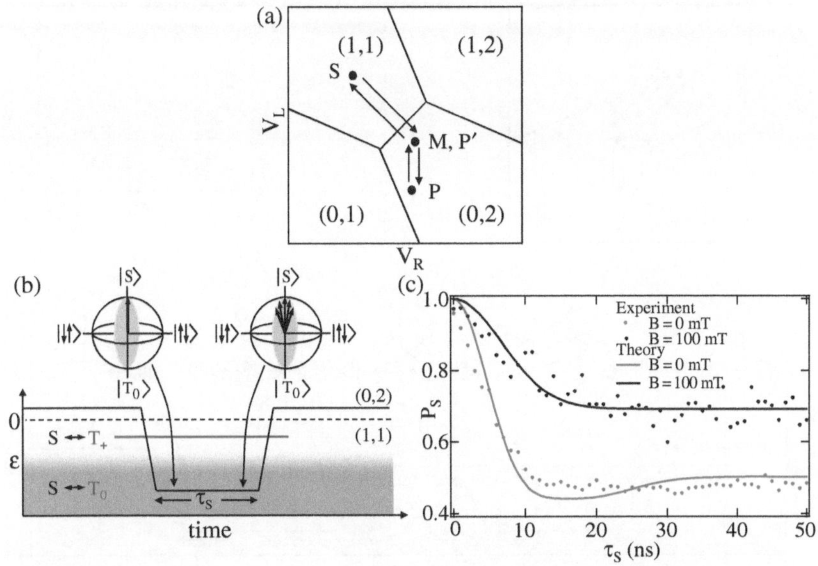

Figure 15.10 (a) Pulse sequence used to measure T_2^*. The system is initialized in the (0, 2)S state at positive detuning. A voltage pulse that is adiabatic with respect to the interdot tunneling rate brings the system to negative detuning where $J(\epsilon) \sim 0$. Hyperfine fields drive rotations between S and T_0. The spin state is projected onto (0,2)S after a free evolution time, τ_S. (b) Detuning as a function of time. Hyperfine fields drive random spin rotations as depicted in the Bloch sphere representation. (c) Singlet state return probability, P_S, as a function of separation time, τ_S.

the charge stability diagram. An intermediate stop at point P′ reduces pulse overshoot. The electrons are held in the (1,1) region of the charge stability diagram for a "separation time" τ_S. After the separation time, the device is pulsed into the (0,2) region for a measurement of the final singlet state probability using spin-to-charge conversion.

In the Bloch sphere representation, the initially prepared spin singlet state is located at the North pole. Once the spins are separated, hyperfine fields drive a random σ_x rotation. In an ensemble averaged picture, the singlet dephases into a mixture of singlet and triplet states, denoted by the fanning out of the Bloch vector. The measured singlet state probability, P_S, exhibits a Gaussian decay with a characteristic timescale $T_2^* \sim 10$ ns. At $B =$ 100 mT, the S state can mix only with T_0. At $B = 0$ mT, S can mix with all three triplet states. As a result, the steady-state value for P_S is smaller at $B = 0$ mT. In theory, one would expect P_S to approach 1/2 at long separation times for $B = 100$ mT. The deviation is due to spin relaxation during the measurement interval.

15.4.7 Spin echo – T_2

Since the random hyperfine fields vary slowly compared to a single cycle of the pulse sequence, nuclear spin dynamics are often treated in the quasi-static approximation, which

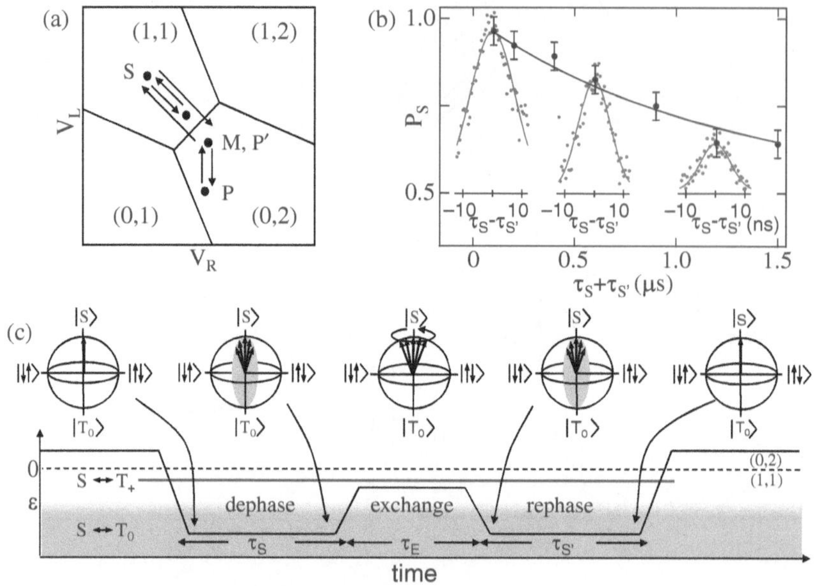

Figure 15.11 The spin echo pulse sequence used to measure T_2 is superimposed on the charge stability diagram in (a). Spin-echo extends the coherence time to $T_2 = 1.2$ μs, as shown in (b). Slight differences between τ_S and $\tau_{S'}$ reduce the echo efficiency, with a sensitivity set by T_2^*. Singlet state evolution is depicted in (c) for a perfect spin-echo. The lower panel of (c) shows the detuning as a function of time for this pulse sequence.

assumes B_{nuc} is frozen during the time span of a single pulse cycle, but varies from one pulse cycle to the next [40]. The fluctuations in the hyperfine fields can be probed using a spin-echo pulse sequence [33]. The pulse sequence is based on the T_2^* measurement, with a key difference being that the free evolution time is interrupted by a σ_z exchange pulse. Details are given in Fig. 15.11.

The system is initialized into the (0,2)S state at point P in the (0,2) region of the charge stability diagram. The singlet is then separated by moving to point S in the (1,1) region of the charge stability diagram via point P′ to reduce pulse overshoot. The electrons are held in the (1,1) region of the charge stability diagram for a "separation time" τ_S. An exchange pulse is applied by shifting the detuning to a more positive value of ϵ for a time τ_E. The system is then shifted back to point S for a time $\tau_{S'}$. Spin-to-charge conversion is used to determine the final singlet state probability, P_S.

If the exchange pulse is selected to execute a perfect σ_z π-pulse, then the singlet state dephasing that took place in the first free evolution period (with duration τ_S) will be reversed, leading to singlet state refocusing. Imperfect refocusing occurs if $\tau_S \neq \tau_{S'}$, as shown in Fig. 15.11b. A reduced echo will also be observed if the hyperfine fields vary during the pulse sequence. Early experiments demonstrated a T_2 on the order of 1 μs [33, 22].

Technical improvements in pulse-fidelity, combined with more sophisticated refocusing pulse sequences, have extended the coherence time out to 200 μs [7, 6].

15.4.8 Single qubit rotations

In this section we describe several methods of single qubit control. It is important to make a distinction between single-spin qubits and two-spin qubits. The Loss–DiVincenzo proposal is based on single-spin qubits. Here coherent rotations between $|\uparrow\rangle$ and $|\downarrow\rangle$ are driven using electron spin resonance. Two-qubit gates are implemented via the exchange interaction. In contrast, the dynamics of the two-electron singlet–triplet qubit are governed by Eqn (15.1). In the basis of $|S\rangle$ and $|\mathrm{T}_0\rangle$, field gradients drive σ_x rotations and exchange drives σ_z rotations, leading to full control of the Bloch vector [27]. Proposals exist for coupling two singlet–triplet qubits via a capacitive interaction [39].

Exchange interaction

Exchange gates were first implemented in a GaAs DQD [33]. As illustrated in the energy level diagram (see Fig. 15.7), hybridization of the (0,2)S and S state leads to a detuning dependent energy splitting, $J(\epsilon)$, between S and T_0. The pulse sequence used to implement the exchange gate is illustrated in Fig. 15.12a. It is very similar to the spin-echo pulse sequence. The main difference relates to the transitions across the (0,2)–(1,1) charge transition. A fast detuning pulse shifts the DQD potential from point P′ to just beyond the S–T_+ anti-crossing. The pulse speed is then made adiabatic with respect to the nuclear fields, shifting the device to point S and loading the system into the ground state of the hyperfine fields, $|\uparrow\downarrow\rangle$. The rise time of the pulses is illustrated in Fig. 15.12c. As in the spin-echo, the exchange pulse drives a σ_z rotation. The ramp back to (0,2) for spin-to-charge conversion is the reverse of the initial ramp into (1,1). Coherent oscillations are shown in Fig. 15.12b. Exchange results in extremely fast ~180 ps $\sqrt{SWAP}$ operations.

Electron spin resonance

In conventional ESR, an external dc magnetic field is applied along the z-axis, resulting in Zeeman splitting of the spin states, $E_z = g\mu_B B$. A perpendicular ac magnetic field is applied to drive Rabi oscillations between the $|\uparrow\rangle$ and $|\downarrow\rangle$ spin states. In Section 15.4.3, we described how spin-to-charge conversion can be used to distinguish a singlet from a triplet state. There, a spin-dependent interdot charge transition is measured using a QPC charge sensor. Similar measurements can be performed in dc transport in the Pauli blockade regime. The double dot is configured in the two-electron regime near the (1,1)–(0,2) charge transition. At negative bias, electrons flow from right to left in the sequence (0,1) → (0,2) → (1,1) → (0,1). Owing to the large (0,2) singlet–triplet splitting, the (0,2)S state is loaded at negative bias. The electron is then transferred to the (1,1)S state and out the other side of the DQD, resulting in a finite current. At positive bias, electrons flow from left to right in

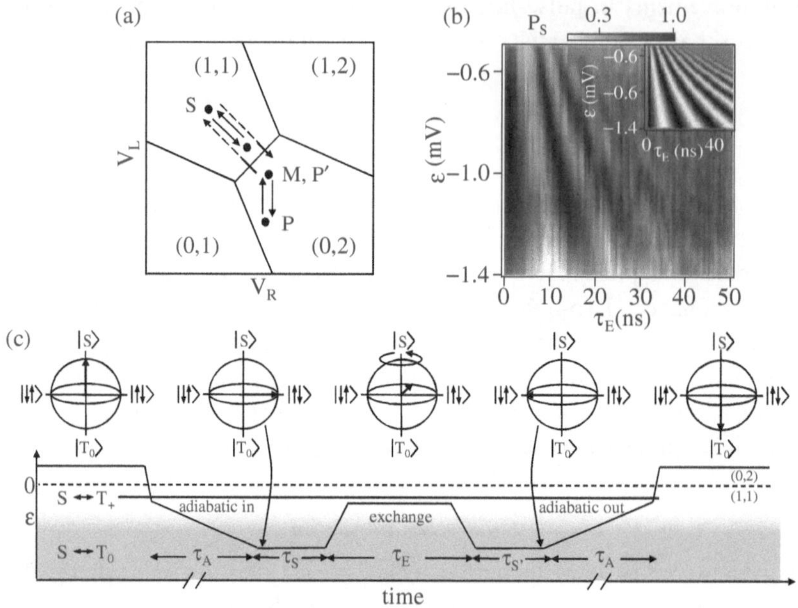

Figure 15.12 (a) The exchange-gate pulse sequence is superimposed on the charge stability diagram. Solid lines correspond to fast rise-time pulses, while dashed lines correspond to slow rise-time pulses. (b) The exchange rotation depends on the pulse length, τ_E, and the detuning, which sets $J(\epsilon)$. (c) The state of the quantum bit at various points in the pulse sequence is illustrated in the Bloch sphere representations. Lower panel: detuning as a function of time for the exchange gate.

the sequence (0,1) → (1,1) → (0,2) → (0,1). Owing to the negligible (1,1) singlet–triplet splitting, a (1,1) triplet state can be loaded. The (1,1) triplet cannot tunnel to (0,2)S due to spin selection rules and the current flow is blocked. This effect, termed Pauli blockade, was first observed in vertical DQDs by Ono *et al.* [32]. A process that flips an electron spin will lift the Pauli blockade. Therefore, a measurement of the Pauli blockade leakage current is a sensitive probe of electron spin resonance and other processes that rotate spins [20, 21].

In 2006, Koppens *et al.* demonstrated single spin ESR on a lateral GaAs DQD [21]. The sample was configured in the two-electron regime under conditions of Pauli blockade. The experiments measured the leakage current in the Pauli blockade regime as a function of external magnetic field and the a.c. frequency, f, of a current applied to a stripline fabricated near the DQD. Effective ESR driving lifts the Pauli blockade, resulting in a finite leakage current when $g\mu_B B = hf$. By sending bursts of microwave radiation to the sample Koppens *et al.* were able to demonstrate coherent control of a single electron spin [21]. The fidelity of the single spin rotations is limited due to the fact that the a.c. microwave field is comparable in amplitude to the fluctuating hyperfine field. Later experiments demonstrated spin-echo, with coherence times of 0.5 μs [22].

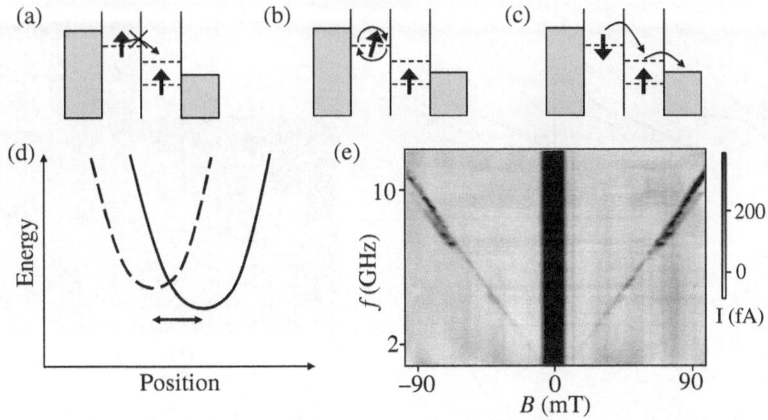

Figure 15.13 Pauli blockade and EDSR. In graphs (a)–(c), a spin-up electron is trapped in the right dot. Transport of other electrons from source to drain is only possible if the electron which tunnels into the left dot has the opposite spin. (d) In EDSR, the position of the electron is modulated by applying an ac voltage to a surface depletion gate. (e) Pauli blockade leakage current measured as a function of magnetic field, B, and the frequency of the a.c. driving, f.

Electric-dipole spin resonance

There is a growing motivation for using electric fields to drive single spin rotations due to the inherent difficulty of generating and localizing gigahertz a.c. magnetic fields on a nanometer scale sample held at millikelvin temperatures. One method for electrical control is called electric-dipole spin resonance (EDSR), where the conventional oscillating magnetic field is replaced by an a.c. electric field applied to one of the depletion gates. The electron trapped inside the dot is driven back and forth by the oscillating gate voltage. Owing to the spin–orbit interaction, the electron feels an effective a.c. magnetic field. As in conventional ESR, coherent spin rotations occur when $g\mu_B B = hf$.

EDSR was first demonstrated in GaAs quantum dots by Nowack *et al.* [31]. The efficiency of the rotations is dependent on the strength of the spin–orbit scattering. Faster Rabi oscillations were recently achieved in an InAs nanowire double dot [30]. Figure 15.13e shows the Pauli blockade leakage current, I, as a function of magnetic field, B, and a.c. driving frequency, f. The large leakage current observed around $B = 0$ is due to the hyperfine fields. Non-zero current is also observed along the two sloped lines that match the resonance condition. Fits to these lines give a g-factor of 8.0. Spin-echo and dynamical decoupling experiments have been achieved using EDSR in InAs nanowire DQDs [30].

Landau–Zener interferometry

A recent experiment demonstrated quantum control of a $|S\rangle$–$|T_+\rangle$ qubit using quantum interference [35]. The process is illustrated in Fig. 15.14 and is based on Landau–Zener transition physics. The DQD is prepared in the (0,2)S state. A voltage pulse is applied to the depletion gates, shifting the system into the (1,1) region of the charge stability diagram. In

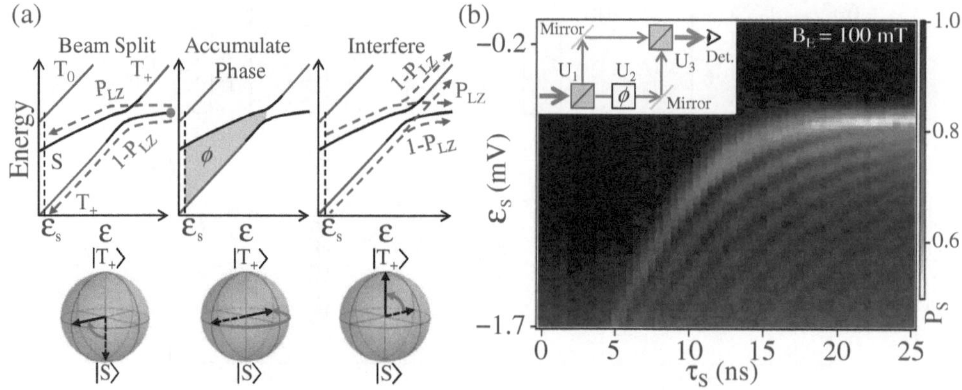

Figure 15.14 Landau–Zener interferometry for spins. (a) Passage through the S–T_+ anti-crossing converts the incoming singlet into a coherent superposition of $|S\rangle$-$|T_+\rangle$. The detuning is maintained at ϵ_S for a time τ_S, resulting in a σ_z rotation. A return sweep through the S–T_+ anti-crossing interferes the two states. (b) Singlet state return probability, P_S, measured as a function of pulse length, τ_S and detuning, ϵ. A complete round-trip through the anti-crossing can be mapped onto an optical interferometer experiment.

contrast to the measurements described above, the voltage pulse is tailored to make the rise-time slow enough that there as an appreciable chance of an adiabatic transition at the S–T_+ anti-crossing. The incoming (0,2)S has some probability amplitude for remaining in the S state, given by $\sqrt{P_{LZ}}$, where $P_{LZ} = e^{\frac{-2\pi\Delta^2}{\hbar\nu}}$ is the Landau–Zener transition probability. Here 2Δ is the energy gap at the anti-crossing and ν is the level velocity. The probability amplitude for a transition to $|T_+\rangle$ is given by $\sqrt{1 - P_{LZ}}$.

Passage through the anti-crossing in the limit of P_{LZ} = 1/2 is equivalent to a 50–50 optical beam splitter. The incoming spin singlet state is converted into a superposition of singlet and triplet states as illustrated in Fig. 15.14a. Detuning is then maintained at a value ϵ_S for a time τ_S. Here the energy difference between S and T_+ results in a σ_z rotation. A sweep back through the S–T_+ anti-crossing interferes the two states and the final singlet state probability, P_S, is determined using spin-to-charge conversion. Resulting interference fringes are shown in Fig. 15.14b. The entire experiment is analogous to an optical interferometer (see inset Fig. 15.14b).

15.5 Outlook

Dramatic progress in quantum science and engineering has taken place over the past ten years. In the case of spin qubits, single electrons have been isolated, detected, and controlled using all-electrical methods. While the dynamics of an isolated electron spin are straightforward, coupling to the environment leads to rich physical complexity that remains to be quantitatively explained. Many challenges exist, both technical and fundamental. We conclude by identifying several areas for future work.

Central-spin problem

An electron spin in a GaAs quantum dot is coupled to $\sim 10^6$ lattice nuclei via the hyperfine interaction, resulting in an Overhauser field of ~2 mT. A typical nuclear spin experiences the Knight field of the electron as well as dipole fields from neighboring nuclei. The large interacting system leads to complicated dynamics, sometimes referred to as the "central-spin problem". Several areas for future work include: (1) developing an accurate model of dynamic nuclear polarization that accounts for longitudinal and transverse hyperfine fields as well as spin–orbit interactions, (2) developing dynamic decoupling pulse sequences that are specifically tailored to the nuclear spin environment, (3) combining conventional nuclear magnetic resonance (NMR) and cooling to actively control nuclear spin dynamics and reach the regime of large polarization, and (4) reducing nuclear spin diffusion.

Scaling and integration

Many of the experiments described here have been performed on different samples, in different labs spread throughout the world. It remains to be seen if all of these methods can be combined into a single experimental setup, allowing the demonstration of simple quantum algorithms using a spin-based quantum processor. The challenge of scaling, while purely technical, is substantial. It necessarily requires coherent coupling of spin qubits.

Spin–orbit control

Recent experiments have demonstrated efficient single-spin control in an InAs nanowire DQD using EDSR [30]. The work suggests that electric fields may be sufficient for quantum control. However, spin-echo resulted in a maximum coherence time of 50 ns, which is several orders of magnitude shorter than the timescales obtained in GaAs quantum dots [30, 31]. The origin of the short coherence time is still not fully understood.

Ultra-coherent materials

Recent developments in the growth of low defect Si/SiGe heterostructures have opened up new possibilities for semiconductor quantum bits. In contrast to GaAs, the naturally abundant isotopes of both Si and Ge possess zero nuclear spin, resulting in a suppressed Overhauser field. Moreover, the spin–orbit coupling is weaker and the crystal is centrosymmetric, which should result in longer spin lifetimes. The few electron regime was recently reached in a lateral quantum dot and a spin lifetime of 6 s was observed in P-doped Si [29].

Acknowledgements

Jason Petta thanks Charlie Marcus for providing the set and setting that enabled many of the early experimental results described here. Petta also acknowledges the support of Arthur Gossard, Loren Pfeiffer, and Mansour Shayegan, who provide the heterostructures upon which these experiments are based. Research at Princeton is supported by the Army

Research Office, DARPA, the National Science Foundation, the Packard Foundation, and the Sloan Foundation.

References

[1] Aleiner, I. L., Brouwer, P. W. and Glazman, L. I. 2002. Quantum effects in Coulomb blockade. *Phys. Rep.*, **358**(5-6), 309–440.

[2] Amasha, S., MacLean, K., Radu, I. P. *et al.* 2008. Electrical control of spin relaxation in a quantum dot. *Phys. Rev. Lett.*, **100**(4), 046803.

[3] Ashoori, R. C., Stormer, H. L., Weiner, J. S. *et al.* 1993. N-electron ground state energies of a quantum dot in a magnetic field. *Phys. Rev. Lett.*, **71**(4), 613–616.

[4] Atatüre, M., Dreiser, J., Badolato, A. *et al.* 2006. Quantum-dot spin-state preparation with near-unity fidelity. *Science*, **312**(5773), 551–553.

[5] Barthel, C., Kjærgaard, M., Medford, J. *et al.* 2010a. Fast sensing of double-dot charge arrangement and spin state with a radio-frequency sensor quantum dot. *Phys. Rev. B*, **81**, 161308.

[6] Barthel, C., Medford, J., Marcus, C. M., Hanson, M. P. and Gossard, A. C. 2010b. Interlaced dynamical decoupling and coherent operation of a singlet–triplet qubit. *Phys. Rev. Lett.*, **105**(26), 266803.

[7] Bluhm, H., Foletti, S., Neder, I. *et al.* 2010. Dephasing time of GaAs electron-spin qubits coupled to a nuclear bath exceeding 200 μs. *Nat. Phys.*, **7**, 109–113.

[8] Ciorga, M., Sachrajda, A. S., Hawrylak, P. *et al.* 2000. Addition spectrum of a lateral dot from Coulomb and spin-blockade spectroscopy. *Phys. Rev. B*, **61**(24), R16315–R16318.

[9] Coish, W. A. and Baugh, J. 2009. Nuclear spins in nanostructures. *Phys. Status Solidi B – Basic Solid State Phys.*, **246**(10), 2203–2215.

[10] D'yakonov, M. I. and Kachorovskii, V. Y. 1986. Spin relaxation of two-dimensional electrons in noncentrosymmetric semiconductors. *Sov. Phys. Semicond.*, **20**, 110–112.

[11] Elzerman, J. M., Hanson, R., Willems van Beveren, L. H. *et al.* 2004. Single-shot read-out of an individual electron spin in a quantum dot. *Nature*, **430**(6998), 431–435.

[12] Field, M., Smith, C. G., Pepper, M. *et al.* 1993. Measurements of Coulomb blockade with a noninvasive voltage probe. *Phys. Rev. Lett.*, **70**(9), 1311–1314.

[13] Fujisawa, T., Austing, D. G., Tokura, Y., Hirayama, Y. and Tarucha, S. 2002. Allowed and forbidden transitions in artificial hydrogen and helium atoms. *Nature*, **419**(6904), 278–281.

[14] Gossard, A. C. 1983. Quantum effects at GaAs/$Al_xGa_{1-x}As$ junctions. *Thin Solid Films*, **104**(3-4), 279–284.

[15] Hanson, R., Witkamp, B., Vandersypen, L. M. K. *et al.* 2003. Zeeman energy and spin relaxation in a one-electron quantum dot. *Phys. Rev. Lett.*, **91**(19), 196802.

[16] Hanson, R., Kouwenhoven, L. P., Petta, J. R., Tarucha, S. and Vandersypen, L. M. K. 2007. Spins in few-electron quantum dots. *Rev. Mod. Phys.*, **79**(4), 1217–1265.

[17] Johnson, A. C., Petta, J. R., Taylor, J. M. *et al.* 2005. Triplet–singlet spin relaxation via nuclei in a double quantum dot. *Nature*, **435**(7044), 925–928.

[18] Khaetskii, A., Loss, D. and Glazman, L. 2003. Electron spin evolution induced by interaction with nuclei in a quantum dot. *Phys. Rev. B*, **67**(19), 195329.

[19] Khaetskii, A. V., Loss, D. and Glazman, L. 2002. Electron spin decoherence in quantum dots due to interaction with nuclei. *Phys. Rev. Lett.*, **88**(18), 186802.

[20] Koppens, F. H. L., Folk, J. A., Elzerman, J. M. *et al.* 2005. Control and detection of singlet-triplet mixing in a random nuclear field. *Science*, **309**(5739), 1346–1350.
[21] Koppens, F. H. L., Buizert, C., Tielrooij, K. J. *et al.* 2006. Driven coherent oscillations of a single electron spin in a quantum dot. *Nature*, **442**(7104), 766–71.
[22] Koppens, F. H. L., Nowack, K. C. and Vandersypen, L. M. K. 2008. Spin echo of a single electron spin in a quantum dot. *Phys. Rev. Lett.*, **100**(23), 4.
[23] Kouwenhoven, L. P., Marcus, C. M., McEuen, L. P. *et al.* 1997. Electron transport in quantum dots. Page 109 of: Sohn, L. L., Kouwenhoven, L. P. and Schön, G. (eds.), *Mesoscopic Electronic Transport*, vol. 345. Springer.
[24] Kroutvar, M., Ducommun, Y., Heiss, D. and Bichler, M. 2004. Optically programmable electron spin memory using semiconductor quantum dots. *Nature*, **432**(November), 81–84.
[25] Kyriakidis, J., Pioro-Ladriere, M., Ciorga, M., Sachrajda, A. S. and Hawrylak, P. 2002. Voltage-tunable singlet-triplet transition in lateral quantum dots. *Phys. Rev. B*, **66**(3), 035320.
[26] Laird, E. A. 2009. *Electrical control of quantum dot spin qubits*. Ph.D. thesis, Harvard University.
[27] Levy, J. 2002. Universal quantum computation with spin-1/2 pairs and Heisenberg exchange. *Phys. Rev. Lett.*, **89**(14), 147902.
[28] Loss, D. and DiVincenzo, D. P. 1998. Quantum computation with quantum dots. *Phys. Rev. A*, **57**(1), 120–126.
[29] Morello, A., Pla, J. J., Zwanenburg, F. A. *et al.* 2010. Single-shot readout of an electron spin in silicon. *Nature*, **467**(7316), 687–691.
[30] Nadj-Perge, S., Frolov, S. M., Bakkers, E. and Kouwenhoven, L. P. 2010. Spin-orbit qubit in a semiconductor nanowire. *Nature*, **468**(7327), 1084–1087.
[31] Nowack, K. C., Koppens, F. H. L., Nazarov, Yu. V. and Vandersypen, L. M. K. 2007. Coherent control of a single electron spin with electric fields. *Science*, **318**(5855), 1430–1433.
[32] Ono, K., Austing, D. G., Tokura, Y. and Tarucha, S. 2002. Current rectification by Pauli exclusion in a weakly coupled double quantum dot system. *Science*, **297**(5585), 1313–1317.
[33] Petta, J. R., Johnson, A. C., Taylor, J. M. *et al.* 2005a. Coherent manipulation of coupled electron spins in semiconductor quantum dots. *Science*, **309**(5744), 2180–2184.
[34] Petta, J. R., Johnson, A. C., Yacoby, A. *et al.* 2005b. Pulsed-gate measurements of the singlet-triplet relaxation time in a two-electron double quantum dot. *Phys. Rev. B*, **72**(16), 161301.
[35] Petta, J. R., Lu, H. and Gossard, A. C. 2010. A coherent beam splitter for electronic spin states. *Science*, **327**(5966), 669–672.
[36] Reilly, D. J., Marcus, C. M., Hanson, M. P. and Gossard, A. C. 2007. Fast single-charge sensing with a rf quantum point contact. *Appl. Phys. Lett.*, **91**(16), 162101.
[37] Schoelkopf, R. J., Wahlgren, P., Kozhevnikov, A. A., Delsing, P. and Prober, D. E. 1998. The radio-frequency single-electron transistor (RF-SET): A fast and ultrasensitive electrometer. *Science*, **280**(5367), 1238–1242.
[38] Schröer, D., Greentree, A. D., Gaudreau, L. *et al.* 2007. Electrostatically defined serial triple quantum dot charged with few electrons. *Phys. Rev. B*, **76**(7), 075306.
[39] Taylor, J. M., Engel, H. A., Dur, W. *et al.* 2005. Fault-tolerant architecture for quantum computation using electrically controlled semiconductor spins. *Nat. Phys.*, **1**(3), 177–183.

[40] Taylor, J. M., Petta, J. R., Johnson, A. C. *et al.* 2007. Relaxation, dephasing, and quantum control of electron spins in double quantum dots. *Phys. Rev. B*, **76**(3), 035315.

[41] van der Wiel, W. G., De Franceschi, S., Elzerman, J. M. *et al.* 2003. Electron transport through double quantum dots. *Rev. of Mod. Phys.*, **75**, 1–22.

[42] Winkler, R., Papadakis, S., De Poortere, E. and Shayegan, M. 2001. Spin–orbit coupling in two-dimensional electron and hole systems. Pages 211–223 of: Kramer, B. (ed.), *Advances in Solid State Physics*. Advances in Solid State Physics, vol. 41. Springer, Berlin/Heidelberg.

16

Theory of electron and nuclear spins in III–V semiconductor and carbon-based dots

H. Ribeiro and G. Burkard

16.1 The magnetic hyperfine Hamiltonian

The dynamics of an electron interacting with an atomic nucleus is mainly governed by the Coulomb interaction. It is this "classical" reasoning that allows us to write the Hamiltonian for the hydrogen atom as the sum of the kinetic energy and the energy of the electrostatic interaction between the electron and the proton. However, this theoretical description of the hydrogen atom cannot entirely explain the observed spectra. Those can only be fully understood in a relativistic model, where for example all the magnetic effects due to the spin of the electron are taken into account. Although the spin-orbit interaction, being one of the relativistic corrections, is important for electron-spin coherence in quantum dots (QDs), we focus on another, equally important, aspect in this chapter: the hyperfine interaction arising from the magnetic coupling between the nuclear and electronic spins.

16.1.1 Electron dynamics in the presence of a nucleus

To understand the origin of the hyperfine Hamiltonian, we will consider as a heuristic guide the hydrogen atom, or the coupling between one nuclear spin and one electron spin. Extended discussions can be found in [1, 2] and a derivation from the Dirac equation in [3].

The dynamics of the two particles with spin, in the absence of an external magnetic field $\boldsymbol{B}$, is described through the Hamiltonian

$$H = \frac{1}{2m}(\boldsymbol{p} - q\boldsymbol{A}_{\mathrm{n}}(\boldsymbol{r}))^2 + qU(r) - g\frac{\mu_{\mathrm{B}}}{\hbar}\boldsymbol{S}\cdot\nabla\times\boldsymbol{A}_{\mathrm{n}}(\boldsymbol{r}), \tag{16.1}$$

where $\boldsymbol{r}$, $\boldsymbol{p}$, and $\boldsymbol{S}$ are, respectively, the position, momentum and spin operator of the electron with mass m_{e} and charge q, g is the Landé factor, and μ_{B} is the Bohr magneton. The quantity $\nabla \times \boldsymbol{A}_{\mathrm{n}}(\boldsymbol{r}) = \boldsymbol{B}_{\mathrm{n}}(\boldsymbol{r})$ is the magnetic field produced by the proton, where the vector potential $\boldsymbol{A}_{\mathrm{n}}(\boldsymbol{r}) = \frac{\mu_0}{4\pi}\frac{\boldsymbol{M}_{\mathrm{n}}\times\boldsymbol{r}}{r^3}$ is expressed using the magnetic dipolar moment of the proton $\boldsymbol{M}_{\mathrm{n}} = \frac{g_{\mathrm{n}}\mu_{\mathrm{n}}}{\hbar}\boldsymbol{I}$. Here g_{n} and μ_{n} are respectively the nuclear Landé factor and magneton, and $\boldsymbol{I}$ the nuclear spin operator.

Since the hyperfine interaction is a small correction to the energy of the hydrogen atom, the terms proportional to $\boldsymbol{A}_{\mathrm{n}}^2(\boldsymbol{r})$ in (16.1) can be safely ignored to obtain the hyperfine Hamiltonian,

Quantum Dots: Optics, Electron Transport and Future Applications, ed. Alexander Tartakovskii. Published by Cambridge University Press.

$$H_{\rm hf} = -\frac{\mu_0 g \mu_{\rm B} g_{\rm n} \mu_{\rm n}}{4\pi\hbar^2}\left(\frac{\boldsymbol{I}\cdot\boldsymbol{L}}{r^3} + 3\frac{(\boldsymbol{I}\cdot\boldsymbol{r})(\boldsymbol{S}\cdot\boldsymbol{r})}{r^5} - \frac{\boldsymbol{I}\cdot\boldsymbol{S}}{r^3} + \frac{8\pi}{3}\boldsymbol{I}\cdot\boldsymbol{S}\delta(\boldsymbol{r})\right). \tag{16.2}$$

The first term in Eq. (16.2) comes from the kinetic energy of Eq. (16.1) and describes the interaction between the orbital motion of the electron, through its angular momentum $\boldsymbol{L} = \boldsymbol{r} \times \boldsymbol{p}$, and the nuclear magnetic moment $\boldsymbol{M}_{\rm n}$. In the literature, it is referred to as *nuclear–orbital interaction*. The other contributions arise from the coupling between the electron spin $\boldsymbol{S}$ and the induced magnetic field of the proton $\boldsymbol{B}_{\rm n}(\boldsymbol{r})$, which can be calculated by choosing $\boldsymbol{M}_{\rm n} = M_{\rm n}\boldsymbol{e}_z$ [1]. The last term of Eq. (16.2) comes from the apparent singularity of $\boldsymbol{B}_{\rm n}(\boldsymbol{r})$ for $\boldsymbol{r} \to 0$ and gives rise to the so-called *contact hyperfine Hamiltonian*. For $|\boldsymbol{r}| > 0$, we obtain a dipole–dipole type interaction between the electronic spin and the nuclear spin, which is referred to as *anisotropic hyperfine interaction*.

16.1.2 Nuclear interactions in quantum dots

A generalization of Eq. (16.2) to the case of many electrons and nuclei can be obtained by summing over all the electrons and nuclei. For simplicity, we consider the case of an electron spin coupled to many nuclear spins. The *nuclear–orbital interaction* becomes

$$H_{\rm hf}^{\rm orb} = -\frac{\mu_0 g \mu_{\rm B}}{4\pi\hbar^2}\sum_k g_{j_k}\mu_{j_k}\frac{\boldsymbol{L}_k\cdot\boldsymbol{I}_k}{|\boldsymbol{r}-\boldsymbol{r}_k|^3}, \tag{16.3}$$

where the sum runs over all the nuclear spins and the index k labels the atomic sites. $\boldsymbol{I}_k$ is the spin operator of the kth nucleus and $\boldsymbol{L}_k = (\boldsymbol{r} - \boldsymbol{r}_k) \times \boldsymbol{p}$ describes the electron orbital angular momentum at the nuclear site $\boldsymbol{r}_k$. The nuclear magneton and Landé factor are written with an index j_k, which labels the nuclear species at site k.

The nuclear–orbital interaction plays an important role in the dynamics only when the electron has a nonzero angular momentum. This typically happens when the electron (or hole) occupies a p-type valence band [4, 5, 6, 7].

The *anisotropic hyperfine interaction* is best written as

$$H_{\rm hf}^{\rm a} = \int {\rm d}\boldsymbol{r}\sum_k \boldsymbol{S}^{\rm T}(\boldsymbol{r})\cdot\boldsymbol{T}_k(\boldsymbol{r})\cdot\boldsymbol{I}_k, \tag{16.4}$$

where the tensor $\boldsymbol{T}_k(\boldsymbol{r})$ is defined component-wise by

$$T_k^{ab}(\boldsymbol{r}) = -\frac{\mu_0 g \mu_{\rm B} g_{j_k}\mu_{j_k}}{4\pi\hbar^2}\left(\frac{-\delta_{ab}}{|\boldsymbol{r}-\boldsymbol{r}_k|^3} + 3\frac{(r^a - r_k^a)(r^b - r_k^b)}{|\boldsymbol{r}-\boldsymbol{r}_k|^5}\right), \tag{16.5}$$

where $a, b \in \{x, y, z\}$ and we have introduced the electron spin density operator

$$\boldsymbol{S}(\boldsymbol{r}) = \frac{1}{2}\sum_{s,s'=\uparrow,\downarrow}\psi_s^\dagger(\boldsymbol{r})\boldsymbol{\sigma}_{ss'}\psi_{s'}(\boldsymbol{r}), \tag{16.6}$$

written in terms of field operators $\psi_s(\boldsymbol{r}) = \sum_l \varphi_l(\boldsymbol{r})c_{ls}$, with c_{ls} the annihilation operator of an electron with spin s and single-particle orbital $\varphi_l(\boldsymbol{r})$, and $\boldsymbol{\sigma}_{ss'} = \langle s|\boldsymbol{\sigma}|s'\rangle$, where $\boldsymbol{\sigma}$ is a vector of Pauli matrices.

The anisotropic hyperfine Hamiltonian, along with the orbital hyperfine Hamiltonian, constitute the main interaction between the spin of electrons occupying a p-orbital and a nuclear spin. As we will see later, it is also the relevant interaction for electrons occupying the π-orbital of graphene and carbon nanotubes [8, 9].

The remaining term, the *hyperfine contact interaction*, plays a major role in the dynamics of materials where the electron occupies a s-type orbital. This interaction is responsible for the loss of coherence of a conduction band electron-spin confined in GaAs quantum dots (QD), as well as for the leakage current in transport experiments in double quantum dot (DQD) setups.

The generalization to many nuclear spins is obtained again by substituting the spin operator in Eq. (16.2) by the spin density operator in Eq. (16.6), which yields

$$H_{\mathrm{hf}}^{\mathrm{c}} = \frac{8\pi}{3}\frac{\mu_0 g \mu_{\mathrm{B}}}{4\pi\hbar^2}\sum_k g_{j_k}\mu_{j_k}\boldsymbol{S}(\boldsymbol{r}_k)\cdot\boldsymbol{I}_k. \tag{16.7}$$

Since a confined electron in a III–V semiconductor quantum dot consists of s-wave atomic orbitals, an effective spin Hamiltonian describing the coupling between one electron spin and many nuclear spins can be derived from Eq. (16.7). This is possible if the electron orbital levels are such that $\Delta E^{\mathrm{orb}} \gg k_{\mathrm{B}}T$, $\langle H_{\mathrm{hf}}\rangle$, where $\langle H_{\mathrm{hf}}\rangle$ denotes the magnitude of the hyperfine coupling. The condition $\Delta E^{\mathrm{orb}} \gg k_{\mathrm{B}}T$ is fulfilled at low temperatures where the wave function of a single electron can be expressed in the envelope-function approximation as $\varphi_0(\boldsymbol{r}) = \sqrt{v_0}u(\boldsymbol{r})\psi_0(\boldsymbol{r})$, when neglecting any possible valley degeneracy. Here v_0 is the atomic volume, $u(\boldsymbol{r})$ is the $\boldsymbol{k}=0$ Bloch amplitude, and $\psi_0(\boldsymbol{r})$ is the ground-state envelope function. The projection of Eq. (16.7) onto the orbital ground-state $\varphi_0(\boldsymbol{r})$ provides the effective spin Hamiltonian

$$H_{\mathrm{eff,\,hf}}^{\mathrm{c}} = \langle\varphi_0|H_{\mathrm{hf}}^{\mathrm{c}}|\varphi_0\rangle = \sum_k A_k\boldsymbol{S}\cdot\boldsymbol{I}_k = \boldsymbol{S}\cdot\boldsymbol{h}, \tag{16.8}$$

which is commonly used to describe the electron–nuclear spin system dynamics in the conduction band of III–V semiconductors. In Eq. (16.8) we introduce the Overhauser field operator $\boldsymbol{h} = \sum_k A_k\boldsymbol{I}_k$ with inhomogeneous hyperfine coupling constant

$$A_k = A_{j_k} v_0\,|\psi_0(\boldsymbol{r}_k)|^2\,, \tag{16.9}$$

where $A_{j_k} = \frac{8\pi}{3}\frac{\mu_0 g\mu_{\mathrm{B}} g_{\mathrm{n}_{j_k}}\mu_{\mathrm{n}_{j_k}}}{4\pi\hbar^2}|u(\boldsymbol{r}_k)|^2$ is the total hyperfine coupling constant for a nucleus of species j at position $\boldsymbol{r}_k$. Since the nuclear Landé factor $g_{\mathrm{n}_{j_k}}$ depends on the nuclear species, it can either be positive or negative and consequently the hyperfine coupling constant can take either sign.

In the literature, Eq. (16.9) often reads $A_k = Av_0|\psi_0(\boldsymbol{r}_k)|^2$, where A is an average hyperfine coupling constant. For GaAs, this is usually $A \approx 90\,\mu\mathrm{eV}$.

16.2 Nuclear – nuclear interactions

The interaction among nuclei results from the coupling of a given nuclear spin at position $\boldsymbol{r}_k$ to the effective magnetic field generated by the remaining nuclear spins. The derivation of the interaction Hamiltonian relies on an expansion technique known as *multipole expansion* (see [2, 10][1]).

16.2.1 Nuclear dipolar interaction

The dipole–dipole Hamiltonian for nuclear spins is given by

$$H_{\mathrm{dd}}^{\mathrm{n}} = \sum_{k \neq l} \boldsymbol{I}_k^{\mathrm{T}} \cdot \boldsymbol{T}_{kl}^{\mathrm{n}} \cdot \boldsymbol{I}_l, \tag{16.10}$$

where the components of the tensor $\boldsymbol{T}_{kl}^{\mathrm{n}}$ are

$$T_{kl}^{\mathrm{n},\, ab} = \frac{\mu_0 g_{j_k} g_{j_l} \mu_{j_k} \mu_{j_l}}{4\pi \hbar^2} \frac{1}{2\,|\boldsymbol{r}_k - \boldsymbol{r}_l|^3} \left(\delta_{ab} - 3 \frac{(r_k^a - r_l^a)(r_k^b - r_l^b)}{|\boldsymbol{r}_k - \boldsymbol{r}_l|^2} \right). \tag{16.11}$$

It has been shown [11, 12] that the nuclear dipolar interaction in combination with the hyperfine contact interaction is also a source of decoherence for the electron spin through spectral diffusion. This can be easily understood if an external magnetic field of a few milliteslas is applied. In this case the secular approximation can be made to derive an effective Hamiltonian from Eqs. (16.10) and (16.11),

$$H_{\mathrm{sec,\,dd}}^{\mathrm{n}} = \frac{\mu_0}{4\pi \hbar^2} \left(\sum_{k \neq l} \Xi_{kl} I_k^z I_l^z - \frac{1}{2} \sum_{\substack{k \neq l \\ (j_k = j_l)}} \Xi_{kl} I_k^+ I_l^- \right), \tag{16.12}$$

where $\Xi_{kl} = g_{j_k} g_{j_l} \mu_{j_k} \mu_{j_l} \frac{1 - 3\cos^2(\theta_{kl})}{2|\boldsymbol{r}_k - \boldsymbol{r}_l|^3}$. The angle θ_{kl} is taken between the external magnetic field and the vector $\boldsymbol{r}_k - \boldsymbol{r}_l$, and $I_k^{\pm} = I_k^x \pm \mathrm{i} I_k^y$ are the ladder operators of the kth nuclear spin. The condition $j_k = j_l$ in the second sum restricts the summation over pairs of sites with the same nuclear species. Even though Eq. (16.12) preserves the total magnetization of the nuclear system, it describes fluctuations that affect the coherence of an electron spin through the so-called flip-flop terms $I_k^+ I_l^-$. These phenomena are discussed in more detail in [13].

16.2.2 Nuclear quadrupolar interaction

Nuclei with spin $I > 1/2$ can have a finite electric moment. This means that their spin couples to the gradient of a nonuniform electrostatic potential $V(\boldsymbol{r})$ [2, 10] and gives rise to the quadrupolar Hamiltonian

[1] In [10], the multipole expansion is illustrated by considering a scalar potential, but the concept is applicable to the vector potential.

$$H_{\text{quad}}^{\text{n}} = \sum_k \sum_{a,b} \left[\left\langle \frac{\partial^2 V(\boldsymbol{r})}{\partial x^a \partial x^b} \right\rangle_{\text{e}} \right]_{\boldsymbol{r}=\boldsymbol{r}_k} Q_k^{ab}, \tag{16.13}$$

where $\langle \cdots \rangle_{\text{e}}$ denotes the expectation value with respect to the electron system evaluated at $\boldsymbol{r} = \boldsymbol{r}_k$ and the quadrupolar tensor is given by

$$Q_k^{ab} = q Q_{j_k} \frac{\frac{3}{2}\left(I_k^a I_k^b + I_k^b I_k^a\right) - \delta_{ab} I_{j_k}\left(I_{j_k}+1\right)}{6 I_{j_k}\left(2 I_{j_k} - 1\right)}. \tag{16.14}$$

The quantities I_{j_k} are the total spin values of the nuclear species at site $\boldsymbol{r}_k$, q the electron charge, and Q_{j_k} is the quadrupole moment.

The quadrupolar interaction is important in strained systems, doped systems or systems with defects where the gradient of the electric field is nonzero at $\boldsymbol{r}_k$. It results in a level splitting in nanostructures that allows for quantum transitions with a change in magnetization $\Delta m = \pm 2$ [14, 15]. It has also been measured in a GaAs 2DEG as a shift of the nuclear spin resonance line of $\gtrsim$10 kHz [16].

Assuming that the potential $V(\boldsymbol{r})$ has an axial symmetry along $\hat{\boldsymbol{n}}$ [2], an effective quadrupolar Hamiltonian in the presence of an external magnetic field can be derived by using the secular approximation,

$$H_{\text{sec, quad}}^{\text{n}} = \frac{1}{4} \sum_k A_{\text{Q},k} f(\theta) \left((I_k^z)^2 - \frac{1}{3} I_{j_k}(I_{j_k}+1) \right). \tag{16.15}$$

Here $A_{\text{Q},k} = \frac{3q\langle \hat{\boldsymbol{n}}\cdot\nabla(\hat{\boldsymbol{n}}\cdot\nabla V(\boldsymbol{r}))\rangle|_{\boldsymbol{r}=\boldsymbol{r}_k} Q_{j_k}}{8 I_{j_k}(2I_{j_k}-1)}$ is the quadrupolar coupling strength and $f(\theta) = 3\cos^2(\theta) - 1$. Even though the magnitude of the quadrupolar interaction given by Eq. (16.15) is small compared to the other couplings discussed earlier, its effect is non-negligible since in a magnetic field it gives rise to non-equidistant spacings of the nuclear Zeeman levels. This property has been used to implement full control of a single spin, since it allows selective addressing of the transitions $m = 1/2 \leftrightarrow -1/2$ and $m = 1/2 \leftrightarrow 3/2$ with different excitation frequencies and design of quantum gates [17]. It can also lead to the suppression of the flip-flop processes due to the dipolar coupling (Eq. (16.12)) when the difference of the quadrupolar coupling strengths between two nuclear spins at sites k and l is bigger than the dipolar coupling strength [18].

16.3 Hyperfine interaction in semiconductor-based quantum dots

Quantum dots have become an attractive field of research for quantum information processing, after the proposal by Loss and DiVincenzo [19] to encode a qubit (the quantum counterpart of a classical bit [20]) in the spin of an electron confined in a quantum dot. Although astonishing experimental progress has been made in measuring and controlling electronic spins [21, 22], a quantum processor with a large number of spin qubits has yet to be built. Qubits in semiconductor-based quantum dots, as in any other real system, suffer from decoherence.[2] Decoherence is a physical phenomenon through which a quantum system, which is not isolated from the environment, loses its capacity to coherently interfere and starts behaving like a classical system [23]. In the case of an electron spin confined in a semiconductor-based quantum dot, the main cause of decoherence is the hyperfine interaction with the host nuclear spins [24, 25].

16.3.1 The double quantum dot system

The Hamiltonian describing confined electrons in a double quantum dot defined in a two-dimensional electron gas (2DEG) in the presence of a magnetic field perpendicular to the 2DEG plane is given by

$$H_0 = \sum_{\substack{i=1,2\\ \sigma=\uparrow,\downarrow}} \left(\varepsilon_i + \frac{1}{2} g^* \mu_B \sigma\right) c_{i\sigma}^\dagger c_{i\sigma} + u \sum_i c_{i\uparrow}^\dagger c_{i\uparrow} c_{i\downarrow}^\dagger c_{i\downarrow} + \tau \sum_\sigma \left(c_{1\sigma}^\dagger c_{2\sigma} + \text{h.c.}\right). \tag{16.16}$$

The indices $i = 1, 2$ and $\sigma = \uparrow, \downarrow \equiv \pm 1$ label the dot number and spin respectively. The operators $c_{i\sigma}$ and $c_{i\sigma}^\dagger$ annihilate and create an electron in dot i with spin σ, g^* denotes the effective Landé g-factor. The first term of H_0 accounts for the single-particle energy of the confined electrons in the presence of a magnetic field, the second for the intra-dot Coulomb energy u when two electrons occupy the same QD, and the last for the electron tunneling with strength τ between the dots.[3]

In most recent experiments the DQD system is operated in a regime with at most two electrons, which is achieved by tuning external gate voltages to adjust the confinement potential and the chemical potential of the leads, thus we first concentrate on the states with exactly two electrons. The projection of Eq. (16.16) into the subspace spanned by the charge configurations (0, 2), (2, 0), and (1, 1) and its diagonalization leads to six low-energy states, which are the superpositions of the singlets S(0, 2), S(2, 0), and S(1, 1) as well as the triplets $T_0(1, 1)$, $T_+(1, 1)$, and $T_-(1, 1)$, where (l, r) denotes the number of electrons in the left and right dot. The triplet states with two particles in the same dot must have electrons occupying different orbitals levels (Pauli principle). Consequently their energies are much higher than the six previously listed states and they can be safely neglected in the following discussions.

The degeneracy between the spin singlet S and spin triplet T_0 is lifted due to the exchange interaction [26]. It has been shown that the exchange coupling allows to generate a SWAP gate (exchanges the states of two qubits, c.f. Section 15.4.8). In addition, it is also possible to build a $\sqrt{\text{SWAP}}$ gate, which in combination with proper single qubit operations allows to construct a CNOT gate (controlled-NOT) [19]. The degeneracy of the singlets S(1, 1) and S(2, 0) at $\varepsilon = u$, as well as between S(1, 1) and S(0, 2) at $\varepsilon = -u$, is lifted due to tunneling. The detuning $\varepsilon = \varepsilon_1 - \varepsilon_2$ of the DQD can be adjusted with external

[2] See also Sections 15.4.4, 15.4.5, 15.4.6 and 15.4.7.
[3] See also Section 15.4.1.

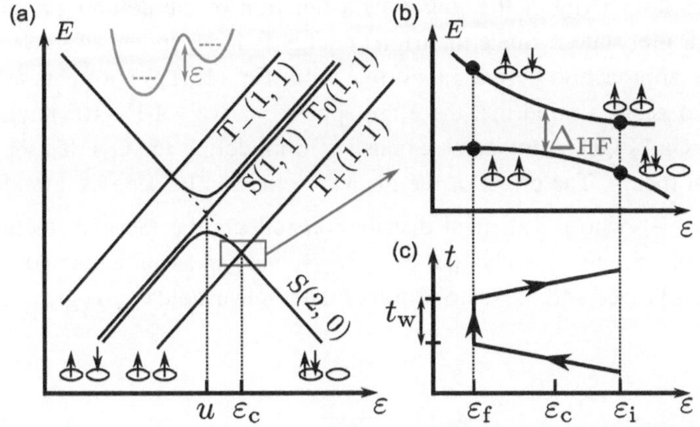

Figure 16.1 (a) Schematic representation of the double quantum dot energy diagram as a function of the detuning $\varepsilon = \varepsilon_1 - \varepsilon_2$. (b) The avoided crossing of the electronic singlet S and T_+ due to the hyperfine interaction with the host nuclei. (c) Perfect square pulse used to manipulate the singlet S and T_+ states (c.f. Section 16.3.5). Figure from Ref. [52].

gate voltages. A schematic representation of the energy levels is shown in Fig. 16.1, where we already emphasize the particular point $\varepsilon = \varepsilon_c$ corresponding to the avoided crossing of the singlet and triplet T_+.

16.3.2 Transport through a double quantum dot system

Let us consider the transport cycle $(0, 1) \to (1, 1) \to (0, 2) \to (0, 1)$, which has been experimentally realized e.g. in [27] and many others, and theoretically studied e.g. in [28]. The first and last processes are incoherent tunnel transitions from and to the leads respectively. The process of interest is the transition $(1, 1) \to (0, 2)$ due to the tunnel coupling between the dots. Such a transition can only occur if the (1, 1) state is the singlet S(1, 1), assuming that tunneling preserves the spin and the only available spin state with a (0, 2) charge configuration is the singlet S(0, 2). If the system is in any (1, 1)-triplet state, there is no transition possible to the final (0, 2) state and thus no current flows through the DQD. This mechanism is known as spin blockade [29].

Even though this reasoning is correct, it does not correspond to the experimentally observed average current, which is nonzero when the applied external magnetic field does not exceed $\approx 10\,\mathrm{mT}$ (in GaAs DQDs). The leakage current arises from the mixing of the triplet states with the singlet states due to the contact hyperfine interaction of the electron spin with the nuclear spins, which was omitted in our previous attempt of an explanation.

The effective Hamiltonian describing the DQD can be written as [28]

$$H_{\mathrm{st}} = E_{\mathrm{S}}(\varepsilon)|\mathrm{S}\rangle\langle\mathrm{S}| + \sum_{i=+,0,-} E_{\mathrm{T}_i}|\mathrm{T}_i\rangle\langle\mathrm{T}_i| + \sqrt{2}\tau(|\mathrm{S}(1,1)\rangle\langle\mathrm{S}(0,2)| + \mathrm{h.c.}) + \sum_{j=1,2} \boldsymbol{S}_j \cdot \boldsymbol{h}_j, \tag{16.17}$$

where $E_S(\varepsilon)$ is the energy of the singlet as a function of the detuning ε, and E_{T_i} is the energy of the triplet state i. Since the triplet energy is independent of ε, we set $E_{T_0} = 0$. A microscopic approach to derive the current from Eq. (16.17) is of course possible, but one has then to keep in mind that there are approximatively 10^5–10^6 nuclear spins in a typical lateral GaAs QD. Another way consists in modeling the Overhauser fields within a semiclassical theory. The effect of the nuclear spins can be seen as a random magnetic field $\mathbf{B}_{n,j} = \frac{\boldsymbol{h}_j}{g^*\mu_B}$ whose statistical distribution reflects the thermal fluctuations of the nuclear ensemble. Since typically $k_B T \gg g_n \mu_n B$, one can assume the nuclear spins to be completely unpolarized and the distribution of the nuclear fields to be given by a Gaussian distribution

$$p(B_{n,i}) = \frac{1}{\sqrt{2\pi}\sigma_n} e^{-\frac{B_{n,i}^2}{2\sigma_n^2}}, \tag{16.18}$$

where $\sigma_n \propto \sqrt{N}$ is the standard deviation of the random field $B_{n,i}$ created by N nuclear spins.

To further simplify the problem, one can assume that the dynamics of the nuclear fields is frozen. The timescale at which the nuclear fields vary is on the order of the nuclear spin relaxation, $\tau_{relax} \simeq 1$ s. This is a few orders of magnitude larger than the timescales associated with the electron dynamics in the transport regime.

In this framework, the effective Hamiltonian describing the spin transport is given by [28]

$$\begin{aligned} H_{st}^{eff} = {} & E_S(\varepsilon)|S\rangle\langle S| + \sum_{i=+,-} E_{T_i}|T_i\rangle\langle T_i| + \sqrt{2}\tau(|S(1,1)\rangle\langle S(0,2)| + \text{h.c.}) \\ & + g^*\mu_B \Big(\delta B_n^z |S(1,1)\rangle\langle T_0| + \frac{B_n^+}{\sqrt{2}}|T_0\rangle\langle T_+| + \frac{B_n^-}{\sqrt{2}}|T_0\rangle\langle T_-| \\ & - \frac{\delta B_n^+}{\sqrt{2}}|S(1,1)\rangle\langle T_+| + \frac{\delta B_n^-}{\sqrt{2}}|S(1,1)\rangle\langle T_-| + \text{h.c.} \Big). \end{aligned} \tag{16.19}$$

Here, the scalar product appearing in Eq. (16.17) has been expanded and the effective difference field $\delta\mathbf{B} = (\mathbf{B}_{n,1} - \mathbf{B}_{n,2})/2$ and average field $\mathbf{B} = (\mathbf{B}_{n,1} + \mathbf{B}_{n,2})/2$ have been introduced, as well as $B_n^{\pm} = B_n^x \pm iB_n^y$ and $\delta B_n^{\pm} = \delta B_n^x \pm i\delta B_n^y$. The energy of the triplet states include the z-component of the total Overhauser field, $E_{T_\pm} = \pm g^*\mu_B B_{tot}^z$.

The current as a function of the external magnetic field B can be computed from Eq. (16.19) with the help of a master equation for the density matrix [30, 31, 32]. The expression for the average current is [28]

$$\begin{aligned} \langle I\rangle = q\Gamma_{in} \Bigg[& \frac{2}{3}\left(2\frac{B_n^2}{B^2} - \frac{B_n^4}{B^4}\right) + \frac{2\sqrt{2\pi}}{3\sqrt{3}}\left(\frac{B_n^5}{B^5} - \frac{B_n^3}{B^3}\right) e^{-\frac{3B^2}{2B_n^2}} \operatorname{erfi}\left(\frac{\sqrt{3}}{\sqrt{2}}\frac{B}{B_n}\right) \\ & - \frac{\pi}{9}\frac{B_n^6}{B^6} e^{-\frac{3B^2}{B_n^2}} \operatorname{erfi}^2\left(\frac{\sqrt{3}}{\sqrt{2}}\frac{B}{B_n}\right) \Bigg], \end{aligned} \tag{16.20}$$

where q is the charge of the electron, Γ_{in} is the rate for an inelastic transition between $|\text{S}(0, 2)\rangle$ and $|\text{S}(1, 1)\rangle$, $\text{erfi}(x)$ is the imaginary error function, and $g^*\mu_\text{B} B_\text{n} = \langle \boldsymbol{h}_{\text{n},i}\rangle_{\text{rms}} = \sqrt{\langle \boldsymbol{h}^2_{\text{n},i}\rangle - \langle \boldsymbol{h}_{\text{n},i}\rangle^2}$ is the root-mean-square expectation value of the Overhauser field operator. The functional form of Eq. (16.20) shows that if the ratio $\frac{B_\text{n}}{B} \to 0$ then the current is suppressed, $\langle I\rangle \to 0$, because the mixing mechanism between the triplet and singlet states becomes inefficient due to the dominant Zeeman splitting. But if the Zeeman splitting is small compared to the hyperfine energy scale, $\frac{B_\text{n}}{B} \gtrsim 1$, there is admixture of the triplet and singlet states resulting in $\langle I\rangle \neq 0$.

16.3.3 Decoherence of the S $-$ T$_0$ *qubit*

After the original proposal to use one electronic spin confined in a QD as a qubit [19], it was subsequently proposed to encode a qubit with two electron spins confined in a DQD [33]. This setup can be operated in a regime of detunings, where the singlet S and triplet $\text{T}_0(1, 1)$ can be seen to a good approximation as an isolated two-level system, whose manipulation can be achieved via the exchange coupling. Since the total magnetization of this subspace is zero, $S^z_{\text{tot}} = S^z_1 + S^z_2 = 0$, it was thought that a DQD with symmetric dots would exhibit a reduced coupling to the environment and consequently longer coherence times than the original proposed qubit. But it was shown experimentally [34] and theoretically [35] that because of the uncorrelated noise produced by the nuclear spins in each QD this system still suffers from short coherence times.

To obtain the effective qubit Hamiltonian, Eq. (16.16) is projected onto the subspace spanned by the four (1, 1) spin states. The resulting Hamiltonian including hyperfine interaction is given up to a constant by [35]

$$H^{\text{eff}}_{(1,\,1)} = g^*\mu_\text{B} B S^z + \frac{J}{2}\boldsymbol{S}\cdot\boldsymbol{S} + \boldsymbol{S}\cdot\boldsymbol{h} + \delta\boldsymbol{S}\cdot\delta\boldsymbol{h}, \tag{16.21}$$

where $\boldsymbol{S} = \boldsymbol{S}_1 + \boldsymbol{S}_2$, $\boldsymbol{h} = \boldsymbol{h}_1 + \boldsymbol{h}_2$, $S^z = S^z_1 + S^z_2$, $\delta\boldsymbol{S} = \boldsymbol{S}_1 - \boldsymbol{S}_2$, and $\delta\boldsymbol{h} = \boldsymbol{h}_1 - \boldsymbol{h}_2$. Here J is the Heisenberg exchange coupling between two spins, which can be assumed positive without loss of generality. If the magnetic field is such that $B \gg \max\{\langle\delta\boldsymbol{h}\rangle_{\text{rms}}/g^*\mu_\text{B}, \langle\boldsymbol{h}\rangle_{\text{rms}}/g^*\mu_\text{B}\}$ and $B \gg J/g^*\mu_\text{B}$, then Eq. (16.21) is to a good approximation block diagonal. Each block corresponds to an eigenspace of the total spin S^z. For an unpolarized nuclear state in a GaAs DQD, a magnetic field $B \approx 10\,\text{mT}$ easily fulfills the previous requirements. The projection of Eq. (16.21) into the $S^z = 0$ subspace is given for $B \gg B_\text{n}$ by $H^{\text{eff}}_{\text{S}-\text{T}_0} = \frac{J}{2}\boldsymbol{S}\cdot\boldsymbol{S} + \delta h^z \delta S^z$ [35], which can be written using the vector of Pauli matrices $\boldsymbol{\tau} = (\boldsymbol{\tau}_x, \boldsymbol{\tau}_y, \boldsymbol{\tau}_z)$ for the qubit states $|\text{S}\rangle \equiv |\tau_z = -1\rangle$ and $|\text{T}_0\rangle \equiv |\tau_z = +1\rangle$ as

$$H_{\text{qubit}} = \frac{J}{2}(1 + \tau_z) + \delta h^z \tau_x. \tag{16.22}$$

A remarkable property of Eq. (16.22) is that its eigenstates are product states between a superposition of the electronic states and a nuclear eigenstate of the operator δh^z. This implies that within the $B \gg B_{\mathrm{n}}$ approximation the electron does not induce any dynamics to the nuclei.

The hyperfine-induced electron spin dynamics can be best studied through the correlator [35]

$$C_{\mathrm{T_0S}}(t) = \sum_{\mathrm{n}} \rho_{\mathrm{I}}(n) \left| \langle n| \otimes \langle \mathrm{T_0}| \mathrm{e}^{-\mathrm{i}H_{\mathrm{qubit}}t} |\mathrm{S}\rangle \otimes |n\rangle \right|^2, \tag{16.23}$$

which gives the probability to find at time $t > 0$ the electron spins in the state $|\mathrm{T_0}\rangle$ knowing that at time $t = 0$ the system was in the state $|\psi(0)\rangle = |\mathrm{S}\rangle \otimes |\psi_{\mathrm{I}}\rangle$, with $|\psi_{\mathrm{I}}\rangle$ an arbitrary superposition of δh^z eigenstates and $\rho_{\mathrm{I}}(n)$ the probability of being in the eigenstate $|n\rangle$.

The correlator (16.23) has been evaluated for different nuclear states in [35]. We discuss here the case where the eigenvalues of δh^z have a Gaussian distribution with mean δh_0 and variance σ_0. The asymptotic behavior of $C_{\mathrm{T_0S}}$ saturates to a finite value [35]

$$C_{\mathrm{T_0S}}(\infty) \sim \begin{cases} \frac{1}{2} - \frac{1}{8}\left(\frac{J}{\delta h_0}\right)^2, & \sigma_0,\ J \ll \delta h_0, \\ 2\left(\frac{\delta h_0}{J}\right)^2, & \sigma_0 \ll \delta h_0 \ll J, \end{cases} \tag{16.24}$$

which differs from a semiclassical calculation, $C_{\mathrm{T_0S}}^{\mathrm{semicl}} = 1/2$. The short-time behavior $t \ll J/4\sigma_0^2$ of the correlator exhibits a Gaussian decay with a timescale $t_{\mathrm{G}} \approx J/4\delta h_0 \sigma_0$ when $J \gg \max(\delta h_0, \sigma_0)$ [35]

$$C_{\mathrm{T_0S}}(t) \sim C_{\mathrm{T_0S}}(\infty) - 2\frac{\sigma_0^2 + \delta h_0^2}{J^2} \mathrm{e}^{-\frac{1}{2}\left(\frac{t}{t_{\mathrm{G}}}\right)^2} \cos\left[\left(J + \frac{2\delta h_0^2}{J}\right)t\right]. \tag{16.25}$$

The long-time behavior $t \gg J/4\sigma_0^2$, also only valid for $J \gg \max(\delta h_0, \sigma_0)$, is not governed by a Gaussian, but a power law decay [35]

$$C_{\mathrm{T_0S}}(t) \sim C_{\mathrm{T_0S}}(\infty) - \mathrm{e}^{-\left(\frac{\delta h_0}{\sqrt{2}\sigma_0}\right)^2} \frac{\cos\left(Jt + \frac{3\pi}{4}\right)}{4\sigma_0\sqrt{J}t^{\frac{3}{2}}}. \tag{16.26}$$

The acquired universal phase shift $3\pi/4$ of the correlator is consistent with the experimental findings reported in [36]. The results presented in [35] show that even in the case of a static nuclear system, there is decoherence due to quantum fluctuations of the nuclear spins.

16.3.4 Nuclear spin state manipulation

Different methods have been developed to reduce the fluctuations of the nuclear state. Ideas have been put forward to project it into approximate Overhauser eigenstates, i.e. eigenstates of the operator h_z or δh_z, either via electrical or optical control [37, 38]. Another method consists in polarizing the nuclear spins [26, 39], but it was shown that a sizeable enhancement of the coherence time would only be possible for a degree of polarization bigger than 99% [25], which has not been achieved yet.

In a DQD setup, the avoided crossing between the singlet and triplet state T_+ (c.f. Fig. 16.1) *a priori* allows to design a nuclear spin polarization method. An adiabatic transition from a spin singlet S(2, 0) to a spin triplet $T_+(1, 1)$ is possible due to the hyperfine interaction. Since angular momentum is conserved, the electronic spin-flip must be accompanied by a nuclear spin-flop. The following cycle $S(2,0) \xrightarrow{\text{adiabatic}} T_+(1,1) \xrightarrow{\text{non-adiabatic}} \tilde{T}_+$, where $\tilde{T}_+$ is the triplet state T_+ for a detuning $\varepsilon > \varepsilon_c$, has been experimentally studied [40]. However, as it will be explained below and experimentally observed, it is impossible to be fully adiabatic while going through the avoided crossing, which readily limits the achievable polarization. Nevertheless, the cycling itself induces a weak measurement which narrows the nuclear spin state, and thus prolongs the coherence time [41]. The microscopic theory describing such transitions in semiconductor quantum dots can be built from the Landau–Zener–Stückelberg–Majorana theory for partially adiabatic transitions.[4] The original formulation of the problem considers a two-level system whose energy depends on a time-varying controllable external parameter. As a function of time, the energy levels exhibit an avoided crossing, due to a stationary interaction H_{int}, through which the system can be swept from an initial state. This passage can be accompanied by a change of populations and relative phase between the states. The LZSM formula, $P_{\text{LZSM}} = \exp\left(-2\pi\, |\langle 1|H_{\text{int}}|0\rangle|^2 / \alpha\hbar\right)$, expresses the non-adiabatic transition probability for a system driven from $t_{\text{i}} = -\infty \to t_{\text{f}} = \infty$ and in which the difference in energy between the states is linearly proportional to time, $\Delta E = \alpha t$. The parameter α is known as the Landau–Zener (LZ) speed, and corresponds to the speed at which the levels are detuned. To describe the nuclear polarizing cycle, there are two important extensions that have to be made to the original model. First, a finite-time theory [46] is more suitable to describe the system, because the pulses have a finite duration, and because of the fluctuations of the Overhauser fields, the position of the avoided crossing, which sets $t = 0$, is not well defined. Second, the theory must be extended to a many-level problem with degeneracies [47]. The number of nuclear spins per quantum dot is about a million, which results in a highly degenerate nuclear state.[5] A review about LZSM physics with emphasis on superconducting qubits can be found in [48].

The projection of Eq. (16.16) near the avoided crossing into the subspace spanned by $T_+(1, 1)$, $S(1, 1)$, and $S(2, 0)$ yields the effective electronic Hamiltonian [52]

$$H_{\text{S,T}_+} \simeq \begin{pmatrix} g^*\mu_{\text{B}}B & 0 & 0 \\ 0 & 0 & \sqrt{2}\tau \\ 0 & \sqrt{2}\tau & u-\varepsilon \end{pmatrix}. \tag{16.27}$$

Here an additive term $\propto \varepsilon \mathbb{1}$ has been neglected. Only the lowest hybridized singlet state $|\text{S}\rangle = c(\varepsilon)|\text{S}(1,1)\rangle + \sqrt{1-c(\varepsilon)^2}|\text{S}(2,0)\rangle$ plays a role in the dynamics. Thus, Eq. (16.27) can be further reduced to [52]

[4] The problem is only widely known in literature as Landau–Zener problem, but it was historically solved at the same time by Landau [42], Zener [43], Stückelberg [44], and Majorana [45] (LZSM).

[5] The nuclear state is degenerate if one neglects the nuclear Zeeman, the longitudinal hyperfine, and the quadrupolar splittings, which are orders of magnitude smaller than the other energy scales considered.

$$H_0(\varepsilon) = E_S(\varepsilon)|S\rangle\langle S| + E_{T_+}|T_+\rangle\langle T_+|, \tag{16.28}$$

where $|S\rangle\langle S|$ is the projector onto the lowest hybridized singlet state obtained by diagonalization of Eq. (16.27), with $c(\varepsilon) = (-u + \varepsilon - \eta)/\sqrt{8\tau^2 + (u - \varepsilon + \eta)^2}$, energy $E_S(\varepsilon) = (u - \varepsilon - \eta)/2$, $\eta = \sqrt{8\tau^2 + (u - \varepsilon)^2}$, and $E_{T_+} = g^*\mu_B B$. Including the hyperfine interaction to Eq. (16.28), a time-dependent bias $\varepsilon = \varepsilon(t)$, and describing the nuclear state with angular momentum eigenstates $|j_1, m_1, j_2, m_2\rangle$ where j_i is the total nuclear magnetic quantum number in dot i and m_i its projection along B, we obtain a microscopic Hamiltonian describing the dynamics at the avoided S$-$T$_+$ crossing [41]

$$H(t) = \sum_{q,\chi} E_{q,\chi}(\varepsilon(t))|q,\chi\rangle\langle q,\chi| + \frac{1}{2}\sum_{i=1,2}\left(S_i^+ h_i^- + S_i^- h_i^+\right), \tag{16.29}$$

with $q = $ S, T$_+$, $|\chi\rangle = |j_1, m_1, j_2, m_2\rangle$, $E_{S,\chi}(\varepsilon) = E_S(\varepsilon)$, and $E_{T_+,\chi}(\varepsilon) = g^*\mu_B B + \langle h_1^z + h_2^z\rangle_\chi/2$.

The dynamics induced by the Hamiltonian, Eq. (16.29), allows us to study the evolution of the fluctuations of the difference Overhauser field operator δh^z responsible for the S$-$T$_0$ qubit decoherence. It was identified in [41] by computing the time evolution of the nuclear spin bath with a generalized LZSM theory that, in addition to a small polarization, a weak measurement [49, 50] of the nuclear state takes place while the DQD detuning is swept through the avoided crossing and the charge is exposed to a strong measurement. This concept can be understood from a simple toy model where there is a single nuclear spin in dot 1, none in dot 2, and the evolution is determined by the LZSM model. After one cycle, the singlet return probability is measured such that the initial incoherent mixture of nuclear spin-up and spin-down evolves to [41]

$$\rho_n^{(0)} = \begin{pmatrix} \frac{1}{2} & 0 \\ 0 & \frac{1}{2} \end{pmatrix} \to \rho_n^{(1)} = \frac{1}{e^{-\frac{\pi A^2}{\alpha\hbar}} + 1}\begin{pmatrix} e^{-\frac{\pi A^2}{\alpha\hbar}} & 0 \\ 0 & 1 \end{pmatrix}. \tag{16.30}$$

The mean of the operator δh^z and its variance, which characterizes the fluctuations of the spin bath, for the initial and final states are respectively $\langle\delta h^z\rangle_0 = 0$, $\sigma_0^z = \sqrt{\langle(\delta h^z)^2\rangle_0 - \langle\delta h^z\rangle_0^2} = \frac{A}{2}$ and $\langle\delta h^z\rangle_1 = -\frac{A}{2}\tanh\left(\frac{\pi A^2}{2\alpha\hbar^2}\right)$, $\sigma_1^z = \frac{A}{2}\cosh^{-1}\left(\frac{\pi A^2}{2\alpha\hbar}\right)$. In the limit where the pulse is infinitely fast, $\alpha \to \infty$, $\sigma_1^z \to \sigma_0^z$ and the variance is unchanged. The slow pulse limit, $\alpha \to 0$, produces $\sigma_1^z = 0$. This actually corresponds to a strong measurement of the nuclear spin, since in this case $\rho_{n,22}^{(1)} = 1$ (fully adiabatic passage). For actual pulses $0 < \alpha < \infty$, we have $\sigma_1^z < \sigma_0^z$. After measuring that the electronic system is in a singlet state, the knowledge about the nuclear spin state is modified. Due to the singlet outcome it is more likely to have a higher population of nuclear spin-up than spin-down. The phenomenon is known as a weak measurement and can be used to prepare the nuclear spin state.

16.3.5 Electronic spin manipulation via the hyperfine coupling

Another interesting phenomenon is what happens to the electronic states when the initialized singlet S(2, 0) is pulsed through the avoided $\mathrm{S}-\mathrm{T}_{+}$ crossing.[6] It was demonstrated experimentally [51] and theoretically [52] that these two states can implement a new type of qubit, which can be manipulated through LZSM interferometry. The coherent dynamics is tested by preparing a singlet state at some initial position ε_{i}, pulse it to some final position ε_{f} and wait in the external magnetic field for a time t_{w} and come back to ε_{i} where the singlet return probability is measured (c.f. Fig. 16.1c).

The dynamics can be described with the effective Hamiltonian [52]

$$H_{\mathrm{eff}}(t) = H_0(\varepsilon(t)) + \frac{1}{2} g^* \mu_{\mathrm{B}} \sum_i \left(S_i^+ B_{\mathrm{n},\,i}^- + S_i^- B_{\mathrm{n},\,i}^+ \right), \tag{16.31}$$

where $H_0(\varepsilon)$ is given by Eq. (16.28), and in contrast to Eq. (16.29) the Overhauser field operators are treated in a semiclassical fashion as in Section 16.3.2. The singlet return probability obtained in this model is [52]

$$P_{\mathrm{S}} = \int \prod_{k=1,2} \mathrm{d}\mathbf{B}_{\mathrm{n},k}\, p(\mathbf{B}_{\mathrm{n},k}) \left| \langle \mathrm{S} | U(\mathbf{B}_{\mathrm{n},\,1},\, \mathbf{B}_{\mathrm{n},\,2}) | \mathrm{S} \rangle \right|^2, \tag{16.32}$$

with the probability distribution (16.18) and the electron spin propagator [52]

$$U(\mathbf{B}_{\mathrm{n},1}, \mathbf{B}_{\mathrm{n},2}) = U_{\mathrm{b}}(\mathbf{B}_{\mathrm{n},1}, \mathbf{B}_{\mathrm{n},2}) U_{\mathrm{w}}(B_{\mathrm{n}}^z) U_{\mathrm{f}}(\mathbf{B}_{\mathrm{n},1}, \mathbf{B}_{\mathrm{n},2}), \tag{16.33}$$

where $U_{\mathrm{b,\,f}} = \mathrm{T}\exp[-\mathrm{i}\int_{t_{\mathrm{i}}}^{t_{\mathrm{f}}} \mathrm{d}t\, H_{\mathrm{eff}}(\varepsilon(t))/\hbar]$ are the backward and forward LZSM propagators and $U_{\mathrm{w}} \simeq \mathrm{T}\exp[-\mathrm{i}\int_0^{t_{\mathrm{w}}} \mathrm{d}t\, H_0(\varepsilon_{\mathrm{f}})/\hbar]$ describes the evolution of the system during the waiting time t_{w} at the final detuning position ε_{f} with $|\varepsilon_{\mathrm{f}} - \varepsilon_{\mathrm{c}}| \gtrsim \Delta_{\mathrm{HF}}$. The results are presented in Fig. 16.2.

The interference patterns present in P_{S} are characteristic of a quantum mechanically coherent process. Interestingly, for $\varepsilon_{\mathrm{f}} > \varepsilon_{\mathrm{c}}$, i.e. the case in which the system is not

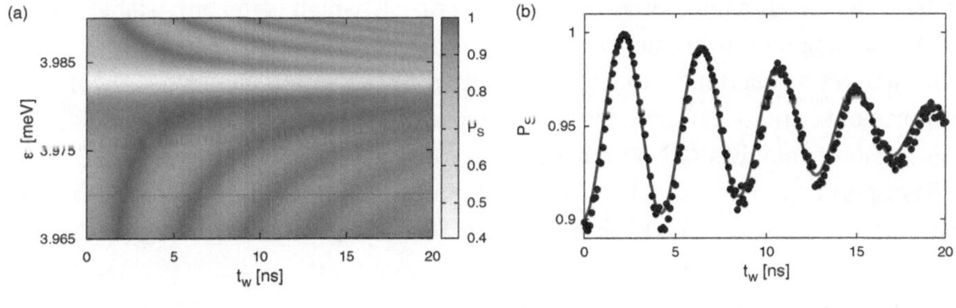

Figure 16.2 (a) The singlet return probability P_{S} as a function of the waiting time t_{w} and final detuning ε_{f}. (b) P_{S} as a function of t_{w} for $\varepsilon_{\mathrm{f}} = 3.97\,\mathrm{meV}$ (horizontal cut along dotted line in (a)). Figure adapted from Ref. [52].

[6] See also Section 15.4.8.

swept through the avoided crossing, the dynamics is also coherent. This result cannot be interpreted with the usual asymptotic LZSM model, but requires the finite-time LZSM model. This phenomenon has not been observed in [51] presumably because of charge dephasing.[7]

The LZSM transition can be viewed as a rotation by an angle θ around an axis $\hat{\boldsymbol{n}}$ in the $\mathrm{S}-\mathrm{T}_+$ space, $U_{\mathrm{LZSM}} = \mathrm{e}^{\mathrm{i}\boldsymbol{\sigma}\cdot\hat{\boldsymbol{n}}\,\theta/2}$. Any single qubit operation can, for instance, be achieved either by engineering a sequence of suitable LZSM transitions, or by using the Euler angle method. To implement the Euler angle method one needs two perpendicular axes around which rotations can be performed. Rotations around the z-axis can be realized with help of the Larmor precession. The LZSM transition provides a rotation around an axis $\hat{\boldsymbol{n}}_{xy}$ that lies in the xy-plane [52] if the propagation times are equal, $-t_{\mathrm{i}} = t_{\mathrm{f}}$. A π-rotation about $\hat{\boldsymbol{n}}_{xy}$ would require exponentially long times, since it corresponds to an adiabatic transition from S to T_+. But the problem can be circumvented by using a sequence of LZSM transitions until the desirable final angle is achieved. A single qubit operation can therefore generally be expressed, by using Eq. (16.33), as [52]

$$\mathcal{D}(\theta,\,\varphi) = \prod_{i=1}^{l} U_{\mathrm{b}}(\theta_{\mathrm{b}}^{(i)})U_z(\varphi^{(i)})U_{\mathrm{f}}(\theta_{\mathrm{f}}^{(i)}), \tag{16.34}$$

where l numbers the necessary forward sweep–wait–backward sweep operations to reach the angles θ and φ. Moreover, a conditional gate can be implemented by capacitively coupling qubits confined in different DQDs. The effect of the coupling results in a shift of the avoided crossing for the target qubit, such that it is possible to perform an effective rotation when the control qubit is in a (1, 1) charge state, whereas if the control qubit is in a (0, 2) charge state nothing happens to the target.

16.4 Hyperfine interaction in carbon-based quantum dots

Since carbon is the only constituent of graphene, hyperfine interactions are expected to be weak due to the small natural abundance ($\approx$1%) of ^{13}C which is the only stable isotope of carbon possessing a nonzero nuclear spin $I = 1/2$. Although the realization of graphene QDs has not yet reached its maturity owing to the fundamental challenges involved with the confinement of massless particles (see, however, [53]), other graphene-based structures such as carbon nanotubes (CNT) have been widely studied experimentally and theoretically. Transport experiments realized in DQDs embedded in CNTs revealed the energy of the hyperfine interaction to be approximatively 100 μeV [54] while theory predicted it to be approximatively 1 μeV [8, 55]. This discrepancy demonstrates that hyperfine interaction in graphene nanostructures is not yet fully understood. New theories show that hyperfine interactions not only couple the electron spin with the nuclear spins but that the latter also couple to the valley degree of freedom [9].

[7] The charge coherence time of a (1, 1)−(2, 0) superposition state has been measured to be ~1ns.

16.4.1 Effective spin-valley hyperfine Hamiltonian

The derivation of the effective hyperfine Hamiltonian for a monolayer graphene-based QD follows the one for semiconductor-based quantum dots. But since electrons occupy π-orbitals, the relevant interaction is given by the Hamiltonian (16.4). The projection onto the lowest orbital level of the QD produces the effective hyperfine Hamiltonian [9].

The lowest orbital of an electron confined in a monolayer graphene QD defined electrostatically and with a finite gap can be obtained via a tight-binding model [56] and reads

$$(\psi_{s\tau})_{l\sigma} = \sqrt{\Omega_{\text{cell}}}\, \mathrm{e}^{\mathrm{i}\tau \boldsymbol{K}\cdot\boldsymbol{r}_{l\sigma}} \Psi_\sigma^\tau(\boldsymbol{r}_{l\sigma})\chi_s, \tag{16.35}$$

where $s \in \{\uparrow, \downarrow\} = (+, -)$ and $\tau \in \{K, K'\} = (+, -)$ are respectively the spin and valley quantum number, σ labels the A and B sites of the sublattices, Ω_{cell} is the area of the unit cell, $\boldsymbol{r}_{l\sigma}$ is the position of a carbon atom on sublattice σ in the lth unit cell, and $\chi_{s=+,-}$ are spinors describing the two possible spin states. The envelope wave function Ψ_σ^τ is obtained by solving the Dirac-type equation

$$\left[v_{\text{F}}\left(\boldsymbol{\sigma}_x p_x + \tau\boldsymbol{\sigma}_y p_y\right) + V_{\text{conf}}(y)\right]\begin{pmatrix}\Psi_A^\tau\\ \Psi_B^\tau\end{pmatrix} = E\begin{pmatrix}\Psi_A^\tau\\ \Psi_B^\tau\end{pmatrix}. \tag{16.36}$$

Here σ_i are the Pauli matrices corresponding to the sublattice degree of freedom, p_i is the momentum operator in the i direction, and V_{conf} is a smooth confinement potential. The main contribution of Eq. (16.4) in the tight-binding approach comes from the on-site matrix element at the location of the nuclear spin, such that we can approximate the hyperfine interaction by $\left(H_{\text{hf}}^{\text{tb}}\right)_{l\sigma,l'\sigma'} = \mathbf{S}^{\text{T}} \cdot \mathbf{T} \cdot \boldsymbol{I}_{l\sigma}\delta_{ll'}\delta_{\sigma\sigma'}$. The projection onto the four levels given by Eq. (16.35) leads to [9]

$$H_{\text{hf,gr}}^{\text{eff}} = \boldsymbol{S} \cdot \left(\boldsymbol{h}^{(0)}\tau_0 + \sum_{i=x,y,z} \boldsymbol{h}^{(i)}\tau_i\right). \tag{16.37}$$

The Overhauser field operators are given by $\boldsymbol{h}^{(k)} = \Omega_{\text{cell}} A \sum_{l\sigma} F_{l\sigma}^{(k)} \boldsymbol{I}_{l\sigma}$, where $F_{l\sigma}^0 = \sum_\tau f_{l\sigma}^\tau/2$, $F_{l\sigma}^x = \text{Re}\left[\mathrm{e}^{-2\mathrm{i}\boldsymbol{K}\cdot\boldsymbol{r}_{l\sigma}}\Psi_\sigma^{(+)*}(\boldsymbol{r}_{l\sigma})\Psi_\sigma^{(-)}(\boldsymbol{r}_{l\sigma})\right]$, $F_{l\sigma}^y = \text{Im}\left[\mathrm{e}^{-2\mathrm{i}\boldsymbol{K}\cdot\boldsymbol{r}_{l\sigma}}\Psi_\sigma^{(+)*}(\boldsymbol{r}_{l\sigma})\Psi_\sigma^{(-)}(\boldsymbol{r}_{l\sigma})\right]$, $F_{l\sigma}^z = \sum_\tau \tau f_{l\sigma}^\tau/2$, and $f_{l\sigma}^\tau = \left|\Psi_\sigma^\tau(\boldsymbol{r}_{l\sigma})\right|^2$.

Since the valley degree of freedom is part of Eq. (16.37), it implies that the hyperfine Hamiltonian is also a source of relaxation and decoherence for the valley degrees of freedom. More importantly it mixes them and therefore influences the behavior of the leakage current through any carbon-based DQD.

16.4.2 Transport through a carbon-based double quantum dot

As in Section 16.3.2, the transport cycle follows the steps $(0, 1) \to (1, 1) \to (0, 2) \to (0, 1)$. But this time, the system is subject to two magnetic fields: an in-plane magnetic field B which causes a spin splitting, $\Delta_s = g\mu_{\text{B}}B$, and an out-of-plane magnetic field B_z which

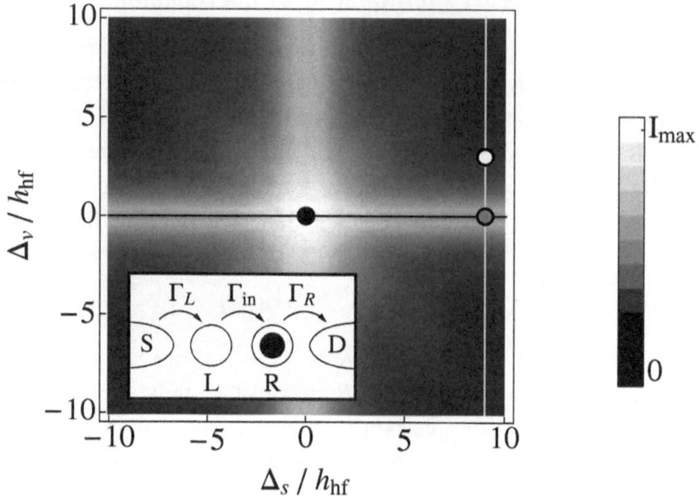

Figure 16.3 Leakage current through a graphene DQD as a function of the Zeeman splitting Δ_s and valley splitting Δ_v. (inset) Schematic representation of the transport cycle through the graphene DQD with the different tunneling rates. Figure from Ref. [9].

in addition introduces a valley splitting, $\Delta_v(B_z)$. The current is numerically calculated as a function of B and B_z using the same formalism as in [28] (c.f. Section 16.3.2), and the results are shown in Fig. 16.3.

A qualitative understanding of the result can be obtained when expressing the (1, 1)-charge states with the total spin quantum number $s = 0, 1$, magnetization m_s, total valley quantum number $\nu = 0, 1$, and its projection along B_z, m_ν. When $\Delta_s = \Delta_v = 0$, the 16 possible states are all degenerate and the mixing induced by Eq. (16.37) is most effective and results in a maximal current. If the magnetic field B is increased, the states with a different magnetization m_s are split, but each m_s subspace is still valley degenerate. This leads to a reduction of the current, but not to its full suppression. If in addition B_z is switched on, but such that $\Delta_s \not\approx \Delta_v$, the states with different m_ν also split. The current therefore becomes strongly suppressed, but does not vanish. It saturates at a finite value. The case where $\Delta_s \approx \Delta_v$ brings the states $|1, 1, 1, -1\rangle$ and $|1, -1, 1, 1\rangle$ into resonance with the four degenerate state $|s, 0, \nu, 0\rangle$ and enhances the current, as it can been seen in Fig. 16.3 along $\Delta_s = \pm\Delta_\nu$.

We acknowledge funding from the DFG within SPP 1285, FOR 912 and SFB 767. We are grateful to A. Auer and J. Hildmann for their careful reading of the manuscript.

References

[1] Cohen-Tannoudji, C., Diu, B. and Laloë, F. 2000. An application of perturbation theory: the fine and hyperfine structure of the hydrogen atom. Pages 1209–1282 of: *Quantum Mechanics Vol. 2*. Wiley-Interscience.

[2] Abragam, A. 1961. *The Principles of Nuclear Magnetism*. Oxford.
[3] Coish, W. A. 2008. Spin in quantum dots: hyperfine interaction, transport, and coherent control. Ph.D. thesis, University of Basel.
[4] Yafet, Y. 1961. Hyperfine interaction due to orbital magnetic moment of electrons with large *g* factors. *J. Phys. Chem. Solids*, **21**, 99–104.
[5] Gr'ncharova, E. I. and Perel', V. I. 1977. Relaxation of nuclear spins interacting with holes in semiconductors. *Sov. Phys. Semicond.*, **11**, 997.
[6] Fisher, J., Coish, W. A., Bulaev, D. V. and Loss, D. 2008. Spin decoherence of a heavy hole coupled to nuclear spins in a quantum dot. *Phys. Rev. B*, **78**, 155329.
[7] Eble, B., Testelin, C., Desfonds, P. *et al.* 2009. Hole–nuclear spin interaction in quantum dots. *Phys. Rev. Lett.*, **102**, 146601.
[8] Fisher, J., Trauzettel, B. and Loss, D. 2009. Hyperfine interaction and electron-spin decoherence in graphene and carbon nanotube quantum dots. *Phys. Rev. B*, **80**, 155401.
[9] Palyi, A. and Burkard, G. 2009. Hyperfine-induced valley mixing and the spin-valley blockade in carbon-based quantum dots. *Phys. Rev. B*, **80**, 201404(R).
[10] Cohen-Tannoudji, C., Diu, B. and Laloë, F. 2000. Electric multipole moments. Pages 1059–1071 of: *Quantum Mechanics Vol. 2*. Wiley-Interscience.
[11] Klauder, J. R. and Anderson, P. W. 1962. Spectral diffusion decay in spin resonance experiments. *Phys. Rev.*, **125**, 912–932.
[12] de Sousa, R. and Das Sarma, S. 2003. Theory of nuclear-induced spectral diffusion: Spin decoherence of phosphorus donors in Si and GaAs quantum dots. *Phys. Rev. B*, **68**, 115322.
[13] Coish, W. A. and Baugh, J. 2009. Nuclear spins in nanostrucutres. *Phys. Status Solidi B*, **246**, 2203–2215.
[14] Salis, G., Fuchs, D. T., Kikkawa, J. M. *et al.* 2001. Optical manipulation of nuclear spin by a two-dimensional electron gas. *Phys. Rev. Lett.*, **86**, 2677.
[15] Salis, G., Awschalom, D. D., Ohno, Y. and Ohno, H. 2001. Origin of enhanced dynamic nuclear polarization and all-optical nuclear magnetic resonance in GaAs quantum wells. *Phys. Rev. B*, **64**, 195304.
[16] Yusa, G., Muraki, K., Takashina, K., Hashimoto, K. and Hirayama, Y. 2005. Controlled multiple quantum coherences of nuclear spins in a nanometre-scale device. *Nature*, **434**, 1001.
[17] Leuenberger, M. N., Loss, D., Poggio M. and Awschalon, D. 2002. Quantum information processing with large nuclear spins in GaAs semiconductors. *Phys. Rev. Lett.*, **89**, 207601.
[18] Maletinsky, P., Kroner, M. and Imamoğlu, A. 2009. Breakdown of the nuclear-spin–temperature approach in quantum-dot demagnetization experiments. *Nature Phys.*, **5**, 407–411.
[19] Loss, D. and DiVincenzo, D. P. 1998. Quantum computation with quantum dots. *Phys. Rev. A*, **57**, 120.
[20] Mermin, N. D. 2007. *Quantum Computer Science*. Cambridge University Press.
[21] Hanson, R., Kouwenhoven, L. P., Petta, J. R., Tarucha, S. and Vandersypen, L. M. K. 2007. Spins in few-electron quantum dots. *Rev. Mod. Phys*, **79**, 1217.
[22] Hanson, R. and Awschalom, D. D. 2008. Coherent manipulation of single spins in semiconductors. *Nature*, **453**, 1043.
[23] Chirolli, L. and Burkard, G. 2008. Decoherence in solid-state qubits. *Adv. Physics*, **57**, 225–285.

[24] Khaetskii, A. V., Loss, D. and Glazman, L. 2002. Electron spin decoherence in quantum dots due to interaction with nuclei. *Phys. Rev. Lett*, **88**, 186802.
[25] Coish, W. A. and Loss, D. 2004. Hyperfine interaction in a quantum dot: non-markovian electron spin dynamics. *Phys. Rev. B*, **70**, 195340.
[26] Burkard, G., Loss, D. and DiVincenzo, D. P. 1999. Coupled quantum dots as quantum gates. *Phys. Rev. B*, **59**, 2070–2078.
[27] Koppens, F. H. L., Folk, J. A., Elzerman, J. M. *et al.* 2005. Control and detection of singlet–triplet mixing in a random nuclear field. *Science*, **309**, 1346.
[28] Jouravlev, O. N. and Nazarov, Y. V. 2006. Electron transport in a double quantum dot governed by a nuclear magnetic field. *Phys. Rev. Lett*, **96**, 176804.
[29] Ono, K., Austing, G. K., Tokura, Y. and Tarucha, S. 2002. Current rectification by Pauli exclusion in a weekly coupled double quantum dot system. *Science*, **297**, 1313.
[30] Stoof, T. H. and Nazarov, Y. V. 1996. Time-dependent resonant tunneling via two discrete states. *Phys. Rev. B*, **53**, 1050.
[31] Breuer, H. P. and Petruccione, F. 2002. Quantum master equations. Pages 105–209 of: *The Theory of Open Quantum Systems*. Oxford University Press.
[32] Nazarov, Y. V. and Blanter, Y. M. 2009. *Quantum Transport*. Cambridge University Press.
[33] Levy, J. 2002. Universal quantum computation with spin-1/2 pairs and Heisenberg exchange. *Phys. Rev. Lett.*, **89**, 147902.
[34] Petta, J. R., Johnson, A. C., Taylor, J. M. *et al.* 2005. Coherent manipulation of coupled electron spins in semiconductor quantum dots. *Science*, **309**, 2180.
[35] Coish, W. A. and Loss, D. 2005. Singlet–triplet decoherence due to nuclear spins in a double quantum dot. *Phys. Rev. B*, **72**, 125337.
[36] Laird, E. A., Petta, J. R., Johnson, A. C. *et al.* 2005. Effect of exchange interaction on spin dephasing in a double quantum dot. *Phys. Rev. Lett.*, **97**, 056801.
[37] Klauser, D., Coish, W. A. and Loss D. 2006. Nuclear spin state narrowing via gate-controlled Rabi oscillations in a double quantum dot. *Phys. Rev. B*, **73**, 205302.
[38] Stepanenko, D., Burkard, G., Giedke, G. and Imamoğlu, A. 2006. Enhancement of electron spin coherence by optical preparation of nuclear spins. *Phys. Rev. Lett.*, **96**, 136401.
[39] Petta, J. R., Taylor, J. M., Johnson, A. C. *et al.* 2008. Dynamic nuclear polarization with single electron spins. *Phys. Rev. Lett.*, **100**, 067601.
[40] Reilly, D. J., Taylor, J. M., Petta, J. R. *et al.* 2008. Suppressing spin qubit dephasing by nuclear state preparation. *Science*, **321**, 817.
[41] Ribeiro, H. and Burkard, G. 2009. Nuclear state preparation via Landau–Zener–Stückelberg transitions in double quantum dots. *Phys. Rev. Lett.*, **102**, 216802.
[42] Landau, L. D. 1932. On the theory of transfer of energy at collisions II. *Phys. Z. Sowjetunion*, **2**, 46.
[43] Zener, C. 1932. Non-adiabatic crossing of energy levels. *Proc. R. Soc. Lond. A*, **137**, 696.
[44] Stückelberg, E. C. G. 1932. Theory of inelastic collisions between atoms (Theory of inelastic collisions between atoms, using two simultaneous differential equations). *Helv. Phys. Acta*, **5**, 369.
[45] Majorana, E. 1932. Oriented atoms in a tunable magnetic field. *Nuovo Cimento*, **9**, 43.
[46] Vitanov, N. V. and Garraway, B. M. 1996. Landau–Zener model: effects of finite coupling duration. *Phys. Rev. A*, **53**, 4288.
[47] Vasilev, G. S., Ivanov, S. S. and Vitanov, N. V. 2007. Degenerate Landau–Zener model: analytical solution. *Phys. Rev. A*, **75**, 013417.

[48] Shevchenko, S. N., Ashhab, S. and Nori, F. 2010. Landau–Zener–Stückelberg interferometry. *Physics Reports*, **492**, 1–30.
[49] Breuer, H. P. and Petruccione, F. 2002. Quantum probability. Pages 57–101 of: *The Theory of Open Quantum Systems*. Oxford University Press.
[50] Preskill, J. 2002. Foundations II: Measurement and Evolution. *Lecture Notes on Quantum Computation*. http://www.theory.caltech.edu/ preskill/ph229.
[51] Petta, J. R., Lu, H. and Gossard, A. C. 2010. A coherent beam splitter for electronic spin states. *Science*, **327**, 669.
[52] Ribeiro, H., Petta, J. R. and Burkard, G. 2010. Harnessing the GaAs quantum dot nuclear spin for quantum control. *Phys. Rev. B*, **82**, 115445.
[53] Trauzettel, B., Bulaev, D. V., Loss, D. and Burkard, G. 2007. Spin qubits in graphene quantum dots. *Nat. Phys.*, **3**, 192.
[54] Churchill, H. O. H., Bestwick, A. J., Harlow, J. W. *et al.* 2005. Electron–nuclear interaction in ^{13}C nanotube double quantum dots. *Nat. Phys.*, **5**, 321.
[55] Yazyev, O. V. 2008. Hyperfine interactions in graphene and related carbon nanostructures. *Nano Lett.*, **8**, 1011.
[56] Castro Neto, A. H., Guinea, F., Peres, N. M. R., Novosolev K. S. and Geim A. K. 2009. Electronic properties of graphene. *Rev. Mod. Phys.*, **81**, 109.

17

Graphene quantum dots: transport experiments and local imaging

S. Schnez, J. Guettinger, F. Molitor, C. Stampfer, M. Huefner, T. Ihn and K. Ensslin

17.1 Introduction

This chapter gives an overview of our recent experimental research on graphene quantum dots. We will focus on two aspects, namely transport experiments revealing the existence of excited states and how they can be used to detect the electron–hole crossover and scanning-gate microscopy to reveal local information on localized states.

One of the extraordinary properties of carbon atoms is the ability to be synthesized into solid structures of any dimensionality: three-dimensional *graphite* and *diamond*, one-dimensional *carbon nanotubes*, and zero-dimensional *fullerenes*. *Graphene*, a two-dimensional carbon allotrope, was the last to be added to this list in 2004, when K. Novoselov *et al.* published their results [32]. Graphene has been stimulating the physicists' imagination since then: remarkable mechanical and electronic properties make it a promising candidate for future technological advances and breakthroughs in fundamental research [10]. In particular, the linear dispersion relation of graphene (for small energies) and an additional degree of freedom due to the valley degeneracy led to new physics not observed before in a condensed-matter environment. In fact, impressive results on the so-called unconventional quantum Hall effect [33, 55] and Klein tunneling [54] were obtained. The significance of the discovery of graphene and the first experiments were acknowledged and the discoverers of graphene were rewarded the Nobel Prize in physics 2010.

It was not obvious from the very beginning of graphene research whether it is easily possible to build more complex nanostructures like quantum point contacts or quantum dots (QDs). Charge carriers in graphene are described by the linear *massless Dirac equation* $H\psi = E\psi$ with the Hamiltonian $H = \hbar c_\star \boldsymbol{\sigma} \cdot \boldsymbol{k}$, $c_\star \approx 10^6$ m/s, and $\boldsymbol{\sigma} = (\sigma_x, \sigma_y)$ being Pauli's spin matrices [52, 43]. This leads to valence and conduction bands which touch each other at the nonequivalent K- and K'-points without a band gap. It is therefore impossible to confine charge carriers using electrostatic gates as in conventional semiconductor two-dimensional electron gases. It turned out that the most straightforward way to circumvent this problem is to cut graphene into small stripes, so-called *nanoribbons* or *constrictions*, thus opening a transport gap [16, 4, 26, 47]. On the other hand, theoretical

Quantum Dots: Optics, Electron Transport and Future Applications, ed. Alexander Tartakovskii. Published by Cambridge University Press.

findings suggested that spin qubits based on graphene QDs will have superior coherence times compared to QDs made of other materials because of a rather small spin–orbit interaction (small atomic charge of carbon atoms) and a very small hyperfine interaction (nuclear spin of zero for ^{12}C-atoms) [50]. The experimental verification of this prediction is one of the driving forces for research on graphene quantum dots.

Graphene is a material where the type of charge carrier can easily be tuned from electrons to holes through a point of nominally zero charge-carrier density – the so-called *Dirac point* or *charge-neutrality point* (CNP). Close to the CNP, electron–hole puddles form so that there is a finite local charge density [25, 56]. Moreover, it turned out that localized states form in nanoribbons. All these effects strongly influence charge transport through graphene nanostructures. A detailed understanding is crucial to construct better-performing nanostructures and can be obtained by employing local imaging tools like scanning-gate microscopy (SGM).

Regarding the fabrication process of our single-layer graphene nanostructures based on the standard mechanical exfoliation procedure, we refer the reader to the more specialized literature for detailed information [46, 32, and references therein]. The outline of this chapter is the following. In Section 17.2 we will introduce the theoretical background information that is necessary to understand why graphene QDs are different from QDs with a parabolic dispersion. Section 17.3 then presents excited-state and magnetic-field spectroscopy. It culminates in our attempts to identify the electron–hole crossover. The technique of scanning-gate microscopy (SGM) and how we use it to locate different states along a QD device is introduced in Section 17.4. In the final section, we will give an outlook on what the next steps in graphene QD research may be.

17.2 Theoretical background

17.2.1 The constant-interaction model

A QD is a small island that can trap charge carriers. For this reason, it is coupled to source and drain leads with electrochemical potentials $\mu_{S,D}$ via tunnel barriers with tunnel coupling rates $\Gamma_{S,D}$. A real graphene QD will be presented in the next section. A bias voltage V_{bias} can be applied between source and drain contacts so that $eV_{bias} = \mu_S - \mu_D$.

Under suitable experimental conditions, QDs show Coulomb blockade. Following the constant-interaction model [1], the electrochemical potential of the QD is given by

$$\mu_N = \epsilon_N + \frac{e^2}{C_\Sigma}\left(N - \frac{1}{2}\right) + e\sum_{i=1}^{n} \frac{C_{0i}}{C_\Sigma}\left(V_i - V_i^{(0)}\right) \tag{17.1}$$

in this regime. The summation runs over all gates of the device to which a voltage V_i is applied; $V_i^{(0)}$ is an offset voltage of that gate. The quantity N is the number of charge carriers trapped on the dot, and $C_\Sigma := -\sum_{i=1}^{n} C_{0i}$ is the total capacitance of the QD to its environment. Quantum mechanical contributions to μ_N are summarized in ϵ_N; this will be discussed in more detail below. The capacitance coefficients C_{0i} are the capacitances

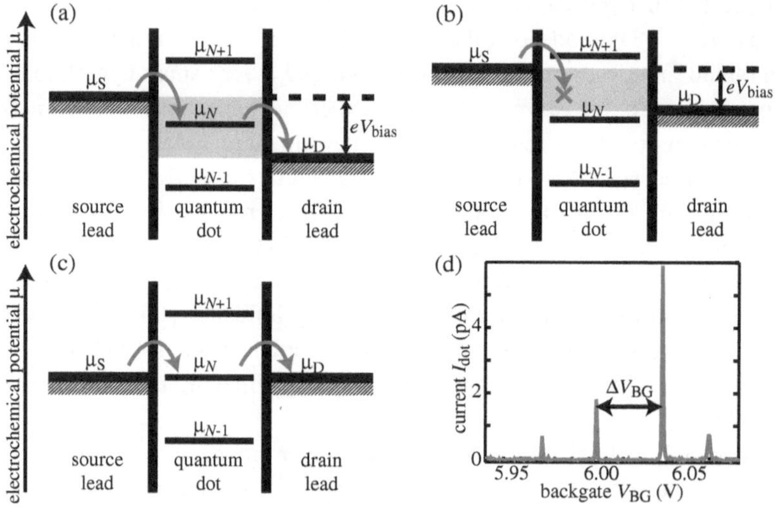

Figure 17.1 Energy diagram of a QD for finite bias $eV_{\text{bias}} = \mu_S - \mu_D$ when current flows (a) or when current is blocked (b). The light gray shaded areas mark the width of the *bias windows*. Whenever one of the discrete energy states μ_N lies within such an area, current through the dot can flow. This is depicted with blue arrows in (a). In (c), the zero-bias case is shown when the dot is *in resonance* with source and drain. Current can then flow, which results in a Coulomb resonance peak in the current I_{dot} through the dot. An experimental trace of I_{dot} as a function of gate voltage and for a small bias voltage of $V_{\text{bias}} = 16\,\mu\text{V}$ is depicted in (d). The distance between two Coulomb resonances is proportional to the addition energy with the lever arm of that gate being the proportionality constant, $\Delta E_C = e\alpha_{\text{BG}} V_{\text{BG}}$. This trace was measured with the device presented in the next section.

between gate i and the QD; the ratio $\alpha_i := -C_{0i}/C_\Sigma$ is called the *lever arm* of gate i. Current flows whenever μ_N is in the bias window of the electrochemical potentials of source and drain, i. e. $\mu_D \leq \mu_N \leq \mu_S$ (or reversed bias). This situation is schematically depicted in Fig. 17.1a. Otherwise, the current is blocked due to Coulomb repulsion, Fig. 17.1b.

According to Eq. (17.1), the electrochemical potential of the QD can be continuously swept by tuning a gate with voltage V_i. If the current through the quantum dot is recorded as a function of gate voltage and for small bias, $eV_{\text{bias}} < k_B T$, *Coulomb resonances* occur whenever the electrochemical potentials of source, dot and drain are in resonance, $\mu_S = \mu_N = \mu_D$ (Fig. 17.1c). An experimental trace of Coulomb resonances is shown in Fig. 17.1d.

The energy difference $\mu_{N+1} - \mu_N$ between two discrete energy states of the QD is called the *addition energy* ΔE_C. It can easily be determined in transport experiments as explained below, and it is given by

$$\Delta E_C := \mu_{N+1} - \mu_N = \frac{e^2}{C_\Sigma} + \Delta(N), \tag{17.2}$$

where $\Delta(N) = \epsilon_{N+1} - \epsilon_N$. The capacitance C_Σ cannot be calculated analytically in general. For a rough estimate of a lower bound of C_Σ, the QD can be considered as a metallic disc embedded in a material with relative dielectric constant ε and infinite extent. Then, C_Σ is approximately equal to the self-capacitance of a disc, $C_\Sigma \approx 8\varepsilon\varepsilon_0 r$. Typical addition energies are in the range of a few millielectronvolts.

This section provides an overview of the constant-interaction model relevant to this chapter. Extensive discussions can be found in [1], [18] and [31].

17.2.2 Graphene quantum dots and quantum confinement

Whenever a quantum mechanical particle is confined to a certain region in space, discrete energy states are a consequence of the wave nature of the particle. This, of course, is also true for charge carriers trapped on a QD and leads to the quantum mechanical contribution ϵ_N to the electrochemical potential in Eq. (17.1). If we neglect any interactions, the total confinement energy can be estimated by

$$E_{\mathrm{conf}}(N) = A \int_0^{E_\mathrm{F}} \mathrm{d}E\, E\mathcal{D}(E) = \frac{2}{3}\hbar c_* \sqrt{\frac{\pi}{A}} N^{3/2}, \qquad (17.3)$$

where E_F is the Fermi energy in the system, A is the area of the QD and $\mathcal{D}(E) = 2E/\left(\pi\hbar^2 c_\star^2\right)$ is the density of states of graphene. At this point, the specific characteristics of graphene enter, namely the linear dispersion relation leading to a linear density of states and the effective velocity $c_\star \approx 10^6$ m/s.

The contribution to the electrochemical potential can be calculated as the difference of the confinement energies for $N+1$ and N charge carriers, namely as $\epsilon_N = E_{\mathrm{conf}}(N+1) - E_{\mathrm{conf}}(N)$. The *single-particle level spacing*, which is the relevant quantity for transport experiments and which also enters the addition energy ΔE_C, Eq. (17.2), is then

$$\Delta(N) = \epsilon(N+1) - \epsilon(N) \approx \frac{\hbar c_*}{2}\sqrt{\frac{\pi}{A}}\frac{1}{\sqrt{N}}. \qquad (17.4)$$

In order to arrive at this expression, we used the fact that $N \gg 1$ in general and neglected higher order terms in N. In other words, the above expression must not be used very close to the CNP where one expects N to be around zero.

In the previous section, we described how current can flow through a QD as charge carriers hop on and off the discrete states of the dot without explicitly considering the quantum-mechanical contributions. Analogously, charge transport can also occur through the excited states spaced by $\Delta(N)$ due to quantum confinement. In Fig. 17.2a, one possibility of how this can happen is depicted. *Inelastic co-tunneling* is a second-order process that can also take place and is shown in Fig. 17.2b. Initially, the QD state with electrochemical potential μ_N is occupied. Borrowing "energy from the vacuum", the charge carrier tunnels out into the drain (process 1) and the system remains in a virtual state. In the next step, a charge carrier tunnels from the source into an excited state (process 2). Processes 1 and 2 have to occur on a time scale given by Heisenberg's uncertainity relation to conserve energy. The complete cycle is referred to be *inelastic* because the final state of the QD is

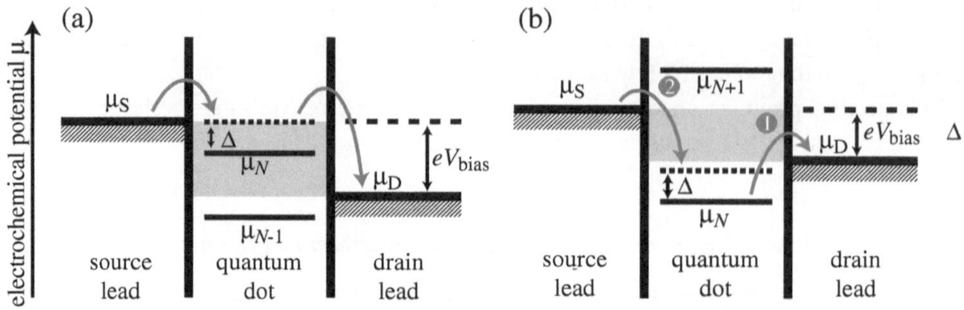

Figure 17.2 Energy diagram of a QD where current transport takes place through excited states. (a) Current through an excited state which is in resonance with the source lead. This provides an additional transport channel (in addition to transport through the ground state μ_N) and leads to enhancement of the conductance. (b) One possible inelastic co-tunneling process. Here, the QD remains in a virtual state after process 1. In order to fulfill Heisenberg's uncertainty relationship for energy and time, process 2 has to occur on a corresponding time scale. The whole process is called *inelastic* because the final state is higher in energy than the initial one by an amount Δ.

higher in energy than the initial state by the amount Δ. This energy is provided by the source–drain bias. Therefore, inelastic co-tunneling is observable only when $\Delta \leq eV_{\text{bias}}$.

17.2.3 Bias spectroscopy

We can experimentally deduce the relevant energy scales of a QD when we perform *bias spectroscopy*, i. e. when the bias voltage is swept and the plunger gate is stepped (or vice versa). A schematic measurement outcome is shown in Fig. 17.3. The white regions of suppressed current are called *Coulomb-blockade diamonds*. Their extent in bias direction is a direct measure of the addition energy, $\Delta E_C = e\Delta V_{\text{bias}}$. The lever arm of the plunger gate is given by the ratio $\alpha_{\text{PG}} = \Delta V_{\text{bias}}/\Delta V_{\text{PG}}$.

Excited states spaced by the single-particle level spacing can be seen as lines of high conductance running in parallel to the edge of Coulomb diamonds. They are depicted as dashed, black lines outside the diamond with N charge carriers. Transport within the diamond-shaped region is forbidden for first-order processes. However, transport employing virtual intermediate states is possible as discussed above. The onset of inelastic co-tunneling is shown in diamond N with a dashed, black line where the condition $\Delta = eV_{\text{bias}}$ is fulfilled. The inelastic co-tunneling onset is aligned with an excited-state line at the edge of the diamond.

17.2.4 Graphene quantum dots in a perpendicular magnetic field

Magnetic-field spectroscopy is a powerful tool to investigate confinement-induced effects in QDs. It is therefore a natural question to ask: what will happen if a graphene QD is

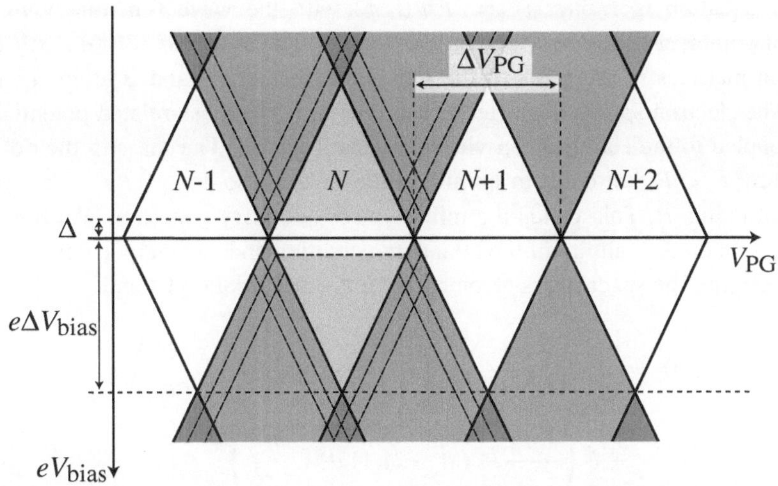

Figure 17.3 Schematic result of a bias spectroscopy on a QD. The conductance is given in color code where darker colors denote higher conductance. The QD is symmetrically biased, $\mu_S = -\mu_D = eV_{bias}/2$. In the white regions, the so-called *Coulomb-blockade diamonds*, the current through the dot is suppressed due to Coulomb repulsion and the number of charge carriers on the dot is fixed. The addition energy can directly be read off from the extent ΔV_{bias} of a diamond in bias direction. The lever arm of the plunger gate is given by the ratio $\alpha_{PG} = \Delta V_{bias}/\Delta V_{PG}$. Excited states appear as lines of higher conductance running in parallel to the edge of diamond N or inside the diamond at the onset of inelastic co-tunneling.

exposed to a perpendicular magnetic field? The contribution to the addition energy, which is determined by the geometry of the QD, e^2/C_Σ, stays constant. The single-particle level spacing Δ, on the other hand, does depend on the magnetic field. This dependence $\Delta(B)$ can be calculated analytically in the case of QDs with a parabolic dispersion and a parabolic confinement potential. The so-called *Fock–Darwin spectrum* shows how energy levels evolve as a function of magnetic field [8, 5] and has been verified under suitable experimental conditions [48].

In the following, we want to establish a corresponding energy spectrum for graphene QDs in a magnetic field and look at the transition from confinement-dominated to magnetic field-dominated behavior of the charge carriers. In particular, we expect that the hole energy levels shift in the opposite direction to the electron energy levels because of the electron–hole symmetry in graphene. This would enable experimentalists to determine the electron–hole crossover by employing magnetic-field spectroscopy.

Several theoretical analyses have been done for this problem [3, 36, 20, 19, 53]. The analysis which we present here is based on the *infinite-mass boundary condition* introduced by Berry and Mondragon [2] as a toy model for two-dimensional neutrinos. We couple an external perpendicular magnetic field B in symmetric gauge to the Hamiltonian expressed in cylindrical coordinates which then reads

$$H = c_\star (\boldsymbol{p} + e\boldsymbol{A}) \cdot \boldsymbol{\sigma} + \tau V(r)\sigma_z. \tag{17.5}$$

The Dirac equation is $H\psi(r,\phi) = E\psi(r,\phi)$ with the wave function being a two-component spinor, $\psi(r,\phi) = (\psi_1(r,\phi), \psi_2(r,\phi))$. The vector $\boldsymbol{\sigma} = (\sigma_x, \sigma_y)$ contains Pauli's spin matrices in the basis of the two sublattices of A- and B-atoms as its components. The electron spin is neglected in our analysis. The mass-related potential energy $V(r)$ is coupled to the Hamiltonian via the σ_z Pauli matrix. The mass in the dot is zero, $V(r) = 0$ for $r < R$, but tends to infinity at the edge of the dot, $V(R) \to \infty$, for a circular dot of radius R. This yields the infinite-mass boundary condition. We refer to [38] for a discussion on the validity of the boundary condition and the analytic derivation of the energy spectrum. The spectrum is given by the following implicit formula

$$0 = \left(1 - \tau \frac{k l_B}{R/l_B}\right) L\left(\frac{k^2 l_B^2}{2} - (m+1), m, \frac{R^2}{2l_B^2}\right) + L\left(\frac{k^2 l_B^2}{2} - (m+2), m+1, \frac{R^2}{2l_B^2}\right). \tag{17.6}$$

The parameter τ takes the two values ± 1 for distinguishing the two valleys K and K', and the magnetic length is defined according to $l_B = \sqrt{\hbar/(eB)}$. The functions $L(a,b,x)$ are generalized Laguerre polynomials which are oscillatory functions. Hence, there is an infinite number of wave vectors k_n for given B, angular momentum quantum number m, and τ which fulfill the above equation. This defines the radial quantum number n which labels the roots of the left part of Eq. (17.6). The energy spectrum $E(n,m,\tau) = \hbar c_\star k_n$ of charge carriers confined to a circular graphene QD which is exposed to a perpendicular magnetic field is determined through Eq. (17.6). The relation $-E(n,m,\tau) = E(n,m,-\tau)$ is a manifestation of the electron–hole symmetry.

This rather complicated formula can be simplified considerably in the two limits $B \to 0$ and $R/l_B \to \infty$ [38]. For $B \to 0$, we retrieve the result of [2], namely

$$\tau J_m(kR) = J_{m+1}(kR), \tag{17.7}$$

where $J_m(x)$ is a Bessel function of the first kind. It can be used to estimate the number of charge carriers confined on a graphene dot when the energy of an excited state is measured. This is an alternative to the density of states argument derived in the previous section. Moreover, we can deduce that $E(n,m,\tau) = E(n,-m-1,-\tau)$ for $B = 0$, derived from the property $J_m(x) = (-1)^m J_{-m}(x)$. Consequently, pairs of states are degenerate at zero magnetic field. There is no state at zero magnetic field and zero energy. This leads to an energy gap between negative and positive energy states.

In contrast, in the limit $R/l_B \to \infty$ we retrieve

$$E_m = \hbar c_\star k_m = \pm c_\star \sqrt{2e\hbar B(m+1)}, \tag{17.8}$$

which are the well-known Landau levels for graphene. Therefore, there will be a transition, governed by the parameter R/l_B, from a regime where the energies of the charge carriers

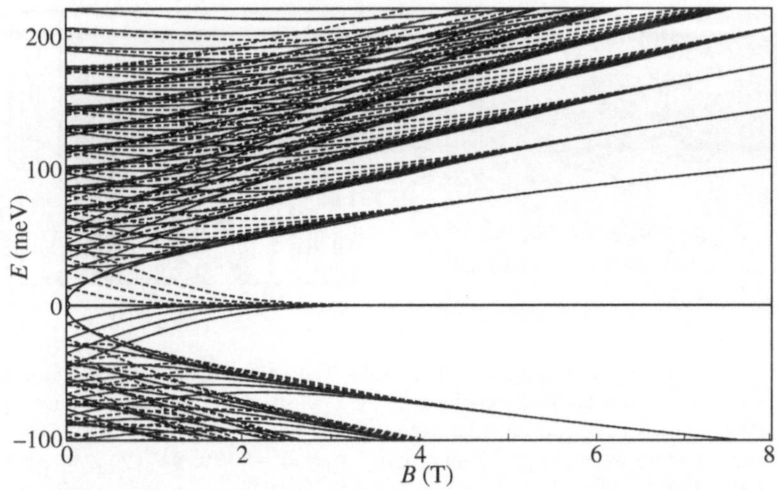

Figure 17.4 Energy spectrum of a QD with $R = 70\,\text{nm}$ which was numerically calculated from Eq. (17.6). The formation of the lowest Landau levels can be seen as predicted by Eq. (17.8). Energy states for $\tau = +1$ are drawn with solid lines, those for $\tau = -1$ with dashed lines. Figure adapted from Ref. [38].

are dominated by confinement (Eq. (17.7)) to Landau levels (Eq. (17.8)). This transition, including these two limiting cases, is described by Eq. (17.6).

We evaluate Eq. (17.6) for a QD of radius $R = 70\,\text{nm}$, which is about the same size as the device presented in the next section. The energy spectrum as a function of magnetic field is shown in Fig. 17.4 for $m = -4, \ldots, 4$ and $n = 1, \ldots, 6$. For $B = 0$, the energy states are not equidistant. For higher magnetic fields, we can see the formation of Landau levels according to the previous discussion. The zero-energy Landau level is formed by states with quantum number $\tau = -1$ and $E > 0$ and those with $\tau = +1$ and $E < 0$.

The lowest conduction-band state has an energy of about 4 meV for $B = 0$. This yields an energy gap of around 8 meV between electron and hole states. Since the energy gap to the next excited state is much lower, the electron–hole transition may be detected experimentally by a confinement-enhanced energy gap. Additionally, the first resonances on both sides of the electron–hole crossover have opposite slopes and merge into the zeroth Landau level. The experimental observation of this effect would clearly identify the crossover.

17.3 Transport experiments

17.3.1 Bias spectroscopy and excited states

A real QD is presented as an atomic-force micrograph under ambient conditions in Fig. 17.5a.[1] Raman spectroscopy was used to confirm the single-layer quality of the

[1] Parts of this section and some figures are reprinted with permission from [39]. Copyright 2009, American Institute of Physics.

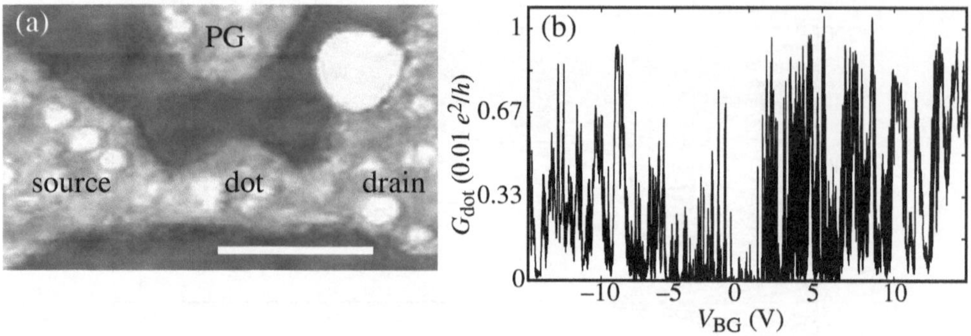

Figure 17.5 (a) Atomic-force micrograph of the QD. The central island which forms the QD is connected to source (S) and drain (D) contacts by two constrictions. A nearby plunger gate (PG) tunes the QD, whereas a back gate sets the overall Fermi energy. The scale bar is 200 nm. (b) A back-gate sweep shows a transport gap from roughly $V_{BG} = -8$ V to 8 V ($V_{bias} = 3$ mV). Figure adapted from Ref. [39].

graphene flake [12]. The highly resistive barriers are formed by two graphene constrictions about 60 nm and 70 nm wide. They separate source and drain contacts from the graphene island (diameter 140 nm). Two gates can be used to tune the device, namely a highly doped silicon back gate (BG) to adjust the overall Fermi level and a plunger gate (PG) to tune the QD itself. In order to increase the device quality, we annealed it for about 24 h in vacuum at 400 K directly before cooling down. The experiments described in the following were carried out in a dilution refrigerator at a base temperature of 40 mK. In a first step, we determined the transport gap by measuring the current I_{dot} through the QD as a function of back gate voltage V_{BG} [45]. It extends roughly from $V_{BG} = -8$ to $+8$ V (Fig. 17.5b). The fact that the gap is centered around zero back-gate voltage is interpreted as evidence for little doping of the graphene device. Current is not completely suppressed in the gap region, but characteristic peaks – interpreted as Coulomb resonances – appear as already shown in Fig. 17.1d. From their width, we estimate an upper bound for the electronic temperature of about 200 mK [18]. Since we were interested in the behavior close to the CNP, we set the back-gate voltage to zero for the following measurements.

As described in the previous section, we performed bias spectroscopy as follows. We plot the differential conductance $G_{dot} = dI_{dot}/dV_{bias}$ in color code as a function of bias voltage V_{bias} and plunger-gate voltage V_{PG} [39]. The outcome of such a measurement is shown in Fig. 17.6. Addition energies are of the order of $\Delta E_C \approx 10$ meV. The estimate employing the disk model according to Eq. (17.2) yields $\Delta E_C \approx 12$ meV and agrees reasonably well with the experimentally found data.[2]

Excited states appear in Fig. 17.6a as distinct lines of increased conductance running in parallel to the edge of the Coulomb diamonds. The single-particle-level spacing is

[2] We assumed the effective dielectric constant including vacuum and the SiO_2 to be $\varepsilon_{eff} = \left(\varepsilon_{vac} + \varepsilon_{SiO_2}\right)/2 = 2.5$.

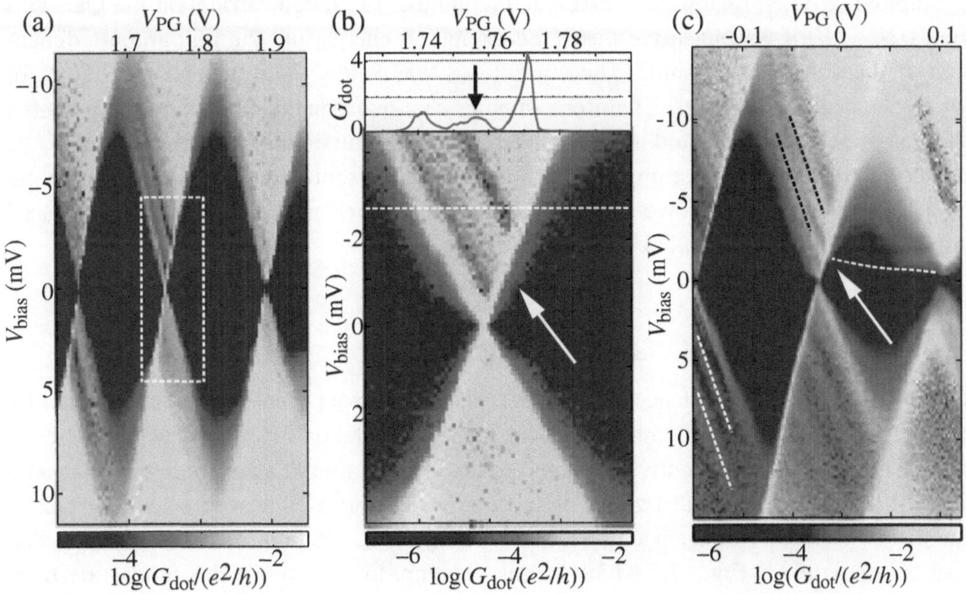

Figure 17.6 Bias spectroscopy of a graphene QD. (a) The differential conductance G_{dot} is plotted in color code as a function of source–drain voltage V_{bias} and plunger-gate voltage V_{PG}. (b) The lower panel is a zoom of the framed area in (a). An excited state is clearly visible (white arrow). The line cut along the dashed line at $V_{\text{bias}} = -2.87$ mV is shown in the upper panel (here, G_{dot} is measured in units of $10^{-3} e^2/h$ and was smoothed over four points). (c) At a smaller plunger-gate regime, several excited states are visible as shown by dashed lines. In the upper part of the right diamond, regions of higher conductance can be seen. This is the onset of inelastic co-tunneling in a graphene quantum dot (see arrow). In all measurements, the back-gate voltage was set to $V_{\text{BG}} = 0$ V and the electronic temperature was around 200 mK. Figures adapted from Ref. [39].

$\Delta \approx 1.6$ meV which we extract from Fig. 17.6b. It is a close up of Fig. 17.6a. The broadening of the excited-state line significantly exceeds thermal broadening as it is obvious from the line cut at $V_{\text{bias}} = 2.78$ mV (upper panel of Fig. 17.6b). We attribute this to an energy-dependent coupling of the excited state to the graphene leads.

We tune the QD closer to the CNP by going to lower plunger-gate voltages as shown in Fig. 17.6c. The pairs of dashed lines highlight more than one excited state. The excitations are found at energies of around 1.6 meV and 3.3 meV (black dashed lines) and 2.1 meV and 4.2 meV (white dashed lines), respectively. More importantly, we also observe the onset of inelastic co-tunneling at lower bias. We distinguish between regions of suppressed and slightly elevated conductance separated by the dotted line inside the right Coulomb diamond. The edge of this conductance step is aligned with the first excited state outside the diamond at an energy of 1.6 meV as it is expected from our theoretical considerations before. This also rules out other explanations for the lines of higher conductance outside the diamonds as, for example, a peak in the density of states in the constriction.

Employing Eq. (17.4), we can estimate the number of charge carriers on the QD. This yields $N \sim 10$ if we put in the measured excitation energy and the lithographic dimension of the dot ($d = 140\,\text{nm}$). The current I_{dot} was below measurement resolution for lower plunger-gate voltages. Therefore, smaller charge-carrier numbers and the potential electron–hole crossover could not be studied. The Coulomb diamonds shown in Fig. 17.6c scatter significantly more strongly in size than those presented in Fig. 17.6a. This might be a consequence of the lower number of charge carriers on the QD and the increased importance of quantum-confinement effects.

17.3.2 Magnetic-Field Spectroscopy

The magnetic-field dependence of different states can experimentally be investigated by monitoring the shift of Coulomb resonances in the B-field. For this reason, we measure the energy shifts of nine consecutive Coulomb peaks as a function of a perpendicular magnetic field [38]. Electrostatic contributions to the energy are subtracted as we will explain below, and the plunger-gate voltage was converted to energy with a lever arm of $\alpha_{\text{PG}} = 0.075$. The result is presented in Fig. 17.7 which is a first attempt to observe the theoretical spectrum of Fig. 17.4 experimentally.

According to Eq. (17.1), the ground-state energy of an N-particle QD can be written as the sum of the single-particle energies $\epsilon_i(B)$ plus the electrostatic contribution to the charging energy Ne^2/C_Σ plus additional gate-dependent terms. The gate voltage V_{PG} tunes the ground-state energy. In such a way, we measured the chemical potential of the

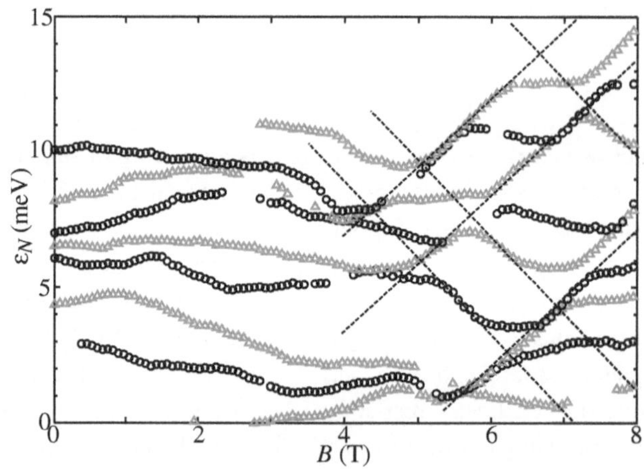

Figure 17.7 Energy spectrum of the QD as obtained by magnetic-field spectroscopy. Consecutive Coulomb resonances are labeled alternatingly with gray triangles and black circles. For small magnetic fields, we recognize the single-level fluctuations, whereas a regular pattern with characteristic linear slopes of ± 2.5 meV/T evolves (see dashed lines) starting around $B = 4\,\text{T}$. This is the transition which is qualitatively predicted by Eq. (17.6). Figure adapted from Ref. [39].

Nth Coulomb resonance if we use V_{PG} to compensate for energetic shifts induced by the magnetic field. The single-particle energy $\epsilon_N(B)$ of the Nth Coulomb resonance is therefore determined by $\epsilon_N(B) = e\alpha_{\mathrm{PG}} V_{\mathrm{PG}}^{\mathrm{res}}(N, B) + Ne^2/C_\Sigma + \text{constant}$, with $V_{\mathrm{PG}}^{\mathrm{res}}$ being the plunger-gate voltage on the resonance. We subtract the constant part and the electrostatic contribution such that consecutive peaks touch each other (shown as triangles and circles, respectively, in Fig. 17.7). The energy spectrum evolves from single-level fluctuations for small B-fields into a regular pattern with characteristic slopes. These slopes are most probable indications of the formation of Landau levels and cannot be explained by the Zeeman effect. In that case, the slopes should be around $g\mu_B = 116$ μeV/T, assuming a g-factor of $g = 2$, which is more than one order of magnitude smaller than observed. This transition from single-level fluctuations to Landau levels is qualitatively desribed by Eq. (17.6).

For this particular sample, it turned out to be impossible to detect the electron–hole crossover employing magnetic-field spectroscopy. The reason for this is that for lower plunger-gate voltages where the crossover was expected to occur the current through the dot was suppressed so strongly that no current signal could be recorded. Therefore, we performed corresponding measurements on another graphene QD where the constrictions could be individually tuned by extra in-plane gates [14]. Figure 17.8 presents the main outcome: Coulomb resonances are recorded over a large plunger-gate range and a magnetic field up to 13 T. Many of the resonances are strongly broadened; the mechanism behind that is still under debate. However, the data also indicate that the states wiggle around for small magnetic field and then tend to bend to the right for lower (e. g. the boxed states to the left) or to the left for higher plunger-gate voltages (e. g. the boxed states to the right). This is exactly what we expect for states heading into the zeroth Landau level according to Fig. 17.4. We therefore conclude that in the regime presented in Fig. 17.8 the electron–hole crossover takes place somewhere for $3\,\mathrm{V} < V_{\mathrm{BG}} < 6\,\mathrm{V}$. Numerical tight-binding calculations are in good quantitative agreement with the data presented here [14].

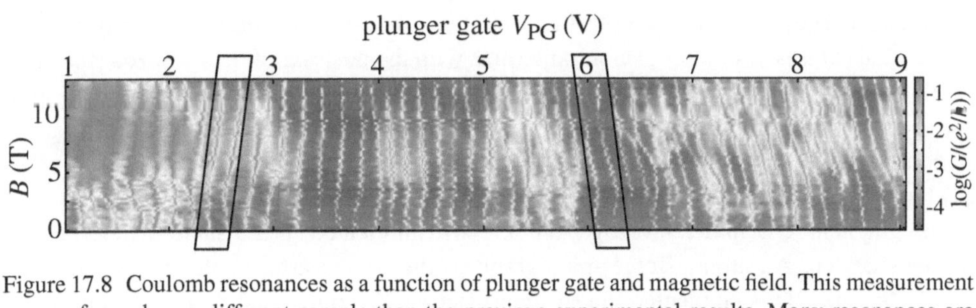

Figure 17.8 Coulomb resonances as a function of plunger gate and magnetic field. This measurement was performed on a different sample than the previous experimental results. Many resonances are broadened due to strong gate- and magnetic-field-dependent coupling to the leads. However, states for lower plunger-gate voltages seem to bend to the right whereas states for higher plunger-gate voltages tend to bend to the left. This is behavior which is expected for states that head into the zeroth Landau level and may indicate that the electron–hole crossover takes place in the presented plunger-gate regime. Figure adapted from Ref. [14].

17.4 Scanning-gate microscopy

It is an experimental fact that we can artificially introduce a so-called transport gap in graphene by cutting it into stripes. If current through the nanoribbon or constriction is then recorded as a function of back gate, current can be pinched off in the regime around the CNP. The simplest and most intuitive explanation goes back to lateral confinement leading to quantization of the lateral wave vector. However, the first experimental study demonstrated that this explanation cannot account for the measured size of the transport gap [16]. By now, many papers – both theoretical and experimental in nature – have been published to improve our understanding of the nature of the gap [30, 17, 9, 49, 26, 47, 29, 44, 7, 24, 34, 37, 42, 51]. It is beyond the scope of this chapter to review the results on nanoribbons and constrictions.

However, the behaviour of constrictions is very complex, e. g. they show Coulomb blockade of overlapping, strongly localized states. Therefore, it is surprising to see how well behaved graphene QDs are, considering the fact that they are connected by two constrictions to source and drain contacts. In conventional transport experiments – such as the one we presented in the previous section – it is very difficult to disentangle influences of the constrictions and effects of the appointed lithographic QD. For these reasons, we will present scanning-gate measurements of a different graphene QD in this section. This technique allows us to locate localized states directly in real space, in such a way distinguishing the different influences of constrictions and QD states on transport.

In scanning-gate microscopy (SGM), the conducting tip of an atomic-force microscope is used to induce a potential perturbation in the nanostructure. It thus alters the conductance of the particular sample as a function of tip position. A scanning-gate measurement is then performed by recording the conductance of the sample while scanning the tip over the nanostructure. In other words, the tip is nothing else than a movable gate. Whereas in conventional transport experiments we change the voltage applied to a gate, in SGM we move the gate, thus altering its lever arm. If we perform SGM on a QD tuned to the Coulomb-blockade regime, Coulomb resonances are lines of constant electrochemical potential. They form closed loops that we refer to as *Coulomb rings*. Intuitively, we can identify the centres of one set of Coulomb rings with the position of the localized state giving rise to the Coulomb blockade. If there is more than one localized state, they will give rise to more sets of intersecting Coulomb rings with different centres. In the following, we demonstrate this technique on a graphene QD where localized states not only form in the QD but also in the constrictions [41].

The graphene QD under investigation here was fabricated in the same manner as the previous device. An atomic-force micrograph of the device after cooldown is shown in Fig. 17.9a: the QD is connected via two constrictions to source and drain; three in-plane gate – the left and right side gates LG and RG and the plunger gate PG – can be used to tune the QD and the constrictions. The QD has a lithographic radius of $r_{\mathrm{dot}} \approx 110\,\mathrm{nm}$; the two constrictions have a width of 30–40 nm.

Conventional transport measurements confirmed that the device was working properly. In particular, as we can see in Fig. 17.9b, the CNP was found to be around $V_{\mathrm{BG}} \approx 30\,\mathrm{V}$ in

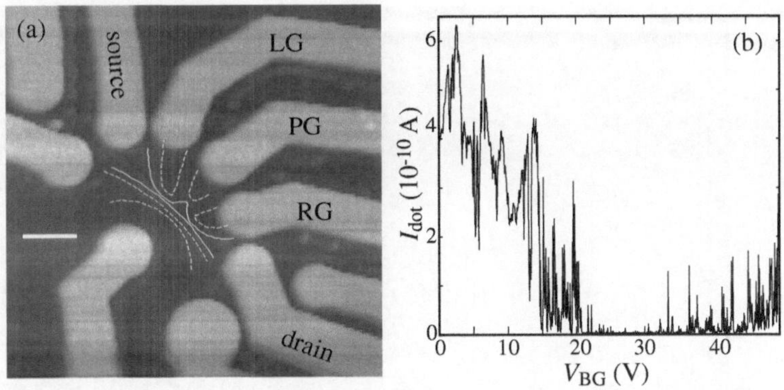

Figure 17.9 (a) *In-situ* atomic-force micrograph of the sample after cooldown at $T \approx 2.6$ K. This image was taken after positioning the tip of our home-built low-temperature AFM [11] above the sample. The outline of the graphene QD is shown with white, continuous lines; white, dashed lines denote nearby in-plane gates. The scale bar is 1 μm. (b) Current through the QD as a function of back-gate voltage. Due to unintentional doping, the CNP is shifted to approximately 30 V; the transport gap ranges from roughly 15 to 45 V. Figure adapted from Ref. [41].

back-gate voltage[3] and, from Coulomb-blockade diamonds, an addition energy of $\Delta E_C^{dot} =$ 3.5 meV was deduced. No signatures of excited states were observed; this is in agreement with previous experiments on graphene QDs of similar size [45].

17.4.1 Scanning-gate measurements

A first series of scanning-gate images were recorded in the Coulomb-blockaded hole regime. The conductance maps shown in Fig. 17.10 have an area of $1.4 \times 1.4\,\mu\text{m}^2$. We can identify three sets of concentric rings. The centers of each are marked with a cross in panel (a). The rings whose centers are labeled "CR" are identified as Coulomb resonances of the QD as verified by the presence of Coulomb diamonds when sweeping the tip and bias voltage. More strikingly, we can identify two more resonant states labeled "R1" and "R2". Their centres are well separated from the QD. If we align the lithographic outline of the QD device with the scanning-gate images such that Coulomb rings are centred at the lithographic QD,[4] states R1 and R2 are located in the source and drain constrictions. While the Coulomb resonances of the QD show the expected behavior, we are more interested in the unintended states in the constrictions. The white ellipse in panels (a) – (d) trace the same resonance of R1. The fact that its diameter increases with increasing left-gate voltage shows that the state in the constriction can be tuned in a similar manner as the QD. Since

[3] This large shift is presumably the result of charged impurities on or near the graphene surface.
[4] In fact, the center of the Coulomb rings is offset from the topographic centre of the QD by around 240 nm. This is a frequently observed behavior [40].

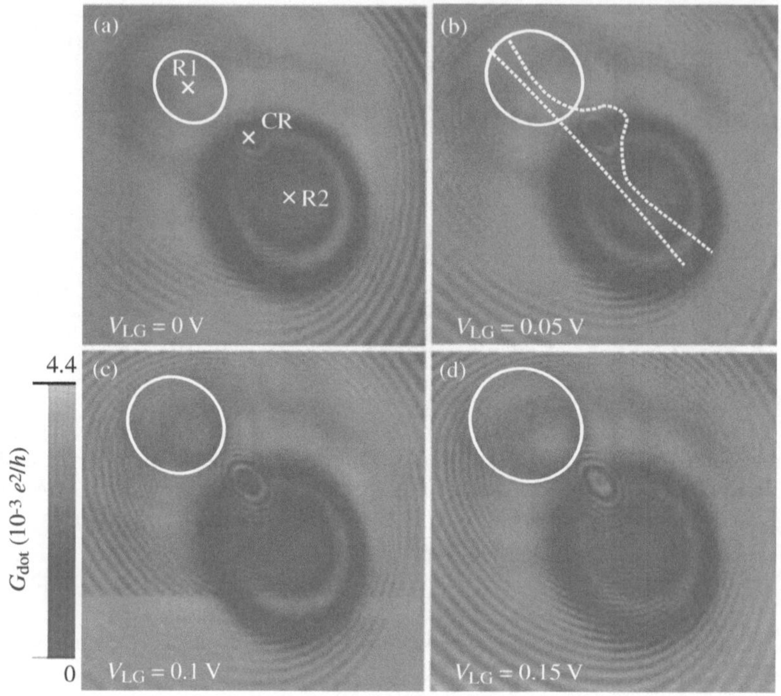

Figure 17.10 A series of conductance maps of an area of $1.4 \times 1.4\,\mu m^2$ where the left-gate voltage was changed according to the values given in each images. Gate voltages are $V_{BG} = 12$ V and $V_{tip} = 2$ V, and the sample was cooled down to around 2.6 K. A symmetric bias voltage of $V_{bias} = 300\,\mu$V was applied in order to drive a current through the QD. The tip was scanned at a constant height of 120 nm. We observe three sets of resonances whose centers are denoted with crosses in panel (a): one set denoted as "CR" are identified as Coulomb resonances of the QD, the other two sets labeled "R1" and "R2" are resonances at a different location. Alignment of the scanning-gate image with the topographic outline of the QD structure (white, dashed lines in panel (b)) shows that resonances R1 and R2 are located in the constrictions. The white, continuous ellipses in all four panels trace the same resonant state in the source constrictions. The fact that its diameter increases in the series shows that the gates tune not only Coulomb resonances of the QD but also the resonant states in the constrictions. The conductance G_{dot} does not drop to zero between two Coulomb rings because the measurements were done at the edge of the transport gap in back-gate voltage where the coupling of dot states to source and drain is rather strong. Figures adapted from Ref. [41].

there is only one apparent set of resonances R1 and R2 per constriction, we conclude that there is only one resonant state in each constriction at this value of the Fermi energy.

We want to analyse the behaviour of the states in the constrictions in more detail. For this reason, we performed SGM at a base temperature of $T = 85$ mK. The result is presented in Fig. 17.11a, where Coulomb rings are now sharper because of the lower temperature. The two white arrows point to resonances of the drain constriction. They show different characteristics than before in Fig. 17.10 because the resonances here are accompanied by avoided crossings of the Coulomb resonances. This is particularly well observable at

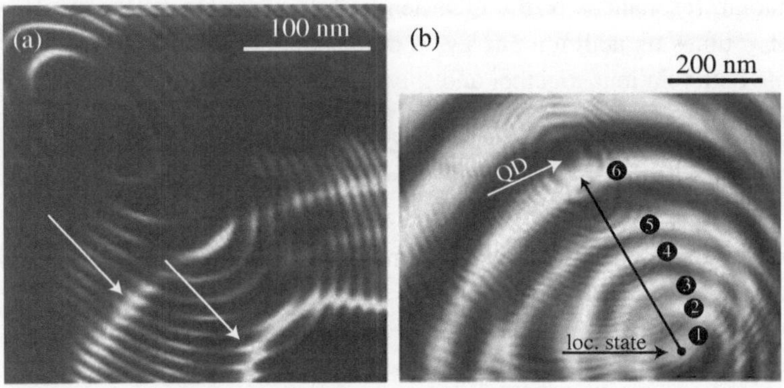

Figure 17.11 SGM in the hole (a) and in the electron regime (b). In (a), the sample was cooled down to 85 mK so that Coulomb rings of the QD (centred in the upper left corner) are sharper than in the previous measurement ($V_{\rm bias} = 35\,\mu$V, $V_{\rm BG} = 12$ V, $V_{\rm tip} = -100$ mV, and tip height 40 nm). Resonances of the state in the drain constriction are accompanied by avoided crossings of the Coulomb resonances (see right white arrow). This is interpreted as charging of a strongly localized state in the constriction with a discrete charge. (b) In the electron regime ($V_{\rm BG} = 50$ V, $V_{\rm RG} = 2$ V, $V_{\rm LG} = 12$ V, $T = 1.76$ K), qualitatively similar patterns are observed: Coulomb rings of the QD (white arrow labeled "QD") are strongly modulated by resonances of the state in the constriction (black arrow labeled "loc. state"). The black line starts in the centre of the localized state at the first resonance and ends at its sixth resonance. This line has a length of $\Delta = 423$ nm and the tip moving along the line shifts the energy of the localized state in the constriction by five times its addition energy.

the inner resonance. This avoided crossing occurs because the state in the constriction is strongly localized such that it can be charged with discrete charges. We will therefore refer to these states now as localized states. A charging event of such a localized state will then shift the electrochemical potential of the QD because of mutual capacitive coupling. This is similar to a double QD where the mutual capacitive coupling also leads to avoided crossings of Coulomb resonances. The precise tuning of the localized state determines whether it shows charge quantization as in Fig. 17.11a or not, Fig. 17.10.

An addition energy $\Delta E_{\rm C}^{\rm loc}$ can be attributed to localized states showing charging effects in the same manner as for QDs. In order to determine the addition energy, we focus on Fig. 17.11b, which presents a scanning-gate image in the electron regime. We see a qualitatively similar pattern as before: Coulomb rings centered around the QD (white arrow labeled "QD"), and resonances centered around the localized state in the drain constriction (black arrow labeled "loc. state"). At the center, we identify the first resonance and then count up to six resonances along the black line. In others words, moving the tip along that line shifts the electrochemical potential of the localized state by five time its addition energy,

$$n_{\rm loc}\Delta E_{\rm C}^{\rm loc} = \left(\alpha_{\rm loc}\left(\boldsymbol{r}_{\rm tip} = \boldsymbol{r}_{\rm loc}\right) - \alpha_{\rm loc}\left(\boldsymbol{r}_{\rm tip} = \boldsymbol{r}_{\rm loc} + \boldsymbol{\Delta}\right)\right) eV_{\rm tip}, \tag{17.9}$$

where $r_{\rm tip}$ and $r_{\rm loc}$ are the positions of the tip and the localized state, $\boldsymbol{\Delta}$ is the vector pointing along the black line, and $n_{\rm loc} = 5$. Along the same black line, we cross about

26 ± 3 Coulomb resonances of the QD, hence shifting its electrochemical potential by $n_{\mathrm{dot}} = 25 \pm 3$ times its addition energy. A corresponding relation for $n_{\mathrm{dot}} \Delta E_{\mathrm{C}}^{\mathrm{dot}}$ can be given. Putting both relations together and solving for $\Delta E_{\mathrm{C}}^{\mathrm{loc}}$ yields

$$\Delta E_{\mathrm{C}}^{\mathrm{loc}} = \underbrace{\frac{\Delta\alpha_{\mathrm{loc}}}{\Delta\alpha_{\mathrm{dot}}}}_{\sim 1} \frac{n_{\mathrm{dot}}}{n_{\mathrm{loc}}} \Delta E_{\mathrm{C}}^{\mathrm{dot}}. \tag{17.10}$$

The first factor is the ratio of the tip lever arms on the localized state and the QD, and we expect this to be of the order of one. With the known addition energy $\Delta E_{\mathrm{C}}^{\mathrm{dot}} = 3.5\,\mathrm{meV}$, this yields $\Delta E_{\mathrm{C}}^{\mathrm{loc}} \approx 18\,\mathrm{meV}$ or an extent of $r_{\mathrm{loc}} = r_{\mathrm{dot}} / \left(\Delta E_{\mathrm{C}}^{\mathrm{loc}} / \Delta E_{\mathrm{C}}^{\mathrm{dot}}\right) \approx 20\,\mathrm{nm}$. A slightly different method for extracting the extent of the localized state in the source constriction yields 10 – 13 nm [41].

In summary, we deduce the following properties of the constrictions forming the QD: localized states form in the constrictions; we observed typically one per constriction at a given Fermi energy. However, this is probably caused by the short length of the constriction. The localized states interact with the QD capacitively. They are coupled to source or drain leads via tunnel barriers. The tunnel coupling strength depends strongly on the Fermi energy giving rise to broad resonances for strong coupling (cf. Fig. 17.10) and to narrow resonances for weak coupling. Below the threshold of e^2/h, the coupling is so weak that charge quantization becomes relevant. This is then visible by avoided crossings of Coulomb resonances (cf. Fig. 17.11a). We determined a radius of 10 – 20 nm for the localized states. This number is in agreement with scanning-tunneling studies on extended graphene samples where similar sizes could be determined for electron–hole puddles [56, 6].

17.5 Summary and outlook

The use of the spins of charge carriers trapped on QDs for information processing has been proposed [23]. Graphene might be particularly interesting because of the predicted long spin-coherence times due to weak spin–orbit coupling and weak hyperfine interaction [50]. The experimental verification of this prediction, in particular a measurement of coherence times, is one of the big goals of graphene QD research. Huge progress has been made during the past few years, starting with the measurement of the Coulomb blockade in graphene QDs [35, 46, 22]. The observation of excited states reported here was the next essential step. Well-defined spin states are expected to form in the few charge-carrier regime, i.e. close to the electron–hole crossover. The identification of the crossover turned out to be trickier than expected. A technique of how it can be identified was explained here and applied to two different graphene QD samples. For the second example, a clear signature of the crossover region was presented. For this particular sample, spin states could consequently be determined [15].

On the other hand, graphene double dots have been fabricated and characterized in transport experiments [21, 27, 28]. Even though transport through excited states was observed, well-defined spin states have not been identified so far; in particular spin-blockade was not observed in either of the two groups. As in the case of single QDs, it is presumably necessary to go to the few charge-carrier regime for such a measurement. The next milestone in graphene QD research is therefore the identification of the electron–hole crossover in double dots (at least the crossover region) and the observation of singlet and triplet states and of Pauli spin blockade. Charge-detection techniques have already been demonstrated in graphene [13] so that the measurement of coherence times will come into reach then.

Our SGM technique has provided insight into the microscopic properties of charge transport in graphene QDs for the first time. The observation of localized states confirmed earlier speculations about the existence of such states in nanoribbons [44, 17, 34, 49, 26, 47]. Although it is impossible to say at this stage whether the localized states are formed due to edge roughness or bulk disorder, our measurements demonstrate how strongly they influence and potentially hamper charge transport through QDs. A more refined SGM study on graphene nanoribbons might shed light on the precise localization mechanism, helping to improve existing graphene nanostructures.

References

[1] Beenakker, C. W. J. 1991. Theory of Coulomb-blockade oscillations in the conductance of a quantum dot. *Phys. Rev. B*, **44**(4), 1646–1656.

[2] Berry, M. V. and Mondragon, R. J. 1987. Neutrino billiards: time-reversal symmetry-breaking without magnetic fields. *Proc. R. Soc. Lond. A*, **412**(1842), 53–74.

[3] Chen, H.-Y., Apalkov, V. and Chakraborty, T. 2007a. Fock–Darwin states of dirac electrons in graphene-based artificial atoms. *Phys. Rev. Lett.*, **98**(18), 186803.

[4] Chen, Z., Lin, Y.-M., Rooks, M. J., and Avouris, P. 2007b. Graphene nano-ribbon electronics. *Physica E: Low-dimensional Systems and Nanostructures*, **40**(2), 228–232. International Symposium on Nanometer-Scale Quantum Physics.

[5] Darwin, C. G. 1930. The diamagnetism of the free electron. *Proc. Cambridge Philos. Soc.*, **27**(86).

[6] Deshpande, A., Bao, W., Miao, F., Lau, C. N. and LeRoy, B. J. 2009. Spatially resolved spectroscopy of monolayer graphene on SiO_2. *Phys. Rev. B*, **79**(20), 205411.

[7] Evaldsson, M., Zozoulenko, I. V., Xu, H. and Heinzel, T. 2008. Edge-disorder-induced Anderson localization and conduction gap in graphene nanoribbons. *Phys. Rev. B*, **78**(16), 161407.

[8] Fock, V. 1928. Bemerkung zur Quantelung des harmonischen Oszillators im Magnetfeld. *Z. Physik*, **47**, 446.

[9] Gallagher, P. Todd, K. and Goldhaber-Gordon, D. 2010. Disorder-induced gap behavior in graphene nanoribbons. *Phys. Rev. B*, **81**(11), 115409.

[10] Geim, A. K., and Novoselov, K. S. 2007. The rise of graphene. *Nat Mater*, **6**(3), 183–191.

[11] Gildemeister, A. E., Ihn, T., Barengo, C., Studerus, P. and Ensslin, K. 2007. Construction of a dilution refrigerator cooled scanning force microscope. *Rev. of Sci. Instrum.*, **78**(1), 013704–6.

[12] Graf, D., Molitor, F., Ensslin, K. *et al.* 2007. Spatially resolved Raman spectroscopy of single- and few-layer graphene. *Nano Lett.*, **7**(2), 238–242.

[13] Guttinger, J., Stampfer, C., Hellmuller, S. *et al.* 2008. Charge detection in graphene quantum dots. *Appl. Phys. Lett.*, **93**(21), 212102.
[14] Guttinger, J., Stampfer, C., Libisch, F. *et al.* 2009. Electron–hole crossover in graphene quantum dots. *Phys. Rev. Lett.*, **103**(4), 046810.
[15] Güttinger, J., Frey, T., Stampfer, C., Ihn, T. and Ensslin, K. 2010. Spin states in graphene quantum dots. *Phys. Rev. Lett.*, **105**(11), 116801.
[16] Han, M. Y., Özyilmaz, B., Zhang, Y. and Kim, P. 2007. Energy band-gap engineering of graphene nanoribbons. *Phys. Rev. Lett.*, **98**(20), 206805.
[17] Han, M. Y., Brant, J. C. and Kim, P. 2010. Electron transport in disordered graphene nanoribbons. *Phys. Rev. Lett.*, **104**(5), 056801.
[18] Ihn, T. 2010. *Semiconductor Nanostructures*. Oxford: Oxford University Press.
[19] Libisch, F., Rotter, S., Güttinger, J., Stampfer, C. and Burgdörfer, J. 2010. Transition to Landau levels in graphene quantum dots. *Phys. Rev. B*, **81**(24), 245411.
[20] Libisch, F., Stampfer, C. and Burgdörfer, J. 2009. Graphene quantum dots: Beyond a Dirac billiard. *Phys. Rev. B*, 115423.
[21] Liu, R.-B., Yao, W. and Sham, L. J. 2010. Quantum computing by optical control of electron spins. *Advances Phys.*, **59**(5), 703–802.
[22] Liu, X., Oostinga, J. B., Morpurgo, A. F. and Vandersypen, L. M. K. 2009. Electrostatic confinement of electrons in graphene nanoribbons. *Phys. Rev. B*, **80**(12), 121407.
[23] Loss, D. and DiVincenzo, D. P. 1998. Quantum computation with quantum dots. *Phys. Rev. A*, **57**(1), 120–126.
[24] Martin, I. and Blanter, Y. M. 2009. Transport in disordered graphene nanoribbons. *Phys. Rev. B*, **79**(23), 235132.
[25] Martin, J., Akerman, N., Ulbricht, G. *et al.* 2008. Observation of electron–hole puddles in graphene using a scanning single-electron transistor. *Nat Phys*, **4**(2), 144–148.
[26] Molitor, F., Jacobsen, A., Stampfer, C. *et al.* 2009a. Transport gap in side-gated graphene constrictions. *Phys. Rev. B*, **79**(7), 075426.
[27] Molitor, F., Dröscher, S., Güttinger, J. *et al.* 2009b. Transport through graphene double dots. *Appl. Phys. Lett.*, **94**(22), 222107.
[28] Molitor, F., Knowles, H., Dröscher, S. *et al.* 2010. Observation of excited states in a graphene double quantum dot. *EPL (Europhysics Letters)*, **89**(6), 67005.
[29] Mucciolo, E. R., Neto, A. H. C. and Lewenkopf, C. H. 2009. Conductance quantization and transport gaps in disordered graphene nanoribbons. *Phys. Rev. B*, **79**(7), 075407.
[30] Nakada, K., Fujita, M., Dresselhaus, G. and Dresselhaus, M. S. 1996. Edge state in graphene ribbons: nanometer size effect and edge shape dependence. *Phys. Rev. B*, **54**(24), 17 954–17 961.
[31] Nazarov, Y. V. and Blanter, Y. M. 2009. *Quantum Transport: Introduction to Nanoscience*. Cambridge: Cambridge University Press.
[32] Novoselov, K. S., Geim, A. K., Morozov, S. V. *et al.* 2004. Electric field effect in atomically thin carbon films. *Science*, **306**(5696), 666–669.
[33] Novoselov, K. S., McCann, E., Morozov, S. V. *et al.* 2006. Unconventional quantum Hall effect and Berry's phase of 2[pi] in bilayer graphene. *Nat. Phys.* **2**(3), 177–180.
[34] Oostinga, J. B., Sacépé, B., Craciun, M. F. and Morpurgo, A. F. 2010. Magnetotransport through graphene nanoribbons. *Phys. Rev. B*, **81**(19), 193408.
[35] Ponomarenko, L. A., Schedin, F., Katsnelson, M. I. *et al.* 2008. Chaotic Dirac billiard in graphene quantum dots. *Science*, **320**(5874), 356–358.

[36] Recher, P., Nilsson, J., Burkard, G. and Trauzettel, B. 2009. Bound states and magnetic field induced valley splitting in gate-tunable graphene quantum dots. *Phys. Rev. B*, **79**(8), 085407.

[37] Ritter, K. A. and Lyding, J. W. 2009. The influence of edge structure on the electronic properties of graphene quantum dots and nanoribbons. *Nat. Mater.*, **8**(3), 235–242.

[38] Schnez, S., Ensslin, K., Sigrist, M. and Ihn, T. 2008. Analytic model of the energy spectrum of a graphene quantum dot in a perpendicular magnetic field. *Phys. Rev. B*, **78**(19), 195427.

[39] Schnez, S., Molitor, F., Stampfer, C. *et al.* 2009. Observation of excited states in a graphene quantum dot. *Appl. Phys. Lett.*, **94**(1), 012107.

[40] Schnez, S. 2010. *Transport Properties and Local Imaging of Graphene Quantum Dots*. Ph.D. thesis, ETH Zurich.

[41] Schnez, S., Güttinger, J. Huefner, M. *et al.* 2010. Imaging localized states in graphene nanostructures. *Phys. Rev. B*, **82**(16), 165445.

[42] Schubert, G., Schleede, J. and Fehske, H. 2009. Anderson disorder in graphene nanoribbons: a local distribution approach. *Phys. Rev. B*, **79**(23), 235116.

[43] Semenoff, G. W. 1984. Condensed-matter simulation of a three-dimensional anomaly. *Phys. Rev. Lett.*, **53**(26), 2449–2452.

[44] Sols, F., Guinea, F. and Neto, A. H. 2007. Coulomb blockade in graphene nanoribbons. *Phys. Rev. Lett.*, **99**(16), 166803.

[45] Stampfer, C., Guttinger, J., Molitor, F. *et al.* 2008a. Tunable Coulomb blockade in nanostructured graphene. *Appl. Phys. Lett.*, **92**(1), 012102.

[46] Stampfer, C., Schurtenberger, E., Molitor, F. *et al.* 2008b. Tunable Graphene Single Electron Transistor. *Nano Letters*, **8**(8), 2378–2383.

[47] Stampfer, C., Güttinger, J., Hellmüller, S. *et al.* 2009. Energy gaps in etched graphene nanoribbons. *Phys. Rev. Lett.*, **102**(5), 056403.

[48] Tarucha, S., Austing, D. G., Honda, T., van der Hage, R. J. and Kouwenhoven, L. P. 1996. Shell filling and spin effects in a few electron quantum dot. *Phys. Rev. Lett.*, **77**(17), 3613–3616.

[49] Todd, K., Chou, H.-T., Amasha, S. and Goldhaber-Gordon, D. 2008. Quantum dot behavior in graphene nanoconstrictions. *Nano Letters*, **9**(1), 416–421.

[50] Trauzettel, B., Bulaev, D. V., Loss, D. and Burkard, G. 2007. Spin qubits in graphene quantum dots. *Nat. Phys.*, **3**(3), 192–196.

[51] Wakabayashi, K., Takane, Y., Yamamoto, M., and Sigrist, M. 2009. Electronic transport properties of graphene nanoribbons. *New J. Phys.*, **11**(9), 095016.

[52] Wallace, P. R. 1947. The band theory of graphite. *Phys. Rev.*, **71**(9), 622–634.

[53] Wimmer, M., Akhmerov, A. R. and Guinea, F. 2010. Robustness of edge states in graphene quantum dots. *ArXiv e-prints*, Mar.

[54] Young, A. F. and Kim, P. 2009. Quantum interference and Klein tunnelling in graphene heterojunctions. *Nat. Phys.*, **5**(3), 222–226.

[55] Zhang, Y., Tan, Y.-W., Stormer, H. L. and Kim, P. 2005. Experimental observation of the quantum Hall effect and Berry's phase in graphene. *Nature*, **438**(7065), 201–204.

[56] Zhang, Y., Brar, V. W., Girit, C., Zettl, A., and Crommie, M. F. 2009. Origin of spatial charge inhomogeneity in graphene. *Nat. Phys.*, **5**(10), 722–726.

Part VI

Single dots for future telecommunications applications

18

Electrically operated entangled light sources based on quantum dots

R. M. Stevenson, A. J. Bennett and A. J. Shields

18.1 Introduction

Quantum information technology promises to offer incredible advantages over current digital systems, allowing intractable problems in science and engineering to be tackled almost instantaneously through quantum computing, and unconditionally secure communication over long distances using quantum key distribution. Many schemes have been developed to implement quantum computing, including using linear optics [28]. The linear optical approach has proved popular due to the limited decoherence of photons with the environment, and accessibility of the components required for simple experiments. At the heart of an optical quantum computer, or extended range quantum key distribution using quantum relays or repeaters [14, 8, 24], lie entangled photons. The characteristics of the sources that create entangled photons, and their properties, are therefore central to realizing the full potential of such applications.

Quantum dots are one technology with which entangled light sources can be built [6]. Although first realised only relatively recently [49], they in principle offer key fundamental and practical advantages over other entangled photon sources. In the fundamental sense, quantum dots can be triggered, so that no more than one entangled photon pair is emitted at a time. This is in stark contrast to Poissonian entangled light sources [47, 27, 13], including the most widely used parametric down-conversion, where zero or multiple photon-pairs are usually emitted due to their probabilistic nature. Furthermore quantum dots have the potential to operate with high efficiency, with current experiments reporting up to 72% collection efficiency for the first and second photon [9, 12]. Operation on-demand, where close to exactly one entangled photon pair is collected, is important to reduce errors in quantum logic operations and becomes increasingly essential as the size (and usefulness) of a quantum processor increases.

However, useful optical quantum computers with the potential to outperform their digital counterparts require a large number of qubits, gates, and entangled photons. Therefore the practical properties of entangled photon sources are equally important in order to realise such machines. In this regard, quantum dots have great potential, as they are based on commercial semiconductor technology, and many thousands of quantum dot entangled light sources can be integrated upon a single chip.

Quantum Dots: Optics, Electron Transport and Future Applications, ed. Alexander Tartakovskii. Published by Cambridge University Press.

Perhaps the most appealing of all practical advantages offered by quantum dots is their potential to be controlled electrically. All pre-existing sources of entangled-photon pairs require a laser for optical excitation [2, 13, 47, 27, 49], so large-scale optical quantum information applications would require complex optical control fields to address each entangled photon pair source. In this chapter we review our recent work which demonstrates the first electrically driven entangled light source, and additionally the first photon pair source where the entanglement of the emitted photons can be activated or deactivated electrically. Furthermore we present analysis of our results that reveals the entangled photon pairs to be fully coherent, indicating that quantum dots generate fundamentally high quality entangled states. This insight and electrical operation shows the true potential of quantum dots for large scale, useful quantum computing.

18.1.1 Principles of entangled light generation with quantum dots

Quantum dots have discrete energy spectra for each charge configuration, allowing a single optical transition, such as the neutral exciton to the ground state, to be isolated spectrally from all others. This forms the basis of quantum dot single-photon emitters [37, 38], which can also be driven electrically [59] and have been shown to create entangled photon pairs probabilistically through two-photon interference [16].

Advantageously, however, single quantum dots are capable of producing entangled photons directly, through the radiative decay of the biexciton state of two electrons and holes to the ground state [6]. The process is shown schematically in Figure 18.1a. Initially the biexciton (XX) state is excited using for example an electrical current or laser pulse. Owing to the Pauli exclusion principle, the biexciton ground state contains two electrons and two holes of opposing spin. In our system, the holes are heavy with spin $\pm 3/2$ as the light holes lie to much higher energy due to the strain within the InGaAs material system, and are not confined. Thus selection rules dictate that the electrons with spin $\pm 1/2$ must recombine with a hole of opposite spin sign in order to emit a photon with angular momentum ± 1, corresponding to left-hand (L) and right-hand (R) circular polarisation. This leads to two recombination paths via the intermediate exciton state X to the ground state G, producing a left-hand (L) followed by right-hand (R) circularly polarised photon, or vice versa, dependent on the spin of the first exciton to recombine.

Assuming the alternate decay paths are otherwise indistinguishable, the emitted photon pair is expected to be in the maximally entangled two-photon state $\Psi^+ = (|L_{XX}R_X\rangle + |R_{XX}L_X\rangle)/\sqrt{2}$ [6]. However, early experiments revealed this not to be the case [53, 45, 55], due to observation of linear polarisation dependent photon energies, caused by significant fine-structure-splitting of the intermediate exciton state. The fine-structure-splitting is caused by asymmetrical structural properties of the quantum dot such as physical elongation, variation in alloy composition, and strain, and has two important consequences. The first is that the exciton spin states are hybridised, resulting in linearly polarised eigenstates, which we denote as V (vertical) and H (horizontal). The second is that a phase

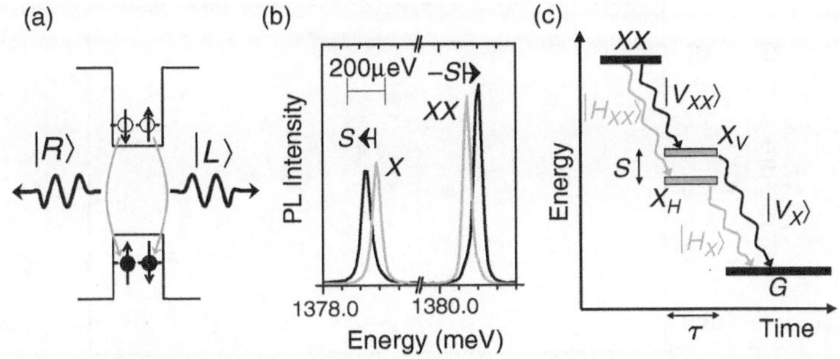

Figure 18.1 Entangled light generation in quantum dots. (a) Schematic of radiative decay of the biexciton (XX) state in a quantum dot. Open and closed discs indicate confined electrons and holes, with arrows indicating spin. Emitted photons are left (L) or right (R) circularly polarised. (b) Linear polarisation splitting of the emission lines in typical quantum dots. Black and grey lines denote vertical (V) and horizontal (H) polarisation. (c) Time-resolved picture of the radiative decay of XX in a quantim dot with fine structure splitting S. Figures adapted from Refs. [44, 53, 51].

difference is acquired between the two alternate decay paths whilst in the intermediate exciton state [51], as shown in Figure 18.1c, resulting in emission of a time-evolving entangled state $\Psi^+ = (|H_{XX}H_X\rangle + e^{iS\tau/\hbar}|V_{XX}V_X\rangle)/\sqrt{2}$. Thus averaged over time τ, little or no entanglement is detected in typical experiments.

18.1.2 Controlling fine-structure-splitting

Since identification of the fine-structure-splitting as a barrier to producing entangled photon pairs, a vast multitude of schemes has been devised by many groups across the world. These include control of physical properties using growth or annealing [57, 15, 39], application of external magnetic [50], strain [46] or electric [20, 30, 5] fields, post-selection in time [1] or energy [51], and optical effects such as enhancement of the spontaneous emission rate [12] or interaction with laser fields [40].

The first scheme reported that resulted in quantum dots with negligible fine-structure-splitting [57], and also the first observation of entangled light emission by a quantum dot [49], was by controlling the size of the quantum dots during growth. This scheme is summarised by Figure 18.2, which plots the fine-structure-splitting against the emission energy for many quantum dots. A clear trend is observed which shows reducing FSS with increasing emission energy. Crucially, the FSS reaches an average of zero around 1.4 eV, then becomes negative, indicating reversal of the energetic ordering of the H and V eigenstates. The reduction and subsequent inversion of the FSS is attributed to poorer confinement of the electron and holes states as the quantum dot becomes smaller, emitting with higher emission energy. To realise an entangled light sources, one must therefore control the growth parameters to produce dots emitting near 1.4 eV.

Entangled photon pair emission from quantum dots has now been observed by using a variety of approaches [49, 1, 21, 51, 40, 39, 12]. However, the above method for controlling

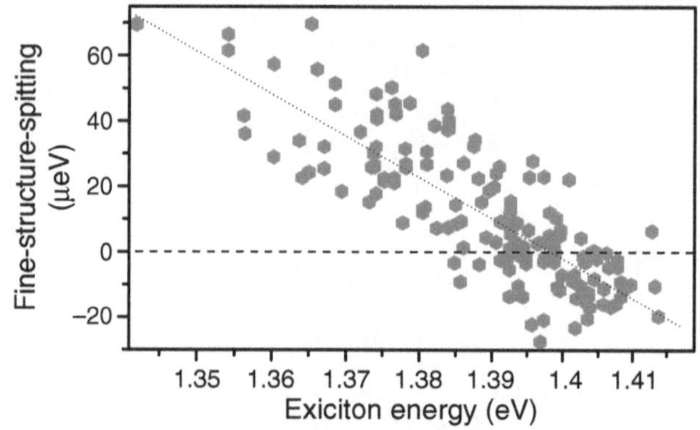

Figure 18.2 Fine-structure-splitting (FSS) versus quantum dot emission energy for many quantum dots. The average FSS is zero for quantum dots with emission energy around 1.4 eV. Figure adapted from Ref. [57].

FSS is our preferred method, as no additional processing or treatment is required to produce entangled light sources.

18.2 Electrically driven entangled light generation

18.2.1 Design of an entangled-light-emitting diode

Our electrically driven entangled light source relies on a quantum dot within its active region. However, unlike all previous experiments which employed optical excitation, the biexciton state is formed by capture of two electrons and two holes via nonresonant electrical injection (Figure 18.3a). A single layer of quantum dots is embedded in a p–i–n-doped planar microcavity, grown by molecular beam epitaxy (MBE). Two features of the design are crucial to successful operation as an entangled-LED. The first is to minimize the fine-structure-splitting by creating dots emitting around 1.4 eV, as described above. The second is an unusually thick cavity (2λ) and intrinsic region ($\sim$400 nm). This is to suppress electrons tunnelling into the dot from the n-doped region during biexciton decay, thereby minimising destruction of entanglement after emission of the first photon from charging of the intermediate exciton state.

We studied a quantum dot in our device which had a fine-structure-splitting of $0.4\pm0.1\,\mu$eV, significantly smaller that the time-bandwidth limited linewidth of $\sim$1 μeV. Its electroluminescence spectrum (recorded at 5 K) is shown in Figure 18.3b.

18.2.2 Entangled light generation under d.c. current

A d.c. current of 31 nAμm^{-2} was passed through the ELED, and photons from the X and XX transitions were resolved in polarisation and in time in order to measure the co- and cross-polarised second-order pair-correlation functions $g^{(2)}(\tau)$. Figure 18.4

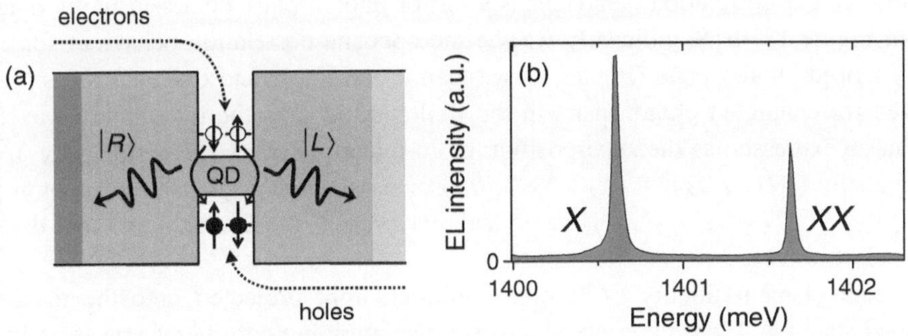

Figure 18.3 (a) Active region of the entangled-light-emitting diode (ELED). Carriers are injected from the n-type (dark) and p-type (light) regions of the diode to emit an entangled photon-pair. (b) CCD-recorded electroluminescence (EL) spectrum of the quantum dot measured here. Figures adapted from Ref. [44].

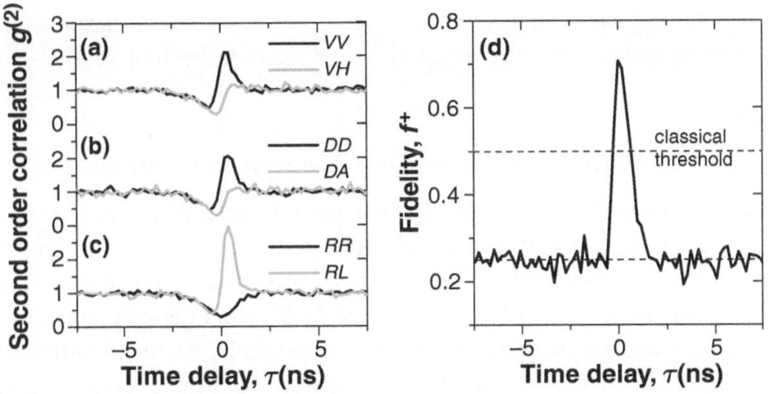

Figure 18.4 Polarised correlation results from d.c. electrical injection into the LED, measured in the (a) rectilinear, (b) diagonal and (c) circular polarisation bases, as a function of the time delay between the first and second photon, τ. Correlations measured for photons of the same (orthogonal) polarisation are shown in black (grey). (d) Fidelity, f^+, as a function of time delay between the first and second photon, τ. Figures adapted from Ref. [44].

shows the measurement of co- and cross-polarised correlations in the rectilinear (a), diagonal (b) and circular (c) polarisation bases. The characteristic shape of a radiative cascade is shown for example by the co-polarised linear correlation labelled VV. Emission of an biexciton photon is followed by an exciton photon of the same polarisation. As a result, $g^{(2)}$ is increased for small positive delays, with a corresponding dip in the cross-polarised correlation labelled HV. Both traces are suppressed for small negative delays owing to the unlikely detection of exciton photon emission closely followed by biexciton photon emission. For times away from $\tau = 0$, $g^{(2)}$ tends to 1 owing to uncorrelated emission events, caused by re-excitation between detection of the first and second photons. Similar correlation behaviour is seen for photon pairs polarised in the diagonal basis in Figure 18.4b. Significantly we see the opposite correlation behaviour for circularly polarised photon pairs (Figure 18.4c) than in the linear and diagonal cases. This is expected for entangled photon pairs in the Bell state Ψ^+, as the two-photon wavefunction can be expressed as the superposition of co-linearly $((|H_{XX}H_X\rangle + |V_{XX}V_X\rangle)/\sqrt{2})$, co-diagonally $((|D_{XX}D_X\rangle + |A_{XX}A_X\rangle)/\sqrt{2})$ or cross-circularly polarised photon pairs $((|L_{XX}R_X\rangle + |R_{XX}L_X\rangle)/\sqrt{2})$. Here D and A represent the diagonal and anti-diagonal linear polarisations.

The entanglement fidelity (f^+) of the emitted light projected onto the maximally entangled state $|\Psi^+\rangle$ was determined directly by combining correlations measured in the rectilinear, diagonal and circular polarisation bases [22]. This is plotted in Figure 18.4d for light emitted in d.c. mode. The peak at $\tau = 0$ gives a maximum d.c. fidelity $f^+ = 0.71 \pm 0.02$. This exceeds the 0.5 threshold for a source emitting a classically polarisation-correlated state by 9 standard deviations, proving that entangled photons have been electrically generated for the first time [44]. The measured fidelity is limited by several factors [22, 52], including the timing jitter on the photon detectors and re-excitation of the quantum dot part way through the cascade, and is discussed in further detail in Section 18.4.

18.2.3 Entangled light generation under a.c. current

Often applications benefit or require operation in pulsed mode, which we achieve by application of an a.c. driving current. Figure 18.5 shows the results of polarised photon-pair correlation experiments conducted when the device was driven with sub-nonosecond current pulses at a repetition rate of 80 MHz. Figure 18.5 shows the measurement of $g^{(2)}$ in the rectilinear (a), diagonal (b) and circular (c) polarisation bases. In both Figure 18.5a and (b) the dominant peak belongs to the co-polarised trace, and in (c) to the cross-polarised trace as expected. From this data we extract a maximum a.c. fidelity of $f^+ = 0.79 \pm 0.02$

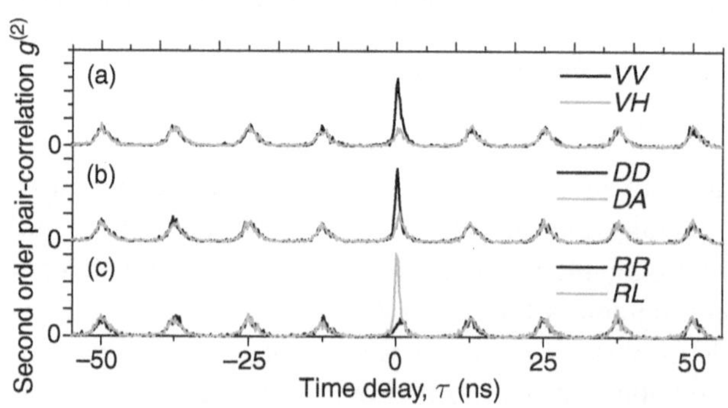

Figure 18.5 Polarised correlation measurements using a.c. electrical injection into the diode, measured in the (a) rectilinear, (b) diagonal and (c) circular polarisation bases. Figures adapted from Ref. [44].

[44]. This a.c. fidelity is larger than the d.c. fidelity presented above. It is likely that this is the result of the lower level of re-excitation of the quantum dot when driven by a short voltage pulse.

18.2.4 Temporal gating and violation of Bell's inequality

As the emitted light state evolves over time, the entanglement fidelity available to an application may be enhanced by limiting the time the detectors remain active. We explore this possibility by changing the gate width used in the measurement of f^+. For the zero period, gate width is defined as the maximum permitted time delay between the two photons to register a detection event, as shown schematically in Figure 18.6a. This changes the proportion of the detected biphotons to be accepted in the measurement of f^+. The fidelity as a function of the gate width and biphoton intensity is plotted in Figure 18.6b. Without rejecting any detected photon pairs (i.e. biphoton intensity = 1.0), $f^+ = 0.530 \pm 0.010$, which is above the classical threshold of 0.5, proving entangled light emission. The fidelity rises quickly as the proportion of biphoton intensity accepted is reduced. This leads to the maximum a.c. fidelity of 0.83±0.03, when 6% of biphoton coincidences closest to $\tau = 0$ are accepted.

Using these results it is also possible to determine Bell's parameters within a quantum mechanics framework in three planes of the Poincaré sphere [58].

Figure 18.6c displays these three Bell parameters measured as a function of the biphoton intensity (bottom axis) and gate width (top axis). Bell's parameters increase as the gate width decreases due to an improving degree of correlation in all three polarisation bases.

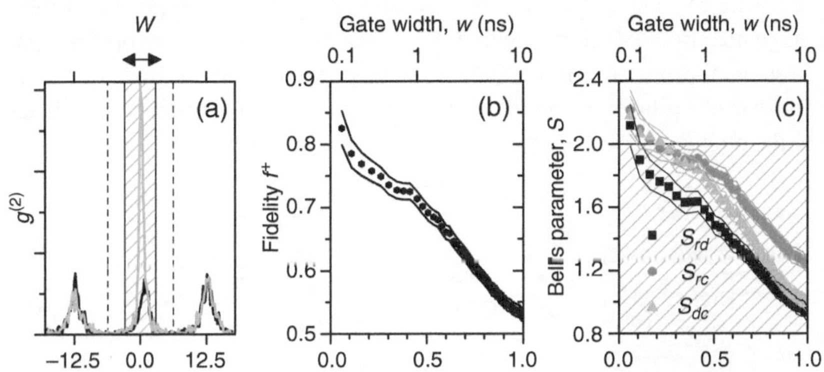

Figure 18.6 (a) Schematic of the gating scheme. Photon pairs are post-selected when their temporal separation τ lies within a gate region as shown. (b) Fidelity f^+ at $\tau = 0$ as a function of the proportion of the total biphoton intensity that is analysed, varied by changing the gate width w (top axis). (c) Bell's parameters at $\tau = 0$ as a function of the proportion of the total biphoton intensity that is analysed. Shaded region indicate values accessible by classical sources. Figures adapted from Ref. [44].

S_{rd} is less than S_{rc} and S_{dc}, as they include correlations in the circular basis which have the highest degree of correlation. At the smallest gate width of 0.1 ns we find; $S_{rd} = 2.12 \pm 0.13$, $S_{dc} = 2.18 \pm 0.12$ and $S_{rc} = 2.22 \pm 0.12$. The latter two violate Bell's inequality by more than one standard deviation, which is bound by two as indicated by the shaded region of Figure 18.6c. Overcoming this threshold shows that the entangled-LED is of high enough quality for applications such as a quantum repeater or relay [8, 24], entanglement-based quantum key distribution [14], and entanglement swapping [60, 41].

18.3 Electrical control of entangled light

18.3.1 Exposing quantum dots to large electric fields

To demonstrate electrically driven entangled light emission from quantum dots the barriers on either side of the dot must be engineered so neutral states dominate the spectrum. In the previous section this was achieved by increasing the thickness of the GaAs intrinsic region at the centre of the diode. We now discuss a device that uses high-energy barriers to suppress the tunnelling of carriers out of the dots.

In previous work, studying InGaAs/GaAs dots electrons tunnel out of the dots in a moderate vertical electric field of tens of kilovolts per centimetre [17] resulting in a rapid intensity reduction and broadening of the photoluminescence emission. Thus, the largest wavelength shift that can be achieved in the photoluminescence of a single InGaAs/GaAs dot is a fraction of a millielectronvolt.

In contrast, the samples we study here contain similar dots grown within a 10 nm thick GaAs quantum well and clad with 75% AlGaAs (Figure 18.7a). This alloy has a bandgap of ~2.2 eV and acts as a barrier to tunnelling for both types of carrier, allowing very large electric fields to be applied in reverse bias. However, by placing the dots in GaAs we retain all properties of the QDs we have previously studied. This layer structure forms a half-wavelength-thick optical spacer between two Bragg mirrors, the planar cavity enhancing the photon extraction efficiency in the vertical direction.

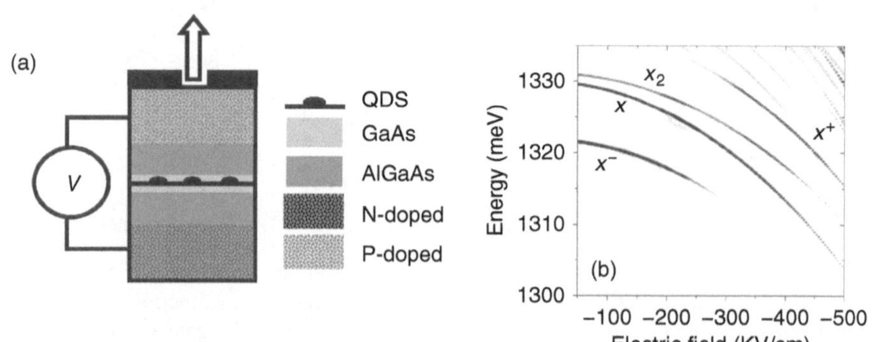

Figure 18.7 (a) Schematic of the centre of the device. (b) The d.c. Stark shift of the excitonic states as a function of field. Figures adapted from Ref. [5].

Figure 18.7b shows the photoluminescence excited with a 850 nm laser from a single dot as a function of the bias across it. The dot displays a characteristic arrangement of emission lines that can easily be identified as exciton (X), biexciton (XX) and charged excitons (X^-/X^+). The energy, E, of the quantum-confined states is shifted by the Stark effect according to the formula $E = E_0 - pF + \beta F^2$, where E_0 is the energy at zero field, p the "permanent dipole" moment and β the "polarisability" (which is negative). Surprisingly, these values of p and β are in good agreement with previously published data [4], albeit with those measurements taken at much smaller fields and/or using photocurrent techniques. However, by applying fields covering a range of 500 kV/cm we can observe Stark shifts of 25 meV in the X state photoluminescence, many times what has previously been possible.

18.3.2 Electrical tuning of the fine-structure-splitting

Thus far, most success in demonstrating entangled light emission from quantum dots has come from selecting those dots that emit near 885 nm. Some of these have a very small fine-structure-splitting (FSS, $|s|$) and many of the rest can be tuned to $|s| = 0$ with an in-plane magnetic field or rapid thermal annealing [49, 57]. Some success at tuning the FSS has been reported using strain [46] but this did not reach zero. Other tuning techniques such as magnetic field [49], strong coherent lasers [25, 40]. Of most relevance to the work here are reports on tuning the FSS with lateral electric field [32, 19, 56] and vertical electric field [5, 31, 35]. These have shown a limited tuning range. Within this plethora of techniques those few that have been able to minimise the FSS have seen the states cross [49, 25, 40, 32, 19].

Arguably, tuning the FSS with vertical electric field is the most practical technique yet reported, but the low confinement energies have limited the fields that can be applied to a few tens of kilovots per centimetre, so relatively small changes in FSS were observed [31, 35]. With the device here much larger electric fields can be applied. Shown in Figure 18.8b are data recorded for 100 dots, showing clearly that there is a large variation in FSS within the ensemble. In fact the mean FSS for QDs emitting at this wavelength is 109 μeV at 50 kV/cm with a standard deviation of 67 μeV. In contrast to this large variation in FSS, every dot displays a remarkably similar linear change of FSS with field, several examples of which are shown in Figure 18.8c. Even the unusual dot we have identified which has its eigenstates at right angles to the rest of those in the ensemble, and is so shown in Figure 18.8b with a negative $|s|$, has the same rate of change of FSS with electric field.

It is not clear what the mechanism is for the change with vertical field, but a detailed theoretical investigation may uncover this. The most elegant explanation is that the vertical dipole moments of the X_H and X_Y eigenstates differ due to anisotropy in the plane of the dot. The fact that the change in FSS with field is linear means the polarisability of the two dipoles is equal. A measured linear shift in FSS of 0.285 ± 0.019 eV.cm/kV suggests a few percent difference in dipole moments of the two exciton eigenstates. In each dot the field can induce a linear change in the FSS of over 100 μeV.

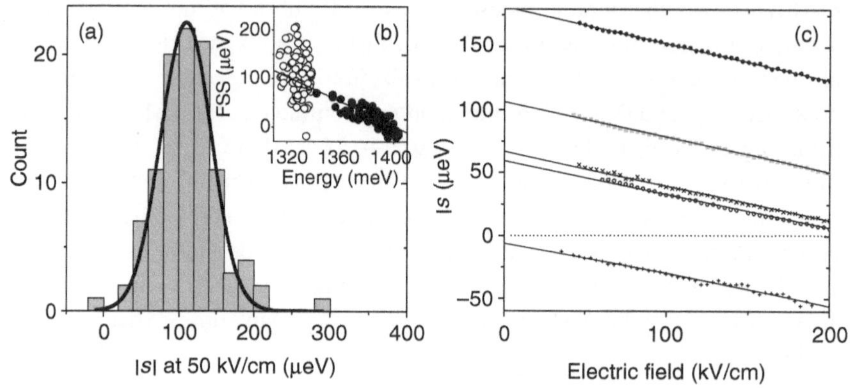

Figure 18.8 (a) Distribution of values of fine structure splitting (FSS, open data points) at a field of 50 kV/cm; (b) a plot of FSS versus emission energy compared to that of Young *et al.* [57] (black circles); and (c) variation in the FSS of with field for six dots. Figures adapted from Ref. [5].

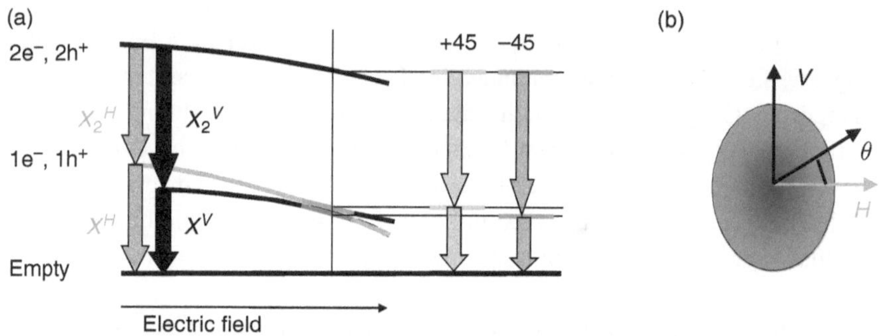

Figure 18.9 (a) Cartoon of the energy levels in dot and (b) the orientation of the eigenstates at some arbitrary field, relative to the horizontal (H) and vertical (V) laboratory axes. Figures adapted from Ref. [5].

18.3.3 Electrically induced coherent coupling of exciton states

Given this range of tuning and the mean FSS approximately half of the dots in the sample can be tuned to a minimum $|s|$. However, we have discovered that in many dots this minimum is not zero, and that we observe an anti-crossing between the states, indicative of coherent coupling between them. Near the minimum splitting the eigenstates rotate in the plane of the sample, being orientated at 45° when the splitting is smallest (Figure 18.9a). Recently, theoretical work using million atom pseudo-potential calculations has suggested that for realistic strain-tuned dots a minimum in the FSS (s_0) will be observed of the order of 3 μeV, due to the symmetry of the crystal [48]. The experimental data can also be explained with a more intuitive model with a single parameter, $|s_0|$, characterising the coupling between the states. The equations are the same as those that describe any system of coupled oscillators:

$$E\begin{pmatrix}\cos\theta\\ \sin\theta\end{pmatrix}=\begin{pmatrix}E_0 & s_0/2\\ s_0/2 & E_0-\gamma(F-F_0)\end{pmatrix}\begin{pmatrix}\cos\theta\\ \sin\theta\end{pmatrix}, \qquad (18.1)$$

where F is the field, and F_0 the field where the eigenstate splitting is smallest. In the absence of any coupling ($|s_0|=0$), the energy between these states varies as $E=\gamma(F-F_0)$ and so are degenerate at field F_0. For a finite coupling the minimum $|s|=|s_0|$. At the anti-crossing the eigenstates are mixtures of H and V, with weighing coefficients $\sin\theta$ and $\cos\theta$. In this particular system θ is a real angle, which gives the orientation of the eigenstates relative to the crystal axes (as shown in Figure 18.9b).

The solutions of this equation are the eigenstates and eigenenergies:

$$E_{\pm}=E_0-\frac{\gamma(F-F_0)}{2}\pm\frac{1}{2}\sqrt{\gamma^2(F-F_0)^2+s_0^2} \qquad (18.2)$$

$$\theta=\pm tan^{-1}\left[\frac{s_0}{\gamma(F-F_0)\pm(E_--E_+)}\right]. \qquad (18.3)$$

Experimentally, we can access θ and $|s|$ by recording the spectrum from the neutral cascade at a range of angles. By fitting the binding energy as a function of angle we see a variation that can be fitted with an amplitude and phase. Data recorded from three different dots is shown in Figure 18.10. We have noticed that some dots prefer to rotate in one direction as they near the anti-crossing and some rotate in the opposite direction, but we chose to present data in Figure 18.10 for dots that all rotate in the same direction. The origin of this handedness remains to be determined, but study of several dozen dots in this sample suggests there is no preferred direction within the ensemble. As can be seen there is a good agreement between the experiment and theory, with only $|s_0|$ and F_0 as free parameters.

Despite the fact the rate of change of $|s|$ appears remarkably similar for all dots, we observe a large variation in the size of the anti-crossing in the fine-structure ranging from $|s_0|=0.3$ to over 40 μeV. If the large variation in measured $|s_0|$ were due to variations in

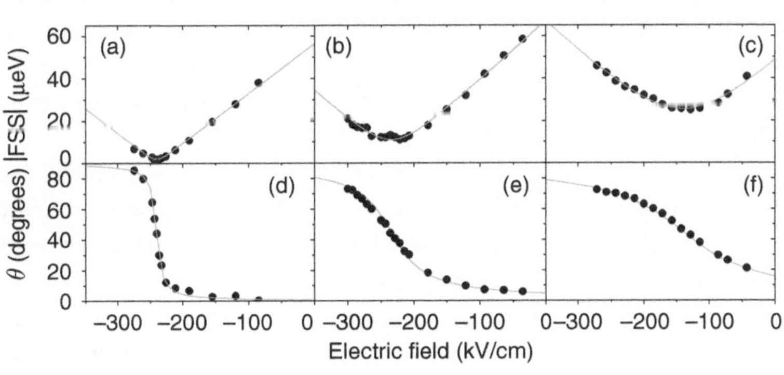

Figure 18.10 (a)–(c) Variation in the magnitude of the fine-structure splitting as a function of field for three dots with differing values of s_0, and (d)–(f) corresponding variation in eigenstate orientation. The grey line shows a fit using the model described in the text. Figures adapted from Ref. [5].

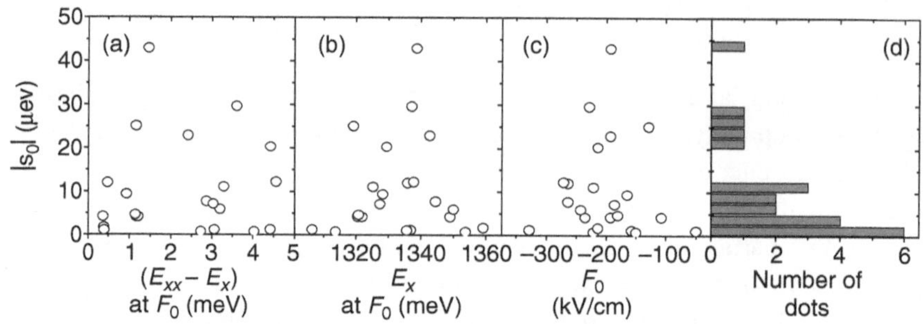

Figure 18.11 A study of 22 dots as a function of various parameters. (a)–(c) The minimum FSS as a function of various parameters for each dot and (d) the frequency of minimum splittings observed. Figures adapted from Ref. [5].

composition, it would likely lead to a strong correlation with emission energy or binding energy. The data in Figure 18.11a–c shows plots of $|s_0|$ and the binding energy ($E_{XX} - E_X$) at F_0, the exciton energy E_X (at F_0) and F_0, respectively. Any trend linking these parameters is not significant in comparison to scatter in the values we determine from the ensemble. However, in a plot of the distribution of $|s_0|$ (Figure 18.11d) it is apparent that smaller values are more frequently observed. In a sizeable fraction of dots we observe $|s_0|$ below 3 μeV, the predicted minimum value of splitting for these dots [48]. Growth of samples with a lower mean $|s|$ at low fields should enable the anti-crossing to be observed in all dots in a sample.

18.3.4 Entangled light emission activated by electric field

In this section we shall present data from one particular dot that shows a minimum $|s|$ of 1.4 μeV at a field of –240 kV/cm (Figure 18.10a). Such a value of $|s|$ is sufficient to demonstrate entangled photon pair emission.

Firstly, polarised cross-correlation measurements were recorded from this dot under CW excitation. At the minimum $|s|$ correlations were recorded in three orthogonal bases: rectilinear (horizontal and vertical), diagonal (+45° and −45°) and circular (left and right). For an unpolarised source these measurements can determine the fidelity of emission with the $(|H_{XX}H_X\rangle + |V_{XX}V_X\rangle)/\sqrt{2}$ state [22] to be 71%.

Cross-correlation measurements were also carried out with pulsed optical excitation at 80 MHz at the same field. In the limit where we only consider those exciton photons emitted immediately after the biexciton, the fidelity reaches 64%. This demonstrates once again that entangled photon pairs are being emitted.

Other CW measurements at increased $|s|$ (Figure 18.12d) appear to show the ability of electrical tuning to “turn-on” and “turn-off” the entangled light emission. More precisely, when a finite splitting in introduced the stored-exciton/biexciton–photon entangled system evolves in time at a different rate, which leads to a lower fidelity with the detectors that we use here [51]. This raises the interesting possibility of using the electric field to control the

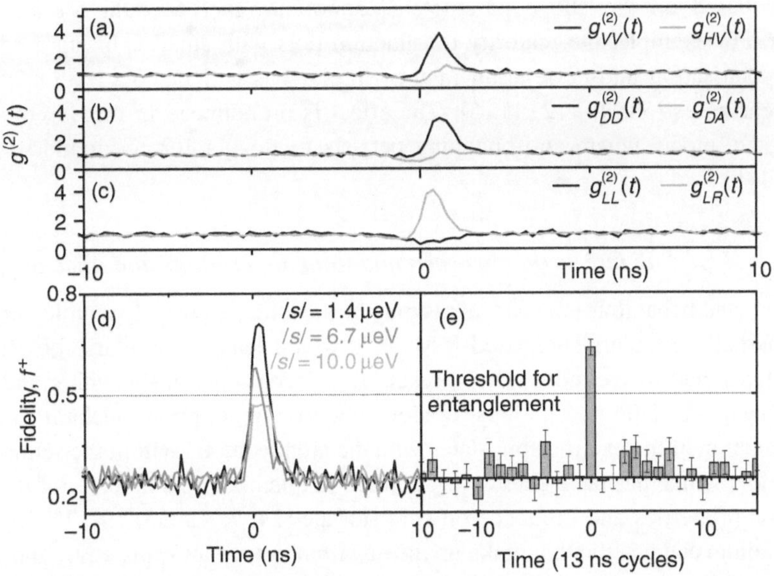

Figure 18.12 (a)–(c) Polarised cross-correlations between the neutral states at the minimum splitting $|s_0|$ with CW excitation. (d) The fidelity of entanglement with CW excitation at various splittings, varied by changing the external field and (e) the fidelity under pulsed excitation at the minimum FSS. In both measurements at the minimum FSS the fidelity exceeds that required to prove entanglement. Figures adapted from Ref. [5].

entangled state in real-time whilst the exciton remains within the dot, something we have achieved on an exciton superposition stored in the dot [10]. Dynamic modulation of the entangled biexciton–photon/stored-exciton system may allow for rapid adjustment of the entangled state to overcome finite dephasing effects, so producing a greater fidelity.

18.4 Interaction of entangled exciton–photon states with nuclei

Previous experiments using pulsed optical excitation have already revealed a great deal about the entanglement dynamics of photon pairs emitted by quantum dots [49, 22, 51]. Effects previously considered include background light from other regions of the sample (such as the wetting layer), detector characteristics such as jitter and dark counts, and properties of the quantum dot itself such as radiative lifetimes, the fine-structure-splitting between the optically active exciton levels, exciton-spin scattering and cross-dephasing. The latter term describes the process whereby the coherence between the two intermediate photon–exciton eigenstates is lost, without affecting the polarisation (spin) properties of the states, and was found previously to have only a weak effect [22]. In the work presented here, decoherence of this type was found to be negligible so we shall disregard its effect from the start to simplify the discussion, though we will show that certain coherent effects can give rise to similar apparent dephasing when averaged over time.

Models based on the above parameters cannot however account for all observed behaviour. For example, the majority of quantum dots with small splitting tend to show stronger polarisation anti-correlation in the circular basis, than polarisation correlation in the linear bases [58, 44, 12, 11, 5]. The effect is pronounced in measurements of an ELED in d.c. mode, where anti-bunching persists for longer for co-circularly polarised photons [44].

18.4.1 Hyperfine interaction between quantum dot excitons and nuclear spins

The unpredicted behaviour in experiments may be attributed to weakly coupled eigenstates that are partially circularly polarised [58]. This would lead to circularly polarised photon pair states that evolve more slowly, resulting in persistent anti-bunching and stronger time-averaged correlations. The most obvious underlying coupling mechanism responsible is interaction of the exciton spin states with the atomic nuclei within the semiconductor quantum dot , which has been extensively studied in quantum dots in terms of its strength, polarisation properties and influence on spin-storage [34, 7, 42, 33, 19]. The isotopes of gallium, indium and arsenic that make up the quantum dot are all spin active, and each randomly oriented nuclear spin interacts with the exciton states via the hyperfine interaction. For the purposes of our analysis, we adopt the convention of approximating the effect of the nuclear population as a net magnetic field B_N (Overhauser field). The net nuclear field fluctuates on a timescale of $\sim$100 μs, has random orientation, and normally distributed strength with mean zero and standard deviation of tens of milliteslas [34, 26].

The fine-structure properties of single quantum dot excitons in the presence of external magnetic fields are strongly orientation dependent [3, 49]. When the field is applied in-plane, hybridisation of optically active with inactive states results in no change in the polarisation of the exciton eigenstates, and a weak, quadratic change in the fine-structure splitting. For our quantum dots the magnitude of the change is 2 μeV/T^{-2} [49], so the expected modification of the fine structure is of the order 1 neV. This is far smaller than the typical splitting of entanglement-optimised quantum dots of the order 1 μeV. The effect of nuclear fields in the plane can therefore be neglected for this analysis.

When applied in the normal direction, a magnetic field lifts the degeneracy between exciton states of opposing spin, resulting in a significant linear change in the fine-structure-splitting of $\sim$200 μeV/T [54]. Thus we expect fluctuations of the fine-structure-splitting of the order of a few microelectronvolts owing to the net nuclear field in this direction, which is comparable to the typical splitting of quantum dots. Therefore for our analysis, we must include the effect of fluctuating nuclear magnetic fields in the normal direction. In addition, diamagnetic effects are neglected as they are polarisation independent [3, 49, 54].

18.4.2 Photon pair states modified by the hyperfine interaction

The effect of a normally oriented magnetic field on the energies of the upper (X_u) and lower (X_l) exciton levels are shown schematically in Figure 18.13a. The horizontal axis

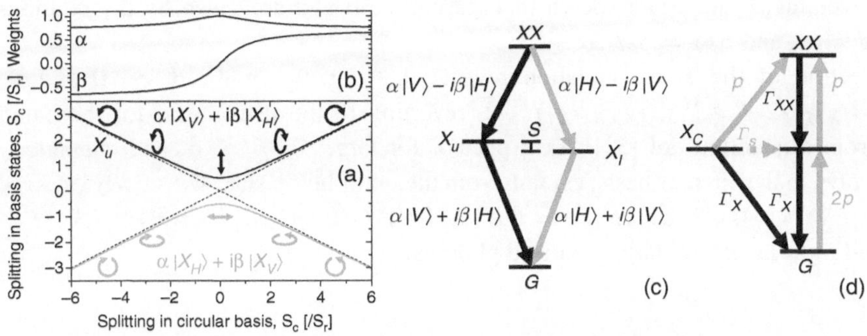

Figure 18.13 (a) Exciton eigenstates (X_u) and (X_l) as a function of the circularly polarised component S_c of the fine-structure-splitting S. The polarisation of light emitted by the excitons is indicated by arrows. (b) Weights α and β of the linearly polarised components of the eigenstates shown in (a). (c) Schematic of radiative decay of the biexciton (XX) state via one of the exciton eigenstates X_u and X_l to the ground state G, with photon polarisation as indicated. (d) Four-level model includes biexciton state (XX), coherent and mixed exciton states (X_C and X_S), and ground state (G). Arrows indicate excitation by electrical current at rate p, radiative decay at rates Γ_{XX} and Γ_X, and potentially spin-scattering at rate Γ_S.

represents the circularly polarised component of the fine-structure-splitting, which is proportional to the magnetic field. The plotted energies and depicted polarisations of the states X_u and X_l are solutions to the following Hamiltonian expressed in the X_H, X_V basis, which denote exciton states that emit horizontal (H) and vertical (V) polarised photons

$$\mathbf{H} = \begin{pmatrix} -S_r & iS_c \\ -iS_c & iS_r \end{pmatrix}. \tag{18.4}$$

Here S_c and S_r denote the circular and rectilinear components of the fine-structure-splitting S between the optically active exciton eigenstates, given by $S = \sqrt{S_c{}^2 + S_r{}^2}$. Significant modification of the splitting is predicted for $S_c \geq S_r$; the polarisation of photons emitted by the exciton eigenstates is expected to be linear for $S_c \ll S_r$, circular for $S_c \gg S_r$, and elliptical for $S_c \sim S_r$, as indicated by arrows in Figure 18.13a.

The modification in fine-structure-splitting and polarisation of the exciton states by fluctuating S_c, introduced by the net nuclear field, consequently changes the form of the entangled photon pair state emitted by the decay of the biexciton state (XX) in a quantum dot. This is shown schematically in Figure 18.13c. The decay proceeds via either X_l or X_u, in each case emitting a pair of photons. The polarisation of each photon is shown, and in the general case are all found to be non-equivalent, elliptically polarised states. Including the phase difference acquired during the time τ spent in the intermediate entangled exciton–photon state [22], the emitted entangled photon pair state is

$$\begin{aligned}\Psi_i = {} & (\alpha|H_{XX}\rangle - i\beta|V_{XX}\rangle)(\alpha|H_X\rangle + i\beta|V_X\rangle)/\sqrt{2} \\ & + e^{iS\tau/\hbar}(\alpha|V_{XX}\rangle - i\beta|H_{XX}\rangle)(\alpha|V_X\rangle + i\beta|H_X\rangle)/\sqrt{2},\end{aligned} \tag{18.5}$$

where weights α and β are shown in Figure 18.13b and are given by the relationships $\alpha^2 + \beta^2 = 1$, and $\beta/\alpha = S_c/(S_r + S)$.

Inspection of the above equation reveals that for $S_c \sim 0$, $\beta \sim 0$ and $\Psi_i \approx (|H_{XX}H_X\rangle/\sqrt{2} + e^{iS\tau/\hbar}(|V_{XX}V_X\rangle)/\sqrt{2}$, reverting to the usual form for the entangled state from a quantum dot [51]. As expected, for large S_c, $\beta \sim \alpha$, and entangled photons evolve in the circular basis, evident from the two-photon state $\Psi_i \approx (|R_{XX}L_X\rangle/\sqrt{2} + e^{iS\tau/\hbar}(|L_{XX}R_X\rangle)/\sqrt{2}$, where $|L\rangle = (|H\rangle + i|V\rangle)/\sqrt{2}$ and $|R\rangle = (|H\rangle - i|V\rangle)/\sqrt{2}$ are left- and right-hand circularly polarised photons.

18.4.3 Modelling entangled photon pair emission

Since the exciton basis states of the quantum dot are dependent on the fluctuating nuclear magnetic field, polarisation and/or spin is a poor choice to differentiate between states, so instead we model our quantum dot levels in terms of the number of excitons in the dot, and their coherent properties, as shown in Figure 18.13d. The resulting levels are the initial coherent exciton state X_C, produced by emission of a previously detected coherent biexciton photon, and the incoherent ground (G), exciton (X_S) and biexciton (XX) states. The levels are linked by re-excitation at rate proportional to p, radiative decay at rates Γ_{XX} and Γ_X and spin-scattering at rate Γ_S, indicated by arrows. Rate equations are constructed and solved by using standard methods, and the density matrix that describes the intensity and polarisation of the two-photon state ρ is computed as follows:

$$\rho(S_r, S_c, \tau) \propto k(X_C(\tau)\rho_e(S_r, S_c, \tau) + X_S(\tau)\rho_m) + (1-k)\rho_m, \qquad (18.6)$$

where τ is the time delay between emission of the biexciton and exciton photons, k is the fraction of light originating exclusively from the dot, ρ_e is the density matrix for entangled light $|\Psi\rangle\langle\Psi|$, and ρ_m the density matrix for maximally mixed light equal to the identity matrix $I_4/4$. Similar methods are used to construct the density matrix for negative time delays, which correspond to emission of an XX photon after detection of an X photon.

To account for fluctuations in the nuclear magnetic field, a weighted average of ρ is numerically obtained from a normally distributed S_c with standard deviation σ and mean zero, as S_c is proportional to the normally distributed nuclear magnetic field.

18.4.4 Coherent entangled photon pair emission

The model was used to fit the second-order cross correlation measurements presented in [44], for a d.c. electrically excited quantum dot. Second-order cross-correlations were calculated as a function of time delay τ, by projection of the calculated density matrix onto the corresponding polarised photon pair states. Finally, the traces were convolved with a Gaussian approximation of the actual detector response functions to account for finite detector resolution. The corresponding experimental and fitted correlations are shown in Figure 18.14 for photon pairs measured in the rectilinear (a), diagonal (b) and circular (c) polarisation bases. Excellent qualitative agreement is observed, revealing nuclear spin effects to be responsible for much of the characteristic shape of photon correlations.

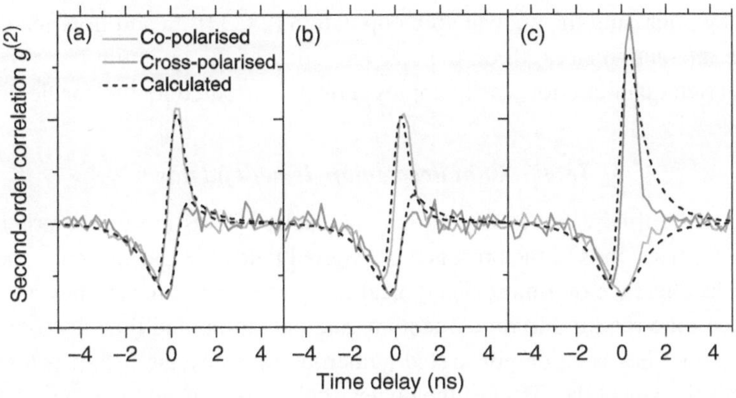

Figure 18.14 Comparison between experimental and simulated correlations from an ELED. (a–c) Second-order correlation measurements as a function of the delay between XX and X photons, $g^{(2)}(\tau)$. Grey and black curves are co- and cross-polarised experimental measurements from [44], performed in the (a) rectilinear, (b) diagonal, and (c) circular polarisation bases. Dashed black curves are corresponding simulated behaviour [52] with parameters as described in the text. Figures adapted from Ref. [44].

The fitted parameters were $k = 0.855$, $\sigma = 2.47\mu\text{eV}$, and $\Gamma_S = 0$, where k represents the fraction of photon pairs originating from the dot and is found to be similar to previous measurements [51]. The fact that no spin-scattering term is required is surprising, as it suggests that photon pair emission from the quantum dot does not suffer dephasing and is fully coherent.

The fitted standard deviation σ of the circular component of the fine-structure-splitting S_c is equivalent to the standard deviation of the energy fluctuations caused by the nuclear spins $\mu_B g_e \Delta B_N$, where g_e is the electron g-factor. Our quantum dots are expected to be quite small owing to their high emission energy, and to consist of $\sim 10^4$ atoms [23]. For dots of this size $\mu_B g_e \Delta B_N$ is $\sim 1.5\,\mu\text{eV}$ in GaAs and $\sim 3.4\,\mu\text{eV}$ in InAs, the latter stronger owing to the larger value for the indium nuclear spin (9/2) compared to gallium (3/2) [7, 36, 29]. Thus the extracted value of $\sigma = 2.47\,\mu\text{eV}$ is consistent with a small InAs quantum dot intermixed with GaAs, as expected in our samples.

The physical effect of nuclear fields can clearly be seen in the correlation traces. For dots with very small splitting such as this (0.4 μeV), the quantum dots have strongly circularly polarised eigenstates owing to the nuclear magnetic field and thus oscillations of $g^{(2)}$ with τ are expected in the rectilinear and diagonal bases as the phase between X_L and X_R evolves. Averaging over the distribution of S results in an additional damping term, resulting in faster convergence of the correlated and anti-correlated traces in the rectilinear and diagonal bases, as observed experimentally and theoretically. The finite temporal resolution of our measurement system means that the same damping effect reduces the maximum and minimum observable correlations due to averaging, leading to stronger correlations in the circular basis, common for dots with small splitting.

For some experiments longer than ~100 μs, it may be advantageous to suppress the hyperfine interaction. This is commonly achieved by applying external magnetic fields

[18, 43], or by polarising the nuclear spin population [23, 33]. In addition, the dependence of hyperfine interaction strength on the size and alloy of the quantum dot suggests using larger, GaAs rich quantum dots, will suppress effects attributed to nuclear fields.

18.4.5 Enhanced entanglement fidelity

We stress that the timescale for fluctuations of nuclear fields are of the order ~ 100 μs [34] up to several seconds in the presence of external fields [43], so although they can lead to apparent decoherence on timescales typical of present experiments, they are effectively static over the time required to do a quantum gate operation, and thus do not cause decoherence. This fact has been demonstrated numerous times by the use of echo techniques which effectively cancel the effect of the nuclear field experienced in each individual cycle [42, 33]. Therefore the entangled state emitted by a quantum dot is expected to be stable for $\sim 10^4$–10^8 cycles, when operating at 100 MHz, which in principle is sufficient for many applications, negating the need to control the hyperfine interaction at all.

To estimate the entanglement properties of the ELED accessible to future applications, we begin with the measured and calculated time-averaged fidelity of the entangled light as a function of the time delay between photons, plotted in Figure 18.15. The measured (black) and calculated (dashed) curves match very well, with peak fidelities of 0.71 ± 0.02 and 0.73 in agreement (within error). Next, we remove the effect of the fluctuating nuclear field, (dark grey curve), which improves the peak fidelity significantly from 0.73 to 0.82. In addition, the time over which entangled light is emitted, defined by fidelities >0.5, is significantly extended from 1.0 ns to 1.8 ns, as the damping effect described above is removed.

Finally, we remove the effect of detector jitter (light grey curve) to reveal an underlying fidelity of 0.9 for this particular device, limited only by the dark counts and background

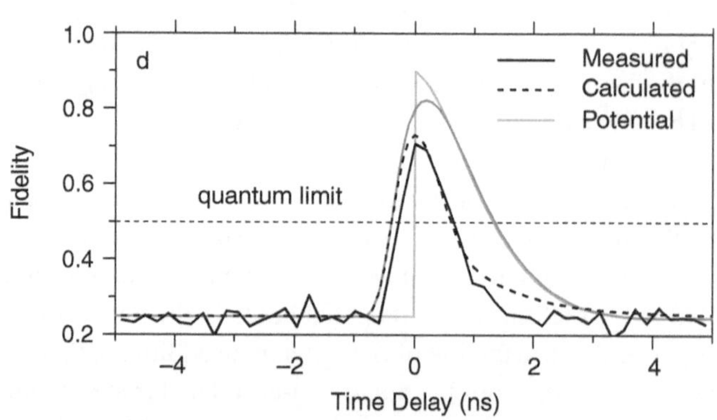

Figure 18.15 Fidelity of photon pair emission with the Bell state $(|H_{XX}H_X\rangle + |V_{XX}V_X\rangle)/\sqrt{2}$, as a function of the time delay τ. The black curve is experimental result from [44], and the dashed curve is the corresponding simulation from [52]. Predicted behaviour for short measurements is shown by the dark grey line, and additionally with fast detectors by the light grey line. Figures adapted from Ref. [44].

light, which define the parameter k. Attempts to increase k by better spectral and spatial filtering of quantum dot light could improve the fidelity further.

18.5 Conclusion

Our results on electrically operated light sources reviewed above represent a great step towards the promised practicality of entangled light sources based on quantum dots. We have shown emission of high-fidelity entangled photons from an electrically driven LED structure, which is the first electrically driven entangled light source. By using a different device we have enabled electric fields of unprecedented magnitude to be applied to single quantum dots, allowing diverse control over the eigenstates in the dot, and voltage-activation and deactivation of entanglement in optically excited photon pairs. By incorporating multiple QD entangled light sources on an electrically operated single semiconductor chip, the realisation of a useful quantum computer based on linear optics will become much more feasible.

Furthermore, our experiments have revealed interesting fundamental properties of quantum dots that could help improve designs for future quantum dot entangled light sources and systems, as well as stimulate new scientific research. Analysis of the time-dependent characteristics of the photon-pair state has identified that fidelity is not affected by any decoherence processes inherent to the quantum dot, previously presumed to affect the intermediate entangled exciton–photon state. This is perhaps surprising as solid-state systems generally suffer from environmental decoherence more severely than photons. Instead, we show that the fidelity is limited by coherent interaction with nuclear spins that could be compensated for, in conjunction with experimentally controllable factors such as re-excitation rate, background light rejection and detector jitter. Finally by application of large electric–fields across QDs, we reveal coherent coupling between exciton eigenstates with varying strengths. This intriguing phenomenon lies in contrast to previous experimental techniques to tune the fine-structure-splitting, which did not reveal the same effect. Its existence leads to new possibilities such as single qubit operations on a stored exciton state.

References

[1] Akopian, N., Lindner, N. H., Poem, E. *et al.* 2006. Entangled photon pairs from semiconductor quantum dots. *Phys. Rev. Lett.*, **96**(13), 130501.

[2] Aspect, A., Grangier, P. and Roger, G. 1982. Experimental realization of Einstein Podolsky Rosen Bohm Gedankenexperiment: a new violation of Bell's inequalities. *Phys. Rev. Lett.*, **49**(2), 91–94.

[3] Bayer, M., Ortner, G., Stern, O. *et al.* 2002. Fine structure of neutral and charged excitons in self-assembled In(Ga)As/(Al)GaAs quantum dots. *Phys. Rev. B*, **65**(19), 195315.

[4] Bennett, A. J., Patel, R. B., Skiba-Syzmanska, J. *et al.* 2010a. Giant Stark shift in the emission of single semiconductor quantum dots. *Phys. Rev. B.*, **97**, 031104.

[5] Bennett, S. D., Cockins, L., Miyahara, Y., Grütter, P., and Clerk, A. A. 2010b. Strong electromechanical coupling of an atomic force microscope cantilever to a quantum dot. *Phys. Rev. Lett.*, **104**(1), 2–5.

[6] Benson, O., Santori, C., Pelton, M. and Yamamoto, Y. 2000. Regulated and entangled photons from a single quantum dot. *Phys. Rev. Lett.*, **84**(11), 2513–6.

[7] Braun, P.-F., Marie, X., Lombez, L. *et al.* 2005. Direct observation of the electron spin relaxation induced by nuclei in quantum dots. *Phys. Rev. Lett.*, **94**(11), 116601.

[8] Briegel, H.-J., Dür, W., Cirac, J. I. and Zoller, P. 1998. Quantum repeaters: the role of imperfect local operations in quantum communication. *Phys. Rev. Lett.*, **81**(26), 5932–5935.

[9] Claudon, J., Bleuse, J., Malik, N. S. *et al.* 2010. A highly efficient single-photon source based on a quantum dot in a photonic nanowire. *Nature Photonics*, **4**, 174.

[10] de la Giroday, A. B., Bennett, A. J., Pooley, M. A. *et al.* 2010. All-electrical coherent control of the exciton states in a single quantum dot. *Phys. Rev. B.*, **82**, 241301.

[11] Dousse, A., Suffczyski, J., Krebs, O. *et al.* 2010a. A quantum dot based bright source of entangled photon pairs operating at 53 K. *Appl. Phys. Lett.*, **97**(8), 081104.

[12] Dousse, A., Suffczynski, J., Beveratos, A. *et al.* 2010b. Ultrabright source of entangled photon pairs. *Nature*, **466**, 217–220.

[13] Edamatsu, K., Oohata, G., Shimizu, R., and Itoh, T. 2004. Generation of ultraviolet entangled photons in a semiconductor. *Nature*, **431**, 167–170.

[14] Ekert, A. K. 1991. Quantum cryptography based on Bell's theorem. *Phys. Rev. Lett.*, **67**(6), 661–663.

[15] Ellis, D. J. P., Stevenson, R. M., Young, R. J. *et al.* 2007. Control of fine-structure splitting of individual InAs quantum dots by rapid thermal annealing. *Appl. Phys. Lett.*, **90**(1), 011907.

[16] Fattal, D., Inoue, K., Vučković, *et al.* 2004. Entanglement formation and violation of bell's inequality with a semiconductor single photon source. *Phys. Rev. Lett.*, **92**(3), 037903.

[17] Finley, J. J., Sabathil, M., Vogl, P. *et al.* 2004. Quantum-confined Stark shifts of charged exciton complexes in quantum dots. *Phys. Rev. B*, **70**(20), 201308.

[18] Gammon, D., Efros, A. L, Kennedy, T. *et al.* 2001. Electron and nuclear spin interactions in the optical spectra of single GaAs quantum dots. *Phys. Rev. Lett.*, **86**(22), 5176–5179.

[19] Gerardot, B. D., Seidl, S., Dalgarno, P. A. *et al.* 2007a. Contrast in transmission spectroscopy of a single quantum dot. *Appl. Phys. Lett.*, **90**(22), 221106.

[20] Gerardot, B. D., Seidl, S., Dalgarno, P. A. *et al.* 2007b. Manipulating exciton fine structure in quantum dots with a lateral electric field. *Appl. Rev. Lett.*, **90**, 041101.

[21] Hafenbrak, R., Ulrich, S. M., Michler, P. *et al.* 2007. Triggered polarization-entangled photon pairs from a single quantum dot up to 30 K. *New J. Phys.*, **9**(9), 315.

[22] Hudson, A. J., Stevenson, R. M., Bennett, A. J. *et al.* 2007. Coherence of an entangled exciton–photon state. *Phys. Rev. Lett.*, **99**, 266802.

[23] Imamoglu, A. 2003. Are quantum dots useful for quantum computation? *Physica E: Low-dimensional Systems and Nanostructures*, **16**(1), 47–50.

[24] Jacobs, B. C., Pittman, T. B. and Franson, J. D. 2002. Quantum relays and noise suppression using linear optics. *Phys. Rev. A*, **66**(5), 052307.

[25] Jundt, G., Robledo, L., Högele, A. Fält, S. and Imamoğlu, A. 2008. Observation of dressed excitonic states in a single quantum dot. *Phys. Rev. Lett.*, **100**(17), 177401.

[26] Khaetskii, A. V., Loss, D. and Glazman, L. 2002. Electron spin decoherence in quantum dots due to interaction with nuclei. *Phys. Rev. Lett.*, **88**(18), 186 802–186 806.

[27] Kiess, T. E. Shih, Y. H., Sergienko, A. V. and Alley, C. O. 1993. Einstein–Podolsky–Rosen–Bohm experiment using pairs of light quanta produced by type-II parametric down-conversion. *Phys. Rev. Lett.*, **71**(24), 3893–3897.

[28] Knill, E., Laflamme, R. and Milburn, G. J. 2001. A scheme for efficient quantum computation with linear optics. *Nature*, **409**, 46–52.

[29] Koppens, F. H. L. 2005. Control and detection of singlet–triplet mixing in a random nuclear field. *Science*, **309**, 1346–1350.

[30] Kowalik, K., Krebs, O., Lemaitre, A. *et al.* 2005. Influence of an in-plane electric field on exciton fine structure in InAs-GaAs self-assembled quantum dots. *Appl. Phys. Lett.*, **86**(4), 041907.

[31] Kowalik, K., Krebs, O., Senellart, P. *et al.* 2006. Stark spectroscopy of coulomb interactions in individual InAs.GaAs self-assembled quantum dots. *Phys. Stat. Sol. (c).*, **3**, 3980–3984.

[32] Kowalik, K., Krebs, O., Lemaitre, A. *et al.* 2007. Monitoring the electrically driven cancellation of exciton fine structure in a semiconductor quantum dot by optical orientation. *Appl. Rev. Lett.*, **91**, 183104.

[33] Lai, C. W., Maletinsky, P., Badolato, A. and Imamoğlu, A. 2006. Knight-field-enabled nuclear spin polarization in single quantum dots. *Phys. Rev. Lett.*, **96**(16), 167 403–167 407.

[34] Maletinsky, P. and Imamoğlu, A. 2009. *Single Semiconductor Quantum Dots*. Springer. Chap. 5, pages 145–184.

[35] Marcet, S., Ohtani, K. and Ohno, H. 2010. Vertical electric field tuning of the exciton fine structure splitting and photon correlation measurements of GaAs quantum dot. *Appl. Phys. Lett.*, **96**, 101117.

[36] Merkulov, I. A. 2002. Electron spin relaxation by nuclei in semiconductor quantum dots. *Phys. Rev. B*, **65**(20), 205309.

[37] Michler, P, Kiraz, a, Becher, C. *et al.* 2000a. A quantum dot single-photon turnstile device. *Science*, **290**(5500), 2282–2285.

[38] Michler, P., Kiraz, A., Becher, C. *et al.* 2000b. A quantum dot single-photon turnstile device. *Science*, **290**, 2282.

[39] Mohan, A., Felici, M., Gallo, P. *et al.* 2010. Polarization-entangled photons produced with high-symmetry site-controlled quantum dots. *Nature Photon.*, **4**, 302–306.

[40] Muller, A., Fang, W., Lawall, J. and Solomon, G. S. 2009. Creating polarization-entangled photon pairs from a semiconductor quantum dot using the optical stark effect. *Phys. Rev. Lett.*, **103**(21), 217402.

[41] Pan, J.-W., Bouwmeester, D., Weinfurter, H. and Zeilinger, A. 1998. Experimental entanglement swapping: entangling photons that never interacted. *Phys. Rev. Lett.*, **80**(18), 3891–3894.

[42] Petta, J. R., Johnson, A. C., Taylor, J. M. *et al.* 2005. Coherent manipulation of coupled electron spins in semiconductor quantum dots. *Science*, **309**(5744), 2180–2184.

[43] Press, D., Greve, K. De, McMahon, P. L. *et al.* 2010. Ultrafast optical spin echo in a single quantum dot. *Nature Photon.*, **4**, 367.

[44] Salter, C. L., Stevenson, R. M., Farrer, I. *et al.* 2010. An entangled-light-emitting diode. *Nature*, **465**, 594–597.

[45] Santori, C., Fattal, D., Vucković, J., Solomon, G. S. and Yamamoto, Y. 2002. Indistinguishable photons from a single-photon device. *Nature*, **419**(6907), 594–597.

[46] Seidl, S., Kroner, M., Hogele, A. *et al.* 2006. Effect of uniaxial stress on excitons in a self-assembled quantum dot. *Appl. Rev. Lett.*, **88**, 203113.

[47] Shih, Y. H. and Alley, C. O. 1988. New type of Einstein–Podolsky–Rosen–Bohm experiment using pairs of light quanta produced by optical parametric down conversion. *Phys. Rev. Lett.*, **61**(26), 2921–2924.

[48] Singh, R. and Bester, G. 2010. Lower bound for the excitonic fine structure splitting in self assembled quantum dots. *Phys. Rev. Lett.*, **104**, 196803.

[49] Stevenson, R., Young, R., See, P. *et al.* 2006a. Magnetic-field-induced reduction of the exciton polarization splitting in InAs quantum dots. *Phys. Rev. B*, **73**(3), 1–4.

[50] Stevenson, R., Young, R. P., Gevaux, D. *et al.* 2006b. Magnetic-field-induced reduction of the exciton polarization splitting in InAs quantum dots. *Phys. Rev. B*, **73**(3), 1–4.

[51] Stevenson, R., Hudson, A., Bennett, A. *et al.* 2008. Evolution of entanglement between distinguishable light states. *Phys. Rev. Lett.*, **101**(17), 1–4.

[52] Stevenson, R. M., Salter, C. L., de la Giroday, A. *et al.* 2011. Coherent entangled light generated by quantum dots in the presence of nuclear magnetic fields. *arXiv:1103.2969v1* [quant-ph].

[53] Stevenson, R. M., Thompson, R. M., Shields, A. J. *et al.* 2002. Quantum dots as a photon source for passive quantum key encoding. *Phys. Rev. B*, **66**(8), 081302.

[54] Stevenson, R. M., Young, R. J., See, P. *et al.* 2004. Time-resolved studies of single quantum dots in magnetic fields. *Physica E*, **21**, 381–384.

[55] Ulrich, S. M., Strauf, S., Michler, P., Bacher, G. and Forchel, A. 2003. Triggered polarization-correlated photon pairs from a single CdSe quantum dot. *Appl. Phys. Lett.*, **83**(9), 1848.

[56] Vogel, M. M., Ulrich, S. M., Hafenbrak, R. *et al.* 2007. Influence of lateral electric fields on multiexcitonic transitions and fine structure of single quantum dots. *Appl. Rev. Lett.*, **91**, 051904.

[57] Young, R. J., Stevenson, R. M., Shields, A. J. *et al.* 2005. Inversion of exciton level splitting in quantum dots. *Phys. Rev. B*, **72**, 113305.

[58] Young, R. J., Stevenson, R. M., Hudson, A. J. *et al.* 2009. Bell-inequality violation with a triggered photon-pair source. *Phys. Rev. Lett.*, **102**, 030406.

[59] Yuan, Z., Kardynal, B. E., Stevenson, R. M. *et al.* 2002. Electrically driven single-photon source. *Science*, **295**, 102.

[60] Żukowski, M., Zeilinger, A. Horne, M. A. and Ekert, A. K. 1993. "Event-ready-detectors" Bell experiment via entanglement swapping. *Phys. Rev. Lett.*, **71**(26), 4287–4290.

19

Deterministic single quantum dot cavities at telecommunication wavelengths

D. Dalacu, K. Mnaymneh, J. Lapointe, G. C. Aers, P. J. Poole,
R. L. Williams and S. Hughes

Scalability requirements in future device application of self-assembled quantum dots for non-classical light generation necessitate control of the quantum dot nucleation site. In this chapter we discuss a site-control technique based on directed self-assembly of InAs/InP quantum dots emitting at telecommunication wavelengths. The site-control method preserves the high optical quality inherent in self-assembled quantum dots and the characteristic signatures of a strongly confined system are observed in the emission spectra. The efficacy of site-control manifests in the coupling of single quantum dots to microcavities required for the fabrication of efficient devices. The *a priori* knowledge of the quantum dot position is used to deterministically couple single dots to high-finesse microcavities with the assurance that one and only one quantum dot is coupled to each cavity. Such devices form the basis of efficient sources of single photons and entangled photon pairs for telecommunications applications that can be manufactured in a scalable manner using conventional semiconductor processing.

19.1 Introduction

Self-assembled quantum dots possess the two-level emitter characteristics required for non-classical light generation [37] in quantum information processing and quantum key distribution. The performance of a quantum dot-based single photon source or entangled photon pair source will depend on how well the dot can be coupled to a high-quality factor Q, small volume V_{eff} microcavity [44]. The cavity is required to channel photons from the exciton decay into an optical mode that can be collected by an external optical system. In addition, the cavity serves to manipulate the characteristic radiative decay rate τ of an excited dipole emitter, enhancing device efficiency. It has been recognized since the work of Purcell in the 1940s [32] that τ is not an immutable property of the emitter, but is a function of its local electromagnetic environment or equivalently the local density of optical states. In a quasi-closed cavity environment, the Purcell factor, F_P, giving the ratio of the spontaneous emission rate within the cavity to the rate within a homogeneous medium with refractive index n, can be expressed [1] as

Quantum Dots: Optics, Electron Transport and Future Applications, ed. Alexander Tartakovskii. Published by Cambridge University Press.

$$\frac{\tau_c}{\tau} = \frac{3Q(\lambda/n)^3}{4\pi^2 V_{eff}} \frac{\delta\omega_c^2}{4(\omega_c - \omega_x)^2 + \delta\omega_c^2} \frac{|\,\mathbf{d}\cdot\mathbf{f}(\mathbf{r_x})\,|^2}{|\,\mathbf{d}\,|^2}. \tag{19.1}$$

Equation (19.1) highlights the difficulties in coupling self-assembled quantum dots to microcavities. By the very nature of the self-assembled growth process, the characteristics of individual dots can vary widely and their spatial location is generally uncontrolled. This makes it difficult to take advantage of the Purcell effect, which requires spectral matching of the emitter ω_x and cavity ω_c frequencies as well as spatial overlap of the emitter's transition dipole moment, $\mathbf{d}$, with an antinode in the cavity mode's electric field distribution, $\mathbf{f}(\mathbf{r_x})$.

The straightforward approach to spectral matching entails careful cavity design and fabrication to tailor the cavity to a specific dot's emission spectrum with sufficient precision so that temperature [27, 45, 35] or electric field [12] can be used to tune the dot into resonance. The precision required is dictated by tuning ranges available, typically ~1 meV, which places stringent tolerances on the cavity fabrication. These tolerances can be relaxed by using a class of cavities based on photonic crystal membranes [25] for which there exists post-fabrication tuning methods [15, 7] that cover a wide spectral range.

The spatial matching of self-assembled quantum dots to microcavities is less trivial. One approach is to use randomly nucleated dot ensembles combined with various methods [3, 21, 10] to align cavities to the dots. A second approach aims to control the quantum dot nucleation site and to use standard alignment methods available to lithography systems to couple cavities to the site-controlled dots [42, 14, 36]. Site-control typically involves some form of surface modification [4, 38, 18, 2] and the concern, whenever there is manipulation of the epitaxial growth surface, is the possibility of adverse effects on the optical quality of the dot. In this chapter, we discuss a site-control technique based on the directed self-assembly of InAs quantum dots using InP nanotemplates [30]. Site-control through the nanotemplate process preserves the optical quality inherent to self-assembled dots, and optical spectroscopy on single dots shows state-filling of the strongly confined system with atomic-like level spacings and narrow linewidths [6]. The single site-controlled dots emit efficiently in the wavelength range 1.3 μm to 1.55 μm and can be coupled to high-Q optical microcavities by using standard semiconductor growth and processing techniques [8]. The site-controlled quantum dot-cavity device is a scalable solution for efficient non-classical light generation for fibre-based communications.

19.2 Directed self-assembly

Control of the quantum dot nucleation site is based on selective area epitaxy [30] where a patterned substrate is used to limit growth of the semiconductor material to areas exposed in a dielectric mask. The substrate is a (001) InP wafer that has been coated with a thin (20 nm) SiO_2 layer. The SiO_2 mask is patterned by using electron-beam (e-beam) lithography and wet etching to produce openings in the oxide, exposing the InP surface as shown in Fig. 19.1a. The wet etch is done by using a buffered HF solution which does not damage

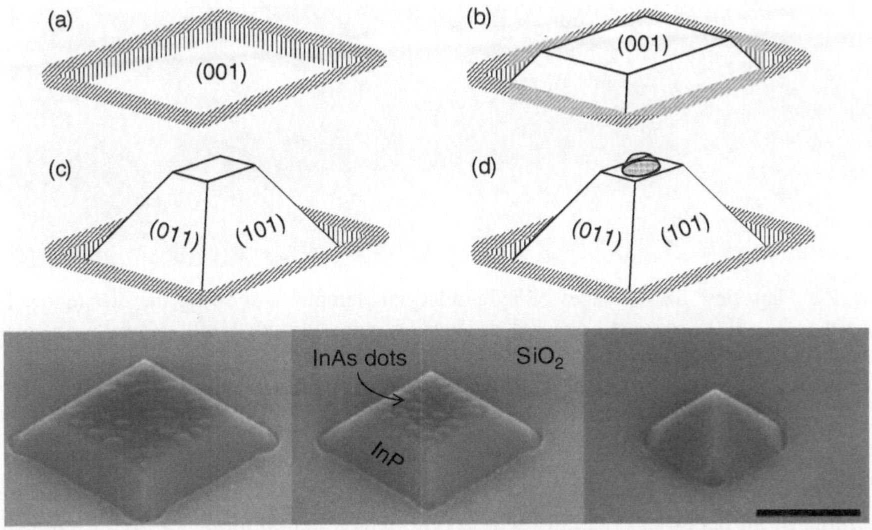

Figure 19.1 Upper panel: schematic illustration of the directed self-assembly process. In (a), the underlying (001) InP substrate is patterned with openings in an otherwise continuous SiO_2 masking layer. (b) Growth occurs exclusively on the exposed InP surface producing pyramidal nanotemplates with (001) top apexes. In (c), further growth results in a controlled reduction of the available (001) top surface area. In (d), deposition of InAs on the nanotemplate produces a single quantum dot nucleated at the pyramid apex. Lower panel: SEM images of nanotemplates viewed at 45° for different size openings. The rightmost image shows a single InAs quantum dot nucleated at the apex of the pyramid. Scale bar indicates 300 nm.

the InP surface, followed by an extensive solvent cleaning step to strip the e-beam resist and leave the InP surface ready for epitaxial growth.

Growth on the patterned substrate is done with chemical-beam epitaxy [29] using trimethyl-indium and cracked PH_3 sources. Growth conditions are chosen that result in growth exclusively on the exposed InP. Growth in the openings takes the form of pyramids with (001) top facets and sloping sidewalls as shown in Fig. 19.1b. With continued InP deposition, the pyramid height increases and the area of the top facet decreases, as in Fig. 19.1c. In this manner, templates whose original dimension is of the order of a few hundred nanometers can be controllably grown until their top dimension approaches $50 \times 50\,\text{nm}^2$, or the order of a typical quantum dot dimension.

The formation of the pyramids occurs for specific crystallographic orientations of the openings. Certain directions promote the formation of higher index facets through increased In adatom migration lengths compared to the (001) top facet that result in a relative incorporation rate on the (001) surface that far exceeds that on the sidewalls. In other words, since material is deposited almost exclusively on the (001) top apex, facets naturally develop and which facets form will depend on the orientation of the opening in the oxide. For example, Fig. 19.2a shows a plan view SEM image of a $< 100 >$ oriented square discussed in Fig. 19.1 for which the sidewalls are $\{110\}$ facets. Alternatively, $< 110 >$

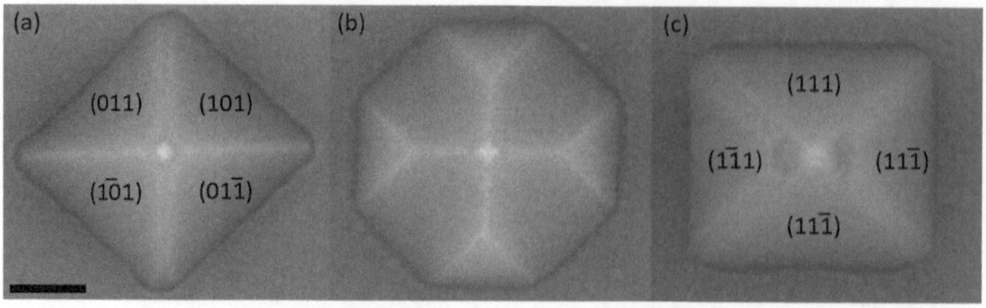

Figure 19.2 Plan view SEM images of pyramidal nanotemplates showing the different possible geometries. (a) < 100 > oriented opening produces a pyramid with {110} sidewalls. (b) octagon opening produces a nanotemplate with four {110} and four {111} sidewalls (c) < 110 > oriented opening produces a pyramid with {111} sidewalls. Scale bar indicates 150 nm.

oriented squares will produce {111} sidewalls shown in Fig. 19.2c whilst Fig. 19.2b shows an example of an octagonal opening which produces a nanotemplate with all eight of the aforementioned facets.

Since it is the group III species that dictates the incorporation in III–V epitaxy, the deposition of InAs (switching precursors from PH_3 to AsH_3) proceeds as does the InP, with the In adatoms migrating to the top (001) facet. Dot formation on the (001) apex (Fig. 19.1d) occurs as on planar substrates, via the Stranski–Krastanow growth mode [41] in which 3D islands nucleate spontaneously after a few monolayers of normal 2D growth of compressively strained layers. The lower panel in Fig. 19.1 shows a series of InP nanotemplates with InAs dots grown in the same run, hence the amount of deposited material is the same. In this case, the controlled reduction of the (001) top apex is achieved through a reduction, from left to right, in the size of the opening. The pyramid in the left image possesses a large (001) top facet which can accommodate many quantum dots. As the size of the opening is reduced, the area of the top facet decreases until only a single quantum dot is nucleated, as in the image on the right. Subsequent capping of the the single dot with InP produces a high-quality, optically active InAs/InP quantum dot whose position is known with high precision.

The beauty of this technique stems from the fact that the InP nanotemplate controlling quantum dot nucleation is prepared *in situ*, in an ultrahigh vacuum environment, with a size and geometry that is determined by the crystallography of the zincblende materials and by the quantity of InP deposited during nanotemplate growth. If the size and orientation of the original opening in the oxide is well controlled then the properties of the individual dots that are nucleated should also be well controlled. Figure 19.3 illustrates the level of control that can be achieved with the directed self-assembly technique. From left to right the dimensions of the opening have been increased in steps of 2.5 nm which corresponds to an increase in the area of the top facet. This increase translates directly to an increase in the size of the quantum dot as shown in the upper panel of the figure. Alternatively, by changing the growth conditions during InAs deposition, one can switch from nucleating a

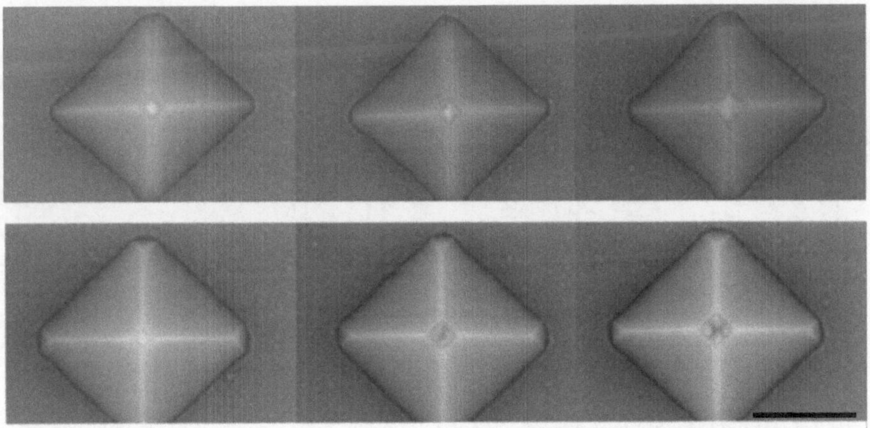

Figure 19.3 Upper panel: plan view SEM images showing control of the quantum dot size by increasing the dimensions of the oxide opening in steps of 2.5 nm from left to right. Lower panel: same but with InAs growth conditions changed to nucleate multiple dots for the same top facet area. Scale bar indicates 300 nm.

single dot to two or four dots for the same top facet area, as shown in the lower panel of the figure.

19.3 Spectroscopy of site-controlled single quantum dots

Since InAs quantum dot formation at the apex of the InP pyramid occurs as on planar substrates, via the Stranski–Krastanow growth mode, capped InP pyramids containing single InAs dots behave optically as their planar counterparts and similar to the more prevalent InAs/GaAs quantum dot system. In particular, the site-controlled quantum dots possess the discrete energy-level spectrum and narrow linewidths reminiscent of atomic systems [22]. A important distinction between the InAs/InP quantum dot system and the InAs/GaAs system [28] is the emission energy: InAs/InP quantum dots emit in the wavelength range relevant for fibre-based applications [29], between 1300 nm and 1550 nm.

With position controlled by lithography and high visibility in an optical microscope, the nanotemplate structures can be individually addressed in a micro-photoluminescence (PL) setup. Typical luminescence measurements for two InAs/InP quantum dots are shown in Fig. 19.4. The emission was measured at 4.2 K in a continuous-flow helium cryostat using non-resonant above-bandgap excitation through a microscope objective. The luminescence is collected through the same microscope objective, dispersed using a single grating spectrometer and detected using a liquid-nitrogen-cooled InGaAs diode array. Figure 19.4a shows the pump power-dependent PL spectra for a dot emitting at 825 meV. The characteristic signatures of a 0-dimensional system with an atomic-like density of states are evident:

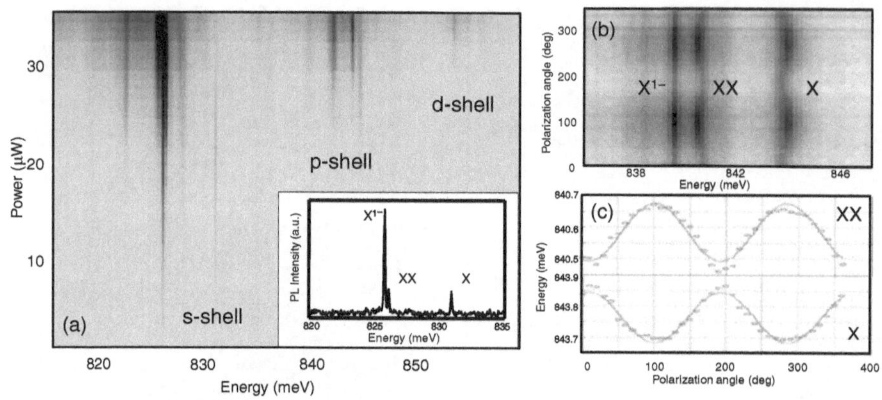

Figure 19.4 (a) Power-dependent PL spectra showing bound s, p, and d orbitals for an InAs on an InP nanotemplate with a ground state energy of 825 meV. The inset shows a low excitation spectrum. (b) Polarization-dependent ground state emission for a dot emitting at 840 meV. (c) Anisotropic exchange splittings in the exciton and biexciton emission extracted from the polarization measurements.

narrow ground state line-width (50 μeV, resolution-limited) at low pump power and bound s, p, and d orbitals evident with increasing excitation.

Owing to the statistical nature of carrier capture processes for nonresonant excitation, one can observe the coexistence of different charge states in the ground-state emission. For the InAs/InP system, the ground state emission (inset of Fig. 19.4a) is typically from a neutral exciton, X, and a singly charged exciton or trion, X^- (electron–hole recombination in the presence of an additional electron). The inset also shows biexciton (XX) emission (single exciton annihilation in the presence of a second exciton of opposite spin) that appears with increasing excitation. Polarization measurements can be used to differentiate the emission from X, XX, and X^-. The neutral exciton emission is split by anisotropic exchange [5] into two orthogonal peaks separated by 20–200 μeV whilst the biexciton emission is split by the same energy but with opposite polarity. This splitting is absent for X^- since the initial state in the transition corresponds to a spin singlet pair of electrons and the final state to an unpaired electron. The exchange splitting for X and XX appears as a shift in emission energy with polarizer angle as shown in Fig. 19.4b for a dot emitting at 840 meV. In Fig. 19.4c the emission energy of X and XX are plotted versus polarization angle, from which an anisotropic exchange splitting of 200 μeV is extracted.

Ideally, one would like to control not only the location, but also the emission energy of the quantum dot. Assuming strain and composition do not change from dot to dot for a given growth condition, energy control entails control of the quantum dot size as in Fig. 19.3. However, sub-millielectronvolt tuning of the quantum dot requires sub-nonometre control of the top apex of the pyramidal template, which is beyond our current

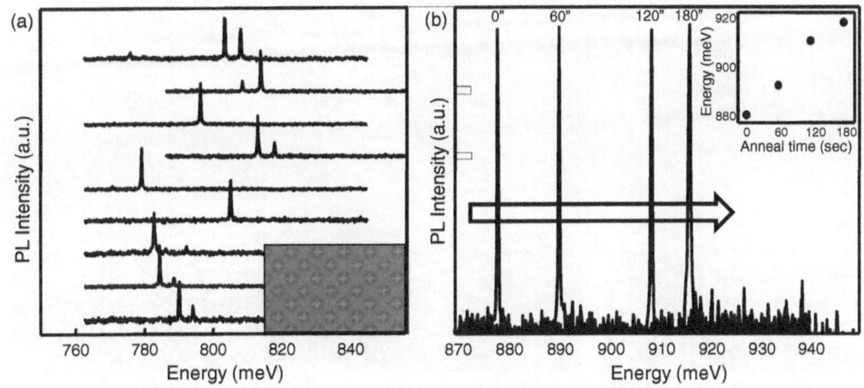

Figure 19.5 Low excitation spectra showing ground-state emission (X and/or X^-) from single dots grown on nominal identical pyramids. (b) Ground-state tuning using RTA at 750 °C. The inset shows the tuning rate of 10 meV/min.

lithography process. The typical spread in ground state emission energy for dots on nominally identical pyramids is shown in Fig. 19.5a and varies by ±20 meV.

An alternative method involves post-growth quantum dot tuning using high-temperature annealing [9]. The compositional changes from As and P interdiffusion will shift the ground-state emission to higher energy as shown in Fig. 19.5b. The tuning rate will depend on temperature and the material system with typical values of 10 meV/min for the InAs/InP system using a rapid thermal anneal (RTA) at 750 °C. Ideally, one would like to tune individual dots on the same chip *in situ*. The temperatures required to promote intermixing have been demonstrated by using a focussed laser [40, 33] leading one to envision a wafer consisting of arrays of site-controlled single emitters, as in the inset of Fig. 19.5a, with identical ground-state emission energies. The annealing process also shows potential for tuning biexciton binding energies and anisotropic exchange splittings [11] which are relevant for the generation of entangled photon pairs [34] through the biexciton–exciton cascade [26].

19.4 Integration of site-controlled quantum dots and cavities

In the following a process is outlined whereby a single InAs quantum dot on an InP pyramid can be spatially coupled to a two-dimensional photonic crystal microcavity [25]. A schematic illustration of the dot–cavity system is shown in the upper panel of Fig. 19.6. The pyramid is located in the membrane of the photonic crystal cavity precisely at the single missing hole defect site where the optical field of the cavity is maximum. To incorporate single pyramidal templates within a photonic crystal membrane the single InAs/InP quantum dot is grown as described in Section 19.2, but in this case the substrate is covered

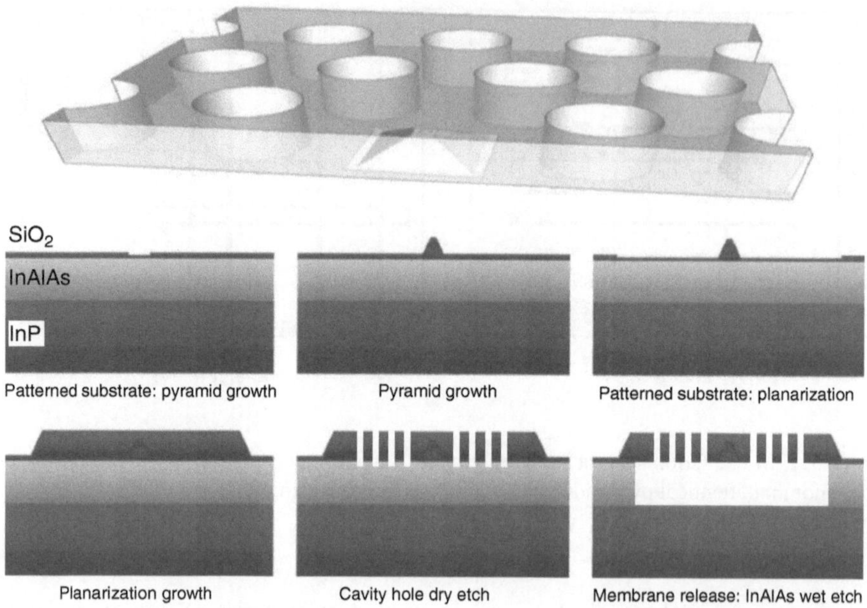

Figure 19.6 Upper panel: schematic of a single site-controlled dot coupled to a photonic crystal microcavity. Lower panel: processing steps involved in the fabrication of the device.

with an InAlAs sacrificial layer, approximately 1 μm thick, that will be removed after cavity fabrication. Also, the SiO_2 opening for the pyramid growth is precisely positioned with respect to alignment marks etched into the substrate that will be used to align the cavity to the dot.

The processing steps for incorporating the site-controlled dot with a photonic crystal cavity are shown in the lower panel of Fig. 19.6. After the initial growth, the single dot pyramids are characterized in emission to identify ground-state energies for subsequent cavity fabrication. The samples are then re-patterned to open up large windows in the SiO_2, centered on the existing pyramids, and re-introduced into the growth chamber for further epitaxial InP growth. In this second growth step, InP is again grown only in the regions of exposed semiconductor, within the large SiO_2 windows. Growth of InP is continued until the pyramidal template is buried within a planarized membrane whose thickness is appropriate for photonic crystal fabrication, typically 200–300 nm and whose lateral dimension, set by the SiO_2 window dimension, is large enough, typically 50–100 μm, to accommodate enough rings of etched photonic crystal holes that the optical loss from the cavity is dominated by vertical loss and not by leakage in the plane. Fabrication of the photonic crystal structure is achieved by depositing a layer of SiO_2 and then a layer of e-beam resist across the planarized dot structure. The electron beam resist is exposed using the alignment marks already etched into the substrate and this is used to etch a hole pattern into the oxide layer. The patterned oxide layer is then used as a mask to transfer the desired hole

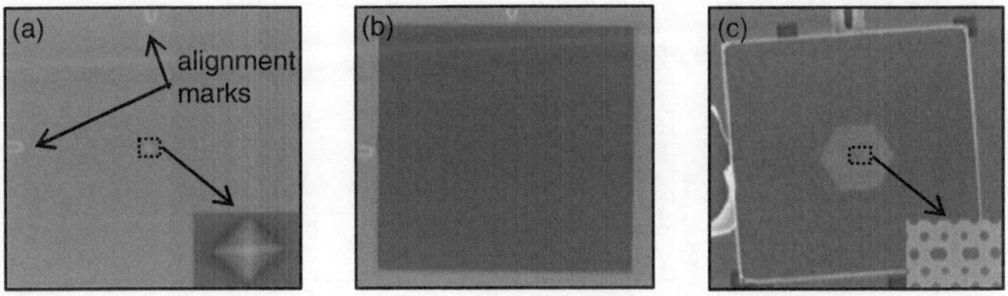

Figure 19.7 (a) SEM image of a single dot pyramid in registry with etched alignment marks. (b) The same pyramid after the planarization growth. (c) Modified single missing hole defect photonic crystal cavity [13] fabricated in the planarized region.

pattern into the InP membrane, which is dry etched using a CH_4, Cl_2, H_2 chemistry into the underlying InAlAs layer. The InAlAs layer is now removed by simple wet chemistry to leave the suspended InP photonic crystal nanocavity membrane with the single pyramidal InP template and single InAs quantum dot precisely located at the optical defect position. Plan view SEM images of the three main processing steps involved in the fabrication the the dot-cavity device are shown in Fig. 19.7.

Spectral matching is achieved using a digital wet-etching technique developed for photonic crystals microcavities [15, 7]. The as-fabricated cavity is designed with the mode on the low-energy side of the dot emission and the cavity is tuned through resonance in 1.7 meV steps using sequential cycles of InP native oxide growth and subsequent removal in a buffered HF solution. The cavity energy as a function of tuning cycles is shown in Fig. 19.8a for a dot emitting at 842 meV. The as-fabricated cavity mode energy was well below 800 meV and emission from the cavity is not observed until the sample has undergone 24 wet etch cycles at which point the cavity mode appears at 816 meV, and 39 cycles takes the cavity one cycle blue of the dot emission. The corresponding spectra are shown in Fig. 19.8b for excitation at saturation which permits tracking of the cavity mode at large detunings due to cavity feeding from excited states of the dot.

The enhancement of the dot emission for zero detuning results from a redirection of the emission into the cavity mode and subsequent vertical leakage, and also from an increase in the emission rate by virtue of the Purcell effect; both of which necessitate spatial matching as dictated by Eqn (19.1). The presence of both effects is elucidated by the integrated intensity versus excitation plot shown in the inset of Fig. 19.8b. At high excitation rates, the decrease in exciton lifetime due to the Purcell effect is seen to shift the onset of saturation of the transition to higher pump powers. The enhanced emission at high excitation rates will therefore include both cavity-mediated modifications to the photon collection efficiency and to the photon emission dynamics.

To interrogate the dot–cavity coupling mechanisms in more detail, the cavity can be tuned back through resonance in arbitrarily fine steps using inert gas condensation [24, 39]. A small volume of N_2 gas is introduced into the cryostat and condenses on the photonic

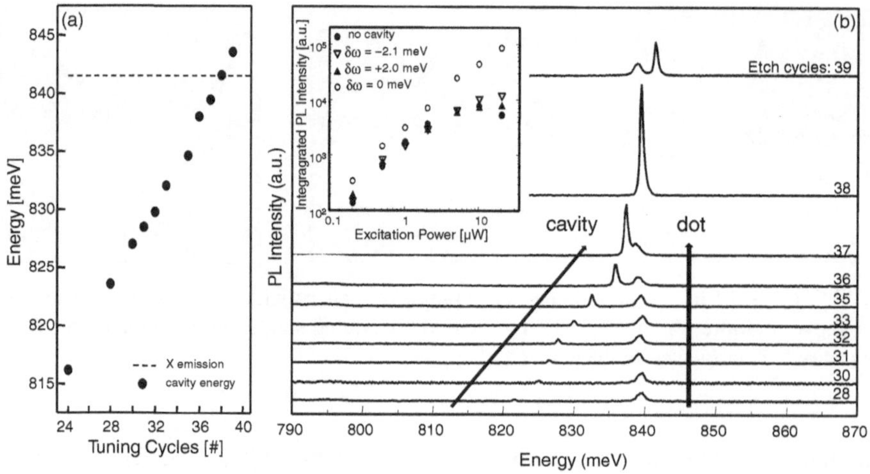

Figure 19.8 (a) Cavity mode energy as a function of oxidation/etch cycles used to tune the cavity through resonance with the exciton emission of the quantum dot. (b) PL spectra as a function of detuning showing the enhanced dot emission at resonance. The inset shows the integrated PL intensity of the dot emission as a function of excitation power before and after cavity fabrication, the latter at and near resonance.

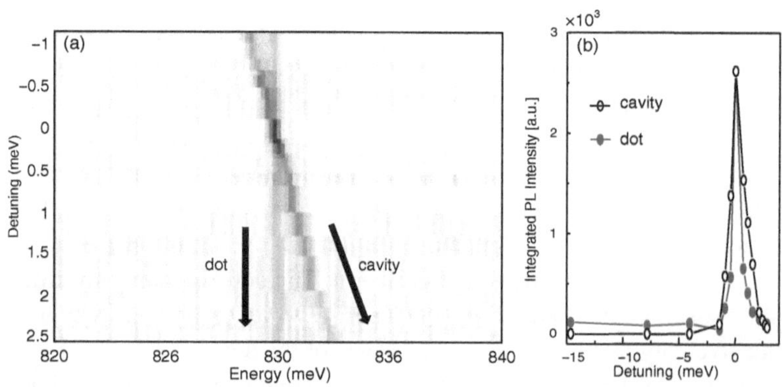

Figure 19.9 (a) PL spectra as a function of detuning using N_2 condensation for a dot emitting at 830 meV pumped well below saturation. (a) Integrated intensity of the dot and cavity under low pump conditions over the entire tuning range.

crystal producing a sub-monolayer film which shifts the cavity resonance to lower energy in steps dictated by the volume and pressure of N_2 injected. The PL spectra as a function of N_2 tuning are shown in Fig. 19.9a for a dot emitting at 830 meV. The spectra were taken using pump powers well below saturation to avoid cavity feeding from higher level excitations. Under these excitation conditions emission from the cavity disappears completely at large detunings (see Fig. 19.9b), as expected from a cavity that contains only one quantum dot and signals the absence of extraneous cavity feeding mechanisms. The onset of cavity emission occurs at a detuning of ~4 meV. This near-resonance cavity emission distinguishes the semiconductor system from its atomic counterpart and highlights the role of the environment in the solid state. In the former, no cavity emission is expected for negligible overlap between the emitter and cavity linewidths. Here, cavity emission is observed for detunings more than an order of magnitude greater than the relevant linewidths and is characteristic of quantum dot–cavity systems [16, 31].

To model the spectra we have used a photon Green function formalism [19] relevant to photonic slab cavities modified to include electron–acoustic phonon scattering [23, 20]. The model describes the emission from a coupled dot–cavity system in terms of emission via radiation modes above the light line, S_r, and emission from the leaky cavity, S_c:

$$S_r(r,\omega) = \Gamma_{rad} \times \left| \frac{\omega_x + \omega}{\omega_x^2 - \omega^2 - i\omega\Gamma_x - \omega\sum_{ph}(\omega) - \frac{4g^2\omega_x\omega_c}{\omega_c^2 - \omega^2 - i\omega\Gamma_c}} \right|^2 \tag{19.2}$$

and

$$S_c(r,\omega) = \Gamma_c \times \left| \frac{\frac{2g\omega_c(\omega_x+\omega)}{\omega_c^2 - \omega^2 - i\omega\Gamma_c}}{\omega_x^2 - \omega^2 - i\omega\Gamma_x - \omega\sum_{ph}(\omega) - \frac{4g^2\omega_x\omega_c}{\omega_c^2 - \omega^2 - i\omega\Gamma_c}} \right|^2, \tag{19.3}$$

where Γ_c is the decay rate of the leaky cavity, $\Gamma_x = \Gamma_{rad} + \Gamma_x^{'}$ is the total exciton decay rate including radiative (Γ_{rad}) and dephasing ($\Gamma_x^{'}$) rates and $g = [\omega_c/2\hbar\epsilon_o]^{1/2}\mathbf{d} \cdot \mathbf{f(r_x)}$ is the dot–cavity coupling rate. The total emission spectrum is given by the sum

$$S_t(r,\omega) = F_c(r)S_c(r,\omega) + F_r(r)S_r(r,\omega), \tag{19.4}$$

where F_r and F_c are geometric factors determined by the collection optics for the emission into radiation and cavity modes, respectively. Phonon-assisted exciton-cavity interactions, responsible for the near-resonance cavity emission [17, 43], are included through the phonon self-energy term $\sum_{ph}(\omega)$.

In Fig. 19.10a, the spectra in Figure 19.9 are replotted normalized to their peak value for clarity. Figure 19.10b shows the corresponding theoretical simulations according to Eqn (19.4). The spectra are calculated using decay rates Γ_c and Γ_x extracted from the measured linewidths and a F_c/F_r ratio equal to 2 estimated from the observed radiation mode emission at large detunings (see Fig. 19.9b). A coupling rate of $g = 130$ μeV is obtained from a fit to the experimental data. To highlight the role of phonon-mediated interactions on the near-resonance cavity emission, the spectra are calculated both with and without the phonon self-energy term $\sum_{ph}(\omega)$. Including phonon interactions enhances the near resonance cavity emission in correspondence with experiment, resulting in excellent agreement between the simulated and measured spectral over the entire tuning range.

The demonstration of deterministic coupling of a single quantum dot to a monolithic microcavity structure is the first step in the implementation of a quantum network for quantum information processing. The next generation of devices for on chip quantum information processing will require waveguide-mediated control of the single dot–cavity

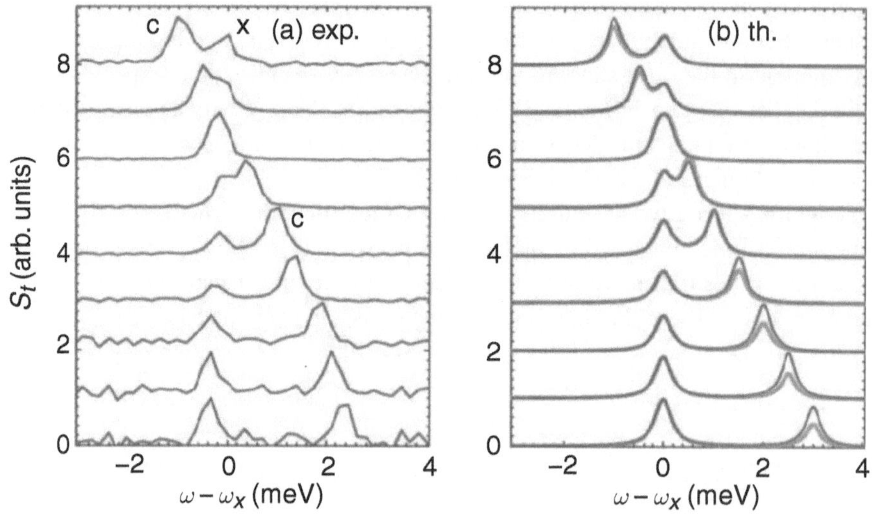

Figure 19.10 (a) Measured spectra from Fig. 19.9a, normalized to highlight the dot and cavity emission as a function of detuning. (b) Spectra calculated from Eqn (19.4) with (dark) and without (light) phonon-mediated exciton–cavity interactions. Figures adapted from Ref. [20].

system. In future architectures we envision nodes of single site-controlled quantum dots in individually addressable microcavities where photons mediate the node to node interaction via connecting photonic crystal waveguides.

19.5 Conclusion

A quantum dot site-control technique based on directed self-assembly has been presented. The approach addresses several key issues relevant to the production of quantum dot-based non-classical light emitters. The process is scalable, potentially capable of producing arrays of identical emitters on the same chip. Spatial alignment of dots and cavities is straightforward using standard alignment techniques compatible with conventional semiconductor processing. The process ensures that each cavity contains only a single dot, eliminating the ambiguity of possible cavity feeding from background emitters. These attributes, together with efficient emission at $\lambda = 1550$ nm, form the basis for a manufacturable technology for the production of quantum dot-based non-classical light sources for fibre-based quantum information processing.

The authors would like to acknowledge the financial support of the Canadian Institute for Advanced Research, the Canadian Institute for Photonic Innovations, QuantumWorks, the Natural Sciences and Engineering Research Council and the Business Development Bank of Canada.

References

[1] Andreani, L. C., Panzarini, G. and Gérard, J.-M. 1999. Strong-coupling regime for quantum boxes in pillar microcavities: Theory. *Phys. Rev. B*, **60**, 13 276.

[2] Atkinson, P., Kiravittaya, S., Benyoucef, M., Rastelli, A. and Schmidt, O. G. 2008. Site-controlled growth and luminescence of InAs quantum dots using in situ Ga-assisted deoxidation of patterned substrates. *Appl. Phys. Lett.*, **93**, 101908.

[3] Badolato, A., Hennessy, K., Atatüre, M. *et al.* 2005. Deterministic coupling of single quantum dots to single nanocavity modes. *Science*, **308**, 1158.

[4] Baier, M. H., Pelucchi, E., Kapon, E. *et al.* 2004. Single photon emission from site-controlled pyramidal quantum dots. *Appl. Phys. Lett.*, **84**, 648.

[5] Bayer, M., Ortner, G., Stern, O. *et al.* 2002. Fine structure of neutral and charged excitons in self-assembled In(Ga)As/(Al)GaAs quantum dots. *Phys. Rev. B*, **65**, 195315.

[6] Chithrani, D., Williams, R. L., Lefebvre, J., Poole, P. J. and Aers, G. C. 2004. Optical spectroscopy of single, site-selected, InAs/InP self-assembled quantum dots. *Appl. Phys. Lett.*, **84**, 978.

[7] Dalacu, D., Frédérick, S., Poole, P. J., Aers, G. C. and Willliams, R. L. 2005. Post-fabrication fine-tuning of photonic crystal microcavities in InAs/InP quantum dot membranes. *Appl. Phys. Lett.*, **87**, 151107.

[8] Dalacu, D., Mnaymneh, K., Sazonova, V. *et al.* 2010. Deterministic emitter-cavity coupling using a single-site controlled quantum dot. *Phys. Rev. B*, **82**, 033381.

[9] Dion, C., Poole, P.J., Raymond, S., Desjardins, P. and Schiettekatte, F. 2007. Tuning of the electronic properties of self-assembled InAs/InP (001) quantum dots using grown-in defect mediated intermixing. *Appl. Phys. Lett.*, **89**, 13105.

[10] Dousse, A., Lanco, L., Suffczyński, J. *et al.* 2008. Controlled light–matter coupling for a single quantum dot embedded in a pillar microcavity using far-field optical lithography. *Phys. Rev. Lett.*, **101**, 267404.

[11] Ellis, D. J. P., Stevenson, R. M., Young, R. J. *et al.* 2007. Control of fine-structure splitting of individual InAs quantum dots by rapid thermal annealing. *Appl. Phys. Lett.*, **90**, 011907.

[12] Faraon, A., Majumdar, A., Kim, H., Petroff, P. and Vučković, J. 2010. Fast electrical control of q quantum dot strongly coupled to a photonic-crystal cavity. *Phys. Rev. Lett.*, **104**, 047402.

[13] Frédérick, S., Dalacu, D., Lapointe, J. *et al.* 2006. Experimental demonstration of high quality factor, x-dipole modes in InAs/InP quantum dot photonic crystal microcvity membranes. *Appl. Phys. Lett.*, **89**, 191115.

[14] Gallo, P., Felici, M., Dwir, B. *et al.* 2008. Integration of site-controlled pyramidal quantum dots and photonic crystal membrane cavities. *Appl. Phys. Lett.*, **92**, 263101.

[15] Hennessy, K., Badolato, A., Tamboli, A. *et al.* 2005. Tuning photonic crystal nanocavity modes by wet chemical digital etching. *Appl. Phys. Lett.*, **87**, 021108.

[16] Hennessy, K., Badolato, A., Winger, M. *et al.* 2007. Quantum nature of a strongly coupled single quantum dot–cavity system. *Nature*, **445**, 896–899.

[17] Hohenester, U. 2010. Cavity quantum electrodynamics with semiconductor quantum dots: role of phonon-assisted cavity feeding. *Phys. Rev. B*, **81**, 155303.

[18] Hsieh, T.-P. and Chyi, J.-I. 2007. Single photon emission from an InGaAs quantum dot precisely position on a nanoplane. *Appl. Phys. Lett.*, **90**, 073105.

[19] Hughes, S. and Yao, P. 2009. Theory of quantum light emission from a strongly-coupled single quantum dot photonic-crystal cavity system. *Opt. Exp.*, **17**, 3322.

[20] Hughes, S., Yao, P., Milde, F. *et al.* 2011. Influence of electron-acoustic phonon scattering on off-resonant cavity feeding within a strongly coupled quantum-dot cavity system. *Phys. Rev. B*, **83**, 165313.
[21] Lee, K. H., Green, A. M., Taylor, R. A. *et al.* 2006. Registration of single quantum dots using cryogenic laser photolithography. *Appl. Phys. Lett.*, **88**, 193106.
[22] Michler, P. 2003. Nonclassical light from single semiconductor quantum dots. *Topics Appl. Phys.*, **90**, 315.
[23] Milde, F., Knorr, A. and Hughes, S. 2008. Role of electron–phonon scattering on the vacuum Rabi splitting of a single-quantum dot and a photonic-crystal-nanocavity. *Phys. Rev. B*, **78**, 035330.
[24] Mosor, S., Hendrickson, J., Richards, B. C. *et al.* 2005. Scanning a photonic crystal slab nanocavity by condensation of xenon. *Appl. Phys. Lett.*, **87**, 141105.
[25] Painter, O. J., Husain, A., Scherer, A. *et al.* 1999. Room temperature photonic crystal defect lasers at near-infrared wavelengths in InGaAsP. *J. Lightwave Technol.*, **17**, 2082.
[26] Pathak, K. and Hughes, S. 2009. Cavity-assisted fast generation of entangled photon pairs through the biexciton-exciton cascade. *Phys. Rev. B*, **80**, 155525.
[27] Peter, E., Senellart, P., Martrou, D. *et al.* 2005. Exciton–photon strong-coupling regime for a single quantum dot embedded in a microcavity. *Phys. Rev. Lett.*, **95**, 067401.
[28] Petroff, P. M. 2003. Epitaxial growth and electronic structure of self-assembled quantum dots. *Topics Appl. Phys.*, **90**, 1–24.
[29] Poole, P. J., McCaffrey, J., Williams, R., Lefebvre, J. and Chitrani, D. 2001. Chemical beam epitaxy of self-assembled InAs/InP quanum dots. *J. Vac. Sci. Technol. B*, **19**, 1467.
[30] Poole, P. J., Dalacu, D., Lefebvre, J. and Williams, R. L. 2010. Selective epitaxy of semiconductor nanopyramids for nanophotonics. *Nanotech.*, **21**, 295302.
[31] Press, D., Götzinger, S. *et al.* 2007. Photon antibunching from a single quantum-dot-microcavity system in the strong coupling regime. *Phys. Rev. Lett.*, **98**, 117402.
[32] Purcell, E. M. 1946. Spontaneous emission probabilities at radio reguencies. *Phys. Rev.*, **69**, 681.
[33] Rastelli, A., Ulhaq, A., Kiravittaya, S. *et al.* 2007. In situ laser microprocessing of single self-assembled quantum dots and optical microcavities. *Appl. Phys. Lett.*, **90**, 073120.
[34] Reimer, M. E., Korkusiński, M., Dalacu, D. *et al.* 2008. Prepositioned single quantum dot in a lateral electric field. *Phys. Rev. B*, **78**, 195301.
[35] Reithmaier, J. P., Sek, G., Löffler, A. *et al.* 2004. Strong coupling in a single quantum dot-semiconductor microcavity system. *Nature*, **432**, 197.
[36] Schneider, C., Heindel, T., Huggenberger, A. *et al.* 2009. Single photon emission from a site-controlled quantum dot-micropillar cavity system. *Appl. Phys. Lett.*, **94**, 111111.
[37] Shields, A. J. 2007. Semiconductor quantum light sources. *Nat. Photonics*, **1**, 215.
[38] Song, H. Z., Usuki, T., Hirose, S. *et al.* 2005. Site-controlled photoluminescence at telecommunication wavelength from InAs/InP quantum dots. *Appl. Phys. Lett.*, **86**, 113.
[39] Srinivasan, K. and Painter, O. 2007. Optical fiber taper coupling and high-resolution wavelength tuning of microdisk resonators at cyrogenic temperatures. *Appl. Phys. Lett.*, **90**, 031114.

[40] Stanowski, R. and Dubowski, J. J. 2009. Laser rapid thermal annealing of quantum semiconductor wafers: a one step bandgap engineering technique. *Appl. Phys. A*, **94**, 667.

[41] Stranski, I. N. and von Krastanow, L. 1939. *Akad. Wiss. Lit. Mainz Math.-Naturwiss Kl. IIb*, **146**, 797.

[42] Sünner, T., Schneider, C., Strauß, M. *et al.* 2008. Scalable fabrication of optical resonators with embedded site-controlled quantum dots. *Opt. Lett.*, **33**, 1759.

[43] Tarel, G. and Savona, V. 2010. Linear spectrum of a quantum dot coupled to a nanocavity. *Phys. Rev. B*, **81**, 075305.

[44] Vahala, K. J. 2003. Optical microcavities. *Nature*, **424**, 839.

[45] Yoshie, T., Scherer, A., Hendrickson, J. *et al.* 2004. Vacuum Rabi splitting with a single quantum dot in a photonic crystal nanocavity. *Nature*, **432**, 200.

Index